The occurrence of singularities is pervasive in many problems in topology, differential geometry and algebraic geometry. This book concerns the study of singular spaces using techniques from a variety of areas of geometry and topology and interactions among them. It contains more than a dozen expository papers on topics ranging from intersection homology, L^2 cohomology and differential operators, to the topology of algebraic varieties, signatures and characteristic classes, mixed Hodge theory, and elliptic genera of singular complex and real algebraic varieties. The book concludes with a list of open problems.

Mathematical Sciences Research Institute
Publications

58

Topology of Stratified Spaces

Mathematical Sciences Research Institute Publications

Volumes 1–4, 6–8, and 10–27 are published by Springer-Verlag

Topology of Stratified Spaces

Edited by

Greg Friedman
Texas Christian University

Eugénie Hunsicker
Loughborough University

Anatoly Libgober
University of Illinois at Chicago

Laurentiu Maxim
University of Wisconsin-Madison

CAMBRIDGE
UNIVERSITY PRESS

Greg Friedman
Department of Mathematics
Texas Christian University
g.friedman@tcu.edu

Eugénie Hunsicker
Department of Mathematical Sciences
Loughborough University
E.Hunsicker@lboro.ac.uk

Anatoly Libgober
Department of Mathematics
University of Illinois at Chicago
libgober@math.uic.edu

Laurentiu Maxim
Department of Mathematics
University of Wisconsin-Madison
maxim@math.wisc.edu

Silvio Levy (*Series Editor*)
Mathematical Sciences Research Institute
Berkeley, CA 94720
levy@msri.org

The Mathematical Sciences Research Institute wishes to acknowledge support by the National Science Foundation and the *Pacific Journal of Mathematics* for the publication of this series.

CAMBRIDGE
UNIVERSITY PRESS

32 Avenue of the Americas, New York NY 10013-2473, USA

Cambridge University Press is part of the University of Cambridge.

It furthers the University's mission by disseminating knowledge in the pursuit of education, learning and research at the highest international levels of excellence.

www.cambridge.org
Information on this title: www.cambridge.org/9781107459472

First published 2011
First paperback edition 2014

A catalogue record for this publication is available from the British Library

Topology of stratified spaces / edited by Greg Friedman ... [et al.].
    p.   cm. – (Mathematical Sciences Research Institute publications ; 58)
Includes bibliographical references.
ISBN 978-0-521-19167-8 (hardback)
1. Singularities (Mathematics) 2. Algebraic spaces. 3. Algebraic topology.
I. Friedman, Greg, 1973–

QA614.58.T67   2011
516.3'5—dc22   2010051108

ISBN 978-0-521-19167-8 Hardback
ISBN 978-1-107-45947-2 Paperback

Topology of Stratified Spaces
MSRI Publications
Volume **58**, 2011

Contents

Preface

This volume is based on lectures given at the Workshop on the Topology of Stratified Spaces that was held at the Mathematical Sciences Research Institute in Berkeley, CA, from September 8 to September 12, 2008. Stratified spaces are usually not quite manifolds — they may possess singularities — but they are composed of manifold layers, the strata. Examples of such spaces include algebraic varieties, quotients of manifolds and varieties by group actions, homotopy stratified spaces, topological and piecewise linear pseudomanifolds, and even manifolds, augmented by filtrations that can arise, for example, via embeddings and their singularities. In recent years, there has been extensive interest and success in expanding to stratified spaces the triumphs of algebraic topology in manifold theory, including the vast progress in the mid-twentieth century on signatures, characteristic classes, surgery theories, and the special homological properties of nonsingular analytic and algebraic varieties, such as the Kähler package and Hodge theories. Such extensions from manifold theory to stratified space theory are rarely straightforward — they tend to involve the discovery and study of subtle interactions between local and global behavior — but vast progress has been made, particularly using such topological tools as intersection homology and the related analytic L^2 cohomology. The goal of the workshop was, and of this proceedings is, to provide an overview of this progress as well as of current research results, with a particular emphasis on communication across the boundaries of the different fields of mathematics that encompass stratified space research. Thus there is an emphasis in this volume on expository papers that give introductions to and overviews of topics in the area of stratified spaces.

Four main areas were featured in the MSRI workshop: L^2 cohomology and Hodge theorems, topology of algebraic varieties, signature theory on singular spaces, and mixed Hodge theory and singularities. For the purpose of giving some organization to the volume, we have grouped the papers roughly into these topics, although some papers overlap more than one area.

There are three papers on analysis and topology. The paper by Dai is an introduction to L^2 cohomology, which discusses some of the analytic considerations that arise in the study of L^2 cohomology and L^2 signatures on stratified spaces as well as relationships between these analytic objects and topological analogues. Building on these basics, the paper of Carron gives an example of how his idea of the exterior derivative on a space having "almost closed range" can be used to calculate the L^2 cohomology in the case of two QALE-type spaces, the Hilbert schemes of two and of three points on $\mathbb{CP}^2$. The paper by Waelder comes from a different angle. It considers "rigid" differential operators on a smooth manifold commuting with an S^1-action, and infinite dimensional analog of the Dirac operator, whose index yields the complex elliptic genus. In this survey, Waelder discusses how rigidity theorems for such operators are related to the problem of defining Chern numbers on singular varieties. The paper of Waelder also fits under the second category: complex algebraic varieties.

The next section, on algebraic varieties, starts with the survey by Kovács and Schwede, which gives an introduction to the study of singularities that are unavoidable in classification problems of smooth algebraic varieties, and especially in the minimal model program, including log-canonical and Du Bois singularities. The paper of Libgober is an overview of the work of the author and his collaborators on extensions of the elliptic genus to singular varieties. It contains a new description of the class of holomorphic functions (quasi-Jacobi forms) that are elliptic genera of complex manifolds (possibly without the Calabi–Yau condition) and includes a survey of recent developments such as the "higher elliptic genera", which contain information on non-simply connected spaces. It also constructs the elliptic genus for singular real algebraic spaces. In their contribution, McCrory and Parusiński make a thorough study of the weight filtration on the Borel–Moore cohomology of real algebraic varieties. The authors give several descriptions of this filtration and show some nice applications of their construction to real algebraic and analytic geometry. Finally, the paper of Maxim gives a new formulation of results about Milnor classes, which generalize Parusiński's Milnor numbers to non-isolated hypersurface singularities. He considers these classes for singular hypersurfaces in complex manifolds and, using his new formulation, gives comparisons to the topologically defined L-classes of Cappell–Shaneson and Goresky–MacPherson.

Three papers discuss aspects of intersection homology and signatures on stratified spaces. The paper by Friedman is an expository survey of perversities in intersection homology, starting with the classical perversities of Goresky and MacPherson and describing various ways in which these have been generalized in the past three decades. It can also be read as a general introduction to intersection homology. Banagl's contribution is an overview of bordism invariant signatures on singular spaces, including the category of Witt space, where the signature on intersection cohomology provides a Witt-bordism invariant signature, and Banagl's new signature on a category of non-Witt spaces under a new type of bordism. Levikov's paper describes the existence of a Wang-type sequence for intersection homology, which has implications for the non-Witt signatures defined by Banagl.

The section on mixed Hodge structures includes a comprehensive survey by Kerr and Pearlstein of recent developments in the theory of normal functions. It begins with the classical theory of Griffiths, Steenbrink, and Zucker, and includes the work of Green, Griffiths, and Kerr on limits of Abel–Jacobi mappings, the equivalence (due to Brosnan, Fang, Nie, and Pearlstein) of the Hodge conjecture to a question about singularities of certain normal functions; work by Brosnan and Pearlstein on the algebraicity of the zero locus of an admissible normal function; and the construction of a Néron model by Green, Griffiths, and Kerr, and by Brosnan, Pearlstein, and Saito. This survey includes several concrete examples and an extensive bibliography, and it concludes with a discussion of open questions in the field. The papers by Yokura and Schürmann, respectively, describe recent work by the authors, together with Brasselet, which gives a positive answer to the question of MacPherson as to whether there exists a unified theory of characteristic classes for singular varieties that is analogous to the classical Hirzebruch theory. Yokura's paper emphasizes the motivic touch of the story, whereas Schürmann's approach relies on Saito's powerful theory of algebraic mixed Hodge modules.

This volume concludes with an annotated list of problems proposed by participants in the workshop.

Neither the workshop nor this volume could have come about without the help of many people. Firstly, MSRI provided funding for the workshop and its staff helped enormously in making it a success. Secondly, Silvio Levy helped arrange publication of this volume and gave us good advice on how to make it as useful a book as possible. Finally, several of our colleagues have been insightful and efficient referees for the various papers herein. We are grateful to everyone for their assistance. We hope that readers will find this volume useful and inspiring.

> Greg Friedman, Eugénie Hunsicker,
> Anatoly Libgober, Laurentiu Maxim
>
> September 2009

Topology of Stratified Spaces
MSRI Publications
Volume **58**, 2011

An introduction to L^2 cohomology

XIANZHE DAI

ABSTRACT. After a quick introduction to L^2 cohomology, we discuss recent
joint work with Jeff Cheeger where we study, from a mostly topological stand-
point, the L^2-signature of certain spaces with nonisolated conical singularities.
The contribution from the singularities is identified with a topological invariant
of the link fibration of the singularities, involving the spectral sequence of the
link fibration.

This paper consists of two parts. In the first, we give an introduction to
L^2 cohomology. This is partly based on [8]. We focus on the analytic aspect
of L^2 cohomology theory. For the topological story, we refer to [1; 22; 31]
and of course the original papers [16; 17]. For the history and comprehensive
literature, see [29]. The second part is based on our joint work with Jeff Cheeger
[11], which gives the contribution to the L^2 signature from nonisolated conical
singularity.

It is a pleasure to thank Eugenie Hunsicker for numerous comments and suggestions.

1. L^2 cohomology: what and why

What is L^2 cohomology? The de Rham theorem provides one of the most useful connections between the topological and differential structure of a manifold.
The differential structure enters the de Rham complex, which is the cochain
complex of smooth exterior differential forms on a manifold M, with the exterior derivative as the differential:

$$0 \to \Omega^0(M) \xrightarrow{d} \Omega^1(M) \xrightarrow{d} \Omega^2(M) \xrightarrow{d} \Omega^3(M) \to \cdots$$

Partially supported by NSF and CNSF..

The de Rham Theorem says that the de Rham cohomology, the cohomology of the de Rham complex, $H^k_{\mathrm{dR}}(M) \overset{\text{def}}{=} \ker d_k / \operatorname{Im} d_{k-1}$, is isomorphic to the singular cohomology:

$$H^k_{\mathrm{dR}}(M) \cong H^k(M; \mathbf{R}).$$

The situation can be further rigidified by introducing geometry into the picture. Let g be a Riemannian metric on M. Then g induces an L^2-metric on $\Omega^k(M)$. As usual, let δ denote the formal adjoint of d. In terms of a choice of local orientation for M, we have $\delta = \pm * d *$, where $*$ is the Hodge star operator. Define the Hodge Laplacian to be

$$\Delta = d\delta + \delta d.$$

A differential form ω is harmonic if $\Delta \omega = 0$.

The great theorem of Hodge then states that, for a closed Riemannian manifold M, every de Rham cohomology class is represented by a unique harmonic form. This theorem provides a direct bridge between topology and analysis of manifolds through geometry, and has found many remarkable applications.

Naturally, then, one would like to extend the theory to noncompact manifolds and manifolds with singularity. The de Rham cohomology is still defined (one would restrict to the smooth open submanifold of a manifold with singularity). However, it does not capture the information at infinity or at the singularity.

One way of remedying this is to restrict to a subcomplex of the usual de Rham complex, namely that of the square integrable differential forms — this leads us to L^2 cohomology.

More precisely, let (Y, g) denote an open (possibly incomplete) Riemannian manifold, let $\Omega^i = \Omega^i(Y)$ be the space of C^∞ i-forms on Y and $L^2 = L^2(Y)$ the L^2 completion of Ω^i with respect to the L^2-metric. Define d to be the exterior differential with the domain

$$\operatorname{dom} d = \{\alpha \in \Omega^i(Y) \cap L^2(Y); \ d\alpha \in L^2(Y)\}.$$

Put

$$\Omega^i_{(2)}(Y) = \Omega^i(Y) \cap L^2(Y).$$

Then one has the cochain complex

$$0 \to \Omega^0_{(2)}(Y) \overset{d}{\to} \Omega^1_{(2)}(Y) \overset{d}{\to} \Omega^2_{(2)}(Y) \overset{d}{\to} \Omega^3_{(2)}(Y) \to \cdots.$$

The L^2-cohomology of Y is defined to be the cohomology of this cochain complex:

$$H^i_{(2)}(Y) = \ker d_i / \operatorname{Im} d_{i-1}.$$

Thus defined, the L^2 cohomology is in general no longer a topological invariant. However, the L^2 cohomology depends only on the quasi-isometry class of the metric.

EXAMPLES. • The real line: For the real line $\mathbb{R}$ with the standard metric,

$$H^i_{(2)}(\mathbb{R}) = 0 \qquad\qquad\qquad \text{if } i = 0,$$
$$H^i_{(2)}(\mathbb{R}) \text{ is infinite-dimensional} \quad \text{if } i = 1.$$

For the first part, this is because constant functions can never be L^2, unless they are zero. For the second part, a 1-form $\phi(x)\,dx$, with $\phi(x)$ having compact support, is obviously closed and L^2, but can never be the exterior derivative of an L^2 function, unless the total integral of ϕ is zero.

• Finite cone: Let $C(N) = C_{[0,1]}(N) = (0, 1) \times N$, where N is a closed manifold of dimension n, with the conical metric $g = dr^2 + r^2 g_N$. Then a result of Cheeger [8] gives

$$H^i_{(2)}(C(N)) = \begin{cases} H^i(N) & \text{if } i < (n+1)/2, \\ 0 & \text{if } i \geq (n+1)/2. \end{cases}$$

Intuitively this can be explained by the fact that some of the differential forms that define classes for the cylinder $N \times (0, 1)$ cannot be L^2 on the cone if their degrees are too big. More specifically, let ω be an i-form on N and extend it trivially to $C(N)$, so ω is constant along the radial direction. Then

$$\int_{C(N)} |\omega|^2_g d\,\mathrm{vol}_g = \int_0^1 \int_N |\omega|_{g_N} r^{n-2i}\,dx\,dr.$$

Thus, the integral is infinite if $i \geq (n+1)/2$.

As we mentioned, the L^2 cohomology is in general no longer a topological invariant. Now clearly, there is a natural map

$$H^i_{(2)}(Y) \longrightarrow H^i(Y, \mathbf{R})$$

via the usual de Rham cohomology. However, this map is generally neither injective nor surjective. On the other hand, in the case when (Y, g) is a compact Riemannian manifold with corner (for a precise definition see the article by Gilles Carron in this volume), the map above is an isomorphism because the L^2 condition is automatically satisfied for any smooth forms.

Also, another natural map is from the compact supported cohomology to the L^2 cohomology:

$$H^i_c(Y) \longrightarrow H^i_{(2)}(Y).$$

As above, this map is also neither injective nor surjective in general.

Instead, the L^2 cohomology of singular spaces is intimately related to the intersection cohomology of Goresky–MacPherson ([16; 17]; see also Greg Friedman's article in this volume for the intersection cohomology). This connection was pointed out by Dennis Sullivan, who observed that Cheeger's local computation of L^2 cohomology for isolated conical singularity agrees with that of

Goresky–MacPherson for the middle intersection homology. In [8], Cheeger established the isomorphism of the two cohomology theories for admissible pseudomanifolds. One of the fundamental questions has been the topological interpretation of the L^2 cohomology in terms of the intersection cohomology of Goresky–MacPherson.

Reduced L^2 cohomology and L^2 harmonic forms. In analysis, one usually works with complete spaces. That means, in our case, the full L^2 space instead of just smooth forms which are L^2. Now the coboundary operator d has well defined strong closure $\bar{d}$ in L^2: $\alpha \in \operatorname{dom} \bar{d}$ and $\bar{d}\alpha = \eta$ if there is a sequence $\alpha_j \in \operatorname{dom} d$ such that $\alpha_j \to \alpha$ and $d\alpha_j \to \eta$ in L^2. (The strong closure is to make $\bar{d}$ a closed operator. There are other notions of closures and extensions, as in [15] for instance.) Similarly, δ has the strong closure $\bar{\delta}$.

One can also define the L^2-cohomology using the strong closure $\bar{d}$. Thus, define

$$H^i_{(2),\#}(Y) = \ker \bar{d}_i / \operatorname{Im} \bar{d}_{i-1} .$$

Then the natural map

$$\iota_{(2)} : \; H^i_{(2)}(Y) \longrightarrow H^i_{(2),\#}(Y)$$

turns out to be always an isomorphism [8].

This is good, but does not produce any new information ... yet! The crucial observation is that, in general, the image of $\bar{d}$ need not be closed. This leads to the notion of reduced L^2-cohomology, which is defined by quotienting out by the closure instead:

$$\bar{H}^i_{(2)}(Y) = \ker \bar{d}_i / \overline{\operatorname{Im} \bar{d}_{i-1}} .$$

The reduced L^2-cohomology is generally not a cohomology theory but it is intimately related to Hodge theory, as we will see.

Now we define the space of L^2-harmonic i-forms $\mathcal{H}^i_{(2)}(Y)$ to be

$$\mathcal{H}^i_{(2)}(Y) = \{\theta \in \Omega^i \cap L^2; d\theta = \delta\theta = 0\}.$$

Some authors define the L^2-harmonic forms differently; compare [31]. The definitions coincide when the manifold is complete. The advantage of our definition is that, when Y is oriented, the Hodge star operator induces

$$* : \mathcal{H}^i_{(2)}(Y) \to \mathcal{H}^{n-i}_{(2)}(Y),$$

which is naturally the Poincaré duality isomorphism.

Now the big question is: Do we still have a Hodge theorem?

Kodaira decomposition, L^2 Stokes and Hodge theorems.   To answer the question, let's look at the natural map, the Hodge map

$$\mathcal{H}^i_{(2)}(Y) \longrightarrow H^i_{(2)}(Y).$$

The question becomes: When is this map an isomorphism? Following Cheeger [8], when the Hodge map is an isomorphism, we will say that the strong Hodge theorem holds.

The most basic result in this direction is the Kodaira decomposition [23] (see also [14]),

$$L^2 = \mathcal{H}^i_{(2)} \oplus \overline{d\Lambda_0^{i-1}} \oplus \overline{\delta\Lambda_0^{i+1}},$$

an orthogonal decomposition which leaves invariant the subspaces of smooth forms. Here subscript 0 denotes having compact support. This result is essentially the elliptic regularity.

It follows from the Kodaira decomposition that

$$\ker \bar{d}_i = \mathcal{H}^i_{(2)} \oplus \overline{d\Lambda_0^{i-1}}.$$

Therefore the question reduces to what the space $\operatorname{Im} \bar{d}_{i-1}$ is in the decomposition. We divide the discussion into two parts:

SURJECTIVITY. If $\operatorname{Im} \bar{d}$ is closed, then $\operatorname{Im} \bar{d} \supset \overline{d\Lambda_0^{i-1}}$. Hence, the Hodge map is surjective in this case.

In particular, this holds if the L^2-cohomology is finite-dimensional.

INJECTIVITY. The issue of injectivity of the Hodge map has to do with the L^2 Stokes theorem. We say that Stokes' theorem holds for Y in the L^2 sense if

$$\langle \bar{d}\alpha, \ \beta \rangle = \langle \alpha, \ \bar{\delta}\beta \rangle$$

for all $\alpha \in \operatorname{dom} \bar{d}$, $\beta \in \operatorname{dom} \bar{\delta}$; or equivalently, if

$$\langle d\alpha, \ \beta \rangle = \langle \alpha, \ \delta\beta \rangle$$

$\alpha \in \operatorname{dom} d$, $\beta \in \operatorname{dom} \delta$.

If the L^2 Stokes theorem holds, one has

$$\mathcal{H}^i_{(2)}(Y) \perp \operatorname{Im} \bar{d}_{i-1},$$

so the Hodge map is injective in this case. Moreover,

$$H^i_{(2)}(Y) = \mathcal{H}^i_{(2)}(Y) \oplus \overline{\operatorname{Im} \bar{d}_{i-1}}/\operatorname{Im} \bar{d}_{i-1}.$$

Here, by the closed graph theorem, the last summand is either 0 or infinite-dimensional. Note also, since it follows that

$$\mathcal{H}^i_{(2)}(Y) \perp \overline{\operatorname{Im} \bar{d}_{i-1}},$$

that
$$\bar{H}^i_{(2)}(Y) \cong \mathcal{H}^i_{(2)}(Y).$$

That is, when the L^2 Stokes theorem holds, the reduced L^2 cohomology is simply the space of L^2 harmonic forms.

Summarizing, if the L^2-cohomology of Y has finite dimension and Stokes' theorem holds on Y in the L^2-sense, then the Hodge theorem holds in this case, and the L^2-cohomology of Y is isomorphic to the space of L^2-harmonic forms. Therefore, when Y is orientable, Poincaré duality holds as well. Consequently, the L^2 signature of Y is well defined in this case.

There are several now classical results regarding the L^2 Stokes theorem. Gaffney [15] showed that the L^2 Stokes theorem holds for complete Riemannian manifolds. On the other hand, for manifolds with conical singularity $M = M_0 \cup C(N)$, the general result of Cheeger [9] says that the L^2 Stokes theorem holds provided that L^2 Stokes holds for N and in addition the middle-dimensional (L^2) cohomology group of N vanishes if dim N is even. In particular, if N is a closed manifold of odd dimension, or $H^{\dim N/2}(N) = 0$ if dim N is even, the L^2 Stokes theorem holds for M.

REMARK. There are various extensions of L^2, for cohomology example, cohomology with coefficients or Dolbeault cohomology for complex manifolds.

2. L^2 signature of nonisolated conical singularities

Nonisolated conical singularities. We now consider manifolds with nonisolated conical singularity whose strata are themselves smooth manifolds. In other words, singularities are of the following type:

(i) The singular stratum consists of disjoint unions of smooth submanifolds.
(ii) The singularity structure along the normal directions is conical.

More precisely, a neighborhood of a singular stratum of positive dimension can be described as follows. Let

$$Z^n \to M^m \xrightarrow{\pi} B^l \tag{2-1}$$

be a fibration of closed oriented smooth manifolds. Denote by $C_\pi M$ the mapping cylinder of the map $\pi : M \to B$. This is obtained from the given fibration by attaching a cone to each of the fibers. Indeed, we have

$$C_{[0,1]}(Z) \to C_\pi M \to B.$$

The space $C_\pi M$ also comes with a natural quasi-isometry class of metrics. A metric can be obtained by choosing a submersion metric on M:

$$g_M = \pi^* g_B + g_Z.$$

Then, on the nonsingular part of $C_\pi M$, we take the metric

$$g_1 = dr^2 + \pi^* g_B + r^2 g_Z. \tag{2-2}$$

The general class of spaces with nonisolated conical singularities as above can be described as follows. A space X in the class will be of the form

$$X = X_0 \cup X_1 \cup \cdots \cup X_k,$$

where X_0 is a compact smooth manifold with boundary, and each X_i (for $i = 1, \ldots, k$) is the associated mapping cylinder $C_{\pi_i} M_i$ for some fibration (M_i, π_i), as above.

More generally, one can consider the iterated construction where we allow manifolds in our initial fibration to have singularities of the type considered above. However, we will restrict ourselves to the simplest situation where the initial fibrations are all modeled on smooth manifolds.

REMARK. An n-dimensional stratified pseudomanifold X is a topological space together with a filtration by closed subspaces

$$X = X_n = X_{n-1} \supset X_{n-2} \supset \cdots \supset X_1 \supset X_0$$

such that for each point $p \in X_i - X_{i-1}$ there is a distinguished neighborhood U in X which is filtered homeomorphic to $C(L) \times B^i$ for a compact stratified pseudomanifold L of dimension $n - i - 1$. The i-dimensional stratum $X_i - X_{i-1}$ is an i-dimensional manifold. A conical metric on X is a Riemannian metric on the regular set of X such that on each distinguished neighborhood it is quasi-isometric to a metric of the type (2-2) with $B = B^i$, $Z = L$ and g_B the standard metric on B^i, g_Z a conical metric on L. Such conical metrics always exist on a stratified pseudomanifold.

L^2 signature of generalized Thom spaces.

A generalized Thom space T is obtained by coning off the boundary of the space $C_\pi M$.

Namely,

$$T = C_\pi M \cup_M C(M)$$

is a compact stratified pseudomanifold with two singular strata: B and a single point (unless B is a sphere).

EXAMPLE. Let $\xi \xrightarrow{\pi} B$ be a vector bundle of rank k. We have the associated sphere bundle

$$S^{k-1} \to S(\xi) \xrightarrow{\pi} B.$$

The generalized Thom space constructed out of this fibration coincides with the usual Thom space equipped with a natural metric.

Now consider the generalized Thom space constructed from an oriented fibration (2-1) of closed manifolds, i.e., both the base B and fiber Z are closed oriented manifolds and so is the total space M. Then T will be a compact oriented stratified pseudomanifold with two singular strata. Since we are interested in the L^2 signature, we assume that the dimension of M is odd (so dim T is even). In addition, we assume the Witt conditions; namely, either the dimension of the fibers is odd or its middle-dimensional cohomology vanishes. Under the Witt conditions, the strong Hodge theorem holds for T. Hence the L^2 signature of T is well defined.

QUESTION. What is the L^2 signature of T?

Let's go back to the case of the usual Thom space.

EXAMPLE (continued). In this case,

$$\text{sign}_{(2)}(T) = -\text{sign}(D(\xi)),$$

the signature of the disk bundle $D(\xi)$ (as a manifold with boundary).

Let Φ denote the Thom class and χ the Euler class. Then the Thom isomorphism gives the commutative diagram

$$
\begin{array}{ccccc}
H^{*+k}(D(\xi), S(\xi)) & \otimes & H^{*+k}(D(\xi), S(\xi)) & \longrightarrow & \mathbb{R} \\
\Big\uparrow{\scriptstyle \pi^*(\cdot)\cup\Phi} & & \Big\uparrow{\scriptstyle \pi^*(\cdot)\cup\Phi} & & \\
H^*(B) & \otimes & H^*(B) & \longrightarrow & \mathbb{R} \\
\phi & & \psi & \longmapsto & [\phi\cup\psi\cup\chi][B].
\end{array}
$$

Thus, $\text{sign}_{(2)}(T)$ is the signature of this bilinear form on $H^*(B)$.

We now introduce the topological invariant which gives the L^2-signature for a generalized Thom space. In [13], in studying adiabatic limits of eta invariants, we introduced a global topological invariant associated with a fibration. (For adiabatic limits of eta invariants, see also [32; 5; 10; 3].) Let (E_r, d_r) be the E_r-term with differential, d_r, of the Leray spectral sequence of the fibration (2-1) in the construction of the generalized Thom space T. Define a pairing

$$E_r \otimes E_r \longrightarrow \mathbb{R}$$

$$\phi \otimes \psi \longmapsto \langle \phi \cdot d_r \psi, \xi_r \rangle,$$

where ξ_r is a basis for E_r^m naturally constructed from the orientation. In case $m = 4k - 1$, when restricted to $E_r^{(m-1)/2}$, this pairing becomes symmetric. We define τ_r to be the signature of this symmetric pairing and put

$$\tau = \sum_{r \geq 2} \tau_r .$$

When the fibration is a sphere bundle with the typical fiber a $(k-1)$-dimensional sphere, the spectral sequence satisfies $E_2 = \cdots = E_k$, $E_{k+1} = E_\infty$ with $d_2 = \cdots = d_{k-1} = 0$, $d_k(\psi) = \psi \cup \chi$. Hence τ coincides with the signature of the bilinear form from the Thom isomorphism theorem. The main result of [11] is this:

THEOREM 1 (CHEEGER–DAI). *Assume either the fiber Z is odd-dimensional or its middle-dimensional cohomology vanishes. Then the L^2-signature of the generalized Thom space T is equal to $-\tau$:*

$$\operatorname{sign}_{(2)}(T) = -\tau.$$

In spirit, our proof of the theorem follows the example of the sphere bundle of a vector bundle. Thus, we first establish an analog of Thom's isomorphism theorem in the context of generalized Thom spaces. In part, this consists of identifying the L^2-cohomology of T in terms of the spectral sequence of the original fibration; see [11] for complete details.

COROLLARY 2. *For a compact oriented space X with nonisolated conical singularity satisfying the Witt conditions, the L^2-signature is given by*

$$\operatorname{sign}_{(2)}(X) = \operatorname{sign}(X_0) + \sum_{i=1}^{k} \tau(X_i) .$$

The study of the L^2-cohomology of the type of spaces with conical singularities discussed here turns out to be related to work on the L^2-cohomology of noncompact hyperkähler manifolds which is motivated by Sen's conjecture; see, for example, [19] and [18]. Hyperkähler manifolds often arise as moduli spaces of (gravitational) instantons and monopoles, and so-called S-duality predicts the dimension of the L^2-cohomology of these moduli spaces (Sen's conjecture). Many of these spaces can be compactified to give a space with nonisolated conical singularities. In such cases, our results can be applied. We also refer the reader to the work [18] of Hausel, Hunsicker and Mazzeo, which studies the L^2-cohomology and L^2-harmonic forms of noncompact spaces with fibered geometric ends and their relation to the intersection cohomology of the compactification. Various applications related to Sen's conjecture are also considered there.

Combining the index theorem of [4] with our topological computation of the L^2-signature of T, we recover the following adiabatic limit formula of [13]; see also [32; 5; 9; 3].

COROLLARY 3. *Assume that the fiber Z is odd-dimensional. Then we have the following adiabatic limit formula for the eta invariant of the signature operator*:

$$\lim_{\varepsilon \to 0} \eta(A_{M,\varepsilon}) = \int_B \mathcal{L}\left(\frac{R^B}{2\pi}\right) \wedge \tilde{\eta} + \tau.$$

In the general case, that is, with no dimension restriction on the fiber, the L^2-signature for generalized Thom spaces is discussed in [21]. In particular, Theorem 1 is proved for the general case in [21]. However, one of the ingredients there is the adiabatic limit formula of [13], rather than the direct topological approach taken here. One of our original motivations was to give a simple topological proof of the adiabatic limit formula. In [20], the methods and techniques in [11] are used in the more general situation to derive a very interesting topological interpretation for the invariant τ_r. On the other hand, in [7], our result on the generalized Thom space, together with the result in [13], is used to derive the signature formula for manifolds with nonisolated conical singularity.

References

[1] A. Borel et al. *Intersection cohomology*. Notes on the seminar held at the University of Bern, Bern, 1983. Reprint of the 1984 edition. Birkhäuser, Boston, 2008.

[2] M. F. Atiyah, V. K. Patodi, and I. M. Singer. Spectral asymmetry and riemannian geometry, I–III. *Math. Proc. Cambridge Philos. Soc.* 77(1975):43–69, 78(1975):405–432, 79(1976):71–99.

[3] J.-M. Bismut and J. Cheeger. η-invariants and their adiabatic limits. *Jour. Amer. Math. Soc.* 2:33–70, 1989.

[4] J.-M. Bismut and J. Cheeger. Remarks on families index theorem for manifolds with boundary. *Differential geometry*, 59–83, Pitman Monogr. Surveys Pure Appl. Math., 52, Longman Sci. Tech., Harlow, 1991 eds. Blaine Lawson and Keti Tenenbaum.

[5] J.-M. Bismut and D. S. Freed. The analysis of elliptic families I, II. *Commun. Math. Phys.* 106:159–167, 107:103–163, 1986.

[6] R. Bott and L. Tu. *Differential forms in algebraic topology*. Graduate Texts in Mathematics, 82. Springer, New York-Berlin, 1982.

[7] J. Brüning. The signature operator on manifolds with a conical singular stratum. *Astérisque*, to appear.

[8] J. Cheeger. On the Hodge theory of Riemannian pseudomanifolds. *Geometry of the Laplace operator*, Proc. Sympos. Pure Math., Amer. Math. Soc., Providence, R.I. 36:91–146, 1980.

[9] J. Cheeger. Spectral geometry of singular Riemannian spaces. *J. Diff. Geom.* 18: 575–657, 1983.

[10] J. Cheeger. Eta invariants, the adiabatic approximation and conical singularities. *J. Diff. Geom.* 26:175–211, 1987.

[11] J. Cheeger and X. Dai. L^2-cohomology of spaces of non-isolated conical singularity and non-multiplicativity of signature, *Riemannian topology and geometric structures on manifolds*, 1–24. Progress in Mathematics, 271, Birkhäuser, Boston, 2009,

[12] J. Cheeger, M. Goresky, and R. MacPherson. L^2-cohomology and intersection homology of singular algebraic varieties. *Seminar on Differential Geometry*, Ann. of Math. Stud., 102, Princeton Univ. Press, Princeton, N.J. pages 303–340, 1982.

[13] X. Dai. Adiabatic limits, nonmultiplicativity of signature, and Leray spectral sequence. *J. Amer. Math. Soc.* 4:265–321, 1991.

[14] G. de Rham. *Differentiable manifolds: forms, currents, harmonic forms*. Grundlehren der Mathematischen Wissenschaften, 266. Springer, Berlin, 1984.

[15] M. Gaffney. A special Stokes's theorem for complete Riemannian manifolds. *Ann. of Math.* (2) 60, (1954). 140–145.

[16] M. Goresky and R. MacPherson. Intersection homology theory. *Topology*, 19:135–162, 1980.

[17] M. Goresky and R. MacPherson. Intersection homology II. *Invent. Math.* 71:77–129, 1983.

[18] T. Hausel, E. Hunsicker, and R. Mazzeo. The Hodge cohomology of gravitational instantons. *Duke Math. J.* 122(3):485–548, 2004.

[19] N. Hitchin. L^2-cohomology of hyperkähler quotients. *Comm. Math. Phys.* 211 (1):153–165, 2000.

[20] E. Hunsicker. Hodge and signature theorems for a family of manifolds with fibration boundary. *Geometry and Topology*, 11:1581–1622, 2007.

[21] E. Hunsicker, R. Mazzeo. Harmonic forms on manifolds with edges. *Int. Math. Res. Not.* 2005, no. 52, 3229–3272.

[22] F. Kirwan. *An introduction to intersection homology theory*. Pitman Research Notes in Mathematics, 187. Longman, Harlow.

[23] K. Kodaira. Harmonic fields in Riemannian manifolds (generalized potential theory). *Ann. of Math.* (2) 50, (1949). 587–665.

[24] E. Looijenga. L^2-cohomology of locally symmetric varieties. *Compositio Math.* 67:3–20, 1988.

[25] R. Mazzeo. The Hodge cohomology of a conformally compact metric. *J. Diff. Geom.* 28:309–339, 1988.

[26] R. Mazzeo and R. Melrose. The adiabatic limit, Hodge cohomology and Leray's spectral sequence for a fibration. *J. Diff. Geom.* 31(1):185–213, 1990.

[27] R. Mazzeo and R. Phillips. Hodge theory on hyperbolic manifolds. *Duke Math. J.* 60:509–559, 1990.

[28] W. Pardon and M. Stern. L^2-$\bar{\partial}$-cohomology of complex projective varieties. *J. Amer. Math. Soc.* 4(3):603–621, 1991.

[29] S. Kleiman. The development of intersection homology theory. *Pure Appl. Math. Q.* 3(1/3):225–282, 2007.

[30] L. Saper and M. Stern. L^2-cohomology of arithmetic varieties. *Ann. of Math.* 132(2):1–69, 1990.

[31] L. Saper and S. Zucker. An introduction to L^2-cohomology. *Several complex variables and complex geometry*, Proc. Sympos. Pure Math., 52, Part 2, Amer. Math. Soc., Providence, RI, 519–534, 1991.

[32] E. Witten. Global gravitational anomalies. *Comm. Math. Phys.* 100:197–229, 1985.

[33] S. Zucker. Hodge theory with degenerating coefficients. L_2 cohomology in the Poincaré metric. *Ann. of Math.* (2) 109:415–476, 1979.

[34] S. Zucker. L_2 cohomology of warped products and arithmetic groups. *Invent. Math.* 70:169–218, 1982.

XIANZHE DAI
MATHEMATICS DEPARTMENT
UNIVERSITY OF CALIFORNIA, SANTA BARBARA
SANTA BARBARA, CA 93106
UNITED STATES
dai@math.ucsb.edu

Topology of Stratified Spaces
MSRI Publications
Volume 58, 2011

The almost closed range condition

GILLES CARRON

ABSTRACT. The almost closed range condition is presented and we explain
how this notion can be used to give a topological interpretation of the space of
L^2 harmonic forms on the Hilbert schemes of 2 and 3 points on $\mathbb{C}^2$.

À Jacques

1. Introduction

When (M, g) is a compact manifold the celebrated theorem of Hodge and
de Rham says that the spaces of L^2 harmonic forms on M are isomorphic to
the cohomology spaces of M; that is, if we denote by

$$\mathcal{H}^k(M, g) = \{\alpha \in L^2_g(\Lambda^k M), \ d\alpha = d^*\alpha = 0\}$$

the space of L^2 harmonic k-forms,[1] then we have a natural isomorphism

$$\mathcal{H}^k(M, g) \simeq H^k(M, \mathbb{R}).$$

When (M, g) is noncompact but complete, the spaces of L^2 harmonic forms
have an interpretation in terms of reduced L^2 cohomology. A general and naive
question is to understand how we can give some topological interpretation for
these spaces of L^2 harmonic forms. There are many results, as well as predictions and conjectures, in this direction. For instance, Zucker's conjecture
[32] about locally symmetric Hermitian spaces, eventually solved by E. Looijenga, L. Saper and M. Stern [18; 27] and extended by A. Nair [22], and the
recent result of L. Saper [25; 26], as well as results for manifolds with flat ends
[6], manifolds with cylindrical end [2], and negatively curved manifolds with
finite volume [17; 30; 31]. Also, L^2 harmonic forms have some significance

---

[1] Here d^* is the formal adjoint of the exterior differentiation operator d for the L^2 structure induced by
the metric g.

in modern physics and there are several predictions based on a duality arising in string theory: for instance there is Sen's conjecture about the moduli space of magnetic monopoles [28] and the Vafa–Witten conjecture about Nakajima's quiver manifolds [29; 13].

When M has a locally finite open covering $M = \bigcup_\alpha U_\alpha$ admitting a partition of unity with bounded gradient such that on any of M, U_α, $U_\alpha \cap U_\beta$, $U_\alpha \cap U_\beta \cap U_\gamma, \ldots$ the L^2-range of d is closed, then we can sometimes use sheaf cohomology to obtain a topological interpretation of the space of L^2 harmonic forms. However, this is not always possible. Several tools have been developed in order to circumvent this difficulty. For instance, pseudodifferential calculi have been used successfully in several situations [19; 20; 21; 14]. Here we present the notion of almost closed range for d which was introduced in [7]. We give some general results (including a Mayer–Vietoris sequence) that are true if this almost closed range condition is satisfied. We also explain that this condition has to be used with some care. In order to illustrate how this notion is used in [7], we explain the arguments (and amongst them the almost closed range condition) leading to the topological interpretation of the space of L^2 harmonic forms on the Hilbert schemes of 2 and 3 points on $\mathbb{C}^2$.

2. L^2 cohomology

We start with basic definitions, to present the setting and fix notation.

2.1. Definitions. Let (M^n, g) be an oriented Riemannian manifold. We endow it with a smooth positive measure $\mu \, d\mathrm{vol}_g$ (where μ is a positive smooth function), so that we can define the space $L^2_\mu(\Lambda^k M)$ of differential k-forms which are in $L^2_\mu(\Lambda^k M)$. This is a Hilbert space when endowed with the norm

$$\|\alpha\|^2_\mu := \int_M |\alpha(x)|^2_g \, \mu \, d\mathrm{vol}_g(x).$$

The associated Hermitian scalar product will be denoted by $\langle \, . \, , . \, \rangle_\mu$.

We introduce the space $Z^k_\mu(M)$ of L^2_μ k-forms that are weakly closed:

$$Z^k_\mu(M) := \{\alpha \in L^2_\mu(\Lambda^k M) : \int_M \alpha \wedge d\varphi = 0 \text{ for all } \varphi \in C^\infty_0(\Lambda^{n-1-k} M)\}$$

The space $Z^k_\mu(M)$ is in fact a subspace of the (maximal) domain of d,

$$Z^k_\mu(M) \subset \mathcal{D}^k_\mu(d),$$

where $\mathcal{D}^k_\mu(d)$ is the maximal domain of d on $L^2_\mu(\Lambda^k M)$. This is the space of $\alpha \in L^2_\mu(\Lambda^k M)$ such that there is a constant C with

$$\left| \int_M \alpha \wedge d\varphi \right| \le C \|\varphi\|_\mu \quad \text{for all } \varphi \in C^\infty_0(\Lambda^{n-1-k} M).$$

When $\alpha \in \mathcal{D}_\mu^k(d)$ we can define $d\alpha \in L_\mu^2(\Lambda^{k+1} M)$ by duality:

$$\int_M d\alpha \wedge \varphi = (-1)^{k+1} \int_M \alpha \wedge d\varphi \quad \text{for all } \varphi \in C_0^\infty(\Lambda^{n-1-k} M).$$

By definition we have $d\mathcal{D}_\mu^{k-1}(d) \subset Z_\mu^k(M)$, but $d\mathcal{D}_\mu^{k-1}(d)$ is not necessarily a closed subspace of $L_\mu^2(\Lambda^k M)$. If we introduce the space $B_\mu^k(M) = d\mathcal{D}_\mu^{k-1}(d)$, then $B_\mu^k(M) \subset Z_\mu^k(M)$ since $Z_\mu^k(M)$ is a closed subspace of $L_\mu^2(\Lambda^k M)$.

DEFINITION 2.1. The k-th reduced L_μ^2 cohomology space is the quotient

$$\mathbb{H}_\mu^k(M) = \frac{Z_\mu^k(M)}{B_\mu^k(M)}.$$

The k-th L_μ^2 cohomology space is the quotient

$$\frac{Z_\mu^k(M)}{d\mathcal{D}_\mu^{k-1}(d)},$$

which is not a Hilbert space but satisfies other good properties of a cohomology theory; for instance, a Mayer–Vietoris sequence holds for L_μ^2 cohomology.

Our aim is to circumvent the fact that in general we have problems computing reduced L_μ^2 from local calculations because the Mayer–Vietoris exact sequence does not hold in the reduced setting.

2.2. Some general properties of reduced L_μ^2 cohomology

Quasi-isometry invariance. The first general fact is a consequence of the definition: L_μ^2 (reduced or not) cohomology spaces depend only on the L_μ^2 topology; hence if g_0 and g_1 are two Riemannian metrics such that $\varepsilon g_0 \leq g_1 \leq g_0/\varepsilon$ for a certain $\varepsilon > 0$, and if μ_0, μ_1 are positive smooth functions such that μ_0/μ_1 and μ_1/μ_0 are bounded, then

$$\mathbb{H}_{\mu_0}^k(M, g_0) = \mathbb{H}_{\mu_1}^k(M, g_1).$$

Smooth forms in L^2 cohomology. Using de Rham's smoothing operator of (see [10] and also [9]), we can show that reduced and nonreduced L_μ^2 cohomology can be computed using only smooth forms; that is,

$$\mathbb{H}_\mu^k(M) \simeq \frac{Z_\mu^k(M) \cap C^\infty(\Lambda^k M)}{d\mathcal{D}_\mu^{k-1}(d) \cap C^\infty(\Lambda^k M) \cap C^\infty(\Lambda^k M)}.$$

This smoothing argument also shows that if M is a closed manifold, reduced and nonreduced L_μ^2 cohomology are both isomorphic to de Rham cohomology.

The smoothing operator gives additional results in the following setting: Assume that M is an open subset in a manifold N such that near every point of the boundary $p \in \partial M = \overline{M} \setminus M$ there is a submersion $x = (x_1, \ldots, x_k) : U \to \mathbb{R}^k$ on a neighborhood of p such that $x(p) = 0$ and $U \cap M = \{x_1 > 0, \ldots, x_k > 0\}$. Such a manifold will be called a manifold with corners. Consider a Riemannian metric on M which extends smoothly to N. Moreover, assume that $(\overline{M}, g)$ is *metrically complete*, that is, for any $o \in M$ and $r > 0$ then the closure $\overline{B(o, r)} \cap \overline{M}$ is compact in $\overline{M}$. This is automatically the case when g extends to a smooth geodesically complete metric on N.

Then we can define two spaces of smooth forms: $C_0^\infty(\Lambda^k M)$ is the set of smooth forms with compact support in M and $C_0^\infty(\Lambda^k \overline{M})$ is the set of smooth forms with compact support in $\overline{M}$. This is illustrated in Figure 1. Then a smoothing argument shows:

PROPOSITION 2.2. *If (M, g) is a Riemannian manifold with corner whose closure is metrically complete, then $C_0^\infty(\Lambda^{k-1} \overline{M})$ is dense in $\mathcal{D}_\mu^{k-1}(d)$ when the domain of d is endowed with the graph norm:*

$$\alpha \mapsto \sqrt{\|\alpha\|_\mu^2 + \|d\alpha\|_\mu^2}.$$

2.3. Harmonic forms and L_μ^2 cohomology. When the Riemannian manifold (M, g) is *geodesically complete* (hence boundaryless), reduced L_μ^2 cohomology has an interpretation in terms of appropriate harmonic L_μ^2 forms. We introduce d_μ^*, the formal L_μ^2 adjoint of the operator d; it is defined through the integration by parts formula

$$\langle d\alpha, \beta \rangle_\mu = \langle \alpha, d_\mu^* \beta \rangle_\mu \quad \text{for all } \alpha \in C_0^\infty(\Lambda^k M) \text{ and } \beta \in C_0^\infty(\Lambda^{k+1} M),$$

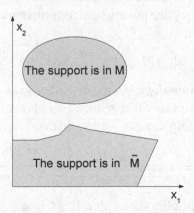

Figure 1. The support of an element of $C_0^\infty(\Lambda^k M)$ and the support of an element of $C_0^\infty(\Lambda^k \overline{M})$. Here $\overline{M}$ is the positive quadrant $\{x_1 \geq 0, x_2 \geq 0\}$ in the plane.

Then we have

$$\mathbb{H}_\mu^k(M) \simeq \{\alpha \in L_\mu^2(\Lambda^k M) : d\alpha = d_\mu^* \alpha = 0\}$$
$$= \{\alpha \in L_\mu^2(\Lambda^k M) : (d_\mu^* d + d d_\mu^*)\alpha = 0\}.$$

3. The almost closed range condition

From now we assume that (M, g) is a manifold with corner whose closure is metrically complete.

3.1. Good primitives. A natural question, which leads to a better understanding of reduced L_μ^2 cohomology, is how an L_μ^2 smooth closed form

$$\alpha \in L^2(\Lambda^k M) \cap C^\infty(\Lambda^k \overline{M})$$

can be zero in reduced L_μ^2 cohomology.

A result of de Rham [10, Theorem 24] implies that for such an α, there is always a $\beta \in C^\infty(\Lambda^k \overline{M})$ such that

$$\alpha = d\beta,$$

but this β will not generally be in L^2. By Proposition 2.2, the vanishing of the reduced L_μ^2 cohomology class of α is equivalent to the existence of a sequence of smooth forms $\beta_j \in C_0^\infty(\Lambda^{k-1} \overline{M})$ such that

$$d\beta = L_\mu^2 \text{-} \lim_{j \to \infty} d\beta_j.$$

Hence the problem is to understand what growth conditions on the primitive β imply the existence of such a sequence of smooth compactly supported forms, (β_j).

It is clear that if $\beta \in L_\mu^2$ then the class of $\alpha = d\beta$ is zero in $\mathbb{H}_\mu^k(M)$. A natural way to obtain a more general condition is to find a sequence of cut-off[2] Lipschitz functions χ_j satisfying the following conditions:

- χ_j tends to 1 uniformly on the compact sets of $\overline{M}$.
- $L_\mu^2 - \lim_{j \to \infty} d\chi_j \wedge \beta = 0$.

We always have $\chi_j \beta \in \mathcal{D}_\mu^{k-1}(d)$ and $d(\chi_j \wedge \beta) = d\chi_j \wedge \beta + \chi_j d\beta$; hence these conditions would imply $\alpha = d\beta = L_\mu^2 - \lim_{j \to \infty} d\beta_j$. The notion of parabolic weights, which we're about to introduce is used to describe the regulation required on the growth of a primitive β at infinity needed for this idea to work.

---

[2]i.e., with compact support on $\overline{M}$.

3.2. Parabolic weights

DEFINITION 3.1. A positive function $w : M \to (0, +\infty)$ is called a *parabolic weight* when there is a function $\psi : (0, +\infty) \to (0, +\infty)$ such that

- for a fixed point $o \in M$ and $r(x) = d(o, x)$ we have $w \geq \dfrac{1}{\psi^2(r)}$, and

- $\displaystyle\int_1^\infty \frac{dr}{\psi(r)} = +\infty.$

LEMMA 3.2. *Assume that* $\alpha \in Z_\mu^k(M)$ *satisfies* $\alpha = d\beta$, *with* $\beta \in L_{w\mu}^2$ *for some parabolic weight* w. *Then the reduced* L_μ^2 *class of* α *is zero.*

We will not give a proof of this result, but we note that the parabolic condition makes possible the choice of a good sequence of cutoff functions as described in the last subsection. We will instead explain where this definition comes from. Parabolicity for a weighted Riemannian manifold (M, g, ν) has several equivalent definitions in terms of Brownian motion, capacity, and existence of positive Green functions [1; 12]. We will only give the following definition:

DEFINITION 3.3. The weighted Riemannian manifold (M, g, ν) is called parabolic if there is a sequence of cut-off Lipschitz functions χ_j such that

- χ_j tends to 1 uniformly on the compact set of $\overline{M}$, and
- $\lim_{j \to \infty} \int_M |d\chi_j|^2 \nu d \operatorname{vol}_g = 0.$

Here is a well known criterion that implies parabolicity.

PROPOSITION 3.4. *Let* o *be a fixed point in the weighted Riemannian manifold* (M, g, ν) *and let*

$$L(r) = \int_{\partial B(o,r)} \nu \, d\sigma_g \quad \text{and} \quad V(r) = \int_{B(o,r)} \nu \, d\operatorname{vol}_g .$$

If

$$\int_1^\infty \frac{dr}{L(r)} = +\infty \quad \text{or} \quad \int_1^\infty \frac{r\,dr}{V(r)} = +\infty,$$

then (M, g, ν) *is parabolic.*

By definition, the parabolicity of the weighted manifold $(M, g, |\beta|^2 \mu)$ implies that the reduced L_μ^2 class of $d\beta = \alpha$ is zero. When w is a parabolic weight, we define the space $\mathcal{C}_{w,\mu}^{k-1}(M)$ to be the set of $\beta \in L_{w\mu}^2(\Lambda^{k-1} M)$ such that the weak differential of β is in L_μ^2:

$$\mathcal{C}_{w,\mu}^{k-1}(M) := \{\beta \in L_{w\mu}^2(\Lambda^{k-1} M) : d\beta \in L_\mu^2\}.$$

Then the parabolicity of w implies that $d : \mathcal{C}_{w,\mu}^{k-1}(M) \to B_\mu^k(M)$ is a bounded operator.

3.3. The almost closed range properties

DEFINITION 3.5. We say that the L^2_μ range of d is *almost closed in degree k with respect to w* when w is a parabolic weight and

$$d\mathcal{C}^{k-1}_{w,\mu}(M) = B^k_\mu(M).$$

In other words, the L^2_μ range of d is almost closed in degree k with respect to w if and only if every L^2_μ closed forms α which is zero in reduced L^2_μ cohomology has a $L^2_{w\mu}$ primitive.

An example. Let Σ be an $(n-1)$-dimensional compact manifold with boundary, endowed with a smooth Riemannian metric h that extends smoothly to $\partial\Sigma$. The truncated cone $C_1(\Sigma) = (1, +\infty) \times \Sigma$ endowed with the conical metric

$$(dr)^2 + r^2 h$$

is a manifold with corner whose closure is metrically complete. Then we have:

PROPOSITION 3.6. *Consider the weight $\mu_a(r, \theta) = r^{2a}$. The $L^2_{\mu_a}$ cohomology of $C_1(\Sigma)$ is given by*

$$\mathbb{H}^k_{\mu_a}(C_1(\Sigma)) = \begin{cases} \{0\} & \text{if } k \le \frac{n}{2} + a, \\ H^k(\Sigma) & \text{if } k > \frac{n}{2} + a. \end{cases}$$

Introduce the two (parabolic) weights $\overline{w} = 1/r^2$ and $w = 1/(r^2 \log^2(r+1))$.

- *If $k \ne n/2 + a$ then the $L^2_{\mu_a}$ range of d is almost closed in degree k with respect to $\overline{w}$.*
- *If $k = n/2 + a$ then in general the $L^2_{\mu_a}$ range of d is almost closed in degree k with respect to w.*
- *If $k = n/2 + a$ and $H^{\frac{n}{2}+a-1}(\Sigma) = \{0\}$, the $L^2_{\mu_a}$ range of d is almost closed in degree k with respect to $\overline{w}$.*

The good news: a Mayer–Vietoris exact sequence. The almost closed range is convenient because it implies a short Mayer–Vietoris exact sequence for reduced L^2 cohomology:

PROPOSITION 3.7. *Assume that $M = U \cup V$ and that U, V and $U \cap V$ are manifolds with corners whose closures are metrically complete. Assume that for a parabolic weight $w : M \to (0, +\infty)$, the L^2_μ range of d is almost closed in degree k with respect to w on M, U, V and $U \cap V$, and that the sequence*

$$\{0\} \to \mathcal{C}^{k-1}_{w,\mu}(M) \xrightarrow{r^*} \mathcal{C}^{k-1}_{w,\mu}(U) \oplus \mathcal{C}^{k-1}_{w,\mu}(V) \xrightarrow{\delta} \mathcal{C}^{k-1}_{w,\mu}(U \cap V) \to \{0\}$$

is exact. Then we have the short Mayer–Vietoris exact sequence

$$\mathbb{H}^{k-1}_{w\mu}(U) \oplus \mathbb{H}^{k-1}_{w\mu}(V) \xrightarrow{\delta} \mathbb{H}^{k-1}_{w\mu}(U \cap V)$$

$$\xrightarrow{b} \mathbb{H}^k_\mu(M) \xrightarrow{r^*} \mathbb{H}^k_\mu(U) \oplus \mathbb{H}^k_\mu(V) \xrightarrow{\delta} \mathbb{H}^k_\mu(U \cap V).$$

We will not give the proof of this result. In fact, the argument is relatively straightforward. We have only to follow the proof of the exactness of the Mayer–Vietoris sequence in de Rham cohomology in the compact case; the hypotheses made here are the ones that are necessary to adapted these classical arguments.

The bad news. There is a difficulty with the assumption

$$\mathcal{C}^{k-1}_{w,\mu}(U) \oplus \mathcal{C}^{k-1}_{w,\mu}(V) \xrightarrow{\delta} \mathcal{C}^{k-1}_{w,\mu}(U \cap V) \to \{0\}.$$

For instance, let $C_1(\Sigma)$ be a truncated cone over a compact manifold with boundary (Σ, h). Now when $\Sigma = \tilde{U} \cup \tilde{V}$, where $\tilde{U}, \tilde{V}, \tilde{U} \cap \tilde{V}$ are open with smooth boundaries, then if we let $U = C_1(\tilde{U})$ and $V = C_1(\tilde{V})$, we have that for the parabolic weight $\overline{w} = 1/r^2$, the sequence

$$\mathcal{C}^{k-1}_{\overline{w},1}(U) \oplus \mathcal{C}^{k-1}_{\overline{w},1}(V) \xrightarrow{\delta} \mathcal{C}^{k-1}_{\overline{w},1}(U \cap V) \to \{0\}$$

is exact. However, for the parabolic weight $w = 1/(r^2 \log^2(r+1))$, the sequence

$$\mathcal{C}^{k-1}_{w,1}(U) \oplus \mathcal{C}^{k-1}_{w,1}(V) \xrightarrow{\delta} \mathcal{C}^{k-1}_{w,1}(U \cap V) \to \{0\}$$

is not (necessarily) exact.

From Proposition 3.6, we see that on a truncated cone we cannot always use only the weight $\overline{w}$. Thus we'll have some difficulties using this exact sequence.

3.4. Comparison with other notions

With the nonparabolicity condition. In [4] and [5], we introduced the notion of nonparabolicity at infinity for the Dirac operator on a complete Riemannian manifold and used it in [6] to compute the L^2 cohomology of manifolds with flat ends. This condition is an extended Fredholmness condition. Specialized to the case of the Gauss–Bonnet operator, $d + d^*_\mu$, this condition is satisfied in the following case:

PROPOSITION 3.8. *Assume (M, g) is a complete Riemannian manifold and there is a weight $w : M \to (0, +\infty)$ and a compact set $K \subset M$ such that*

$$\|\alpha\|_{w\mu} \le \|(d + d^*_\mu)\alpha\|_\mu \quad \text{for all } \alpha \in C^\infty_0(\Lambda^k(M \setminus K)),$$

Then the reduced L^2_μ cohomology of M is finite-dimensional. Moreover, for any $\alpha \in L^2_\mu(\Lambda^k M)$, there exist $h \in L^2(\Lambda^k M)$ such that dh and $d^*_\mu h$ vanish, $\beta \in L^2_{w\mu}(\Lambda^{k-1} M)$, and $\gamma \in L^2_{w\mu}(\Lambda^{k+1} M)$, such that

$$\alpha = h + d\beta + d^*_\mu \gamma.$$

Finally, for any $\alpha \in B^k_\mu(M)$, there is a $\beta \in L^2_{w\mu}(\Lambda^{k-1} M)$ such that $\alpha = d\beta$.

Hence, under the assumptions of the Proposition 3.8, if w is a parabolic weight the L^2_μ range of d is almost closed in degree k with respect to w. There is a closely related result about the almost closed range condition [7]:

PROPOSITION 3.9. *Assume that* (M, g) *is a complete Riemannian manifold and that* $w : M \to (0, +\infty)$ *is a parabolic weight. Suppose there are a positive constant* C *and a compact set* $K \subset M$ *such that*

$$C\|\alpha\|^2_{w\mu} \le \|d\alpha\|^2_\mu + \|d^*_{w\mu}\alpha\|^2_{w\mu} \quad \text{for all } \alpha \in C^\infty_0(\Lambda^k(M \setminus K)). \quad (3\text{-}1)$$

Then

- *the* L^2_μ *range of* d *is almost closed in degree* k *with respect to* w;
- *the* $L^2_{w\mu}$ *range of* d *is almost closed in degree* $k-1$ *with respect to* w;
- *the space* $H^{k-1}_{w\mu}(M)$ *is finite-dimensional.*

Moreover, these three properties imply the existence of a positive constant C *and a compact set* $K \subset M$ *such that the inequality* (3-1) *holds.*

The first proposition is in fact a statement about the operator $d + d^*_\mu$, whereas the second is a statement about d.

With more classical cohomology theory. Let (X, g) be a complete Riemannian manifold, fix a degree k and assume that we have a sequence of weights w_l (which will depend on k in general) such that $w_k = 1$ and, for all degrees l, the $L^2_{w_l\mu}$ range of d is almost closed in degree l with respect to w_{l-1}/w_l. Then we consider the complex

$$\cdots \to C^{l-1}_{w_{l-1}/w_l, w_l\mu}(X) \xrightarrow{d} C^l_{w_l/w_{l+1}, w_{l+1}\mu}(X) \xrightarrow{d} \cdots$$

When the cohomology of this complex can be computed from a local computation (that is, when there is a Poincaré lemma characterizing the cohomology of this complex), then the L^2_μ cohomology of X can be obtained from the degree k cohomology space of this complex. This method has been used successfully by T. Hausel, E. Hunsicker and R. Mazzeo in [14] to obtain a topological interpretation of the $L^2_{\mu=1}$ cohomology of manifolds with fibered cusp ends or with fibered boundary ends. However, in this case the proof is not simple and the authors have to face the same kind of difficulty as the one we encountered on page 20, essentially because the choice of primitive sometimes doesn't lead to a complex whose cohomology follows from local computations. In this paper, the authors had to compare the $L^2_{\mu=1}$ cohomology with two other weighted cohomologies, L^2_μ, $L^2_{1/\mu}$, with $\mu = r^\varepsilon$ or $\mu = e^{\varepsilon r}$ where r is the function given by distance to a fixed point; this comparison is made with an adapted pseudodifferential calculus.

4. The QALE geometry of the Hilbert scheme of 2 or 3 points

We will now describe the QALE geometry of the Hilbert scheme of 2 and 3 points on $\mathbb{C}^2$. The Hilbert scheme of n points on $\mathbb{C}^2$, denoted by $\mathrm{Hilb}_0^n(\mathbb{C}^2)$, is a crepant resolution of the quotient of $(\mathbb{C}^2)_0^n = \{q \in (\mathbb{C}^2)^n, \sum_j q_j = 0\}$ by the action of the symmetric group S_n, which acts by permutation of the indices:

$$\sigma \in S_n, q \in (\mathbb{C}^2)_0^n, \ \sigma.q = (q_{\sigma^{-1}(1)}, q_{\sigma^{-1}(2)}, \dots, q_{\sigma^{-1}(n)}).$$

Hence we have a resolution of singularities map

$$\pi : \mathrm{Hilb}_0^n(\mathbb{C}^2) \to (\mathbb{C}^2)_0^n/S_n.$$

4.1. The case of 2 points. For $n = 2$, we have

$$(\mathbb{C}^2)_0^n = \{(x, -x) : x \in \mathbb{C}^2\};$$

hence

$$(\mathbb{C}^2)_0^2/S_2 \simeq \mathbb{C}^2/\{\pm \mathrm{Id}\}.$$

Now the crepant resolution of $\mathbb{C}^2/\{\pm \mathrm{Id}\}$ is $T^*\mathbb{P}^1(\mathbb{C})$; indeed, we have that the cotangent bundle of $\mathbb{P}^1(\mathbb{C})$ is the set of pairs (L, ξ) where L is a line in $\mathbb{C}^2$ and $\xi : \mathbb{C}^2 \to \mathbb{C}^2$ a linear map such that the range of ξ is contained in L and such that the kernel of ξ contains L. That is, ξ induces a linear map $\bar{\xi} : \mathbb{C}^2/L \to L$. In particular, $T^*\mathbb{P}^1(\mathbb{C}) \setminus \mathbb{P}^1(\mathbb{C})$ is identified with the set

$$\{\xi \in \mathcal{M}_2(\mathbb{C}^2) : \xi \neq 0, \xi \circ \xi = 0\} = \{\begin{pmatrix} a & b \\ c & -a \end{pmatrix} : a^2 = bc, (a, b, c) \neq (0, 0, 0)\},$$

through the identification $\xi \mapsto (\mathrm{Im}\,\xi, \xi)$. This space is diffeomorphic to the quotient $(\mathbb{C}^2 \setminus \{0\})/\{\pm \mathrm{Id}\}$ through the map

$$\pm(x, y) \mapsto \begin{pmatrix} xy & y^2 \\ x^2 & -xy \end{pmatrix}.$$

$T^*\mathbb{P}^1(\mathbb{C})$ carries a remarkable metric, the Eguchi–Hanson metric, which is Kähler and Ricci flat [11; 3]. Moreover, this metric on $T^*\mathbb{P}^1(\mathbb{C}) \setminus \mathbb{P}^1(\mathbb{C}) \simeq \mathbb{C}^2/\{\pm \mathrm{Id}\}$ is asymptotic to the Euclidean metric. Such a metric is called asymptotically locally euclidean (ALE in short).

4.2. The case of 3 points. We can also understand the geometry of $\mathrm{Hilb}_0^3(\mathbb{C}^2)$. Outside a compact set, $(\mathbb{C}^2)_0^3/S_3$ is a truncated cone over $\mathbb{S}^7/S_3$ and the singular set of the quotient $\mathbb{S}^7/S_3$ pulls back to $\mathbb{S}^7$ as a disjoint union of 3 sub-spheres $\mathbb{S}^3$ given by the intersection of $\mathbb{S}^7$ with the collision planes given by

$$P_{i,j} = \{(q_1, q_2, q_3) \in (\mathbb{C}^2)_0^3 : q_i = q_j\}, \quad \text{where } i < j.$$

This is illustrated in Figure 2, left. These three spheres are interchanged by the action of S_3; hence the singular set of $\mathbb{S}^7/S_3$ is a sphere $\mathbb{S}^3$ and the geometry of

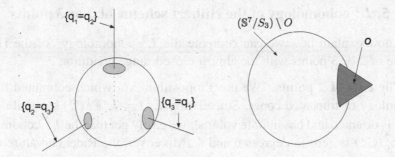

Figure 2. Left: The three collision planes in $(\mathbb{C}^2)_0^3$. Right: $\mathbb{S}^7/S_3$.

$\mathbb{S}^7/S_3$ near the singular set is the one of $\mathbb{B}^2/\{\pm \operatorname{Id}\} \times \mathbb{S}^3$, where $\mathbb{B}^2$ is the unit ball in $\mathbb{C}^2$.

Hence, as illustrated in Figure 2, right, outside a compact set, $(\mathbb{C}^2)_0^3/S_3$ is the union of

- U, a truncated cone over $(\mathbb{S}^7/S_3) \setminus O$, where O is an ε-neighborhood of the singular set (hence homeomorphic to $\mathbb{B}^2/\{\pm \operatorname{Id}\} \times \mathbb{S}^3$),
- $\{(x, v) \in \mathbb{C}^2/\{\pm \operatorname{Id}\} \times \mathbb{C}^2 : |x|^2 + |v|^2 > 1, |x| < \varepsilon|v|\}$.

The geometry at infinity of $\operatorname{Hilb}_0^3(\mathbb{C}^2)$ is the union of two open sets:

- U: a truncated cone over $(\mathbb{S}^7/S_3) \setminus O$, and
- $V = \{(y, v) \in T^*\mathbb{P}^1(\mathbb{C}) \times (\mathbb{C}^2 \setminus \mathbb{B}^2) : |y| < \varepsilon|v|\}$, where $|y|$ is the pullback to $T^*\mathbb{P}^1(\mathbb{C})$ of the Euclidean distance;

that is, there is a compact set $K \subset \operatorname{Hilb}_0^3(\mathbb{C}^2)$ such that $\operatorname{Hilb}_0^3(\mathbb{C}^2) \setminus K = U \cup V$.

In [16], D. Joyce constructed a hyperkähler metric g on $\operatorname{Hilb}_0^3(\mathbb{C}^2)$ (in particular this is a Kähler and Ricci flat metric) which is quasi-asymptotically locally euclidean (QALE) asymptotic to $(\mathbb{C}^2)_0^3/S_3$, which means that

- on U, the truncated cone over $(\mathbb{S}^7/S_3) \setminus O$, we have for all $l \in \mathbb{N}$

$$\nabla^l(g - \mathrm{eucl}) = O\left(\frac{1}{r^{4+l}}\right),$$

where r is the radial function on this cone.

- On V we have, for all $l \in \mathbb{N}$,

$$\nabla^l\big(g - (g_{\operatorname{Hilb}_0^2(\mathbb{C}^2)} + \mathrm{eucl})\big) = O\left(\frac{1}{|y|^{2+l}|v|^2}\right).$$

5. L^2 cohomology of the Hilbert scheme of 2 or 3 points

We now explain how we can compute the L^2 cohomology[3] of the Hilbert Scheme of 2 or 3 points with the almost closed range condition.

5.1. The case of 2 points. We use Proposition 3.6, which computed the L^2 cohomology of truncated cones. Since $\text{Hilb}_0^2(\mathbb{C}^2) = T^*\mathbb{P}^1(\mathbb{C})$ has real dimension 4, is oriented and has infinite volume, we easily get that the L^2 cohomology of $\text{Hilb}_0^2(\mathbb{C}^2)$ is zero in degrees 0 and 4. Moreover, the Ricci curvature of the Eguchi–Hansen metric is zero, so the Bochner formula implies that for any L^2 harmonic 1-form α, we get

$$0 = \int_{\text{Hilb}_0^2(\mathbb{C}^2)} |d\alpha|^2 + |d^*\alpha|^2 = \int_{\text{Hilb}_0^2(\mathbb{C}^2)} |\nabla\alpha|^2.$$

Hence the space of L^2 harmonic 1-forms is trivial and it remains only to compute the L^2 cohomology of $\text{Hilb}_0^2(\mathbb{C}^2)$ in degree 2. The main point is this:

LEMMA 5.1. *The natural map from cohomology with compact support to L^2 cohomology is surjective in degree 2:*

$$H_c^2(\text{Hilb}_0^2(\mathbb{C}^2)) \to \mathbb{H}^2(\text{Hilb}_0^2(\mathbb{C}^2)) \to \{0\}.$$

PROOF. As a matter of fact, if $\alpha \in Z_{\mu=1}^2(\text{Hilb}_0^2(\mathbb{C}^2))$ is a L^2 closed 2-form, its restriction to the neighborhood U of infinity[4] is exact, according to Proposition 3.6. Moreover, for w as in that proposition, we can find $\beta \in \mathcal{C}_{w,1}^1(U)$ a primitive of $\alpha|_U$:

$$\text{on } U, \ \alpha = d\beta.$$

If $\bar{\beta} \in \mathcal{C}_{w,1}^1(\text{Hilb}_0^2(\mathbb{C}^2))$ is an extension of β then because w is a parabolic weight, $\alpha - d\bar{\beta}$ and α have the same L^2 cohomology class and $\alpha - d\bar{\beta}$ has compact support.   □

The Hodge and Poincaré dualities imply that we also have an injective map from L^2 cohomology to absolute cohomology:

$$\{0\} \to \mathbb{H}^2(\text{Hilb}_0^2(\mathbb{C}^2)) \to H^2(\text{Hilb}_0^2(\mathbb{C}^2)).$$

But the natural map from cohomology with compact support to absolute cohomology is an isomorphism in degree 2; hence we have the following isomorphism:

---

[3] L^2 cohomology refers to reduced $L_{\mu=1}^2$ cohomology. From now on we will avoid the subscript 1 when dealing with spaces related to $L_{\mu=1}^2$ cohomology.

[4] The ALE condition says that on U, the Eguchi–Hanson metric and the Euclidean metric (for which U will be a truncated cone over $\mathbb{S}^3/\{\pm \text{Id}\}$) are quasi-isometric. Hence by 2.2, the L^2 cohomology of the two metrics are the same.

THEOREM 5.2. *For* $\mathrm{Hilb}_0^2(\mathbb{C}^2) = T^*\mathbb{P}^1(\mathbb{C})$ *endowed with the Eguchi–Hansen metric, we have*

$$\mathbb{H}^k(\mathrm{Hilb}_0^2(\mathbb{C}^2)) = \begin{cases} \{0\} & \text{if } k \neq 2, \\ \mathbb{R} \simeq H_c^2(\mathrm{Hilb}_0^2(\mathbb{C}^2)) \simeq H^2(\mathrm{Hilb}_0^2(\mathbb{C}^2)) & \text{if } k = 2. \end{cases}$$

REMARK 5.3. In fact, there is a general result about the L^2 cohomology of manifolds with conical ends: *Suppose* (X^n, g) *is a Riemannian manifold with conical ends (meaning that there is a compact set with smooth boundary $K \subset X$ such that $(X \setminus K, g)$ is isometric to the truncated cone $C_1(\partial K)$). Let w be a smooth function on X such that*

$$w(r, \theta) = r^{-2}(1 + \log r)^{-2}. \quad \text{on } X \setminus K \simeq C_1(\partial K).$$

Then, on (X, g), the L^2 range of d is almost closed in any degree with respect to w and the L^2 cohomology of (X, g) is given by

$$\mathbb{H}_\mu^k(X) = \begin{cases} H_c^k(X) & \text{if } k < n/2, \\ \mathrm{Im}\big(H_c^k(X) \to H^k(X)\big) & \text{if } k = n/2, \\ H^k(X) & \text{if } k > n/2. \end{cases}$$

There are different proofs of this result. The first one uses the scattering calculus developed by Melrose; see [21, Theorem 4] and [14, Theorem 1A]. The second uses the almost closed range condition, the computation of the L^2 and weighted L_w^2 cohomologies of a truncated cone, and the Mayer–Vietoris sequence 3.7 [7, Theorem 4.11]. For the case of the Eguchi–Hanson metric, this topological interpretation of the space of L^2 cohomology can also be obtained using explicit computation of harmonic forms because this metric has an SU(2) invariance; hence the harmonic equation reduces to an ODE [15, section 5.5].

5.2. The case of 3 points

A vanishing result outside degree 4. According to [8], the QALE metric on $\mathrm{Hilb}_0^3(\mathbb{C}^2)$ constructed by D. Joyce coincides with the one of H. Nakajima, who showed in [23] that $\mathrm{Hilb}_0^3(\mathbb{C}^2)$ can be endowed with a hyperkähler metric using the hyperkähler reduction of a Euclidean quaternionic space.[5] According to N. Hitchin, the L^2 cohomology of a hyperkähler reduction of a Euclidean quaternionic space is trivial except perhaps for the degree equal to the middle (real) dimension [15]. Hence in our case, we only need to compute the L^2 cohomology of $\mathrm{Hilb}_0^3(\mathbb{C}^2)$ in degree 4.

---

[5]This is a general fact for all the Hilbert schemes of points in $\mathbb{C}^2$, $\mathrm{Hilb}_0^n(\mathbb{C}^2)$, $n \geq 2$.

The result in degree 4. It is again true that for $\mathrm{Hilb}^3_0(\mathbb{C}^2)$, the natural map from cohomology with compact support to absolute cohomology is an isomorphism in degree 4 and moreover these spaces have dimension 1. We have:

LEMMA 5.4. *There exists a compact set $K \subset \mathrm{Hilb}^3_0(\mathbb{C}^2)$ such that $\mathrm{Hilb}^3_0(\mathbb{C}^2)$ retracts on K and such that any L^2 closed 4-form α on $\mathrm{Hilb}^3_0(\mathbb{C}^2) \setminus K$ has a primitive $\beta \in L^2_w$ with $w = 1/(r\log(r+1))^2$. In particular, on $\mathrm{Hilb}^3_0(\mathbb{C}^2) \setminus K$, the L^2 range of d is almost closed in degree 4 with respect to w.*

Using this and the same arguments as in the case of 2 points, we obtain:

THEOREM 5.5. *For $\mathrm{Hilb}^3_0(\mathbb{C}^2)$ endowed with the QALE metric described in (4.2) we have:*

$$\mathbb{H}^k(\mathrm{Hilb}^3_0(\mathbb{C}^2)) = \begin{cases} \{0\} & \text{if } k \neq 4, \\ \mathbb{R} \simeq H^4_c(\mathrm{Hilb}^3_0(\mathbb{C}^2)) \simeq H^4(\mathrm{Hilb}^3_0(\mathbb{C}^2)) & \text{if } k = 4. \end{cases}$$

THE PROOF OF LEMMA 5.4. Let $K \subset \mathrm{Hilb}^3_0(\mathbb{C}^2)$ be a compact set such that $\mathrm{Hilb}^3_0(\mathbb{C}^2) \setminus K = U \cup V$, where

- U is a truncated cone over $(\mathbb{S}^7/S_3) \setminus O$, and
- $V = \{(y, v) \in T^*\mathbb{P}^1(\mathbb{C}) \times (\mathbb{C}^2 \setminus \mathbb{B}^2) : |y| < \varepsilon|v|\}$, where $|y|$ is the pullback to $T^*\mathbb{P}^1(\mathbb{C})$ of the Euclidean distance.

The main point in the proof of Lemma 5.4 is the following result concerning the L^2 cohomology of V:

LEMMA 5.6. $\mathbb{H}^4(V) = \{0\}$ *and* $Z^4(V) = d\mathcal{C}^3_{w,1}(V)$. *That is, any L^2 closed 4-form α on V has a primitive $\varphi \in L^2_w(\Lambda^3 V)$, that is, $\alpha = d\varphi$.*

We will only sketch the proof of this lemma.

Consider $\alpha \in Z^4_{\mu=1}(V)$. The set $U \cap V$ is a truncated cone over the product $\mathbb{S}^3/\{\pm \mathrm{Id}\} \times \mathbb{S}^3 \times (\varepsilon, 2\varepsilon)$. The third Betti number of this product is not zero; hence by Proposition 3.6, there is a $\psi \in L^2_w(\Lambda^3(U \cap V))$ such that

$$\alpha = d\psi \quad \text{on } U \cap V.$$

We cannot extend ψ to V as an element of $\mathcal{C}^3_{w,1}(V)$ but only as an element $\overline{\psi} \in \mathcal{C}^3_{w,\rho}(V)$ where $\rho(y, v) = 1/\log^2(|v| + 1)$. Then

$$\alpha - d\overline{\psi} \in Z^4_\rho(V)$$

and, because this form is zero on $V \cap U$, it can be extended to the whole $T^*\mathbb{P}^1(\mathbb{C}) \times (\mathbb{C}^2 \setminus \mathbb{B}^2)$. This extension will be also denoted by $\alpha - d\overline{\psi}$. It is still a closed form; that is,

$$\alpha - d\overline{\psi} \in Z^4_\rho(T^*\mathbb{P}^1(\mathbb{C}) \times (\mathbb{C}^2 \setminus \mathbb{B}^2)).$$

Now using a Künneth-type argument and the computation of the L_ρ^2 cohomology of $\mathbb{C}^2 \setminus \mathbb{B}^2$, it can be shown that the L_ρ^2 cohomology of $T^*\mathbb{P}^1(\mathbb{C}) \times (\mathbb{C}^2 \setminus \mathbb{B}^2)$ vanishes in degree 4 and that if we introduce the weight $\overline{w}_1(y, v) = 1/(1+|y|)^2$, there are

$$u \in \mathcal{C}_{\overline{w}, \rho}^3(T^*\mathbb{P}^1(\mathbb{C}) \times (\mathbb{C}^2 \setminus \mathbb{B}^2)) \quad \text{and} \quad v \in \mathcal{C}_{\overline{w}_1, \rho}^3(T^*\mathbb{P}^1(\mathbb{C}) \times (\mathbb{C}^2 \setminus \mathbb{B}^2))$$

such that

$$\alpha - d\overline{\psi} = du + dv.$$

If we let $\varphi = \overline{\psi} + u + v$, then because on V we have $\varepsilon\overline{w} \leq \overline{w}_1$, we conclude that $\varphi \in L_w^2(\Lambda^3 V)$ and

$$\alpha = d\varphi.$$

With this result, we can finish the proof of Lemma 5.4: we cannot use the Mayer–Vietoris exact sequence because of the log factor in the weight w (see middle of page 20). However, we will use some of the arguments leading to the proof of the exactness of the Mayer–Vietoris sequence.

Let α be a closed L^2 4-form outside K. Because U is a truncated cone, we know that there is $\varphi_U \in \mathcal{C}_{w,1}^3(U)$ such that $\alpha = d\varphi_U$ on U, and by Lemma 5.6, there is a $\varphi_V \in \mathcal{C}_{w,1}^3(V)$ such that on V

$$\alpha = d\varphi_V.$$

Now on the intersection $U \cap V$ the difference $\varphi_U - \varphi_V$ is a closed L_w^2 3-form. But $U \cap V$ is a truncated cone over $\mathbb{S}^3/\{\pm \operatorname{Id}\} \times \mathbb{S}^3 \times (\varepsilon, 2\varepsilon)$ and there is an analogue of Proposition 3.6 for the L_w^2 cohomology, the threshold now being $n/2-1 = 3$ in our case. But the second Betti number of $\mathbb{S}^3/\{\pm \operatorname{Id}\} \times \mathbb{S}^3 \times (\varepsilon, 2\varepsilon)$ is zero; hence on $U \cap V$, $\varphi_U - \varphi_V$ has a primitive $\eta \in \mathcal{C}_{\overline{w}, w}^2(U \cap V)$ which can be extend to a $\overline{\eta} \in \mathcal{C}_{\overline{w}, w}^2(U)$. Now we can define ψ

$$\psi = \begin{cases} \varphi_U + d\overline{\eta} & \text{on } U, \\ \varphi_V & \text{on } V. \end{cases}$$

By construction, we have $\alpha = d\psi$ and $\psi \in L_w^2(\Lambda^3(\operatorname{Hilb}_0^3(\mathbb{C}^2) \setminus K))$. $\qquad \square$

5.3. Conclusion. In the physics literature, $\operatorname{Hilb}_0^n(\mathbb{C}^2)$ is associated to the moduli space of instantons on noncommutative $\mathbb{R}^4$ [24]. One motivation for the study of L^2 cohomology of $\operatorname{Hilb}_0^n(\mathbb{C}^2)$ comes from a question of C. Vafa and E. Witten: in [29], see also the nice survey of T. Hausel [13], the following conjecture is formulated (note that $2(n-1) = \frac{1}{2} \dim_{\mathbb{R}} \operatorname{Hilb}_0^n(\mathbb{C}^2)$):

$$\mathbb{H}^k = \begin{cases} \{0\} & \text{if } k \neq 2(n-1), \\ \operatorname{Im}(H_c^k(\operatorname{Hilb}_0^n(\mathbb{C}^2)) \to H^k(\operatorname{Hilb}_0^n(\mathbb{C}^2))) & \text{if } k = 2(n-1). \end{cases} \tag{5-1}$$

However, Vafa and Witten have said that "unfortunately, we do not understand the prediction of S-duality on noncompact manifolds precisely enough to fully exploit them." According to N. Hitchin's vanishing result [15], the first part of this conjecture is true. The result above says that this is true for $n = 2, 3$.

6. Other results and perspectives

In [7], the L^2 cohomology of certain QALE spaces is computed. The proof uses the same general idea given in 5.2, but the argumentation is considerably longer and we cannot in general use the same vanishing result. Quasi Asymptotically Locally Euclidean (QALE) geometry is defined by induction. A QALE manifold asymptotic to $\mathbb{C}^n / \Gamma$ (where Γ is a finite subgroup of $SU(n)$) is a manifold whose geometry at infinity is the union of a piece that looks (up to a finite cover) like a subset of the product $Y \times \mathbb{C}^p$, where Y is a QALE manifold asymptotic to $\mathbb{C}^{n-p} / A$ for some A a finite subgroup of $SU(n - p)$. In [7], we computed the L^2 cohomology of QALE spaces where the singular space $\mathbb{C}^n / \Gamma$ has only two singular strata ($\{0\}$ and a finite union of linear subspaces). In order to prove the Vafa–Witten conjecture (5-1), it is important to be able to understand the L^2 cohomology of more general QALE spaces. The almost closed range condition has been a interesting tool for doing this for the case of $\mathrm{Hilb}_0^3(\mathbb{C}^2)$. We hope that it will also be useful in other situations and for the other Hilbert schemes of points.

References

[1] A. Ancona A. *Théorie du potentiel sur des graphes et des variétés*. Lecture Notes in Math. 1427, Springer, Berlin, 1990.

[2] M. F. Atiyah, V. K. Patodi, I. M. Singer, Spectral asymmetry and Riemannian geometry I, *Math. Proc. Camb. Phil. Soc.* **77** (1975), 43–69.

[3] E. Calabi, Métriques kählériennes et fibrés holomorphes, *Ann. Sci. École Norm. Sup.* **12**:2 (1979), 269–294.

[4] G. Carron, Un théorème de l'indice relatif. *Pacific J. Math.* **198** (2001), 81–107.

[5] G. Carron, Théorèmes de l'indice sur les variétés non-compactes, *J. Reine Angew. Math.*, **541** (2001), 81–115.

[6] G. Carron, L^2-cohomology of manifolds with flat ends, *Geom. Funct. Anal.* **13**:2 (2003), 366–395.

[7] G. Carron, Cohomologie L^2 des variétés QALE. arXiv:math.DG/0501290. To appear in *J. Reine Angew. Math*.

[8] G. Carron, On the QALE geometry of Nakajima's metric, arXiv:0811.3870. To appear in *J. Inst. Math. Jussieu*.

[9] J. Cheeger, On the Hodge theory of Riemannian pseudomanifolds. In *Geometry of the Laplace operator* (Honolulu, 1979). Proc. Sympos. Pure Math., XXXVI, Amer. Math. Soc., Providence, 1980, 91–146,

[10] G. de Rham, *Variétés différentiables. Formes, courants, formes harmoniques*. 3rd ed., Hermann, Paris, 1973.

[11] T. Eguchi, A. Hanson, Self-dual solutions to Euclidean gravity. *Ann. Physics*, **120**:1 (1979), 82–106.

[12] A. Grigoryan, Analytic and geometric background of recurrence and non-explosion of the Brownian motion on Riemannian manifolds. *Bull. Amer. Math. Soc. (N.S.)* **36**:2 (1999), 135–249.

[13] T. Hausel, S-duality in hyperkähler Hodge theory, preprint, arXiv:0709.0504. To appear in *The many facets of geometry: a tribute to Nigel Hitchin,* edited by O. Garcia-Prada, J.-P. Bourguignon, and S. Salamon, Oxford University Press.

[14] T. Hausel, E. Hunsicker, R. Mazzeo, Hodge cohomology of gravitational instantons, *Duke Math. J.* **122**:3 (2004), 485–548.

[15] N. Hitchin, L^2-cohomology of hyperkähler quotients, *Comm. Math. Phys.* **211** (2000), 153–165.

[16] D. Joyce, *Compact manifolds with special holonomy*, Oxford University Press (2000) Oxford Mathematical Monographs.

[17] J. Lott. On the spectrum of a finite-volume negatively-curved manifold. *Amer. J. Math.* **123**:2 (2001), 185–205.

[18] E. Looijenga, L^2-cohomology of locally symmetric varieties, *Compositio Math.* **67** (1998), 3–20.

[19] R. Mazzeo, The Hodge cohomology of a conformally compact metric, *J. Differential Geom.* **28**:2 (1988), 309–339.

[20] R. Mazzeo, R. S. Phillips, Hodge theory on hyperbolic manifolds, *Duke Math. J.* **60**:2 (1990), 509–559.

[21] R. Melrose, Spectral and scattering theory for the Laplacian on asymptotically Euclidean spaces. In *Spectral and scattering theory: Proceedings of the Taniguchi international workshop*, edited by M. Ikawa. Basel: Marcel Dekker. Lect. Notes Pure Appl. Math. **161** (1994), 85–130.

[22] A. Nair, Weighted cohomology of arithmetic groups. *Ann. of Math.* **150**:1 (1999), 1–31.

[23] H. Nakajima, *Lectures on Hilbert schemes of points on surfaces*. American Mathematical Society, Providence, RI, 1999.

[24] N. Nekrasov, A. Schwarz, Instantons on noncommutative $\mathbb{R}^4$, and $(2, 0)$ superconformal six-dimensional theory. *Comm. Math. Phys.* **198**:3 (1998), 689–703.

[25] L. Saper, L^2-cohomology of locally symmetric spaces. I, *Pure and Applied Mathematics Quarterly*, **1**:4 (2005), 889–937.

[26] L. Saper, L^2-harmonic forms on locally symmetric spaces, Oberwolfach reports, **2**:3, (2005), 2229–2230.

[27] L. Saper, M. Stern, L_2-cohomology of arithmetic varieties, *Ann. Math.* **132**:1 (1990), 1–69.

[28] A. Sen, Dyon-monopole bound states, selfdual harmonic forms on the multi-monopole moduli space and $Sl(2, \mathbb{Z})$-invariance of string theory, *Phys. Lett.* **B 329** (1994), 217–221.

[29] C. Vafa, E. Witten, A strong coupling test of S-duality. *Nuclear Phys. B* **431**:1–2 (1994), 3–77.

[30] N. Yeganefar, Sur la L^2-cohomologie des variétés à courbure négative, *Duke Math. J.* **122**:1 (2004), 145–180.

[31] N. Yeganefar, L^2-cohomology of negatively curved Kähler manifolds of finite volume. *Geom. Funct. Anal.* **15**:5 (2005), 1128–1143.

[32] S. Zucker, L_2 cohomology of warped products and arithmetic groups, *Inventiones Math.*, **70** (1982), 169–218.

GILLES CARRON
LABORATOIRE DE MATHÉMATIQUES JEAN LERAY (UMR 6629)
UNIVERSITÉ DE NANTES
2, RUE DE LA HOUSSINIÈRE
B.P. 92208
44322 NANTES CEDEX 3
FRANCE
  Gilles.Carron@math.univ-nantes.fr

Topology of Stratified Spaces
MSRI Publications
Volume **58**, 2011

Rigidity of differential operators and Chern numbers of singular varieties

ROBERT WAELDER

ABSTRACT. A differential operator D commuting with an S^1-action is said to be rigid if the nonconstant Fourier coefficients of ker D and coker D are the same. Somewhat surprisingly, the study of rigid differential operators turns out to be closely related to the problem of defining Chern numbers on singular varieties. This relationship comes into play when we make use of the rigidity properties of the complex elliptic genus–essentially an infinite-dimensional analogue of a Dirac operator. This paper is a survey of rigidity theorems related to the elliptic genus, and their applications to the construction of "singular" Chern numbers.

1. Rigidity of elliptic differential operators

Let $D : \Gamma(E) \to \Gamma(F)$ be an elliptic operator mapping sections of a vector bundle E to sections of F. If D commutes with a $T = S^1$ action, then ker D and coker D are finite-dimensional S^1-modules. We define the character-valued index

$$\mathrm{Ind}_T(D) = \ker D - \mathrm{coker}\, D \in R(T)$$

For example, if $D = d + d^* : \Omega^{\mathrm{even}} \to \Omega^{\mathrm{odd}}$ is the de Rham operator on a smooth manifold X with a T action, then by Hodge theory and homotopy invariance of de Rham cohomology, $\mathrm{Ind}_T(D)$ is a trivial virtual T-module of rank equal to the Euler characteristic of X. In general, if $\mathrm{Ind}_T(D)$ is a trivial T-module, we say that D is *rigid*. In the case where D is the de Rham operator, both ker D and coker D are independently trivial T-modules. However, more interesting cases exist where D is rigid, but both ker D and coker D are nontrivial T-modules. For example, if X is a spin manifold and $D : \Gamma(\Delta^+) \to \Gamma(\Delta^-)$ is the Dirac

The author is supported by an NSF postdoctoral fellowship.

operator, then D is rigid. It is instructive to sketch the proof of this fact, which is due to Atiyah and Hirzebruch [3]:

For simplicity, assume that T acts on X with isolated fixed points $\{p\}$, and that the action lifts to the spin bundles $\Delta^{\pm}$. At each fixed point p, $T_p X$ decomposes into a sum of one-dimensional complex representations of T with weights $m_1(p), \ldots, m_n(p)$, where $2n = \dim X$. If we view $\mathrm{Ind}_T(D)$ as a function of $t \in T$, then by the Lefschetz fixed point formula,

$$\mathrm{Ind}_T(D) = \sum_p \frac{1}{\prod_{j=1}^n (t^{m_j/2} - t^{-m_j/2})}$$

A priori, $\mathrm{Ind}_T(D)$ is a function only on the unit circle in $\mathbb{C}$. However, the above formula shows that we can analytically continue $\mathrm{Ind}_T(D)$ to a meromorphic function on S^2, with possible poles restricted to lie on the unit circle. But since $\mathrm{Ind}_T(D)$ is a virtual T-module, and therefore has a finite Fourier decomposition of the form $\mathrm{Ind}_T(D) = \sum a_k t^k$, all such poles on the unit circle must cancel. It follows that $\mathrm{Ind}_T(D)$ is constant. Furthermore, by taking the limit as $t \to \infty$, one sees that the character-valued index is identically zero. A similar proof shows that on a complex manifold, $\bar{\partial} + \bar{\partial}^*$ (whose corresponding index is the arithmetic genus) is rigid with respect to holomorphic torus actions.

The situation becomes more difficult if we investigate the rigidity of the twisted Dirac operators $D \otimes E$, where E is an equivariant vector bundle. For example, if $d_S = D \otimes (\Delta^+ \oplus \Delta^-)$ is the signature operator on a spin manifold, the Lefschetz fixed point formula for the index of $d_S \otimes TX$ gives:

$$\mathrm{Ind}_T(d_S \otimes TX) = \sum_p \prod_{j=1}^n \frac{1 + t^{-m_j(p)}}{1 - t^{-m_j(p)}} \cdot \sum (t^{m_j(p)} + t^{-m_j(p)})$$

Here $\pm m_j(p)$ are the weights of the T-action on the complexified tangent bundle of X at p. The factors $\sum (t^{m_j(p)} + t^{-m_j(p)})$ come from the twisting of the rigid operator d_S by TX. Thus, in this situation, the fixed point formula for $\mathrm{Ind}_T(d_S \otimes TX)$ has poles at 0 and ∞, and we can no longer apply the same argument.

It is therefore astonishing that, based on ideas from physics, Witten predicted the rigidity of an infinite sequence of twisted Dirac operators of this nature on a spin manifold. Witten's insight came from generalizing a quantum mechanics-inspired proof of the Atiyah–Hirzebruch theorem to its analogue in the setting of superstring theory. We briefly sketch this point of view, as given in [17]: In supersymmetric quantum mechanics on a spin manifold X (with one real fermion field), the Hilbert space of states corresponds to the space of square-integrable spinors. Quantization of the supercharge Q_+ yields the Dirac operator. In passing to superstring theory, the Hilbert space of states should be interpreted as spinors on the loop space of X. It therefore makes sense to think of the quantization of the supercharge in this theory as a Dirac operator on the loop space. Now for any manifold X, the loop space of X possesses a natural S^1 action given by rotating the loops. The fixed points of this action correspond to the space of constant loops, which we may identify with X itself. Via formal application of the Atiyah–Bott–Lefschetz fixed point formula one can reduce the S^1 character-valued index of operators constructed out of Q_+ to integrals over X. To give an example, let Δ denote the spin bundle on the loop space. If we quantize a theory with two fermionic fields $\psi_\pm$, the associated Hilbert space becomes $\Delta \otimes \Delta$. Now in finite dimensions, $\Delta \otimes \Delta$ corresponds to the de Rham complex. $\Delta \otimes \Delta$ therefore provides a good candidate for the de Rham complex on the loop space. At the classical level of this theory, one has an involution σ on the space of superfields, sending $\psi_+ \mapsto -\psi_+$ and $\psi_- \mapsto \psi_-$ which preserves the action Lagrangian. When X is spin, this involution descends to the quantum theory; the corresponding action of σ on $\Delta \otimes \Delta$ may be interpreted as the Hodge star operator acting on forms. Consequently, one can construct out of Q_+ and σ a canonical choice of a signature operator on the loop space. By the fixed point formula, its S^1-charactered valued index reduces to the index of

$$
d_S \otimes \bigotimes_{m=1}^{\infty} \Lambda_{q^m} TX \otimes \bigotimes_{m=1}^{\infty} S_{q^m} TX = d_S \otimes \Theta_q
$$

over X. Here, for any vector bundle E, we define

$$
\Lambda_{q^m}(E) = 1 + q^m E + q^{2m} E \wedge E + \cdots
$$

and

$$
S_{q^m}(E) = 1 + q^m E + q^{2m} E^2 + \cdots,
$$

where q^m denote the weights of the induced S^1 action of an S^1-bundle over X. If X itself has an S^1 action, the character-valued index of $d_S \otimes \Theta_q$ as a function of $e^{i\theta}$ may be interpreted as the signatures associated to a family of field theories parameterized by θ. The rigidity of $d_S \otimes \Theta_q$ then follows from a formal application of deformation invariance of the index of Dirac operators on loop spaces. For details, see [16] or [17].

Note that since $d_S \otimes \Theta_q = d_S + 2q d_S \otimes TX + \cdots$, the rigidity of $d_S \otimes TX$ now follows from the rigidity of the $d_S \otimes \Theta_q$. It is interesting to point out that, although $d_S \otimes \Theta_q$ is defined on any oriented manifold, it is only rigid for spin manifolds. Heuristically this makes sense when we view $d_S \otimes \Theta_q$ as the signature operator on the loop space of X. For if X is oriented, the signature operator d_S is easily seen to be rigid. But the the loop space is oriented precisely when X is spin.

Dirac operators on the loop space provide concrete examples of *elliptic genera*. These are homomorphisms $\varphi : \Omega^{SO} \to R$ from the oriented cobordism ring to an auxiliary ring R, whose characteristic power series are defined in terms of certain elliptic integral expressions.

The rigidity theorems of Witten were initially proved under restricted hypotheses by Landweber, Stong, and Ochanine [8; 10], and later proved in complete form by Bott, Taubes, and Liu [6; 9]. The simplest and most direct proof was discovered by Liu, who observed that the modular properties of the elliptic genera implied their rigidity. We will provide a sketch of Liu's argument for the case of the complex elliptic genus, which is defined as the index of $\bar{\partial} \otimes E_{q,y}$ on an almost complex manifold of dimension $2n$, where $E_{q,y}$ is given by

$$E_{q,y} = y^{-n/2} \bigotimes_{m=1}^{\infty} \Lambda_{-yq^{m-1}} T''X \otimes \Lambda_{-yq^m} T'X \otimes \bigotimes_{m=1}^{\infty} S_{q^m} T''X \otimes S_{q^m} T'X$$

Here $TX \otimes \mathbb{C} = T'X \oplus T''X$ denotes the decomposition of the complexified tangent bundle into holomorphic and antiholomorphic components. By Riemann–Roch, the ordinary index of this operator is given by the integral

$$\int_X \prod_{T'X} \frac{x_j \vartheta\left(\frac{x_j}{2\pi i} - z, \tau\right)}{\vartheta\left(\frac{x_j}{2\pi i}, \tau\right)}.$$

Here x_j denote the formal Chern roots of $T'X$, $y = e^{2\pi i z}$ and $q = e^{2\pi i \tau}$. $\vartheta(v, \tau)$ denotes the Jacobi theta function

$$\vartheta(v, \tau) = \prod_{m=1}^{\infty} (1 - q^m) \cdot q^{1/8} 2 \sin \pi v \prod_{m=1}^{\infty} (1 - q^m e^{2\pi i v}) \prod_{m=1}^{\infty} (1 - q^m e^{-2\pi i v})$$

We will frequently refer to $\mathrm{Ind}(\bar{\partial} \otimes E_{q,y})$ as $\mathrm{Ell}(X; z, \tau)$. The almost-complex version of Witten's rigidity theorem for this operator states that the complex elliptic genus of X is rigid provided that $c_1(X) = 0$.

The idea of the proof is as follows: For simplicity, assume that the T-action on X has isolated fixed points $\{p\}$, with equivariant weights $m_j(p)$ on $T'_p X$. Writing $t \in T$ as $t = e^{2\pi i u}$, we have by the Lefschetz fixed point formula,

$$\mathrm{Ind}_T(\bar{\partial} \otimes E_{q,y}) = \sum_p \prod_{j=1}^{n} \frac{\vartheta(m_j(p)u - z, \tau)}{\vartheta(m_j(p)u, \tau)}$$

Write $\mathrm{Ind}_T(\bar{\partial} \otimes E_{q,y}) = F(u, z, \tau)$. It is evident from the fixed point formula that $F(u, z, \tau)$ is a meromorphic function on $\mathbb{C} \times \mathbb{C} \times \mathbb{H}$ which is holomorphic in z and τ. Let $z = \frac{1}{N}$ where N is a common multiple of the weights $m_j(p)$. Then, using the translation formulas:

$$\vartheta(u+1, \tau) = -\vartheta(u, \tau),$$

$$\vartheta(u+\tau, \tau) = q^{-1/2} e^{-2\pi i u} \vartheta(u, \tau),$$

it is easy to see that $F(u+1, \frac{1}{N}, \tau) = F(u, \frac{1}{N}, \tau)$ and that $F(u + N\tau, \frac{1}{N}, \tau) = F(u, \frac{1}{N}, \tau)$. Thus, for each fixed τ, $F(u, \frac{1}{N}, \tau)$ is a meromorphic function on the torus defined by the lattice $\mathbb{Z} \oplus N\mathbb{Z}\tau$. Suppose we could show that $F(u, \frac{1}{N}, \tau)$ was in fact holomorphic in u. Then for each multiple N of the weights $m_j(p)$ and for each $\tau \in \mathbb{H}$, $F(u, \frac{1}{N}, \tau)$ would have to be constant in u. It would follow that $\frac{\partial}{\partial u} F(u, \frac{1}{N}, \tau) \equiv 0$. Since this equation held for an infinite set of (u, z, τ) containing a limit point, it would hold for all (u, z, τ). Hence $F(u, z, \tau)$ would be independent of u, which is precisely the statement of rigidity for the operator $\operatorname{Ind}_T(\bar{\partial} \otimes E_{q,y})$.

Thus, we are reduced to proving $F(u, z, \tau)$ is holomorphic. Let $\left(\begin{smallmatrix} a & b \\ c & d \end{smallmatrix}\right) \in SL_2(\mathbb{Z})$ act on $\mathbb{C} \times \mathbb{C} \times \mathbb{H}$ by the rule

$$(u, z, \tau) \mapsto \left(\frac{u}{c\tau+d}, \frac{z}{c\tau+d}, \frac{a\tau+b}{c\tau+d}\right).$$

Using the transformation formula

$$\vartheta\left(\frac{u}{c\tau+d}, \frac{a\tau+b}{c\tau+d}\right) = \zeta(c\tau+d)^{\frac{1}{2}} e^{\frac{\pi i c u^2}{c\tau+d}} \vartheta(u, \tau),$$

where ζ is an eighth root of unity, one sees that $F\left(\frac{u}{c\tau+d}, \frac{z}{c\tau+d}, \frac{a\tau+b}{c\tau+d}\right)$ is equal to

$$K \cdot \sum_p e^{-2\pi i c \sum_{j=1}^n m_j(p) u z / (c\tau+d)} \prod_{j=1}^n \frac{\vartheta(m_j(p)u - z, \tau)}{\vartheta(m_j(p)u, \tau)}$$

where K is a nonzero holomorphic function of (u, z, τ). Now the Calabi–Yau condition implies that the only possible T-action on K_X is given by multiplication by a constant along the fibers. Since $\sum_{j=1}^n m_j(p)$ is the weight of the T-action induced on K_X^*, it follows that $\sum_{j=1}^n m_j(p)$ is the same constant for all p. We may therefore pull the expression $e^{-2\pi i c \sum_{j=1}^n m_j(p) u z / (c\tau+d)}$ outside of the above summation sign, and conclude that

$$F\left(\frac{u}{c\tau+d}, \frac{z}{c\tau+d}, \frac{a\tau+b}{c\tau+d}\right) = K' F(u, z, \tau),$$

for K' a nonzero holomorphic function.

Now the key observation: First, by the fixed point formula, $F(u, z, \tau)$ has possible poles only for $u = r + s\tau$, where $r, s \in \mathbb{Q}$. Moreover, since $F(u, z, \tau)$

is the character-valued index of an elliptic differential operator, the poles of $F(u, z, \tau)$ must cancel for $u \in \mathbb{R}$, since in that case $F(u, z, \tau)$ admits a Fourier decomposition $\sum b_k(z, \tau) e^{2\pi i k u}$ (in a rigorous treatment of the subject, one must of course deal with convergence issues regarding this summation). Note that this is also the key observation in Bott and Taubes' proof. Thus, for u a possible pole, write $u = (m/\ell)(c\tau + d)$, where c and d are relatively prime. By relative primality, we can find integers a and b so that $ad - bc = 1$, i.e., $\left(\begin{smallmatrix} a & b \\ c & d \end{smallmatrix}\right) \in \mathrm{SL}_2(\mathbb{Z})$. Then

$$K' \cdot F\left(\frac{m}{\ell}(c\tau + d), z, \tau\right) = F\left(\frac{m}{\ell}, \frac{z}{c\tau + d}, \frac{a\tau + b}{c\tau + d}\right)$$

where $K' \neq 0$. It follows that $F(u, z, \tau)$ is holomorphic, which completes the proof.

The above rigidity theorem for the complex elliptic genus on a Calabi–Yau manifold has an interesting analogue for toric varieties, which has applications to the study of singular varieties. Let Σ be a complete fan which corresponds to a smooth toric variety X. This means that Σ is a finite union of cones $\{C_i\}$ inside the real vector space $N \otimes \mathbb{R}$, where N is an integral lattice of rank n. For any two cones C_1, C_2 in Σ, we require that $C_1 \cap C_2$ is a proper subcone, and that the union of the cones in Σ covers all of $N \otimes \mathbb{R}$. The smoothness requirement for Σ means that the k-dimensional cones have k generators, each lying in N. Recall that the data of Σ gives rise to a natural scheme structure as follows: For each cone $C \subset \Sigma$, we define the sheaf of regular functions

$$\Gamma(U_C) = \mathbb{C}[e^f]_{f \in S_C}$$

where S_C is the collection of linear functionals $f \in \mathrm{Hom}(N, \mathbb{Z})$ that are non-negative along C. The toric variety X corresponding to these data is the variety with affine charts given by $U_C = \mathrm{Specm}\, \Gamma(U_C)$.

Note that inclusions of cones $C_1 \subset C_2$ give rise to inclusions of open sets $U_{C_1} \subset U_{C_2}$. In particular, since every cone C contains the point $0 \in N$ as a subcone, every open set U_C contains the open set

$$U_0 = \mathrm{Specm}\, \mathbb{C}[e^{\mathrm{Hom}(N, \mathbb{Z})}] \cong (\mathbb{C}^*)^n.$$

The action of this complex torus on itself is easily seen to extend to all of U_C. In this way, X inherits a natural action by a complex torus $T_{\mathbb{C}}$, with isolated fixed points.

There is a nice relationship between the $T_{\mathbb{C}}$-invariant divisors on a smooth toric variety and combinatorial data of its associated simplicial fan: the $T_{\mathbb{C}}$-invariant divisors on X are in one-one correspondence with piecewise linear functionals on Σ. For example, if f is a piecewise linear functional on Σ, then f is completely determined by its values $f(v_i)$ on the generators $v_i \in N$

of the 1-dimensional rays of Σ. These generators, in turn, define $T_{\mathbb{C}}$-Cartier divisors by the following prescription: If C is a cone containing v_i, we define $\mathcal{O}(v_i)(U_C) = \Gamma(U_C) \cdot e^{v_i^*}$, where v_i^* is the piecewise linear functional which is 1 on v_i and 0 on the remaining 1-dimensional rays of Σ. Otherwise, we set $\mathcal{O}(v_i)(U_C) = \Gamma(U_C)$. In this way, each piecewise linear f gives rise to the divisor $D_f = \sum f(v_i)\mathcal{O}(v_i)$. In terms of this correspondence, it turns out there is a simple criterion for determining whether a $\mathbb{Q}$-divisor D_f is linearly equivalent to zero: namely, $D_f \sim_{\mathbb{Q}} 0$ if and only if $f \in \mathrm{Hom}(N, \mathbb{Q})$.

Now, the canonical divisor $K_X = D_{f_{-1}}$, where f_{-1} is the piecewise linear functional given by $f_{-1}(v_i) = -1$. Clearly if Σ is complete, f_{-1} cannot be given by a globally defined linear functional in $\mathrm{Hom}(N, \mathbb{Z})$. So compact smooth toric varieties are never Calabi–Yau, and consequently we can expect no rigidity properties for their complex elliptic genera. Note, however, that TX is stably equivalent to $\bigoplus_{i=1}^{\ell} \mathcal{O}(v_i)$, where the sum is taken over all the 1-dimensional rays v_i of Σ. Thus, up to a normalization factor, the elliptic genus of X is given by the index of $\overline{\partial} \otimes \xi$, where ξ equals

$$\bigotimes_{i=1}^{\ell} \bigotimes_{n=1}^{\infty} \Lambda_{-yq^{n-1}}\mathcal{O}(v_i)^{-1} \otimes \Lambda_{-y^{-1}q^n}\mathcal{O}(v_i) \otimes \bigotimes_{m=1}^{\infty} S_{q^m}\mathcal{O}(v_i)^{-1} \otimes S_{q^m}\mathcal{O}(v_i)$$

We may view ξ as a function of the $T_{\mathbb{C}}$-line bundle $\otimes_{i=1}^{\ell}\mathcal{O}(v_i)$. In this light, is natural to introduce, for any $T_{\mathbb{C}}$-line bundle $\otimes_{i=1}^{\ell}\mathcal{O}(v_i)^{a_i}$, with $a_i \neq 0$, the following vector bundle, denoted as $\xi(a_1, \ldots, a_{\ell})$:

$$\bigotimes_{i=1}^{\ell} \bigotimes_{n=1}^{\infty} \Lambda_{-y^{a_i}q^{n-1}}\mathcal{O}(v_i)^{-1} \otimes \Lambda_{-y^{-a_i}q^n}\mathcal{O}(v_i) \otimes \bigotimes_{m=1}^{\infty} S_{q^m}\mathcal{O}(v_i)^{-1} \otimes S_{q^m}\mathcal{O}(v_i)$$

We may think of $\overline{\partial} \otimes \xi(a_1, \ldots, a_{\ell})$ as a kind of generalized elliptic genus for the toric variety X. The analogue of the Calabi–Yau condition for this generalized elliptic genus is the triviality of the $\mathbb{Q}$-line bundle $\otimes_{i=1}^{\ell}\mathcal{O}(v_i)^{a_i}$. In fact, if this bundle is trivial, then

$$\mathrm{Ind}_T \overline{\partial} \otimes \xi(a_1, \ldots, a_{\ell}) = 0 \in R(T)[\![q, y, y^{-1}]\!]$$

for any compact torus $T \subset T_{\mathbb{C}}$. To prove this, it suffices to assume that $T = S^1$ and that the T-action on X has isolated fixed points. We can always find such a T by first picking a dense 1-parameter subgroup τ of a maximal compact subtorus of $T_{\mathbb{C}}$, and then letting T be generated by a compact 1-parameter subgroup whose initial tangent direction is sufficient close to that of τ. Then the rigidity of $\overline{\partial} \otimes \xi(a_1, \ldots, a_{\ell})$ follows from Liu's modularity technique discussed above. To see that $\mathrm{Ind}_T \overline{\partial} \otimes \xi(a_1, \ldots, a_{\ell})$ is identically 0, we use the following trick observed by Hattori [7]. Let $F(u, z, \tau) = \mathrm{Ind}_T \bar{\partial} \otimes \xi(a_1, \ldots, a_\ell)$. The modular properties of F imply that $F(u + \tau, z, \tau) = e^{2\pi i c z} F(u, z, \tau)$. Here c is the weight of the T-action on the trivial bundle $\otimes_{i=1}^{\ell} \mathcal{O}(v_i)^{a_i}$. For a generic choice of $T \subset T_{\mathbb{C}}$, this weight will be nonzero. But since $F(u, z, \tau)$ is constant in u, we must have that $F(u, z, \tau) = e^{2\pi i c z} F(u, z, \tau)$. This implies that $F(u, z, \tau) = 0$.

2. Chern numbers of singular varieties

We now turn to the problem of defining Chern numbers on singular varieties, a subject which at first glance appears unrelated to the discussion above. In what follows we will find that rigidity theorems provide a powerful tool in solving these types of problems. We first discuss some background.

If X is a smooth compact almost-complex manifold of dimension $2n$, the Chern numbers of X are the numbers of the form

$$c_{i_1, \ldots, i_n} = \int_X c_1^{i_1} \cdot c_2^{i_2} \cdots c_n^{i_n}$$

where c_i denotes the ith Chern class of $T'X$ and $i_1 + 2i_2 + \cdots + n i_n = \dim X / 2$ (so that the total degree of the integrand is $2n$). Chern numbers are easily seen to be functions on the complex cobordism ring Ω_U^*. Moreover, they completely characterize Ω_U^* in the sense that two almost complex manifolds with the same Chern numbers must be complex cobordant.

Much of algebraic geometry consists of efforts to extend techniques from the theory of smooth manifolds to singular varieties. Minimal model theory suggests that one can approach this problem by working on a smooth (or "nearly smooth") birational model of a given singular variety X. For a special combination of Chern numbers, this approach works without any difficulties: namely, the Chern numbers defining the Todd genus. For if X is a smooth complex manifold, the Todd genus of X is given by the alternating sum

$$\chi_0(X) = \sum_{i=0}^{n} (-1)^i \dim H_{\bar{\partial}}^{i,0}(X).$$

By Hartog's theorem, the space of holomorphic i-forms is birationally invariant, and is therefore well-defined even when X is singular, by passing to a resolution of singularities. On the other hand, if X is smooth, then by Riemann–Roch,

$$\chi_0(X) = \int_X \prod_{i=1}^{n} \frac{x_i}{1 - e^{-x_i}}$$

where x_i denote the formal Chern roots of the holomorphic tangent bundle. The combination of Chern numbers obtained by performing the above integration therefore makes sense for any compact singular variety defined over $\mathbb{C}$.

More generally, we consider the following naive attempt at defining combinations of Chern numbers on X: Suppose we are lucky enough to have a smooth minimal model Y for X. Then define $c_{i_1,\ldots,i_n}(X) = c_{i_1,\ldots,i_n}(Y)$. The main problem with this approach is that, even when smooth minimal models exist, we should not expect them to be unique. In fact, we expect different minimal models for X to be related to each other by codimension-2 surgeries called flips and flops. A priori, it is not at all clear what combinations of Chern numbers will be preserved under such operations.

In [11] Totaro set out to classify the combinations of Chern numbers invariant under classical flops. Here we say that two varieties X_1 and X_2 differ by a classical flop if they are the two small resolutions of an n-fold Y whose singular locus is locally the product of a smooth $n-3$-fold Z and the 3-fold node $xy - zw = 0$. More precisely, X_1 and X_2 are constructed as follows: blowing up along Z defines a resolution of Y whose exceptional set is a $\mathbb{P}^1 \times \mathbb{P}^1$ bundle over Z with normal bundle $\mathcal{O}(-1,-1)$. Here $\mathcal{O}(-1,-1)$ denotes the line bundle over a $\mathbb{P}^1 \times \mathbb{P}^1$-bundle which coincides with the tautological bundle along each $\mathbb{P}^1$ direction. Blowing down along either of these $\mathbb{P}^1$ fibers therefore produces two distinct small resolutions X_1 and X_2 of Y.

Totaro demonstrated that the combinations of Chern numbers invariant under classical flops were precisely the combinations of Chern numbers encoded by the complex elliptic genus in the Riemann–Roch formula. We sketch the first half of his argument–namely, that the complex elliptic genus is invariant under classical flops. As X_1 and X_2 are identical away from their exceptional sets, their difference $X_1 - X_2$ is complex cobordant to a fibration E over Z. In fact, if the exceptional sets of X_1 and X_2 are the $\mathbb{P}^1$-bundles $\mathbb{P}(A)$ and $\mathbb{P}(B)$ corresponding to the rank 2 complex bundles A and B over Z, then as a differentiable manifold, E is simply the $\mathbb{P}^3$ bundle $\mathbb{P}(A \oplus B^*)$ over Z. Now the way that E is actually constructed is by taking a tubular neighborhood of $\mathbb{P}(A) \subset X_1$ and gluing it to a tubular neighborhood of $\mathbb{P}(B) \subset X_2$ along their common boundaries (which are both diffeomorphic to $Z \times S^3$). The crucial point is that the stably almost complex structure on E induced by this construction makes E into an SU-fibration. That is, E is a $\mathbb{P}^3$-bundle whose the stable tangent bundle in the vertical direction has a complex structure satisfying $c_1 = 0$. He calls these fibers "twisted projective space" $\widetilde{\mathbb{P}}_{2,2}$. The fiber-integration formula implies that $\mathrm{Ell}(E; z, \tau) = \int_Z \mathrm{Ell}_T(\widetilde{\mathbb{P}}_{2,2}; z, \tau, x_1, \ldots, x_4) \cdot \mathcal{E}ll(Z; z, \tau)$. Here $\mathcal{E}ll(Z; z, \tau)$ is the cohomology class which appears as the integrand in the Riemann–Roch formula for the elliptic genus of Z. More importantly, $\mathrm{Ell}_T(\widetilde{\mathbb{P}}_{2,2}; z, \tau, x_1, \ldots, x_4)$ denotes the character-valued elliptic genus of $\widetilde{\mathbb{P}}_{2,2}$ with the standard T^4 action, with the generators $u_1, \ldots, u_4$ of the Lie algebra of T^4 evaluated at the Chern roots $x_1, \ldots, x_4$ of $A \oplus B$. Since $\widetilde{\mathbb{P}}_{2,2}$ is an SU-

manifold, by the Witten rigidity theorem, $\text{Ell}_T(\widetilde{\mathbb{P}}_{2,2}; z, \tau, x_1, \ldots, x_4) = \text{const.}$
Thus, the elliptic genus of E is simply the product $\text{Ell}(\widetilde{\mathbb{P}}_{2,2}; z, \tau) \cdot \text{Ell}(Z; z, \tau)$.
Moreover, since $\widetilde{\mathbb{P}}_{2,2}$ is cobordant to $Y_1 - Y_2$, where Y_i are the small resolutions of a 3-fold node, and since classical flopping is symmetric for 3-folds,
$\widetilde{\mathbb{P}}_{2,2} \sim Y_2 - Y_1$. Hence $\widetilde{\mathbb{P}}_{2,2} \sim 0$ in the complex cobordism ring. We therefore
have that $\text{Ell}(X_1; z, \tau) - \text{Ell}(X_2; z, \tau) = \text{Ell}(\widetilde{\mathbb{P}}_{2,2}; z, \tau) \cdot \text{Ell}(Z; z, \tau) = 0$.

An obvious consequence of the above discussion is that for varieties Y whose
singular locus is locally the product of a smooth variety with a 3-fold node, it
makes sense to define the elliptic genus of Y to be the elliptic genus of one of
its small resolutions. However, most singular varieties fail to possess even one
small resolution. It is therefore natural to ask whether one can continue to define
the elliptic genus for a more general class of singularities. The right approach
to answering this question is to expand one's category to include pairs (X, D),
where X is a variety and D is a divisor on X with the property that $K_X - D$
is $\mathbb{Q}$-Cartier. A map $f : (X, D) \rightarrow (Y, \Delta)$ in this category corresponds to a
birational morphism $f : X \rightarrow Y$ satisfying $K_X - D = f^*(K_Y - \Delta)$. The idea
is to first define the elliptic genus for smooth pairs (X, D) in such a way that
$\text{Ell}(X, D; z, \tau)$ becomes functorial with respect to morphisms of pairs. Given
two resolutions $f_i : X_i \rightarrow Y$ of a singular variety Y, with $K_{X_i} - D_i = f^* K_Y$,
we could then find resolutions $g_i : (M, D) \rightarrow (X, D_i)$ making the following
diagram commute:

$$
\begin{array}{ccc}
(M, D) & \xrightarrow{\;\;g_1\;\;} & (X_1, D_1) \\
{\scriptstyle g_2}\big\downarrow & & \big\downarrow{\scriptstyle f_1} \\
(X_2, D_2) & \xrightarrow{\;\;f_2\;\;} & (Y, 0)
\end{array}
$$

Functoriality of the elliptic genus would then imply that

$$\text{Ell}(X_1, D_1; z, \tau) = \text{Ell}(M, D; z, \tau) = \text{Ell}(X_2, D_2; z, \tau).$$

It would then make sense to define $\text{Ell}(Y; z, \tau) \equiv \text{Ell}(X_1, D_1; z, \tau)$.

One can simplify this approach by making two observations. First, by introducing further blow-ups, one can always assume that the exceptional divisors
$D_i \subset X_i$ have smooth components with simple normal crossings. (Such resolutions are called "log resolutions".) Second, by a deep result of Wlodarczyk [1],
the birational map $(X_1, D_1) \dashrightarrow (X_2, D_2)$ may be decomposed into a sequence
of maps

$$(X_1, D_1) = (X^{(0)}, D^{(0)}) \dashrightarrow \cdots \dashrightarrow (X^{(N)}, D^{(N)}) = (X_2, D_2)$$

where each of the arrows are blow-ups or blow-downs along smooth centers
which have normal crossings with respect to the components of $D^{(j)}$. It therefore suffices to define $\text{Ell}(X, D; z, \tau)$ for smooth pairs (X, D), where D is a

simple normal crossing divisor, and prove that $\mathrm{Ell}(X, D; z, \tau)$ is functorial with respect to blow-ups along smooth centers which have normal crossings with respect to the components of D. This procedure has been carried out successfully by Borisov and Libgober in [4], and by Chin-Lung Wang in [15]. They define $\mathrm{Ell}(X, D; z, \tau)$ by the formula

$$
\int_X \prod_j \frac{x_j \vartheta\left(\frac{x_j}{2\pi i} - z, \tau\right)}{\vartheta\left(\frac{x_j}{2\pi i}, \tau\right)} \prod_i \frac{\vartheta\left(\frac{D_i}{2\pi i} - (a_i + 1)z, \tau\right)\vartheta(z, \tau)}{\vartheta\left(\frac{D_i}{2\pi i} - z, \tau\right)\vartheta((a_i + 1)z, \tau)}
\tag{2-1}
$$

Here the x_j denote the formal Chern roots of TX and the D_i denote the first Chern classes of the components D_i of D with coefficients $a_i(X, D)$. Note that since $\vartheta(0, \tau) = 0$, the above expression only makes sense for $a_i \neq -1$. Naturally, this places some restrictions on the types of singularities allowed in the definition of $\mathrm{Ell}(Y; z, \tau)$. For example, at the very least Y must possess a log resolution $(X, D) \to (Y, 0)$ such that none of the discrepancy coefficients $a_i(X, D)$ are equal to -1. In order to ensure that $\mathrm{Ell}(Y; z, \tau)$ does not depend on our choice of a log resolution (X, D), we actually must require that the discrepancy coefficients $a_i(X, D) > -1$. To see why, suppose that (X_1, D_1) and (X_2, D_2) are two log resolutions of Y with discrepancy coefficients $a_i(X_j, D_j) \neq -1$. To prove that $\mathrm{Ell}(X_1, D_1; z, \tau) = \mathrm{Ell}(X_2, D_2; z, \tau)$, we must connect these two resolutions by a sequence of blow-ups and blow-downs, applying functoriality of the elliptic genus of pairs at each stage. But if some of the discrepancy coefficients $a_i(X_1, D_1)$ are greater than -1, and others less than -1, then after blowing up X_1, we may acquire discrepancy coefficients equal to -1. In this case, the elliptic genus of one of the intermediate pairs in the chain of blow-ups and blow-downs will be undefined, and consequently we will have no means of comparing the elliptic genera of (X_1, D_1) and (X_2, D_2). The only obvious way of avoiding this problem is to require $a_i(X_j, D_j) > -1$. This constraint is quite familiar to minimal model theorists; singular varieties Y possessing this property are said to have *log-terminal* singularities.

Functoriality of the elliptic genus provides a nice explanation for the invariance of the elliptic genus under classical flops. For if X_1 and X_2 are related by a classical flop, then there exists a common resolution $f_i : X \to X_i$ with $f_1^* K_{X_1} = f_2^* K_{X_2}$. Two varieties related in this way are said to be *K-equivalent*. One therefore discovers from this approach that the fundamental relation leaving the elliptic genus invariant is not flopping but K-equivalence.

The original proofs of functoriality of the elliptic genus, by Borisov, Libgober, and Wang, are based on an explicit calculation of the push-forward f_* of the integrand in (2-1), where $f : (X, D) \to (X_0, D_0)$ is a blow-down. The obstruction to this push-forward giving the correct integrand on X_0 is given by an elliptic function with values in $H^*(X_0)$. One can then use basic elliptic function theory to show that this function vanishes. In what follows, we will sketch a different proof, similar to the one in [13], that makes use of the rigidity properties of the elliptic genus. This approach has several advantages: the first is that the proof can be easily generalized to more exotic versions of elliptic genera, such as the character-valued elliptic genus for orbifolds. Though the original proofs could be adapted to this situation, their implementation in the most general setting is cumbersome. Another advantage is that some variation of this approach appears to be useful for studying elliptic genera for varieties with non-log-terminal singularities. We will have more to say on this in the following section. Recall though that the rigidity of the elliptic genus for SU-manifolds was the key step in Totaro's proof of the invariance of elliptic genera under classical flops. It is therefore reasonable to expect rigidity phenomena to play a useful role in the study of elliptic genera of singular varieties.

Proceeding with the proof, we let X be a smooth variety and $D = \sum a_i D_i$ a simple normal crossing divisor on X. Let $f : \tilde{X} \to X$ be the blow-up along a smooth subvariety which has normal crossings with respect to the components of D. We let $\tilde{D} = \sum a_i \tilde{D}_i + mE$ be the sum of the proper transforms of D_i and the exceptional divisor E, whose coefficients are chosen so that $K_{\tilde{X}} - \tilde{D} = f^*(K_X - D)$.

To avoid getting bogged down in technical details, assume $f : \tilde{X} \to X$ is the blow-up at a single point $p = D_1 \cap \ldots \cap D_n$, and $D_1, \ldots, D_n$ are the only components of D. Then $T\tilde{X}$ is stably equivalent to $f^*TX \oplus \bigoplus_{i=1}^n \mathcal{O}(\tilde{D}_i) \oplus \mathcal{O}(E)$. From (2-1), it follows immediately that the proof of the blow-up formula for the elliptic genus amounts to proving that

$$\int_{\tilde{X}} f^* \left\{ \prod_{T'X} \frac{\frac{x_j}{2\pi i} \vartheta(\frac{x_j}{2\pi i} - z)\vartheta'(0)}{\vartheta(\frac{x_j}{2\pi i})\vartheta(-z)} \right\} \prod_{i=1}^n \frac{\frac{\tilde{D}_i}{2\pi i}\vartheta(\frac{\tilde{D}_i}{2\pi i} - (a_i + 1)z)\vartheta'(0)}{\vartheta(\frac{\tilde{D}_i}{2\pi i})\vartheta(-(a_i + 1)z)}$$

$$\times \frac{\frac{E}{2\pi i}\vartheta(\frac{E}{2\pi i} - (m + 1)z)\vartheta'(0)}{\vartheta(\frac{E}{2\pi i})\vartheta(-(m + 1)z)}$$

$$= \int_X \prod_{T'X} \frac{\frac{x_j}{2\pi i}\vartheta(\frac{x_j}{2\pi i} - z)\vartheta'(0)}{\vartheta(\frac{x_j}{2\pi i})\vartheta(-z)} \prod_{i=1}^n \frac{\frac{D_i}{2\pi i}\vartheta(\frac{D_i}{2\pi i} - (a_i + 1)z)\vartheta'(0)}{\vartheta(\frac{D_i}{2\pi i})\vartheta(-(a_i + 1)z)}$$

Here, for ease of exposition, we have omitted the dependence of ϑ on τ. Note that $\tilde{D}_i = f^*D_i - E$ in the above expression. Thus, if we expand both sides in the variables f^*D_i, E, and D_i, the blow-up formula is easily seen to hold for integrals of Chern and divisor data not involving E. Note however that in a neighborhood of E, $(\tilde{X}, \tilde{D})$ has the exact same structure as the blow-up of $\mathbb{C}^n$ at the origin, with the divisors $\tilde{D}_1, \ldots, \tilde{D}_n$ corresponding to the proper transforms of the coordinate hyperplanes of $\mathbb{C}^n$. For the purpose of proving the blow-up formula, we may therefore assume that $X \cong (\mathbb{P}^1)^n$ and that $\tilde{X}$ is the blow-up of X at $[0:1] \times \cdots \times [0:1]$. Viewed as a toric variety, X is defined by the fan $\Sigma \subset N \otimes \mathbb{R}$ with 1-dim rays $\mathbb{R}(\pm e_1), \ldots, \mathbb{R}(\pm e_n)$, where $e_1, \ldots, e_n$ form an integral basis for the lattice N. The fan $\tilde{\Sigma}$ of $\tilde{X}$ is obtained from Σ by adding the ray $\mathbb{R}(e_1 + \cdots + e_n)$. The divisors $D_i \subset X$ correspond to the rays $\mathbb{R}e_i$ in Σ; and the divisors $\tilde{D}_i$ and E correspond to the rays $\mathbb{R}e_i$ and $\mathbb{R}(e_1 + \cdots + e_n)$ in $\tilde{\Sigma}$. Using the fact that the tangent bundle of smooth toric variety with $T_{\mathbb{C}}$-invariant divisors D_j, $j = 1, \ldots, \ell$, is stably equivalent to $\bigoplus_{j=1}^{\ell} \mathcal{O}(D_j)$, the blow-up formula for X reduces to proving that

$$
\int_{\tilde{X}} \prod_{k=1}^{n+1} \frac{\frac{\tilde{D}_k}{2\pi i} \vartheta(\frac{\tilde{D}_k}{2\pi i} - (a_k+1)z)\vartheta'(0)}{\vartheta(\frac{\tilde{D}_k}{2\pi i})\vartheta(-(a_k+1)z)} \prod_{k=1}^{n} \frac{\frac{\tilde{D}_{-k}}{2\pi i}\vartheta(\frac{\tilde{D}_{-k}}{2\pi i} - (a_{-k}+1)z)\vartheta'(0)}{\vartheta(\frac{\tilde{D}_{-k}}{2\pi i})\vartheta(-(a_{-k}+1)z)}
$$

$$
= \int_{X} \prod_{j=1}^{n} \frac{\frac{D_j}{2\pi i}\vartheta(\frac{D_j}{2\pi i} - (a_j+1)z)\vartheta'(0)}{\vartheta(\frac{D_j}{2\pi i})\vartheta(-(a_j+1)z)} \prod_{j=1}^{n} \frac{\frac{D_{-j}}{2\pi i}\vartheta(\frac{D_{-j}}{2\pi i} - (a_{-j}+1)z)\vartheta'(0)}{\vartheta(\frac{D_{-j}}{2\pi i})\vartheta(-(a_{-j}+1)z)}
$$

In this formula, D_{-j} denote the $T_{\mathbb{C}}$-divisors on X corresponding to the 1-dim rays $\mathbb{R}(-e_j)$, with coefficients $a_{-j} = 0$. $\tilde{D}_{-j}$ denote their proper transforms, which are simply given by $f^* D_{-j}$, since D_{-j} are defined away from the blow-up locus. For ease of exposition, we also let $\tilde{D}_{n+1} = E$, with $a_{n+1} = m$.

Now the crucial observation is that in the above formula, RHS $-$ LHS is independent of the coefficients a_{-j}. For since $\tilde{D}_{-j}$ are disjoint from E, any divisor intersection data involving $\tilde{D}_{-j}$ will be unchanged after formally setting $E = 0$. Therefore, the parts of RHS $-$ LHS depending a_{-j} will be unchanged after setting $E = 0$. But formally letting $E = 0$ clearly gives RHS $=$ LHS. Consequently, RHS $-$ LHS depends only on $a_1, \ldots, a_n$.

Let us therefore define a_{-j} so that $(1 + a_{-j}) = -(1 + a_j)$. As discussed in the previous section, the set of coefficients $(1 + a_{\pm j})$ assigned to the rays $\mathbb{R}(\pm e_j)$ give rise to a piecewise linear functional $g = g_{1+a_i, 1+a_{-i}}$ on the fan Σ. In fact, g is simply the global linear functional which maps the basis vectors e_i to $(1 + a_i)$. As $g \in \mathrm{Hom}(N, \mathbb{Z})$, it also defines a global linear functional on $\tilde{\Sigma}$, taking the value $\sum_{i=1}^{n}(1+a_i)$ on $e_1 + \cdots + e_n$. Now by the discrepancy formula for blow-ups, $\sum_{i=1}^{n}(1+a_i) = (1+m)$. We see from this that the piecewise linear functional on $\tilde{\Sigma}$ defined by assigning the coefficients $(1+a_{\pm j})$ to $\mathbb{R}(\pm e_j)$ and $(1+m)$ to $\mathbb{R}(e_1 + \cdots + e_n)$ corresponds to this same global linear functional g.

It follows that the bundles

$$
\mathcal{O}(e_1 + \cdots + e_n)^{1+m} \otimes \bigotimes_{i=1}^{n} \mathcal{O}(e_i)^{1+a_i} \otimes \mathcal{O}(-e_i)^{1+a_{-i}}
$$

and $\bigotimes_{i=1}^{n} \mathcal{O}(e_i)^{1+a_i} \otimes \mathcal{O}(-e_i)^{1+a_{-i}}$ are trivial as $\mathbb{Q}$-line bundles on $\tilde{X}$ and X, respectively. Consequently,

$$\text{Ind } \overline{\partial} \otimes \xi(1 + a_i, 1 + m, 1 + a_{-i}) = \text{Ind } \overline{\partial} \otimes \xi(1 + a_i, 1 + a_{-i}) = 0.$$

But, up to a normalization factor, $\text{Ind } \overline{\partial} \otimes \xi(1 + a_i, 1 + m, 1 + a_{-i}) = \text{RHS}$ and $\text{Ind } \overline{\partial} \otimes \xi(1 + a_i, 1 + a_{-i}) = \text{LHS}$ for the given new values of a_{-i}. Thus, $\text{RHS} = \text{LHS}$ for $(1 + a_{-i}) = -(1 + a_i)$, and therefore also for $a_{-i} = 0$.

This completes the proof of the blow-up formula for the case where the blow-up locus is a single point. For completeness, let us outline the case for the blow-up along a subvariety Z with normal crossings with respect to the components of D. This case is handled in much the same way, the only difference being that instead of reducing to the situation where X is toric, we instead reduce to the case where X is a toric fibration, fibered over the blow-up locus Z. Namely, by deformation to the normal cone, we may assume that

$$X = \mathbb{P}(M \oplus 1) \times \mathbb{P}(L_1 \oplus 1) \times \cdots \times \mathbb{P}(L_r \oplus 1).$$

Here, for the components D_i intersecting Z,

$$L_i = \mathcal{O}(D_i)|_Z$$

and M is the quotient of $N_{Z/X}$ by $\oplus L_i$. The product $\times$ is the fiber product of the corresponding projective bundles over Z. We now view D_i as the divisors given by the zero sections of the bundles L_i. Moreover, the zero sections of L_i and M together define a copy of Z in $\mathbb{P}(M \oplus 1) \times \mathbb{P}(L_1 \oplus 1) \times \cdots \times \mathbb{P}(L_r \oplus 1)$ with the same normal bundle $N_{Z/X}$ as in the original space. We let $\tilde{X}$ be the blow-up along this copy of Z. The proof of the blow-up formula then follows the same reasoning as in the toric case, where we now make use of the rigidity of fiberwise analogues of the operators $\overline{\partial} \otimes \xi(\vec{a})$. For example, let us examine how to generalize the bundle $\xi(1 + a_i, 1 + a_{-i})$ on $(\mathbb{P}^1)^n$ to the fibration X.

For each fibration $\pi_i : \mathbb{P}(L_i \oplus 1) \to Z$, we have the exact sequence

$$0 \to S_i \to \pi_i^*(L_i \oplus 1) \to Q_i \to 0$$

of tautological bundles. The vertical tangent bundle to $\mathbb{P}(L_i \oplus 1)$ is stably equivalent to the direct sum of hyperplane bundles $H_i \oplus H_{-i}$, where $H_i = \text{Hom}(\pi_i^* L_i, S_i)$ and $H_{-i} = \text{Hom}(1, S_i)$. Similarly, the vertical tangent bundle to the fibration

$$\pi : \mathbb{P}(M \oplus 1) \to Z,$$

with tautological bundle S is stably equivalent to the direct sum $V \oplus H$ where $V = \text{Hom}(\pi^* M, S)$ and $H = \text{Hom}(1, S)$. All of these bundles extend naturally to the whole fibration X. Recall that if $\alpha_i = -\alpha_{-i}$, then $\overline{\partial} \otimes \xi(\alpha_i, \alpha_{-i})$ defines a elliptic operator on $(\mathbb{P}^1)^n$ with vanishing equivariant index (note that for convenience of notation we have defined $\alpha_i = 1 + a_i$). For the fibration X, we replace $\bar{\partial} \otimes \xi(\alpha_i, \alpha_{-i})$ by the following fiberwise analogue:

$$\bar{\partial} \otimes \overset{\pm r}{\underset{i=\pm 1}{\bigotimes}} \overset{\infty}{\underset{n=1}{\bigotimes}} \Lambda_{-y^{\alpha_i} q^{n-1}} H_i^* \otimes \Lambda_{y^{-\alpha_i} q^n} H_i \otimes \overset{\infty}{\underset{m=1}{\bigotimes}} S_{q^m} H_i^* \otimes S_{q^m} H_i$$

$$\otimes \overset{\infty}{\underset{n=1}{\bigotimes}} \Lambda_{-y q^{n-1}} V^* \otimes \Lambda_{y^{-1} q^n} V \otimes \overset{\infty}{\underset{m=1}{\bigotimes}} S_{q^m} V^* \otimes S_{q^m} V$$

$$\otimes \overset{\infty}{\underset{n=1}{\bigotimes}} \Lambda_{-y^{-d-1} q^{n-1}} H^* \otimes \Lambda_{y^{d+1} q^n} H \otimes \overset{\infty}{\underset{m=1}{\bigotimes}} S_{q^m} H^* \otimes S_{q^m} H.$$

Here $d = \text{rank}(M)$. By performing a fiber integration over X, one can show that the rigidity of this operator with respect to the obvious torus action on the fibers follows directly from the rigidity results obtained for $\bar{\partial} \otimes \xi(\alpha_i, \alpha_{-i})$. Analogously, there exists a natural generalization of the operator $\bar{\partial} \otimes \xi(1 + a_i, 1 + m, 1 + a_{-i})$ to a rigid operator on the fibration $\tilde{X}$. We therefore see that the blow-up formula for the elliptic genus is in all cases a consequence of rigidity phenomena on toric varieties.

Before moving on, we make a simple observation which will prove convenient in the next section. Let X be a smooth toric variety with toric divisors D_i. Since TX is stably equivalent to $\bigoplus_{i=1}^{\ell} \mathcal{O}(D_i)$, the elliptic genus of the pair $(X, \sum a_i D_i)$ is equal to the index of the operator

$$\bar{\partial} \otimes \xi(a_1 + 1, \ldots, a_{\ell} + 1),$$

up to a normalization factor. Moreover, the condition that $\bigotimes_{i=1}^{\ell} \mathcal{O}(D_i)^{a_i+1}$ is trivial is equivalent to the condition that $K_X - \sum a_i D_i = 0$ as a Cartier divisor. In this case, we say that $(X, \sum a_i D_i)$ is a Calabi–Yau pair. Hence, a trivial consequence of the rigidity theorem for the elliptic genus of toric varieties is that $\text{Ell}(X, D; z, \tau) = 0$ whenever (X, D) is a toric Calabi–Yau pair.

3. Beyond log-terminal singularities

As observed above, Borisov, Libgober, and Wang's approach to defining the elliptic genus of a singular variety Y only appears to work when Y has log-terminal singularities. This is due to the division by $\vartheta((a_i + 1)z)$ in the formula for the elliptic genus of the pair (X, D), where (X, D) is a resolution of Y with discrepancy coefficients $a_i(X, D)$. In pursuit of the broader question, "for what class of singularities can we make sense of Chern data?", it is natural to ask whether log-terminality represents an essential constraint. In what follows, we will demonstrate that at the very least, the elliptic genus can be defined for all but a finite class of normal surface singularities.

Since the terms $\vartheta((a_i+1)z)$ do not involve any Chern data, the first thing one might try to do is simply throw these terms away in the definition of the elliptic genus of a pair. However, this approach is of little use since one would lose functoriality with respect to birational morphisms. As a second attempt, one could introduce a perturbation $a_i + \varepsilon b_i$ to each of the discrepancy coefficients a_i of D, and take the limit as $\varepsilon \to 0$. Two obvious difficulties with this approach are (1) the limit does not always exist, and (2), even when the limit exists, it depends on the choice of the perturbation. Moreover, deciding on some fixed perturbation in advance (like letting all $b_i = 1$) runs into problems if we hope to preserve functoriality.

To carry out this perturbation approach, we therefore require a distinguished class of perturbation divisors $\Delta(X, D) = \{\sum \varepsilon b_i D_i\}$ satisfying the following two properties:

(1) For every $D_\varepsilon \in \Delta(X, D)$, the limit as $\varepsilon \to 0$ of $\text{Ell}(X, D + D_\varepsilon; z, \tau)$ exists and is independent of the choice of D_ε.

(2) If $f : (\tilde{X}, \tilde{D}) \to (X, D)$ is a blow-up, then $f^* \Delta(X, D) \subset \Delta(\tilde{X}, \tilde{D})$.

Assuming we have found a set of perturbation divisors satisfying these properties, we could then define the elliptic genus of a singular variety Y by the following procedure: Pick a log-resolution (X, D) of Y, and choose $D_\varepsilon \in \Delta(X, D)$. Then define $\text{Ell}(Y; z, \tau) = \lim_{\varepsilon \to 0} \text{Ell}(X, D + D_\varepsilon; z, \tau)$. The important point is that if $f : (\tilde{X}, \tilde{D}) \to (X, D)$ is a blow-up, and $\tilde{D}_\varepsilon \in \Delta(\tilde{X}, \tilde{D})$, then the answer we get for the elliptic genus of Y is the same, regardless of whether we work with $(X, D + D_\varepsilon)$ or with $(\tilde{X}, \tilde{D} + \tilde{D}_\varepsilon)$. To see why, note that $f^*(K_X - D - D_\varepsilon) = K_{\tilde{X}} - \tilde{D} - f^* D_\varepsilon$. Thus, by functoriality of the elliptic genus with respect to blow-ups, $\text{Ell}(X, D + D_\varepsilon; z, \tau) = \text{Ell}(\tilde{X}, \tilde{D} + f^* D_\varepsilon; z, \tau)$. By property (2), $f^* D_\varepsilon$ lies inside $\Delta(\tilde{X}, \tilde{D})$. Hence, property (1) of $\Delta(\tilde{X}, \tilde{D})$ implies that $\lim_{\varepsilon \to 0} \text{Ell}(\tilde{X}, \tilde{D} + \tilde{D}_\varepsilon; z, \tau) = \lim_{\varepsilon \to 0} \text{Ell}(\tilde{X}, \tilde{D} + f^* D_\varepsilon; z, \tau)$.

For the case of complex surfaces, we have a natural candidate for $\Delta(X, D)$ satisfying the second property; namely the set

$$\{\Delta_\varepsilon : \Delta_\varepsilon D_i = 0 \text{ for all } D_i \subset D \text{ with discrepancy coefficient} = -1\}$$

For if Δ_ε is in this set, and $\tilde{D}_i \subset \tilde{D}$ has coefficient equal to -1, then $f^* \Delta_\varepsilon \tilde{D}_i = \Delta_\varepsilon f_* \tilde{D}_i$. Now, either $\tilde{D}_i$ is the proper transform of a divisor with -1 discrepancy, or it is a component of the exceptional locus of f. In the former case, $\Delta_\varepsilon f_* \tilde{D}_i = 0$ by virtue Δ_ε belonging to the set $\Delta(X, D)$; in the latter case, $f_* \tilde{D}_i = 0$.

We still must verify that the $\varepsilon \to 0$ limit of $\text{Ell}(X, D + D_\varepsilon; z, \tau)$ is well-defined and independent of the choice of $D_\varepsilon \in \Delta(X, D)$ when (X, D) is a resolution of a singular complex surface Y. Unfortunately, it is too much to ask that this property hold for all normal surface singularities. Suppose, however, that (X, D) is a log resolution of a normal surface Y satisfying the following additional property: For every component $D_i \subset D$ with discrepancy coefficient $a_i(X, D) = -1$, $D_i \cong \mathbb{P}^1$ and D_i intersects at most two other components D_{i_1}, D_{i_2} of D at a single point (and $a_{i_k}(D) \neq -1$ for $k = 1, 2$). In other words, we assume that the local geometry in a tubular neighborhood U of D_i is indistinguishable from a tubular neighborhood of a toric divisor, and moreover that we can find disjoint such neighborhoods for every component D_i with a -1 discrepancy. Note that since D_i is an exceptional curve, the adjunction formula implies that $(X, D)|_U$ is a toric Calabi–Yau pair. Under this additional assumption, it turns out that $\lim_{\varepsilon \to 0} \mathrm{Ell}(X, D + D_\varepsilon; z, \tau)$ exists and is independent of the choice of $D_\varepsilon \in \Delta(X, D)$.

To see why the limit exists, note that $\mathrm{Ell}(X, D + D_\varepsilon; z, \tau)$ is a meromorphic function of ε with at most a simple pole at $\varepsilon = 0$. Up to a normalization factor, the residue of $\mathrm{Ell}(X, D + D_\varepsilon; z, \tau)$ at $\varepsilon = 0$ corresponds to $\sum_{a_i(X,D)=-1} \mathrm{Ell}(D_i, D + D_\varepsilon|_{D_i}; z, \tau)$. By adjunction, $(D_i, D + D_\varepsilon|_{D_i})$ are all toric Calabi–Yau pairs, and consequently, the residue of $\mathrm{Ell}(X, D + D_\varepsilon; z, \tau)$ vanishes by the rigidity theorems discussed in the previous section.

It remains to check that this limit is independent of the choice of $D_\varepsilon \in \Delta(X, D)$. Suppose then that $D_\varepsilon, D'_\varepsilon$ are two possible perturbation divisors. Since the $\varepsilon \to 0$ limit of $\mathrm{Ell}(X, D + D_\varepsilon; z, \tau) - \mathrm{Ell}(X, D + D'_\varepsilon; z, \tau)$ depends only on the local geometry near the divisor components D_i with $a_i(X, D) = -1$, we may reduce the calculation to the case where (X, D) is a toric variety. Moreover, since (X, D) is Calabi–Yau in the tubular neighborhoods U_i of the above divisor components, we may further reduce to the situation where (X, D) is a Calabi–Yau pair. By definition, D_ε and D'_ε are trivial over U_i and we may extend them to trivial divisors over all of X without affecting the $\varepsilon \to 0$ limit of $\mathrm{Ell}(X, D + D_\varepsilon; z, \tau)$ or $\mathrm{Ell}(X, D + D'_\varepsilon; z, \tau)$. We have thus reduced the calculation to the case where $(X, D + D_\varepsilon)$ and $(X, D + D'_\varepsilon)$ are both toric Calabi–Yau pairs. The rigidity theorem for the elliptic genus in this case then implies that $\mathrm{Ell}(X, D + D_\varepsilon; z, \tau) = \mathrm{Ell}(X, D + D'_\varepsilon; z, \tau) = 0$ for all ε, which clearly implies that their limits are the same as $\varepsilon \to 0$.

Of course, the above discussion is moot unless one can find a reasonably large class of surface singularities whose resolutions satisfy the additional criterion of being locally toric in a neighborhood of the exceptional curves with -1 discrepancies. Fortunately, as observed by Willem Veys [12], nearly all normal surface singularities satisfy this property. The sole exceptions consist of the normal surfaces with strictly log-canonical singularities. These are surfaces whose resolutions (X, D) satisfy $a_i(X, D) \geq -1$, with some $a_i(X, D) = -1$. A well-known example is the surface singularity obtained by collapsing an elliptic curve to a point. For a complete classification of these singularities, see [2]. Based on this observation, Veys used a limiting procedure similar to the one given here to define Batyrev's string-theoretic Hodge numbers for normal surfaces without strictly log-canonical singularities.

Note that, for dimensionality reasons, the elliptic genus of a smooth surface is a coarser invariant than the surface's collective Hodge numbers. Nevertheless, the approach discussed here affords several advantages. First, the technique of applying the rigidity properties of toric Calabi–Yau pairs is easy to adapt to more complicated invariants, such as the character-valued elliptic genus and the elliptic genus of singular orbifolds. These are finer invariants than the ordinary elliptic genus which are not characterized entirely by Hodge numbers. Second, this approach provides some clues about how to define elliptic genera for higher-dimensional varieties whose singularities are not log-terminal. For example, a possible generalization of the locally toric structure we required of the -1 discrepancy curves is to demand that all -1 discrepancy divisors be toric varieties fibered over some smooth base. The analogue of property (2) for $\Delta(X, D)$ in this case is that $c_1(D_\varepsilon) = 0$ when restricted to each fiber of a -1 discrepancy divisor.

4. Further directions

Singular Chern numbers constructed out of elliptic genera have an interesting interpretation when the singular variety is the quotient of a smooth variety X by a finite group G. In this situation, quantum field theory on orbifolds gives rise to a definition for the elliptic genus of X/G constructed entirely out of the orbifold data of (X, G). This orbifold version of the elliptic genus turns out to be closely related to the singular elliptic genus of X/G: for example, when the G-action has no ramification divisor, the orbifold elliptic genus of (X, G) equals the singular elliptic genus of X/G. This is a specific example, proved by Borisov and Libgober [5], of a much larger interaction between representation theory and topology known as the McKay correspondence.

Note that the log-terminality constraint comes for free in this case, since the germs of quotients $\mathbb{C}^n/G$, where G is a finite subgroup of $GL(n, \mathbb{C})$ are always log-terminal. Suppose however that X itself is singular. By following a procedure similar to the one discussed above for the elliptic genus, one can continue to define a singular analogue of the orbifold elliptic genus of (X, G). At this point it is natural to ask whether the McKay correspondence continues to hold when we allow X to have singularities. When X has log-terminal singularities, this follows directly out of Borisov and Libgober's proof of the McKay correspondence. For more general singularities the answer to this question is not known, although the McKay correspondence has been verified for the cases discussed in the previous section: namely, when X is a normal surface without strictly log-canonical singularities. See, for example, [14].

As we have seen, many of the techniques for studying elliptic genera in birational geometry can be traced back to some rigidity property of the elliptic genus. It is therefore not surprising that most of these techniques (such as functoriality of the elliptic genus of a divisor pair) work equally well for the character-valued elliptic genus. From Totaro's work, we know that the elliptic genus completely determines the collection of Chern numbers invariant under classical flops. An obvious question then is whether the analogous statement holds for equivariant Chern numbers. From the functoriality property of the character-valued elliptic genus, one easily verifies that the equivariant Chern numbers encoded by the character-valued elliptic genus are indeed invariant under equivariant flops. The more difficult question is whether all flop-invariant equivariant Chern numbers factor through the character-valued elliptic genus. It appears that some knowledge of the image of the character-valued elliptic genus over an equivariant cobordism ring must play a role in answering this question.

References

[1] Dan Abramovich, Kalle Karu, Kenji Matsuki, and Jaroslaw Wlodarczyk. "Torification and factorization of birational maps." *J. Amer. Math. Soc.*, 15(3):531–572, 2002.

[2] "Flips and abundance for algebraic threefolds". In *Papers from the Second Summer Seminar on Algebraic Geometry held at the University of Utah*, (Salt Lake City, 1991), Astérisque No. 211, Société Mathématique de France, Paris, 1992.

[3] Michael Atiyah and Friedrich Hirzebruch. "Spin-manifolds and group actions." In *Essays on Topology and Related Topics (Mémoires dédiés à Georges de Rham)*. Springer, New York, 1970, 18–28.

[4] Lev Borisov and Anatoly Libgober. "Elliptic genera of singular varieties." *Duke Math. J.*, 116(2):319–351, 2003.

[5] Lev Borisov and Anatoly Libgober. "McKay correspondence for elliptic genera." *Ann. of Math. (2)*, 161(3):1521–1569, 2005.

[6] Raoul Bott and Clifford Taubes. "On the rigidity theorems of Witten." *J. Amer. Math. Soc.*, 2(1):137–186, 1989.

[7] Akio Hattori. "Elliptic genera, torus orbifolds and multi-fans. II." *Internat. J. Math.*, 17(6):707–735, 2006.

[8] Peter S. Landweber and Robert E. Stong. "Circle actions on spin manifolds and characteristic numbers." *Topology*, 27(2):145–161, 1988.

[9] Kefeng Liu. "On elliptic genera and theta-functions." *Topology*, 35(3):617–640, 1996.

[10] Serge Ochanine. "Genres elliptiques équivariants." In *Elliptic curves and modular forms in algebraic topology*, 107–122. Springer, Berlin, 1988.

[11] Burt Totaro. "Chern numbers of singular varieties and elliptic homology." *Ann. of Math.* (2), 151(2):757–791, 2000.

[12] Willem Veys. "Stringy invariants of normal surfaces." *J. Algebraic Geom.*, 13(1): 115–141, 2004.

[13] Robert Waelder. "Equivariant elliptic genera." *Pacific J. Math.*, 235(2):345–377, 2008.

[14] Robert Waelder. "Singular McKay correspondence for normal surfaces." Preprint, math.AG/0810.3634.

[15] Chin-Lung Wang. "*K*-equivalence in birational geometry and characterizations of complex elliptic genera." *J. Algebraic Geom.*, 12(2):285–306, 2003.

[16] Edward Witten. "The index of the Dirac operator in loop space." In *Elliptic curves and modular forms in algebraic topology*, 161–181. Springer, Berlin, 1988.

[17] Edward Witten. "Index of Dirac operators." In *Quantum fields and strings: a course for mathematicians*, Vol. 1, 475–511. Amer. Math. Soc., Providence, RI, 1999.

ROBERT WAELDER
DEPARTMENT OF MATHEMATICS, STATISTICS, AND COMPUTER SCIENCE
UNIVERSITY OF ILLINOIS AT CHICAGO
410 SCIENCE AND ENGINEERING OFFICES
851 S. MORGAN STREET
CHICAGO, IL 60607-7045
   rwaelder@math.uic.edu

Topology of Stratified Spaces
MSRI Publications
Volume **58**, 2011

Hodge theory meets the minimal model program: a survey of log canonical and Du Bois singularities

SÁNDOR J. KOVÁCS AND KARL E. SCHWEDE

ABSTRACT. We survey some recent developments in the study of singularities related to the classification theory of algebraic varieties. In particular, the definition and basic properties of Du Bois singularities and their connections to the more commonly known singularities of the minimal model program are reviewed and discussed.

1. Introduction

The primary goal of this note is to survey some recent developments in the study of singularities related to the minimal model program. In particular, we review the definition and basic properties of *Du Bois singularities* and explain how these singularities fit into the minimal model program and moduli theory.

Since we can resolve singularities [Hir64], one might ask why we care about them at all. It turns out that in various situations we are forced to work with singularities even if we are only interested in understanding smooth objects.

One reason we are led to study singular varieties is provided by the minimal model program [KM98]. The main goal is the classification of algebraic varieties and the plan is to find reasonably simple representatives of all birational classes and then classify these representatives. It turns out that the simplest objects in a birational class tend to be singular. What this really means is that when choosing a birational representative, we aim to have simple *global* properties and this is often achieved by a singular variety. Being singular means that

Mathematics Subject Classification: 14B05.

Kovács was supported in part by NSF Grant DMS-0554697 and the Craig McKibben and Sarah Merner Endowed Professorship in Mathematics. Schwede was partially supported by RTG grant number 0502170 and by a National Science Foundation Postdoctoral Research Fellowship.

there are points where the *local* structure is more complicated than on a smooth variety, but that allows for the possibility of still having a somewhat simpler global structure and along with it, good local properties at most points.

Another reason to study singularities is that to understand smooth objects we should also understand how smooth objects may deform and degenerate. This leads to the need to construct and understand moduli spaces. And not just moduli for the smooth objects: degenerations provide important information as well. In other words, it is always useful to work with complete moduli problems, i.e., extend our moduli functor so it admits a compact (and preferably projective) coarse moduli space. This also leads to having to consider singular varieties.

On the other hand, we have to be careful to limit the kinds of singularities that we allow in order to be able to handle them. One might view this survey as a list of the singularities that we must deal with to achieve the goals stated above. Fortunately, it is also a class of singularities with which we have a reasonable chance to be able to work.

In particular, we will review Du Bois singularities and related notions, including some very recent important results. We will also review a family of singularities defined via characteristic-p methods, the Frobenius morphism, and their connections to the other set of singularities we are discussing.

Definitions and notation. Let k be an algebraically closed field. Unless otherwise stated, all objects will be assumed to be defined over k. A *scheme* will refer to a scheme of finite type over k and unless stated otherwise, a *point* refers to a closed point.

For a morphism $Y \to S$ and another morphism $T \to S$, the symbol Y_T will denote $Y \times_S T$. In particular, for $t \in S$ we write $X_t = f^{-1}(t)$. In addition, if $T = \operatorname{Spec} F$, then Y_T will also be denoted by Y_F.

Let X be a scheme and $\mathscr{F}$ an $\mathscr{O}_X$-module. The *m-th reflexive power* of $\mathscr{F}$ is the double dual (or reflexive hull) of the m-th tensor power of $\mathscr{F}$:

$$\mathscr{F}^{[m]} := (\mathscr{F}^{\otimes m})^{**}.$$

A *line bundle* on X is an invertible $\mathscr{O}_X$-module. A $\mathbb{Q}$-*line bundle* $\mathscr{L}$ on X is a reflexive $\mathscr{O}_X$-module of rank 1 that possesses a reflexive power which is a line bundle, i.e., there exists an $m \in \mathbb{N}_+$ such that $\mathscr{L}^{[m]}$ is a line bundle. The smallest such m is called the *index* of $\mathscr{L}$.

- For the advanced reader: whenever we mention Weil divisors, assume that X is S_2 [Har77, Theorem 8.22A(2)] and think of a *Weil divisorial sheaf*, that is, a rank 1 reflexive $\mathscr{O}_X$-module which is locally free in codimension 1. For flatness issues consult [Kol08a, Theorem 2].
- For the novice: whenever we mention Weil divisors, assume that X is normal and adopt the definition [Har77, p. 130].

For a Weil divisor D on X, its associated *Weil divisorial sheaf* is the $\mathcal{O}_X$-module $\mathcal{O}_X(D)$ defined on the open set $U \subseteq X$ by the formula

$$\Gamma(U, \mathcal{O}_X(D)) = \left\{ \frac{a}{b} \ \middle| \ \begin{array}{l} a, b \in \Gamma(U, \mathcal{O}_X), \ b \text{ is not a zero divisor anywhere} \\ \text{on } U, \text{ and } D|_U + \mathrm{div}_U(a) - \mathrm{div}_U(b) \geq 0. \end{array} \right\}$$

and made into a sheaf by the natural restriction maps.

A Weil divisor D on X is a *Cartier divisor*, if its associated Weil divisorial sheaf, $\mathcal{O}_X(D)$ is a line bundle. If the associated Weil divisorial sheaf, $\mathcal{O}_X(D)$ is a $\mathbb{Q}$-line bundle, then D is a $\mathbb{Q}$-*Cartier divisor*. The latter is equivalent to the property that there exists an $m \in \mathbb{N}_+$ such that mD is a Cartier divisor. Weil divisors form an abelian group. Tensoring this group with $\mathbb{Q}$ (over $\mathbb{Z}$) one obtains the group of $\mathbb{Q}$-*divisors* on X. (If X is not normal, some unexpected things can happen in this process; see [Kol92, Chapter 16].)

The symbol $\sim$ stands for *linear* and $\equiv$ for *numerical equivalence* of divisors.

Let $\mathcal{L}$ be a line bundle on a scheme X. It is said to be *generated by global sections* if for every point $x \in X$ there exists a global section $\sigma_x \in H^0(X, \mathcal{L})$ such that the germ σ_x generates the stalk $\mathcal{L}_x$ as an $\mathcal{O}_X$-module. If $\mathcal{L}$ is generated by global sections, then the global sections define a morphism

$$\phi_{\mathcal{L}} : X \to \mathbb{P}^N = \mathbb{P}\big(H^0(X, \mathcal{L})^*\big).$$

$\mathcal{L}$ is called *semi-ample* if $\mathcal{L}^m$ is generated by global sections for $m \gg 0$. $\mathcal{L}$ is called *ample* if it is semi-ample and $\phi_{\mathcal{L}^m}$ is an embedding for $m \gg 0$. A line bundle $\mathcal{L}$ on X is called *big* if the global sections of $\mathcal{L}^m$ define a rational map $\phi_{\mathcal{L}^m} : X \dashrightarrow \mathbb{P}^N$ such that X is birational to $\phi_{\mathcal{L}^m}(X)$ for $m \gg 0$. Note that in this case $\mathcal{L}^m$ need not be generated by global sections, so $\phi_{\mathcal{L}^m}$ is not necessarily defined everywhere. We leave it for the reader the make the obvious adaptation of these notions for the case of $\mathbb{Q}$-line bundles.

The *canonical divisor* of a scheme X is denoted by K_X and the *canonical sheaf* of X is denoted by ω_X.

A smooth projective variety X is of *general type* if ω_X is big. It is easy to see that this condition is invariant under birational equivalence between smooth projective varieties. An arbitrary projective variety is of *general type* if so is a desingularization of it.

A projective variety is *canonically polarized* if ω_X is ample. Notice that if a smooth projective variety is canonically polarized, then it is of general type.

2. Pairs and resolutions

For the reader's convenience, we recall a few definitions regarding pairs.

DEFINITION 2.1. A *pair* (X, Δ) consists of a normal[1] quasiprojective variety or complex space X and an effective $\mathbb{Q}$-divisor $\Delta \subset X$. A *morphism of pairs* $\gamma : (\widetilde{X}, \widetilde{\Delta}) \to (X, \Delta)$ is a morphism $\gamma : \widetilde{X} \to X$ such that $\gamma(\mathrm{Supp}(\mathrm{wt}\,\Delta)) \subseteq \mathrm{Supp}(\Delta)$. A morphism of pairs $\gamma : (\widetilde{X}, \widetilde{\Delta}) \to (X, \Delta)$ is called *birational* if it induces a birational morphism $\gamma : \mathrm{wt}\,X \xrightarrow{\sim} X$ and $\gamma(\mathrm{wt}\,\Delta) = \Delta$. It is an *isomorphism* if it is birational and it induces an isomorphism $\gamma : \mathrm{wt}\,X \xrightarrow{\sim} X$.

DEFINITION 2.2. Let (X, Δ) be a pair, and $x \in X$ a point. We say that (X, Δ) is *snc at* x, if there exists a Zariski-open neighborhood U of x such that U is smooth and $\Delta \cap U$ is reduced and has only simple normal crossings (see Section 3B for additional discussion). The pair (X, Δ) is *snc* if it is snc at all $x \in X$.

Given a pair (X, Δ), let $(X, \Delta)_{\mathrm{reg}}$ be the maximal open set of X where (X, Δ) is snc, and let $(X, \Delta)_{\mathrm{Sing}}$ be its complement, with the induced reduced subscheme structure.

REMARK 2.2.1. If a pair (X, Δ) is snc at a point x, this implies that all components of Δ are smooth at x. If instead of the condition that U is Zariski-open one would only require this analytically locally, then Definition 2.2 would define normal crossing pairs rather than pairs with simple normal crossing.

DEFINITION 2.3. A *log resolution* of (X, Δ) is a proper birational morphism of pairs $\pi : (\widetilde{X}, \mathrm{wt}\,\Delta) \to (X, \Delta)$ that satisfies the following four conditions:

- $\widetilde{X}$ is smooth.
- $\mathrm{wt}\,\Delta = \pi_*^{-1}\Delta$ is the strict transform of Δ.
- $\mathrm{Exc}(\pi)$ is of pure codimension 1.
- $\mathrm{Supp}(\widetilde{\Delta} \cup \mathrm{Exc}(\pi))$ is a simple normal crossings divisor.

If, in addition,

- the strict transform $\widetilde{\Delta}$ of Δ has smooth support,

then we call π an *embedded resolution* of $\Delta \subset X$:

In many cases, it is also useful to require that π is an isomorphism over $(X, \Delta)_{\mathrm{reg}}$.

3. Introduction to the singularities of the mmp

Even though we have introduced pairs and most of these singularities make sense for pairs, to make the introduction easier to digest we will mostly discuss the case when $\Delta = \varnothing$. As mentioned in the introduction, one of our goals is to show why we are forced to work with singular varieties even if our primary interest lies with smooth varieties.

---

[1] Occasionally, we will discuss pairs in the nonnormal setting. See Section 3F for more details.

3A. Canonical singularities. For an excellent introduction to this topic the reader is urged to take a thorough look at Miles Reid's Young Person's Guide [Rei87]. Here we will only touch on the subject.

Let us suppose that we would like to get a handle on some varieties. Perhaps we want to classify them or make some computations. In any case, a useful thing to do is to embed the object in question into a projective space (if we can). Doing so requires a (very) ample line bundle. It turns out that in practice these can be difficult to find. In fact, it is not easy to find any nontrivial line bundle on an abstract variety.

One possibility, when X is smooth, is to try a line bundle that is "handed" to us, namely some (positive or negative) power of the *canonical line bundle*; $\omega_X = \det T_X^*$. If X is not smooth but instead normal, we can construct ω_X on the smooth locus and then push it forward to obtain a rank one reflexive sheaf on all of X (which sometimes is still a line bundle). Next we will explore how we might "force" this line bundle to be ample in some (actually many) cases.

Let X be a minimal surface of general type that contains a (-2)-curve (a smooth rational curve with self-intersection -2). For an example of such a surface consider the following.

EXAMPLE 3.1. $\tilde{X} = (x^5 + y^5 + z^5 + w^5 = 0) \subseteq \mathbb{P}^3$ with the $\mathbb{Z}_2$-action that interchanges $x \leftrightarrow y$ and $z \leftrightarrow w$. This action has five fixed points, $[1:1:-\varepsilon^i:-\varepsilon^i]$ for $i = 1, \ldots, 5$, where ε is a primitive fifth root of unity. Hence the quotient $\tilde{X}/\mathbb{Z}_2$ has five singular points, each a simple double point of type A_1. Let $X \to \tilde{X}/\mathbb{Z}_2$ be the minimal resolution of singularities. Then X contains five (-2)-curves, the exceptional divisors over the singularities.

Let us return to the general case, that is, X is a minimal surface of general type that contains a (-2)-curve, $C \subseteq X$. As $C \simeq \mathbb{P}^1$, and X is smooth, the adjunction formula gives us that $K_X \cdot C = 0$. Therefore K_X is not ample.

On the other hand, since X is a minimal surface of general type, it follows that K_X is semi-ample, that is, some multiple of it is base-point free. In other words, there exists a morphism,

$$|mK_X| : X \to X_{\mathrm{can}} \subseteq \mathbb{P}\big(H^0(X, \mathscr{O}_X(mK_X))^*\big).$$

This may be shown in several ways. For example, it follows from Bombieri's classification of pluricanonical maps, but perhaps the simplest proof is provided by Miles Reid [Rei97, E.3].

It is then relatively easy to see that this morphism onto its image is independent of m (as long as mK_X is base point free). This constant image is called the *canonical model* of X, and will be denoted by X_{can}.

The good news is that the canonical line bundle of X_{can} is indeed ample, but the trouble is that X_{can} is singular. We might consider this as the first sign of the necessity of working with singular varieties. Fortunately the singularities are not too bad, so we still have a good chance to work with this model. In fact, the singularities that can occur on the canonical model of a surface of general type belong to a much studied class. This class goes by several names; they are called *du Val singularities*, or *rational double points*, or *Gorenstein, canonical singularities*. For more on these singularities, refer to [Dur79; Rei87].

3B. Normal crossings. These singularities already appear in the construction of the moduli space of stable curves (or if the reader prefers, the construction of a compactificaton of the moduli space of smooth projective curves). If we want to understand degenerations of smooth families, we have to allow normal crossings.

A *normal crossing* singularity is one that is locally analytically (or formally) isomorphic to the intersection of coordinate hyperplanes in a linear space. In other words, it is a singularity locally analytically defined as $(x_1 x_2 \cdots x_r = 0) \subseteq \mathbb{A}^n$ for some $r \leq n$. In particular, as opposed to the curve case, for surfaces it allows for triple intersections. However, triple intersections may be "resolved": Let $X = (xyz = 0) \subseteq \mathbb{A}^3$. Blow up the origin $O \in \mathbb{A}^3$ to obtain $\sigma : \mathrm{Bl}_O \mathbb{A}^3 \to \mathbb{A}^3$, and consider the proper transform of X, $\sigma : \widetilde{X} \to X$. Observe that $\widetilde{X}$ has only double normal crossings.

Another important point to remember about normal crossings is that they are *not* normal. In particular they do not belong to the previous category. For some interesting and perhaps surprising examples of surfaces with normal crossings see [Kol07].

3C. Pinch points. Another nonnormal singularity that can occur as the limit of smooth varieties is the pinch point. It is locally analytically defined as the locus $(x_1^2 = x_2 x_3^2) \subseteq \mathbb{A}^n$. This singularity is a double normal crossing away from the pinch point. Its normalization is smooth, but blowing up the pinch point (i.e., the origin) does not make it any better. (Try it for yourself!)

3D. Cones. Let $C \subseteq \mathbb{P}^2$ be a curve of degree d and $X \subseteq \mathbb{P}^3$ the projectivized cone over C. As X is a degree d hypersurface, it admits a smoothing.

EXAMPLE 3.2. Let $\mathcal{E} = (x^d + y^d + z^d + tw^d = 0) \subseteq \mathbb{P}^3_{x:y:z:w} \times \mathbb{A}^1_t$. The special fiber $\mathcal{E}_0$ is a cone over a smooth plane curve of degree d and the general fiber $\mathcal{E}_t$, for $t \neq 0$, is a smooth surface of degree d in $\mathbb{P}^3$.

This, again, suggests that we must allow some singularities. The question is whether we can limit the type of singularities we must deal with. More particularly to this case, can we limit the type of cones we need to allow?

First we need an auxiliary computation. By the nature of the computation it is easier to use *divisors* instead of *line bundles*.

COMMENTARY 3.3. One of our ultimate goals is to construct a moduli space for canonical models of varieties. We are already aware that the minimal model program has to deal with singularities and so we must allow some singularities on canonical models. We would also like to understand what constraints are imposed if our goal is to construct a moduli space. The point is that in order to construct our moduli space, the objects must have an ample canonical class. It is possible that a family of canonical models degenerates to a singular fiber that has singularities worse than the original canonical models. An important question then is whether we may resolve the singularities of this special fiber and retain ampleness of the canonical class. The next example shows that this is not always possible.

EXAMPLE 3.4. Let W be a smooth variety and $X = X_1 \cup X_2 \subseteq W$ such that X_1 and X_2 are Cartier divisors in W. Then by the adjunction formula we have

$$K_X = (K_W + X)|_X,$$
$$K_{X_1} = (K_W + X_1)|_{X_1},$$
$$K_{X_2} = (K_W + X_2)|_{X_2}.$$

Therefore

$$K_X|_{X_i} = K_{X_i} + X_{3-i}|_{X_i} \tag{3.4.1}$$

for $i = 1, 2$, so we have

$$K_X \text{ is ample} \iff K_X|_{X_i} = K_{X_i} + X_{3-i}|_{X_i} \text{ is ample for } i = 1, 2. \tag{3.4.2}$$

Next, let X be a normal projective surface with K_X ample and an isolated singular point $P \in \operatorname{Sing} X$. Assume that X is isomorphic to a cone $\Xi_0 \subseteq \mathbb{P}^3$ as in Example 3.2, locally analytically near P. Further assume that X is the special fiber of a family Ξ that itself is smooth. In particular, we may assume that all fibers other than X are smooth. As explained in (3.3), we would like to see whether we may resolve the singular point $P \in X$ and still be able to construct our desired moduli space, i.e., that K of the resolved fiber would remain ample. For this purpose we may assume that P is the only singular point of X.

Let $\Upsilon \to \Xi$ be the blowing up of $P \in \Xi$ and let $\tilde{X}$ denote the proper transform of X. Then $\Upsilon_0 = \tilde{X} \cup E$, where $E \simeq \mathbb{P}^2$ is the exceptional divisor of the blowup. Clearly, $\sigma : \tilde{X} \to X$ is the blowup of P on X, so it is a smooth surface and $\tilde{X} \cap E$ is isomorphic to the degree d curve over which X is locally analytically a cone.

We would like to determine the condition on d that ensures that the canonical divisor of Υ_0 is still ample. According to (3.4.2) this means that we need that $K_E + \tilde{X}|_E$ and $K_{\tilde{X}} + E|_{\tilde{X}}$ be ample.

As $E \simeq \mathbb{P}^2$, $\omega_E \simeq \mathcal{O}_{\mathbb{P}^2}(-3)$, so $\mathcal{O}_E(K_E + \tilde{X}|_E) \simeq \mathcal{O}_{\mathbb{P}^2}(d-3)$. This is ample if and only if $d > 3$.

As this computation is local near P the only relevant issue about the ampleness of $K_{\widetilde{X}} + E|_{\widetilde{X}}$ is whether it is ample in a neighborhood of $E_X := E|_{\widetilde{X}}$. By the next claim this is equivalent to asking when $(K_{\widetilde{X}} + E_X) \cdot E_X$ is positive.

CLAIM. *Let Z be a smooth projective surface with nonnegative Kodaira dimension and $\Gamma \subset Z$ an effective divisor. If $(K_Z + \Gamma) \cdot C > 0$ for every proper curve $C \subset Z$, then $K_Z + \Gamma$ is ample.*

PROOF. By the assumption on the Kodaira dimension there exists an $m > 0$ such that $m K_Z$ is effective, hence so is $m(K_Z + \Gamma)$. Then by the assumption on the intersection number, $(K_Z + \Gamma)^2 > 0$, so the statement follows by the Nakai–Moishezon criterium. $\qquad\square$

Observe that, by the adjunction formula,

$$(K_{\widetilde{X}} + E_X) \cdot E_X = \deg K_{E_X} = d(d - 3),$$

as E_X is isomorphic to a plane curve of degree d. Again, we obtain the same condition as above and thus conclude that K_{Υ_0} may be ample only if $d > 3$.

Now, if we are interested in constructing moduli spaces, one of the requirements of being stable is that the canonical bundle be ample. This means that in order to obtain a compact moduli space we have to allow cone singularities over curves of degree $d \leq 3$. The singularity we obtain for $d = 2$ is a rational double point, but the singularity for $d = 3$ is not even rational. This does not fit any of the earlier classes we discussed. It belongs to the one discussed in the next section.

3E. Log canonical singularities. Let us investigate the previous situation under more general assumptions.

COMPUTATION 3.5. Let $D = \sum_{i=0}^{r} \lambda_i D_i$ $(\lambda_i \in \mathbb{N})$, be a divisor with only normal crossing singularities in a smooth ambient variety such that $\lambda_0 = 1$. Using a generalized version of the adjunction formula shows that in this situation (3.4.1) remains true.

$$K_D|_{D_0} = K_{D_0} + \sum_{i=1}^{r} \lambda_i D_i|_{D_0} \qquad (3.5.1)$$

Let $f : \Xi \to B$ a projective family with $\dim B = 1$, Ξ smooth and K_{Ξ_b} ample for all $b \in B$. Further let $X = \Xi_{b_0}$ for some $b_0 \in B$ a singular fiber and let $\sigma : \Upsilon \to \Xi$ be an embedded resolution of $X \subseteq \Xi$. Finally let $Y = \sigma^* X = \widetilde{X} + \sum_{i=1}^{r} \lambda_i F_i$ where $\widetilde{X}$ is the proper transform of X and F_i are exceptional divisors for σ. We are interested in finding conditions that are necessary for K_Y to remain ample.

Let $E_i := F_i|_{\widetilde{X}}$ be the exceptional divisors for $\sigma : \widetilde{X} \to X$ and for the simplicity of computation, assume that the E_i are irreducible. For K_Y to be ample we need $K_Y|_{\mathrm{wt}\, X}$ as well as $K_Y|_{F_i}$ for all i to be ample. Clearly, the important one of these for our purposes is $K_Y|_{\mathrm{wt}\, X}$, for which 3.5.1) gives

$$K_Y|_{\widetilde{X}} = K_{\widetilde{X}} + \sum_{i=1}^{r} \lambda_i E_i.$$

As usual, we may write $K_{\widetilde{X}} = \sigma^* K_X + \sum_{i=1}^{r} a_i E_i$, so we are looking for conditions to guarantee that $\sigma^* K_X + \sum (a_i + \lambda_i) E_i$ be ample. In particular, its restriction to any of the E_i has to be ample. To further simplify our computation let us assume that $\dim X = 2$. Then the condition that we want satisfied is that, for all j,

$$\left(\sum_{i=1}^{r} (a_i + \lambda_i) E_i \right) \cdot E_j > 0. \tag{3.5.2}$$

Write the sum in parentheses as $E_+ - E_-$, where

$$E_+ = \sum_{a_i + \lambda_i \geq 0} |a_i + \lambda_i| E_i \quad \text{and} \quad E_- = \sum_{a_i + \lambda_i < 0} |a_i + \lambda_i| E_i.$$

Choose a j such that $E_j \subseteq \mathrm{Supp}\, E_+$. Then $E_- \cdot E_j \geq 0$ since $E_j \nsubseteq E_-$ and (3.5.2) implies that $(E_+ - E_-) \cdot E_j > 0$. These together imply that $E_+ \cdot E_j > 0$ and then that $E_+^2 > 0$. However, the E_i are exceptional divisors of a birational morphism, so their intersection matrix, $(E_i \cdot E_j)$ is negative definite.

The only way this can happen is if $E_+ = 0$. In other words, $a_i + \lambda_i < 0$ for all i. However, the λ_i are positive integers, so this implies that K_Y may remain ample only if $a_i < -1$ for all $i = 1, \ldots, r$.

The definition of a *log canonical singularity* is the exact opposite of this condition. It requires that X be normal and admit a resolution of singularities, say $Y \to X$, such that all the $a_i \geq -1$. This means that the above argument shows that we may stand a fighting chance if we resolve singularities that are *worse* than log canonical, but have no hope to do so with log canonical singularities. In other words, this is another class of singularities that we have to allow. As we remarked above, the class of singularities we obtained for the cones in the previous subsection belong to this class. In fact, all the normal singularities that we have considered so far belong to this class.

The good news is that by now we have covered most of the ways that something can go wrong and found the class of singularities we must allow. Since we already know that we have to deal with some nonnormal singularities and in fact in this example we have not really needed that X be normal, we conclude that we will have to allow the nonnormal cousins of log canonical singularities. These are called *semi-log canonical singularities*, and we turn to them now.

3F. Semi-log canonical singularities. Semi-log canonical singularities are very important in moduli theory. These are exactly the singularities that appear on stable varieties, the higher dimensional analogs of stable curves. However, their definition is rather technical, so the reader might want to skip this section at the first reading.

As a warm-up, let us first define the normal and more traditional singularities that are relevant in the minimal model program.

DEFINITION 3.6. A pair (X, Δ) is called *log $\mathbb{Q}$-Gorenstein* if $K_X + \Delta$ is $\mathbb{Q}$-Cartier, i.e., some integer multiple of $K_X + \Delta$ is a Cartier divisor. Let (X, Δ) be a log $\mathbb{Q}$-Gorenstein pair and $f : \widetilde{X} \to X$ a log resolution of singularities with exceptional divisor $E = \bigcup E_i$. Express the log canonical divisor of $\widetilde{X}$ in terms of $K_X + \Delta$ and the exceptional divisors:

$$K_{\widetilde{X}} + \mathrm{wt}\, \Delta \equiv f^*(K_X + \Delta) + \sum a_i E_i$$

where $\mathrm{wt}\, \Delta = f_*^{-1}\Delta$, the strict transform of Δ on $\mathrm{wt}\, X$ and $a_i \in \mathbb{Q}$. Then the pair (X, Δ) has

$$
\left.
\begin{array}{r}
terminal \\
canonical \\
plt \\
klt \\
log\ canonical
\end{array}
\right\}
\text{singularities if}
\left\{
\begin{array}{l}
a_i > 0 \\
a_i \geq 0 \\
a_i > -1 \\
a_i > -1 \text{ and } \lfloor \Delta \rfloor \leq 0 \\
a_i \geq -1
\end{array}
\right.
$$

for all log resolutions f and all i. The corresponding definitions for nonnormal varieties are somewhat more cumbersome. We include them here for completeness, but the reader should feel free to skip them and assume that for instance "semi-log canonical" means something that can be reasonably considered a nonnormal version of log canonical.

Suppose that X is a reduced equidimensional scheme that

(i) satisfies Serre's condition S2 (see [Har77, Theorem 8.22A(2)]), and
(ii) has only simple normal double crossings in codimension 1 (in particular X is Gorenstein in codimension 1).[2]

The conditions imply that we can treat the canonical module of X as a divisorial sheaf even though X is not normal. Further suppose that D is a $\mathbb{Q}$-Weil divisor on X (again, following [Kol92, Chapter 16], we assume that X is regular at the generic point of each component in Supp D).

REMARK 3.7. Conditions (i) and (ii) in Definition 3.6 imply that X is seminormal since it is seminormal in codimension 1; see [GT80, Corollary 2.7].

---

[2] Sometimes a ring that is S2 and Gorenstein in codimension 1 is called quasinormal.

Set $\rho : X^N \to X$ to be the normalization of X and suppose that B is the divisor of the conductor ideal on X^N. We denote by $\rho^{-1}(D)$ the pullback of D to X^N.

DEFINITION 3.8. We say that (X, D) is *semi-log canonical* if

(i) $K_X + D$ is $\mathbb{Q}$-Cartier, and
(ii) the pair $(X^N, B + \rho^{-1} D)$ is log canonical.

Actually, the original definition of semi-log canonical singularities (which is equivalent to this one) uses the theory of semi-resolutions. For details, see [KSB88], [Kol92, Chapter 12], and [Kol08b].

4. Hyperresolutions and Du Bois's original definition

A very important construction is Du Bois's generalized De Rham complex. The original construction of Du Bois's complex, $\underline{\Omega}_X^{\cdot}$, is based on simplicial resolutions. The reader interested in the details is referred to the original article [DB81]. Note also that a simplified construction was later obtained in [Car85] and [GNPP88] via the general theory of polyhedral and cubic resolutions. At the end of the paper, we include an appendix in which we explain how to construct, and give examples of cubical hyperresolutions. An easily accessible introduction can be found in [Ste85]. Another useful reference is the recent book [PS08].

In [Sch07] one of us found a simpler alternative construction of (part of) the Du Bois complex, not involving a simplicial resolution; see also Section 6 below. However we will discuss the original construction because it is important to keep in mind the way these singularities appeared, as that explains their usefulness. For more on applications of Du Bois's complex and Du Bois singularities see [Ste83], [Kol95, Chapter 12], [Kov99], and [Kov00b].

The word "hyperresolution" will refer to either simplicial, polyhedral, or cubic resolution. Formally, the construction of $\underline{\Omega}_X^{\cdot}$ is the same regardless the type of resolution used and no specific aspects of either types will be used.

The following definition is included to make sense of the statements of some of the forthcoming theorems. It can be safely ignored if the reader is not interested in the detailed properties of Du Bois's complex and is willing to accept that it is a very close analog of the De Rham complex of smooth varieties.

DEFINITION 4.1. Let X be a complex scheme (i.e., a scheme of finite type over $\mathbb{C}$) of dimension n. Let $D_{\text{filt}}(X)$ denote the derived category of filtered complexes of $\mathcal{O}_X$-modules with differentials of order ≤ 1 and $D_{\text{filt,coh}}(X)$ the subcategory of $D_{\text{filt}}(X)$ of complexes $K^{\cdot}$ such that for all i, the cohomology sheaves of $\text{Gr}_{\text{filt}}^i K^{\cdot}$ are coherent; see [DB81], [GNPP88]. Let $D(X)$ and $D_{\text{coh}}(X)$ denote the derived categories with the same definition except that the complexes are assumed to have the trivial filtration. The superscripts $+, -, b$ carry the usual meaning (bounded below, bounded above, bounded). Isomorphism in these categories is denoted by $\simeq_{\text{qis}}$. A sheaf $\mathscr{F}$ is also considered a complex $\mathscr{F}^{\cdot}$ with $\mathscr{F}^0 = \mathscr{F}$ and $\mathscr{F}^i = 0$ for $i \neq 0$. If $K^{\cdot}$ is a complex in any of the above categories, then $h^i(K^{\cdot})$ denotes the i-th cohomology sheaf of $K^{\cdot}$.

The right derived functor of an additive functor F, if it exists, is denoted by RF and $R^i F$ is short for $h^i \circ RF$. Furthermore, $\mathbb{H}^i$, $\mathbb{H}^i_Z$, and $\mathscr{H}^i_Z$ will denote $R^i \Gamma$, $R^i \Gamma_Z$, and $R^i \mathscr{H}_Z$ respectively, where Γ is the functor of global sections, Γ_Z is the functor of global sections with support in the closed subset Z, and $\mathscr{H}_Z$ is the functor of the sheaf of local sections with support in the closed subset Z. Note that according to this terminology, if $\phi : Y \to X$ is a morphism and $\mathscr{F}$ is a coherent sheaf on Y, then $R\phi_* \mathscr{F}$ is the complex whose cohomology sheaves give rise to the usual higher direct images of $\mathscr{F}$.

THEOREM 4.2 [DB81, 6.3, 6.5]. *Let X be a proper complex scheme of finite type and D a closed subscheme whose complement is dense in X. Then there exists a unique object $\underline{\Omega}^{\cdot}_X \in \mathrm{Ob}\, D_{\mathrm{filt}}(X)$ such that, using the notation*

$$\underline{\Omega}^p_X := \mathrm{Gr}^p_{\mathrm{filt}}\, \underline{\Omega}^{\cdot}_X[p],$$

the following properties are satisfied:

(a) $\underline{\Omega}^{\cdot}_X \simeq_{\text{qis}} \mathbb{C}_X$; *i.e., $\underline{\Omega}^{\cdot}_X$ is a resolution of the constant sheaf $\mathbb{C}$ on X.*

(b) $\underline{\Omega}^{\cdot}_{(\_)}$ *is functorial; i.e., if $\phi : Y \to X$ is a morphism of proper complex schemes of finite type, there exists a natural map ϕ^* of filtered complexes*

$$\phi^* : \underline{\Omega}^{\cdot}_X \to R\phi_* \underline{\Omega}^{\cdot}_Y.$$

Furthermore, $\underline{\Omega}^{\cdot}_X \in \mathrm{Ob}\big(D^b_{\mathrm{filt,coh}}(X)\big)$, and if ϕ is proper, ϕ^* is a morphism in $D^b_{\mathrm{filt,coh}}(X)$.

(c) *Let $U \subseteq X$ be an open subscheme of X. Then*

$$\underline{\Omega}^{\cdot}_X|_U \simeq_{\text{qis}} \underline{\Omega}^{\cdot}_U.$$

(d) *If X is proper, there exists a spectral sequence degenerating at E_1 and abutting to the singular cohomology of X:*

$$E^{pq}_1 = \mathbb{H}^q\big(X, \underline{\Omega}^p_X\big) \Rightarrow H^{p+q}(X^{\mathrm{an}}, \mathbb{C}).$$

(e) *If $\varepsilon_{\cdot} : X_{\cdot} \to X$ is a hyperresolution, then*

$$\underline{\Omega}^{\cdot}_X \simeq_{\text{qis}} R\varepsilon_{\cdot *} \underline{\Omega}^{\cdot}_{X_{\cdot}}.$$

In particular, $h^i\big(\underline{\Omega}^p_X\big) = 0$ for $i < 0$.

(f) *There exists a natural map, $\mathscr{O}_X \to \underline{\Omega}^0_X$, compatible with (4.2.(b)).*

(g) *If X is smooth, then*

$$\underline{\Omega}_X^{\cdot} \simeq_{\mathrm{qis}} \Omega_X^{\cdot}.$$

In particular,

$$\underline{\Omega}_X^p \simeq_{\mathrm{qis}} \Omega_X^p.$$

(h) *If $\phi : Y \to X$ is a resolution of singularities, then*

$$\underline{\Omega}_X^{\dim X} \simeq_{\mathrm{qis}} R\phi_* \omega_Y.$$

(i) *Suppose that $\pi : \widetilde{Y} \to Y$ is a projective morphism and $X \subset Y$ a reduced closed subscheme such that π is an isomorphism outside of X. Let E denote the reduced subscheme of $\widetilde{Y}$ with support equal to $\pi^{-1}(X)$ and $\pi' : E \to X$ the induced map. Then for each p one has an exact triangle of objects in the derived category,*

$$\underline{\Omega}_Y^p \longrightarrow \underline{\Omega}_X^p \oplus R\pi_* \underline{\Omega}_{\widetilde{Y}}^p \xrightarrow{-} R\pi'_* \underline{\Omega}_E^p \xrightarrow{+1} .$$

It turns out that Du Bois's complex behaves very much like the de Rham complex for smooth varieties. Observe that condition (d) says that the Hodge-to-de Rham spectral sequence works for singular varieties if one uses the Du Bois complex in place of the de Rham complex. This has far reaching consequences and if the associated graded pieces, $\underline{\Omega}_X^p$ turn out to be computable, then this single property leads to many applications.

Notice that condition (f) gives a natural map $\mathcal{O}_X \to \underline{\Omega}_X^0$, and we will be interested in situations when this map is a quasi-isomorphism. When X is proper over $\mathbb{C}$, such a quasi-isomorphism will imply that the natural map

$$H^i(X^{\mathrm{an}}, \mathbb{C}) \to H^i(X, \mathcal{O}_X) = \mathbb{H}^i(X, \underline{\Omega}_X^0)$$

is surjective because of the degeneration at E_1 of the spectral sequence in condition (d).

Following Du Bois, Steenbrink was the first to study this condition and he christened this property after Du Bois.

DEFINITION 4.3. A scheme X is said to have *Du Bois singularities* (or *DB singularities* for short) if the natural map $\mathcal{O}_X \to \underline{\Omega}_X^0$ from condition (f) in Theorem 4.2 is a quasi-isomorphism.

REMARK 4.4. If $\varepsilon : X_{\textbf{.}} \to X$ is a hyperresolution of X (see the Appendix for a how to construct cubical hyperresolutions) then X has Du Bois singularities if and only if the natural map $\mathcal{O}_X \to R\varepsilon_{\textbf{.}*}\mathcal{O}_{X_{\textbf{.}}}$ is a quasi-isomorphism.

EXAMPLE 4.5. It is easy to see that smooth points are Du Bois. Deligne proved that normal crossing singularities are Du Bois as well [DJ74, Lemme 2(b)].

We will see more examples of Du Bois singularities in later sections.

5. An injectivity theorem and splitting the Du Bois complex

In this section, we state an injectivity theorem involving the dualizing sheaf that plays a role for Du Bois singularities similar to the role that Grauert–Riemenschneider plays for rational singularities. As an application, we state a criterion for Du Bois singularities related to a "splitting" of the Du Bois complex.

THEOREM 5.1 [Kov99, Lemma 2.2; Sch09, Proposition 5.11]. *Let X be a reduced scheme of finite type over $\mathbb{C}$, $x \in X$ a (possibly nonclosed) point, and $Z = \overline{\{x\}}$ its closure. Assume that $X \setminus Z$ has Du Bois singularities in a neighborhood of x (for example, x may correspond to an irreducible component of the non-Du Bois locus of X). Then the natural map*

$$H^i\left(R\mathcal{H}om_X^{\cdot}(\underline{\Omega}_X^0, \omega_X^{\cdot})\right)_x \to H^i(\omega_X^{\cdot})_x$$

is injective for every i.

The proof uses the fact that for a projective X, $H^i(X^{an}, \mathbb{C}) \to \mathbb{H}^i(X, \underline{\Omega}_X^0)$ is surjective for every $i > 0$, which follows from Theorem 4.2.

It would also be interesting and useful if the following generalization of this injectivity were true.

QUESTION 5.2. Suppose that X is a reduced scheme essentially of finite type over $\mathbb{C}$. Is it true that the natural map of sheaves

$$H^i\left(R\mathcal{H}om_X^{\cdot}(\underline{\Omega}_X^0, \omega_X^{\cdot})\right) \to H^i(\omega_X^{\cdot})$$

is injective for every i?

Even though Theorem 5.1 does not answer Question 5.2, it has the following extremely useful corollary.

THEOREM 5.3 [Kov99, Theorem 2.3; Kol95, Theorem 12.8]. *Suppose that the natural map $\mathcal{O}_X \to \underline{\Omega}_X^0$ has a left inverse in the derived category (that is, a map $\rho : \underline{\Omega}_X^0 \to \mathcal{O}_X$ such that the composition $\mathcal{O}_X \longrightarrow \underline{\Omega}_X^0 \xrightarrow{\ \rho\ } \mathcal{O}_X$ is an isomorphism). Then X has Du Bois singularities.*

PROOF. Apply the functor $R\mathcal{H}om_X(\_, \omega_X^{\cdot})$ to the maps $\mathcal{O}_X \longrightarrow \underline{\Omega}_X^0 \xrightarrow{\ \rho\ } \mathcal{O}_X$. Then by the assumption, the composition

$$\omega_X^{\cdot} \xrightarrow{\ \delta\ } R\mathcal{H}om_X(\underline{\Omega}_X^0, \omega_X^{\cdot}) \longrightarrow \omega_X^{\cdot}$$

is an isomorphism. Let $x \in X$ be a possibly nonclosed point corresponding to an irreducible component of the non-Du Bois locus of X and consider the stalks at x of the cohomology sheaves of the complexes above. We obtain that the natural map

$$H^i\left(R\mathcal{H}om_X(\underline{\Omega}_X^0, \omega_X^{\cdot})\right)_x \to H^i(\omega_X^{\cdot})_x$$

is surjective for every i. But it is also injective by Theorem 5.1. This proves that $\delta : (\omega_X^\bullet)_x \to R\mathcal{H}om_X(\underline{\Omega}_X^0, \omega_X^\bullet)_x$ is a quasi-isomorphism. Finally, applying the functor $R\mathcal{H}om_{\mathcal{O}_{X,x}}(\_, (\omega_X^\bullet)_x)$ one more time proves that X is Du Bois at x, contradicting our choice of $x \in X$ □

This also gives the following Boutot-like theorem for Du Bois singularities (cf. [Bou87]).

COROLLARY 5.4 [Kov99, Theorem 2.3; Kol95, Theorem 12.8]. *Suppose that* $f : Y \to X$ *is a morphism,* Y *has Du Bois singularities and the natural map* $\mathcal{O}_X \to Rf_*\mathcal{O}_Y$ *has a left inverse in the derived category. Then* X *also has Du Bois singularities.*

PROOF. Observe that the composition is an isomorphism

$$\mathcal{O}_X \to \underline{\Omega}_X^0 \to Rf_*\underline{\Omega}_Y^0 \simeq Rf_*\mathcal{O}_Y \to \mathcal{O}_X.$$

Then apply Theorem 5.3.   □

As an easy corollary, we see that rational singularities are Du Bois (which was first observed in the isolated case by Steenbrink in [Ste83, Proposition 3.7]).

COROLLARY 5.5 [Kov99; Sai00]. *If* X *has rational singularities, then* X *has Du Bois singularities.*

PROOF. Let $\pi : \widetilde{X} \to X$ be a log resolution. One has the composition $\mathcal{O}_X \to \underline{\Omega}_X^0 \to R\pi_*\mathcal{O}_{\widetilde{X}}$. Since X has rational singularities, this composition is a quasi-isomorphism. Apply Corollary 5.4.   □

6. Hyperresolution-free characterizations of Du Bois singularities

The definition of Du Bois singularities given via hyperresolution is relatively complicated (hyperresolutions themselves can be rather complicated to compute; see 2). In this section we state several hyperresolution free characterizations of Du Bois singularities. The first such characterization was given by Steenbrink in the isolated case. Another, more analytic characterization was given by Ishii and improved by Watanabe in the isolated quasi-Gorenstein[3] case. Finally the second named author gave a characterization that works for any reduced scheme.

A relatively simple characterization of an affine cone over a projective variety being Du Bois is given in [DB81]. Steenbrink generalized this criterion to all normal isolated singularities. It is this criterion that Steenbrink, Ishii, Watanabe, and others used extensively to study isolated Du Bois singularities.

---

[3]A variety X is quasi-Gorenstein if K_X is a Cartier divisor. It is not required that X is Cohen–Macaulay.

THEOREM 6.1 [DB81, Proposition 4.13; Ste83, 3.6]. *Let (X, x) be a normal isolated Du Bois singularity, and $\pi : \tilde{X} \to X$ a log resolution of (X, x) such that π is an isomorphism outside of $X \setminus \{x\}$. Let E denote the reduced preimage of x. Then (X, x) is a Du Bois singularity if and only if the natural map*

$$R^i \pi_* \mathscr{O}_{\tilde{X}} \to R^i \pi_* \mathscr{O}_E$$

is an isomorphism for all $i > 0$.

PROOF. Using Theorem 4.2, we have an exact triangle

$$\underline{\Omega}^0_X \longrightarrow \underline{\Omega}^0_{\{x\}} \oplus R\pi_* \underline{\Omega}^0_{\tilde{X}} \overset{-}{\longrightarrow} R\pi_* \underline{\Omega}^0_E \overset{+1}{\longrightarrow}.$$

Since $\{x\}$, $\tilde{X}$ and E are all Du Bois (the first two are smooth, and E is snc), we have the following exact triangle

$$\underline{\Omega}^0_X \longrightarrow \mathscr{O}_{\{x\}} \oplus R\pi_* \mathscr{O}_{\tilde{X}} \overset{-}{\longrightarrow} R\pi_* \mathscr{O}_E \overset{+1}{\longrightarrow}.$$

Suppose first that X has Du Bois singularities (that is, $\underline{\Omega}^0_X \simeq_{\mathrm{qis}} \mathscr{O}_X$). By taking cohomology and examining the long exact sequence, we see that $R^i \pi_* \mathscr{O}_{\tilde{X}} \to R^i \pi_* \mathscr{O}_E$ is an isomorphism for all $i > 0$.

So now suppose that $R^i \pi_* \mathscr{O}_{\tilde{X}} \to R^i \pi_* \mathscr{O}_E$ is an isomorphism for all $i > 0$. By considering the long exact sequence of cohomology, we see that $H^i(\underline{\Omega}^0_X)$ is zero for all $i > 0$. On the other hand, $H^0(\underline{\Omega}^0_X)$ is naturally identified with the seminormalization of $\mathscr{O}_X$; see Proposition 7.8 below. Thus if X is normal, then $\mathscr{O}_X \to H^0(\underline{\Omega}^0_X)$ is an isomorphism. $\qquad\square$

We now state a more analytic characterization, due to Ishii and slightly improved by Watanabe. First we recall the definition of the plurigenera of a singularity.

DEFINITION 6.2. For a singularity (X, x), we define the plurigenera $\{\delta_m\}_{m \in \mathbb{N}}$;

$$\delta_m(X, x) = \dim_{\mathbb{C}} \Gamma(X \setminus x, \mathscr{O}_X(mK_X))/L^{2/m}(X \setminus \{x\}),$$

where $L^{2/m}(X \setminus \{x\})$ denotes the set of all $L^{2/m}$-integrable m-uple holomorphic n-forms on $X \setminus \{x\}$.

THEOREM 6.3 [Ish85, Theorem 2.3; Wat87, Theorem 4.2]. *Let $f : \tilde{X} \to X$ be a log resolution of a normal isolated Gorenstein singularity (X, x) of dimension $n \geq 2$. Set E to be the reduced exceptional divisor (the preimage of x). Then (X, x) is a Du Bois singularity if and only if $\delta_m(X, x) \leq 1$ for any $m \in \mathbb{N}$.*

In [Sch07], a characterization of arbitrary Du Bois singularities is given that did not rely on hyperresolutions, but instead used a single resolution of singularities. An improvement of this was also obtained in [ST08, Proposition 2.20]. We provide a proof for the convenience of the reader.

THEOREM 6.4 [Sch07; ST08, Proposition 2.20]. *Let X be a reduced separated scheme of finite type over a field of characteristic zero. Suppose that $X \subseteq Y$ where Y is smooth and suppose that $\pi : \widetilde{Y} \to Y$ is a proper birational map with $\widetilde{Y}$ smooth and where $\overline{X} = \pi^{-1}(X)_{\mathrm{red}}$, the reduced preimage of X, is a simple normal crossings divisor (or in fact any scheme with Du Bois singularities). Then X has Du Bois singularities if and only if the natural map $\mathcal{O}_X \to R\pi_*\mathcal{O}_{\overline{X}}$ is a quasi-isomorphism.*

In fact, we can say more. There is an isomorphism

$$R\pi_*\mathcal{O}_{\overline{X}} \xrightarrow{\;\sim\;} \underline{\Omega}^0_X$$

such that the natural map $\mathcal{O}_X \to \underline{\Omega}^0_X$ can be identified with the natural map $\mathcal{O}_X \to R\pi_*\mathcal{O}_{\overline{X}}$.

PROOF. We first assume that π is an isomorphism outside of X. Then using Theorem 4.2, we have an exact triangle

$$\underline{\Omega}^0_Y \longrightarrow \underline{\Omega}^0_X \oplus R\pi_*\underline{\Omega}^0_{\widetilde{Y}} \xrightarrow{\;-\;} R\pi_*\underline{\Omega}^0_{\overline{X}} \xrightarrow{\;+1\;}.$$

Using the octahedral axiom, we obtain the diagram

$$
\begin{array}{ccccc}
C^{\cdot} & \longrightarrow & \underline{\Omega}^0_Y & \longrightarrow & \underline{\Omega}^0_X \xrightarrow{+1} \\
\downarrow{\scriptstyle\sim} & & \downarrow{\scriptstyle\alpha} & & \downarrow{\scriptstyle\beta} \\
C^{\cdot} & \longrightarrow & R\pi_*\underline{\Omega}^0_{\widetilde{Y}} & \longrightarrow & R\pi_*\underline{\Omega}^0_{\overline{X}} \xrightarrow{+1}
\end{array}
$$

where $C^{\cdot}$ is simply the object in the derived category that completes the triangles. But notice that the vertical arrow α is an isomorphism since Y has rational singularities (in which case each term in the middle column is isomorphic to $\mathcal{O}_Y$). Thus the vertical arrow β is also an isomorphism.

One always has a commutative diagram

$$
\begin{array}{ccc}
\mathcal{O}_X & \longrightarrow & \underline{\Omega}^0_X \\
\downarrow & & \downarrow{\scriptstyle\beta} \\
R\pi_*\mathcal{O}_{\overline{X}} & \xrightarrow{\;\delta\;} & R\pi_*\underline{\Omega}^0_{\overline{X}}
\end{array}
$$

(where the arrows are the natural ones). Observe that $\overline{X}$ has Du Bois singularities since it has normal crossings, thus δ is a quasi-isomorphism. But then the theorem is proven at least in the case that π is an isomorphism outside of X.

For the general case, it is sufficient to show that $R\pi_*\mathcal{O}_{\overline{X}}$ is independent of the choice of resolution. Since any two log resolutions can be dominated by a third, it is sufficient to consider two log resolutions $\pi_1 : Y_1 \to Y$ and $\pi_2 : Y_2 \to Y$ and a map between them $\rho : Y_2 \to Y_1$ over Y. Let $F_1 = (\pi_1^{-1}(X))_{\mathrm{red}}$ and $F_2 = (\pi_2^{-1}(X))_{\mathrm{red}} = (\rho^{-1}(F_1))_{\mathrm{red}}$. Dualizing the map and applying Grothendieck duality implies that it is sufficient to prove that $\omega_{Y_1}(F_1) \leftarrow R\rho_*(\omega_{Y_2}(F_2))$ is a quasi-isomorphism.

We now apply the projection formula while twisting by $\omega_{Y_1}^{-1}(-F_1)$. Thus it is sufficient to prove that

$$R\rho_*(\omega_{Y_2/Y_1}(F_2 - \rho^*F_1)) \to \mathcal{O}_{Y_1}$$

is a quasi-isomorphism. But note that $F_2 - \rho^*F_1 = -\lfloor \rho^*(1-\varepsilon)F_1 \rfloor$ for sufficiently small $\varepsilon > 0$. Thus it is sufficient to prove that the pair $(Y_1, (1-\varepsilon)F_1)$ has klt singularities by Kawamata–Viehweg vanishing in the form of local vanishing for multiplier ideals; see [Laz04, 9.4]. But this is true since Y_1 is smooth and F_1 is a reduced integral divisor with simple normal crossings. $\qquad\square$

It seems that in this characterization the condition that the ambient variety Y is smooth is asking for too much. We propose that the following may be a more natural characterization. For some motivation and for a statement that may be viewed as a sort of converse; see Conjecture 12.5 and the discussion preceding it.

CONJECTURE 6.5. *Theorem 6.4 should remain true if the hypothesis that Y is smooth is replaced by the condition that Y has rational singularities.*

Having Du Bois singularities is a local condition, so even if X is not embeddable in a smooth scheme, one can still use Theorem 6.4 by passing to an affine open covering.

To illustrate the utility and meaning of Theorem 6.4, we will explore the situation when X is a hypersurface inside a smooth scheme Y. In the notation of Theorem 6.4, we have the diagram of exact triangles

$$
\begin{array}{ccccccc}
R\pi_*\mathcal{O}_{\widetilde{Y}}(-\overline{X}) & \longrightarrow & R\pi_*\mathcal{O}_{\widetilde{Y}} & \longrightarrow & R\pi_*\mathcal{O}_{\overline{X}} & \xrightarrow{+1} & \\
\uparrow{\scriptstyle\alpha} & & \uparrow{\scriptstyle\beta} & & \uparrow{\scriptstyle\gamma} & & \\
0 \longrightarrow & \mathcal{O}_Y(-X) & \longrightarrow & \mathcal{O}_Y & \longrightarrow & \mathcal{O}_X & \longrightarrow 0
\end{array}
$$

Since Y is smooth, β is a quasi-isomorphism (as then Y has at worst rational singularities). Therefore, X has Du Bois singularities if and only if the map α is a quasi-isomorphism. However, α is a quasi-isomorphism if and only if the dual map

$$R\pi_*\omega_{\widetilde{Y}}^{\bullet}(\overline{X}) \to \omega_Y^{\bullet}(X) \qquad (6.5.1)$$

is a quasi-isomorphism. The projection formula tells us that Equation 6.5.1 is a quasi-isomorphism if and only

$$R\pi_* \mathscr{O}_{\widetilde{Y}}(K_{\widetilde{Y}/Y} - \pi^* X + \overline{X}) \to \mathscr{O}_X \qquad (6.5.2)$$

is a quasi-isomorphism. Note however that $-\pi^* X + \overline{X} = \lceil -(1-\varepsilon)\pi^* X \rceil$ for $\varepsilon > 0$ and sufficiently close to zero. Thus the left side of Equation 6.5.2 can be viewed as $R\pi_* \mathscr{O}_{\widetilde{Y}}(\lceil K_{\widetilde{Y}/Y} - (1-\varepsilon)\pi^* X \rceil)$ for $\varepsilon > 0$ sufficiently small. Note that Kawamata–Viehweg vanishing in the form of local vanishing for multiplier ideals implies that $\mathscr{J}(Y, (1-\varepsilon)X) \simeq_{\mathrm{qis}} R\pi_* \mathscr{O}_{\widetilde{Y}}(\lceil K_{\widetilde{Y}/Y} - (1-\varepsilon)\pi^* X \rceil)$. Therefore X has Du Bois singularities if and only if $\mathscr{J}(Y, (1-\varepsilon)X) \simeq \mathscr{O}_X$.

COROLLARY 6.6. *If X is a hypersurface in a smooth Y, then X has Du Bois singularities if and only if the pair (Y, X) is log canonical.*

Du Bois hypersurfaces have also been characterized via the Bernstein–Sato polynomial; see [Sai09, Theorem 0.5].

7. Seminormality of Du Bois singularities

In this section we show that Du Bois singularities are partially characterized by seminormality. First we remind the reader what it means for a scheme to be seminormal.

DEFINITION 7.1 [Swa80; GT80]. Suppose that R is a reduced excellent ring and that $S \supseteq R$ is a reduced R-algebra which is finite as an R-module. We say that the extension $i : R \hookrightarrow S$ is *subintegral* if

(i) i induces a bijection on spectra, Spec $S \to$ Spec R, and
(ii) i induces an isomorphism of residue fields over every (possibly nonclosed) point of Spec R.

REMARK 7.2. In [GT80], subintegral extensions are called quasi-isomorphisms.

DEFINITION 7.3 [Swa80; GT80]. Suppose that R is a reduced excellent ring. We say that R is *seminormal* if every subintegral extension $R \hookrightarrow S$ is an isomorphism. We say that a scheme X is *seminormal* if all of its local rings are seminormal.

REMARK 7.4. In [GT80], the authors call R seminormal if there is no proper subintegral extension $R \hookrightarrow S$ such that S is contained in the integral closure of R (in its total field of fractions). However, it follows from [Swa80, Corollary 3.4] that the definition above is equivalent.

REMARK 7.5. Seminormality is a local property. In particular, a ring is seminormal if and only if it is seminormal after localization at each of its prime (equivalently, maximal) ideals.

REMARK 7.6. The easiest example of seminormal schemes are schemes with snc singularities. In fact, a one dimensional variety over an algebraically closed field is seminormal if and only if its singularities are locally analytically isomorphic to a union of coordinate axes in affine space.

We will use the following well known fact about seminormality.

LEMMA 7.7. *If X is a seminormal scheme and $U \subseteq X$ is any open set, then $\Gamma(U, \mathscr{O}_X)$ is a seminormal ring.*

PROOF. We leave it as an exercise to the reader.   □

It is relatively easy to see, using the original definition via hyperresolutions, that if X has Du Bois singularities, then it is seminormal. Du Bois certainly knew this fact (see [DB81, Proposition 4.9]) although he didn't use the word seminormal. Later Saito proved that seminormality in fact partially characterizes Du Bois singularities. We give a different proof of this fact, from [Sch09].

PROPOSITION 7.8 [Sai00, Proposition 5.2; Sch09, Lemma 5.6]. *Suppose that X is a reduced separated scheme of finite type over $\mathbb{C}$. Then $h^0(\underline{\Omega}_X^0) = \mathscr{O}_{X^{sn}}$ where $\mathscr{O}_{X^{sn}}$ is the structure sheaf of the seminormalization of X.*

PROOF. Without loss of generality we may assume that X is affine. We need only consider $\pi_* \mathscr{O}_E$ by Theorem 6.4. By Lemma 7.7, $\pi_* \mathscr{O}_E$ is a sheaf of seminormal rings. Now let $X' = \mathrm{Spec}(\pi_* \mathscr{O}_E)$ and consider the factorization

$$E \to X' \to X.$$

Note $E \to X'$ must be surjective since it is dominant by construction and is proper by [Har77, II.4.8(e)]. Since the composition has connected fibers, so must have $\rho : X' \to X$. On the other hand, ρ is a finite map since π is proper. Therefore ρ is a bijection on points. Because these maps and schemes are of finite type over an algebraically closed field of characteristic zero, we see that $\Gamma(X, \mathscr{O}_X) \to \Gamma(X', \mathscr{O}_{X'})$ is a subintegral extension of rings. Since X' is seminormal, so is $\Gamma(X', \mathscr{O}_{X'})$, which completes the proof.   □

8. A multiplier-ideal-like characterization of Cohen–Macaulay Du Bois singularities

In this section we state a characterization of Cohen–Macaulay Du Bois singularities that explains why Du Bois singularities are so closely linked to rational and log canonical singularities.

We first do a suggestive computation. Suppose that X embeds into a smooth scheme Y and that $\pi : \widetilde{Y} \to Y$ is an embedded resolution of X in Y that is an isomorphism outside of X. Set $\widetilde{X}$ to be the strict transform of X and set $\overline{X}$ to be the reduced preimage of X. We further assume that $\overline{X} = \widetilde{X} \cup E$ where E is a reduced simple normal crossings divisor that intersects $\widetilde{X}$ transversally in another reduced simple normal crossing divisor. Note that E is the exceptional divisor of π (with reduced scheme structure). Set $\Sigma \subseteq X$ be the image of E. We have the short exact sequence

$$0 \to \mathscr{O}_{\widetilde{X}}(-E) \to \mathscr{O}_{\overline{X}} \to \mathscr{O}_E \to 0$$

We apply $R\mathcal{H}om_{\mathscr{O}_Y}(\_, \omega_{\widetilde{Y}}^{\cdot})$ followed by $R\pi_* \_$ and obtain the exact triangle

$$R\pi_* \omega_E^{\cdot} \longrightarrow R\pi_* \omega_{\overline{X}}^{\cdot} \longrightarrow R\pi_* \omega_{\widetilde{X}}(E)[\dim X] \xrightarrow{+1}$$

Using condition (i) in Theorem 4.2, the leftmost object can be identified with $R\mathcal{H}om_{\mathscr{O}_\Sigma}(\underline{\Omega}_\Sigma^0, \omega_\Sigma^{\cdot})$ and the middle object, $R\pi_* \omega_{\overline{X}}^{\cdot}$, can be identified with $R\mathcal{H}om_{\mathscr{O}_X}(\underline{\Omega}_X^0, \omega_X^{\cdot})$. Recall that X has Du Bois singularities if and only if the natural map $R\mathcal{H}om_{\mathscr{O}_X}(\underline{\Omega}_X^0, \omega_X^{\cdot}) \to \omega_X^{\cdot}$ is an isomorphism. Therefore, the object $\pi_* \omega_{\widetilde{X}}(E)$ is closely related to whether or not X has Du Bois singularities. This inspired the following result, which we do not prove.

THEOREM 8.1 [KSS10, Theorem 3.1] . *Suppose that X is normal and Cohen–Macaulay. Let $\pi : X' \to X$ be a log resolution, and denote the reduced exceptional divisor of π by G. Then X has Du Bois singularities if and only if $\pi_* \omega_{X'}(G) \simeq \omega_X$.*

We mention that the main idea in the proof is to show that

$$\pi_* \omega_{X'}(G) \simeq H^{-\dim X}\left(R\mathcal{H}om_{\mathscr{O}_X}(\underline{\Omega}_X^0, \omega_X^{\cdot}) \right).$$

Related results can also be obtained in the nonnormal Cohen–Macaulay case; see [KSS10] for details.

REMARK 8.2. The submodule $\pi_* \omega_{X'}(G) \subseteq \omega_X$ is independent of the choice of log resolution. Thus this submodule may be viewed as an invariant which partially measures how far a scheme is from being Du Bois (compare with [Fuj08]).

As an easy corollary, we obtain another proof that rational singularities are Du Bois (this time via the Kempf-criterion for rational singularities).

COROLLARY 8.3. *If X has rational singularities, then X has Du Bois singularities.*

PROOF. Since X has rational singularities, it is Cohen–Macaulay and normal. Then $\pi_* \omega_{X'} = \omega_X$ but we also have $\pi_* \omega_{X'} \subseteq \pi_* \omega_{X'}(G) \subseteq \omega_X$, and thus $\pi_* \omega_{X'}(G) = \omega_X$ as well. Then use Theorem 8.1. $\square$

We also see immediately that log canonical singularities coincide with Du Bois singularities in the Gorenstein case.

COROLLARY 8.4. *Suppose that X is Gorenstein and normal. Then X is Du Bois if and only if X is log canonical.*

PROOF. X is easily seen to be log canonical if and only if $\pi_* \omega_{X'/X}(G) \simeq \mathcal{O}_X$. The projection formula then completes the proof. $\qquad\qquad\square$

In fact, a slightly improved version of this argument can be used to show that every Cohen–Macaulay log canonical pair is Du Bois; see [KSS10, Theorem 3.16].

9. The Kollár–Kovács splitting criterion

The proof of the following, rather flexible, criterion for Du Bois singularities can be found in the original paper.

THEOREM 9.1 [KK10]. *Let $f : Y \to X$ be a proper morphism between reduced schemes of finite type over $\mathbb{C}$, $W \subseteq X$ an arbitrary subscheme, and $F := f^{-1}(W)$, equipped with the induced reduced subscheme structure. Let $\mathscr{I}_{W \subseteq X}$ denote the ideal sheaf of W in X and $\mathscr{I}_{F \subseteq Y}$ the ideal sheaf of F in Y. Assume that the natural map ϱ*

$$\mathscr{I}_{W \subseteq X} \xrightarrow[\varrho]{\quad\quad} Rf_* \mathscr{I}_{F \subseteq Y}$$

with a dashed arrow ϱ' curving back from $Rf_* \mathscr{I}_{F \subseteq Y}$ to $\mathscr{I}_{W \subseteq X}$

admits a left inverse ϱ', that is, $\rho' \circ \rho = \mathrm{id}_{\mathscr{I}_{W \subseteq X}}$. Then if Y, F, and W all have DB singularities, so does X.

REMARK 9.1.1. Notice that it is not required that f be birational. On the other hand the assumptions of the theorem and [Kov00a, Theorem 1] imply that if $Y \setminus F$ has rational singularities, e.g., if Y is smooth, then $X \setminus W$ has rational singularities as well.

This theorem is used to derive various consequences in [KK10], some of which are formally unrelated to Du Bois singularities. We will mention some of these in the sequel, but the interested reader should look at the original article to obtain the full picture.

10. Log canonical singularities are Du Bois

Log canonical and Du Bois singularities are very closely related as we have seen in the previous sections. This was first observed in [Ish85]; see also [Wat87] and [Ish87].

Recently, Kollár and the first named author gave a proof that log canonical singularities are Du Bois using Theorem 9.1. We will sketch some ideas of the proof here. There are two main steps. First, one shows that the non-klt locus of a log canonical singularity is Du Bois (this generalizes [Amb98] and [Sch08, Corollary 7.3]). Then one uses Theorem 9.1 to show that this property is enough to conclude that X itself is Du Bois. For the first part we refer the reader to the original paper. The key point of the second part is contained in the following Lemma. Here we give a different proof than in [KK10].

LEMMA 10.1. *Suppose* (X, Δ) *is a log canonical pair and that the reduced non-klt locus of* (X, Δ) *has Du Bois singularities. Then* X *has Du Bois singularities.*

PROOF. First recall that the multiplier ideal $\mathcal{J}(X, \Delta)$ is precisely the defining ideal of the non-klt locus of (X, Δ) and since (X, Δ) is log canonical, it is a radical ideal. We set $\Sigma \subseteq X$ to be the reduced subscheme of X defined by this ideal. Since the statement is local, we may assume that X is affine and thus that X is embedded in a smooth scheme Y. We let $\pi : \widetilde{Y} \to Y$ be an embedded resolution of (X, Δ) in Y and we assume that π is an isomorphism outside the singular locus of X. Set $\overline{\Sigma}$ to be the reduced-preimage of Σ (which we may assume is a divisor in $\widetilde{Y}$) and let $\widetilde{X}$ denote the strict transform of X. We consider the diagram of exact triangles

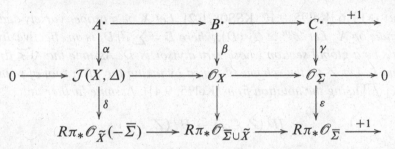

Here the first row is made up of objects in $D^b_{\mathrm{coh}}(X)$ needed to make the columns into exact triangles. Since Σ has Du Bois singularities, the map ε is an isomorphism and so $C^{\cdot} \simeq 0$. On the other hand, there is a natural map $R\pi_* \mathcal{O}_{\widetilde{X}}(-\overline{\Sigma}) \to R\pi_* \mathcal{O}_{\widetilde{X}}(K_{\widetilde{X}} - \pi^*(K_X + \Delta)) \simeq \mathcal{J}(X, \Delta)$ since (X, Δ) is log canonical. This implies that the map α is the zero map in the derived category. However, we then see that β is also zero in the derived category which implies that $\mathcal{O}_X \to R\pi_* \mathcal{O}_{\overline{\Sigma} \cup \widetilde{X}}$ has a left inverse. Therefore, X has Du Bois singularities (since $\overline{\Sigma} \cup \widetilde{X}$ has simple normal crossing singularities) by Theorems 5.3 and 6.4. $\square$

11. Applications to moduli spaces and vanishing theorems

The connection between log canonical and Du Bois singularities have many useful applications in moduli theory. We will list a few without proof.

SETUP 11.1. Let $\phi : X \to B$ be a flat projective morphism of complex varieties with B connected. Assume that for all $b \in B$ there exists a $\mathbb{Q}$-divisor D_b on X_b such that (X_b, D_b) is log canonical.

REMARK 11.2. Notice that it is not required that the divisors D_b form a family.

THEOREM 11.3 [KK10]. *Under the assumptions* 11.1, $h^i(X_b, \mathcal{O}_{X_b})$ *is independent of* $b \in B$ *for all* i.

THEOREM 11.4 [KK10]. *Under the assumptions* 11.1, *if one fiber of* ϕ *is Cohen–Macaulay (resp. S_k for some k), so are all the fibers.*

THEOREM 11.5 [KK10]. *Under the assumptions* 11.1, *the cohomology sheaves* $h^i(\omega_\phi^\bullet)$ *are flat over* B, *where* $\omega_\phi^\bullet$ *denotes the relative dualizing complex of* ϕ.

Du Bois singularities also appear naturally in vanishing theorems. As a culmination of the work of Tankeev, Ramanujam, Miyaoka, Kawamata, Viehweg, Kollár, and Esnault–Viehweg, Kollár proved a rather general form of a Kodaira-type vanishing theorem in [Kol95, 9.12]. Using the same ideas this was slightly generalized to the following theorem in [KSS10].

THEOREM 11.6 [Kol95, 9.12; KSS08, 6.2]. *Let X be a proper variety and $\mathcal{L}$ a line bundle on X. Let $\mathcal{L}^m \simeq \mathcal{O}_X(D)$, where $D = \sum d_i D_i$ is an effective divisor, and let s be a global section whose zero divisor is D. Assume that $0 < d_i < m$ for every i. Let Z be the scheme obtained by taking the m-th root of s (that is, $Z = X[\sqrt[m]{s}]$ using the notation from [Kol95, 9.4]). Assume further that*

$$H^j(Z, \mathbb{C}_Z) \to H^j(Z, \mathcal{O}_Z)$$

is surjective. Then, for any collection of $b_i \geq 0$, the natural map

$$H^j\big(X, \mathcal{L}^{-1}(-\textstyle\sum b_i D_i)\big) \to H^j(X, \mathcal{L}^{-1})$$

is surjective.

This, combined with the fact that log canonical singularities are Du Bois, yields that Kodaira vanishing holds for log canonical pairs:

THEOREM 11.7 [KSS10, 6.6]. *Kodaira vanishing holds for Cohen–Macaulay semi-log canonical varieties: Let (X, Δ) be a projective Cohen–Macaulay semi-log canonical pair and $\mathcal{L}$ an ample line bundle on X. Then $H^i(X, \mathcal{L}^{-1}) = 0$ for $i < \dim X$.*

It turns out that Du Bois singularities appear naturally in other kinds of vanishing theorems. We cite one here.

THEOREM 11.8 [GKKP10, 9.3]. *Let (X, D) be a log canonical reduced pair of dimension $n \geq 2$, $\pi : \tilde{X} \to X$ a log resolution with π-exceptional set E, and $\tilde{D} = \mathrm{Supp}(E + \pi^{-1}D)$. Then*

$$R^{n-1}\pi_* \mathcal{O}_{\tilde{X}}(-\tilde{D}) = 0.$$

12. Deformations of Du Bois singularities

Given the importance of Du Bois singularities in moduli theory it is an important obvious question whether they are invariant under small deformation.

It is relatively easy to see from the construction of the Du Bois complex that a general hyperplane section (or more generally, the general member of a base point free linear system) on a variety with Du Bois singularities again has Du Bois singularities. Therefore the question of deformation follows from the following.

CONJECTURE 12.1. *(cf. [Ste83]) Let $D \subset X$ be a reduced Cartier divisor and assume that D has only Du Bois singularities in a neighborhood of a point $x \in D$. Then X has only Du Bois singularities in a neighborhood of the point x.*

This conjecture was proved for isolated Gorenstein singularities by Ishii [Ish86]. Also note that rational singularities satisfy this property; see [Elk78].

We also have the following easy corollary of the results presented earlier:

THEOREM 12.2. *Assume that X is Gorenstein and D is normal.[4] Then the statement of Conjecture 12.1 is true.*

PROOF. The question is local so we may restrict to a neighborhood of x. If X is Gorenstein, then so is D as it is a Cartier divisor. Then D is log canonical by (8.4), and then the pair (X, D) is also log canonical by inversion of adjunction [Kaw07]. (Recall that if D is normal, then so is X along D). This implies that X is also log canonical and thus Du Bois. □

It is also stated in [Kov00b, 3.2] that the conjecture holds in full generality. Unfortunately, the proof is not complete. The proof published there works if one assumes that the non-Du Bois locus of X is contained in D. For instance, one may assume that this is the case if the non-Du Bois locus is isolated.

The problem with the proof is the following: it is stated that by taking hyperplane sections one may assume that the non-Du Bois locus is isolated. However, this is incorrect. One may only assume that the *intersection* of the non-Du Bois locus of X with D is isolated. If one takes a further general section then it will

---

[4]This condition is actually not necessary, but the proof becomes rather involved without it.

miss the intersection point and then it is not possible to make any conclusions about that case.

Therefore currently the best known result with regard to this conjecture is the following:

THEOREM 12.3 [Kov00b, 3.2]. *Let $D \subset X$ be a reduced Cartier divisor and assume that D has only Du Bois singularities in a neighborhood of a point $x \in D$ and that $X \setminus D$ has only Du Bois singularities. Then X has only Du Bois singularities in a neighborhood of x.*

Experience shows that divisors not in general position tend to have worse singularities than the ambient space in which they reside. Therefore one would in fact expect that if $X \setminus D$ is reasonably nice, and D has Du Bois singularities, then perhaps X has even better ones.

We have also seen that rational singularities are Du Bois and at least Cohen–Macaulay Du Bois singularities are not so far from being rational cf. 8.1. The following result of the second named author supports this philosophical point.

THEOREM 12.4 [Sch07, Theorem 5.1]. *Let X be a reduced scheme of finite type over a field of characteristic zero, D a Cartier divisor that has Du Bois singularities and assume that $X \setminus D$ is smooth. Then X has rational singularities (in particular, it is Cohen–Macaulay).*

Let us conclude with a conjectural generalization of this statement:

CONJECTURE 12.5. *Let X be a reduced scheme of finite type over a field of characteristic zero, D a Cartier divisor that has Du Bois singularities and assume that $X \setminus D$ has rational singularities. Then X has rational singularities (in particular, it is Cohen–Macaulay).*

Essentially the same proof as in (12.2) shows that this is also true under the same additional hypotheses.

THEOREM 12.6. *Assume that X is Gorenstein and D is normal.[5] Then the statement of Conjecture 12.5 is true.*

PROOF. If X is Gorenstein, then so is D as it is a Cartier divisor. Then by (8.4) D is log canonical. Then by inversion of adjunction [Kaw07] the pair (X, D) is also log canonical near D. (Recall that if D is normal, then so is X along D).

As X is Gorenstein and $X \setminus D$ has rational singularities, it follows that $X \setminus D$ has canonical singularities. Then X has only canonical singularities everywhere. This can be seen by observing that D is a Cartier divisor and examining the discrepancies that lie over D for (X, D) as well as for X. Therefore, by [Elk81], X has only rational singularities along D.   □

---

[5] Again, this condition is not necessary, but makes the proof simpler.

13. Analogs of Du Bois singularities in characteristic $p > 0$

Starting in the early 1980s, the connections between singularities defined by the action of the Frobenius morphism in characteristic $p > 0$ and singularities defined by resolutions of singularities started to be investigated, cf. [Fed83]. After the introduction of tight closure in [HH90], a precise correspondence between several classes of singularities was established. See, for example, [FW89; MS91; HW02; Smi97; Har98; MS97; Smi00; Har05; HY03; Tak04; TW04; Tak08]. The second named author partially extended this correspondence in his doctoral dissertation by linking Du Bois singularities with F-injective singularities, a class of singularities defined in [Fed83]. The currently known implications are summarized below.

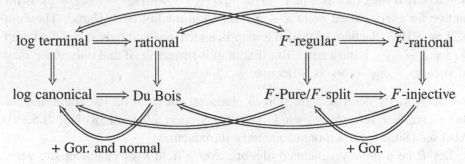

We will give a short proof that normal Cohen–Macaulay singularities of dense F-injective type are Du Bois, based on the characterization of Du Bois singularities given in Section 8.

Note that Du Bois and F-injective singularities also share many common properties. For example F-injective singularities are also seminormal [Sch09, Theorem 4.7].

First however, we will define F-injective singularities (as well as some necessary prerequisites).

DEFINITION 13.1. Suppose that X is a scheme of characteristic $p > 0$ with absolute Frobenius map $F : X \to X$. We say that X is *F-finite* if $F_* \mathcal{O}_X$ is a coherent $\mathcal{O}_X$-module. A ring R is called *F-finite* if the associated scheme $\operatorname{Spec} R$ is F-finite.

REMARK 13.2. Any scheme of finite type over a perfect field is F-finite; see for example [Fed83].

DEFINITION 13.3. Suppose that $(R, \mathfrak{m})$ is an F-finite local ring. We say that R is *F-injective* if the induced Frobenius map $F : H^i_{\mathfrak{m}}(R) \to H^i_{\mathfrak{m}}(R)$ is injective for every $i > 0$. We say that an F-finite scheme is *F-injective* if all of its stalks are F-injective local rings.

REMARK 13.4. If $(R, \mathfrak{m})$ is F-finite, F-injective and has a dualizing complex, then R_Q is also F-injective for any $Q \in \mathrm{Spec}\, R$. This follows from local duality; see [Sch09, Proposition 4.3] for details.

·LEMMA 13.5. *Suppose X is a Cohen–Macaulay scheme of finite type over a perfect field k. Then X is F-injective if and only if the natural map $F_* \omega_X \to \omega_X$ is surjective.*

PROOF. Without loss of generality (since X is Cohen–Macaulay) we can assume that X is equidimensional. Set $f : X \to \mathrm{Spec}\, k$ to be the structural morphism. Since X is finite type over a perfect field, it has a dualizing complex $\omega_X^{\boldsymbol{\cdot}} = f^! k$ and we set $\omega_X = h^{-\dim X}(\omega_X^{\boldsymbol{\cdot}})$. Since X is Cohen–Macaulay, X is F-injective if and only if the Frobenius map $H_x^{\dim X}(\mathscr{O}_{X,x}) \longrightarrow H_x^{\dim X}(F_* \mathscr{O}_{X,x})$ is injective for every closed point $x \in X$. By local duality (see [Har66, Theorem 6.2] or [BH93, Section 3.5]) such a map is injective if and only if the dual map $F_* \omega_{X,x} \to \omega_{X,x}$ is surjective. But that map is surjective, if and only if the map of sheaves $F_* \omega_X \to \omega_X$ is surjective.   □

We now briefly describe reduction to characteristic $p > 0$. Excellent and far more complete references include [HH09, Section 2.1] and [Kol96, II.5.10]. Also see [Smi01] for a more elementary introduction.

Let R be a finitely generated algebra over a field k of characteristic zero. Write $R = k[x_1, \ldots, x_n]/I$ for some ideal I and let S denote $k[x_1, \ldots, x_n]$. Let $X = \mathrm{Spec}\, R$ and $\pi : \widetilde{X} \to X$ a log resolution of X corresponding to the blow-up of an ideal J. Let E denote the reduced exceptional divisor of π. Then E is the subscheme defined by the radical of the ideal $J \cdot \mathscr{O}_{\widetilde{X}}$.

There exists a finitely generated $\mathbb{Z}$-algebra $A \subset k$ that includes all the coefficients of the generators of I and J, a finitely generated A algebra $R_A \subset R$, an ideal $J_A \subset R_A$, and schemes $\widetilde{X}_A$ and E_A of finite type over A such that $R_A \otimes_A k = R$, $J_A R = J$, $\widetilde{X}_A \times_{\mathrm{Spec}\, A} \mathrm{Spec}\, k = X$ and $E_A \times_{\mathrm{Spec}\, A} \mathrm{Spec}\, k = E$ with E_A an effective divisor with support defined by the ideal $J_A \cdot \mathscr{O}_{\widetilde{X}_A}$. We may localize A at a single element so that Y_A is smooth over A and E_A is a reduced simple normal crossings divisor over A. By further localizing A (at a single element), we may assume any finite set of finitely generated R_A modules is A-free (see [Hun96, 3.4] or [HR76, 2.3]) and we may assume that A itself is regular. We may also assume that a fixed affine cover of E_A and a fixed affine cover of $\widetilde{X}_A$ are also A-free.

We will now form a family of positive characteristic models of X by looking at all the rings $R_t = R_A \otimes_A k(t)$ where $k(t)$ is the residue field of a maximal ideal $t \in T = \mathrm{Spec}\, A$. Note that $k(t)$ is a finite, and thus perfect, field of characteristic p. We may also tensor the various schemes X_A, E_A, etc. with $k(t)$ to produce a characteristic p model of an entire situation.

By making various cokernels of maps free A-modules, we may also assume that maps between modules that are surjective (respectively injective) over k correspond to surjective (respectively injective) maps over A, and thus surjective (respectively injective) in our characteristic p model as well; see [HH09] for details.

DEFINITION 13.6. A ring R of characteristic zero is said to have dense F-injective type if for every family of characteristic $p \gg 0$ models with A chosen sufficiently large, a Zariski dense set of those models (over Spec A) have F-injective singularities.

THEOREM 13.7 [Sch09]. *Let X be a reduced scheme of finite type over $\mathbb{C}$ and assume that it has dense F-injective type. Then X has Du Bois singularities.*

PROOF. We only provide a proof in the case that X is normal and Cohen–Macaulay. For a complete proof, see [Sch09]. Let $\pi : \tilde{X} \to X$ be a log resolution of X with exceptional divisor E. We reduce this entire setup to characteristic $p \gg 0$ such that the corresponding X is F-injective. Let $F^e : X \to X$ be the e-iterated Frobenius map.

We have the commutative diagram

$$
\begin{array}{ccc}
F_*^e \pi_* \omega_{\tilde{X}}(p^e E) & \longrightarrow & \pi_* \omega_{\tilde{X}}(E) \\
\rho \downarrow & & \downarrow \beta \\
F_*^e \omega_X & \xrightarrow{\phi} & \omega_X
\end{array}
$$

where the horizontal arrows are induced by the dual of the Frobenius map, $\mathscr{O}_X \to F_*^e \mathscr{O}_X$, and the vertical arrows are the natural maps induced by π. By hypothesis, ϕ is surjective. On the other hand, for $e > 0$ sufficiently large, the map labeled ρ is an isomorphism. Therefore the map $\phi \circ \rho$ is surjective which implies that the map β is also surjective. But as this holds for a dense set of primes, it must be surjective in characteristic zero as well, and in particular, as a consequence X has Du Bois singularities. $\square$

It is not known whether the converse of this statement is true:

OPEN PROBLEM 13.8. If X has Du Bois singularities, does it have dense F-injective type?

Since F-injective singularities are known to be closely related to Du Bois singularities, it is also natural to ask how F-injective singularities deform cf. Conjecture 12.1. In general, this problem is also open.

OPEN PROBLEM 13.9. If a Cartier divisor D in X has F-injective singularities, does X have F-injective singularities near D?

In the case that X (equivalently D) is Cohen–Macaulay, the answer is affirmative, see [Fed83]. In fact, Fedder defined F-injective singularities partly because they seemed to deform better than F-pure singularities (the conjectured analog of log canonical singularities).

Appendix A. Connections with Buchsbaum rings

In this section we discuss the links between Du Bois singularities and Buchsbaum rings. Du Bois singularities are not necessarily Cohen–Macaulay, but in many cases, they are Buchsbaum (a weakening of Cohen–Macaulay).

Recall that a local ring $(R, \mathfrak{m}, k)$ has *quasi-Buchsbaum* singularities if

$$\mathfrak{m} H^i_{\mathfrak{m}}(R) = 0$$

for all $i < \dim R$. Further recall that a ring is called *Buchsbaum* if $\tau^{\dim R} R\Gamma_{\mathfrak{m}}(R)$ is quasi-isomorphic to a complex of k-vector spaces. Here $\tau^{\dim R}$ is the brutal truncation of the complex at the $\dim R$ location. Note that this is not the usual definition of Buchsbaum singularities, rather it is the so-called Schenzel's criterion; see [Sch82]. Notice that Cohen–Macaulay singularities are Buchsbaum (after truncation, one obtains the zero-object in the derived category).

It was proved by Tomari that isolated Du Bois singularities are quasi-Buchsbaum (a proof can be found in [Ish85, Proposition 1.9]), and then by Ishida that isolated Du Bois singularities were in fact Buchsbaum. Here we briefly review the argument to show that isolated Du Bois singularities are quasi-Buchsbaum since this statement is substantially easier.

PROPOSITION A.1. *Suppose that (X, x) is an isolated Du Bois singularity with $R = \mathcal{O}_{X,x}$. Then R is quasi-Buchsbaum.*

PROOF. Note that we may assume that X is affine. Since $\operatorname{Spec} R$ is regular outside its the maximal ideal $\mathfrak{m}$, it is clear that some power of $\mathfrak{m}$ annihilates $H^i_{\mathfrak{m}}(R)$ for all $i < \dim R$. We need to show that the smallest power for which this happens is 1. We let $\pi : \widetilde{X} \to X$ be a log resolution with exceptional divisor E as in Theorem 6.1. Since X is affine, we see that $H^i_{\mathfrak{m}}(R) \simeq H^{i-1}(X \setminus \{\mathfrak{m}\}, \mathcal{O}_X) \simeq H^{i-1}(\widetilde{X} \setminus E, \mathcal{O}_{\widetilde{X}})$ for all $i > 0$. Therefore, it is enough to show that $\mathfrak{m} H^{i-1}(\widetilde{X} \setminus E, \mathcal{O}_{\widetilde{X}}) = 0$ for all $i < \dim X$. In other words, we need to show that $\mathfrak{m} H^i(\widetilde{X} \setminus E, \mathcal{O}_{\widetilde{X}}) = 0$ for all $i < \dim X - 1$.

We examine the long exact sequence

$$\cdots \longrightarrow H^{i-1}(\widetilde{X} \setminus E, \mathcal{O}_{\widetilde{X}}) \longrightarrow H^i_E(\widetilde{X}, \mathcal{O}_{\widetilde{X}}) \longrightarrow H^i(\widetilde{X}, \mathcal{O}_{\widetilde{X}})$$

$$\longrightarrow H^i(\widetilde{X} \setminus E, \mathcal{O}_{\widetilde{X}}) \longrightarrow \cdots$$

Now, $H_E^i(\widetilde{X}, \mathscr{O}_{\widetilde{X}}) = R^i(\Gamma_{\mathfrak{m}} \circ \pi_*)(\mathscr{O}_X)$ which vanishes for $i < \dim X$ by the Matlis dual of Grauert-Riemenschneider vanishing. Therefore

$$H^i(\widetilde{X} \setminus E, \mathscr{O}_{\widetilde{X}}) \simeq H^i(\widetilde{X}, \mathscr{O}_{\widetilde{X}})$$

for $i < \dim X - 1$. Finally, since X is Du Bois, $H^i(\widetilde{X}, \mathscr{O}_{\widetilde{X}}) = H^i(E, \mathscr{O}_E)$ by Theorem 6.1. But it is obvious that $\mathfrak{m}H^i(E, \mathscr{O}_E) = 0$ since E is a *reduced* divisor whose image in X is the point corresponding to $\mathfrak{m}$. The result then follows. $\square$

It is easy to see that isolated F-injective singularities are also quasi-Buchsbaum.

PROPOSITION A.2. *Suppose that $(R, \mathfrak{m})$ is a local ring that is F-injective. Further suppose that $\operatorname{Spec} R \setminus \{\mathfrak{m}\}$ is Cohen–Macaulay. Then $(R, \mathfrak{m})$ is quasi-Buchsbaum.*

PROOF. Since the punctured spectrum of R is Cohen–Macaulay, $H_{\mathfrak{m}}^i(R)$ is annihilated by some power of $\mathfrak{m}$ for $i < \dim R$. We will show that the smallest such power is 1. Choose $c \in \mathfrak{m}$. Since R is F-injective, $F^e : H_{\mathfrak{m}}^i(R) \to H_{\mathfrak{m}}^i(R)$ is injective for all $e > 0$. Choose e large enough so that $c^{p^e} H_{\mathfrak{m}}^i(R)$ is zero for all $i < e$. However, for any element $z \in H_{\mathfrak{m}}^i(R)$, $F^e(cz) = c^{p^e} F^e(z) \in c^{p^e} H_{\mathfrak{m}}^i(R) = 0$ for $i < \dim R$. This implies that $cz = 0$ and so $\mathfrak{m}H_{\mathfrak{m}}^i(R) = 0$ for $i < \dim R$. $\square$

Perhaps the most interesting open question in this area is the following:

OPEN PROBLEM 1.3 (TAKAGI). Are F-injective singularities with isolated non-CM locus Buchsbaum?

Given the close connection between F-injective and Du Bois singularities, this question naturally leads to the next one:

OPEN PROBLEM 1.4. Are Du Bois singularities with isolated non-CM locus Buchsbaum?

2. Cubical hyperresolutions

For the convenience of the reader we include a short appendix explaining the construction of cubical hyperresolutions, as well as several examples. We follow [GNPP88] and mostly use their notation.

First let us fix a small universe to work in. Let Sch denote the category of reduced schemes. Note that the usual fibred product of schemes $X \times_S Y$ need not be reduced, even when X and Y are reduced. We wish to construct the fibred product in the category of reduced schemes. Given any scheme W (reduced or not) with maps to X and Y over S, there is always a unique morphism $W \to X \times_S Y$, which induces a natural unique morphism $W_{\mathrm{red}} \to (X \times_S Y)_{\mathrm{red}}$.

It is easy to see that $(X \times_S Y)_{\text{red}}$ is the fibred product in the category of reduced schemes.

Let us denote by $\underline{1}$ the category $\{0\}$ and by $\underline{2}$ the category $\{0 \to 1\}$. Let n be an integer ≥ -1. We denote by $\square_n^+$ the product of $n + 1$ copies of the category $\underline{2} = \{0 \to 1\}$ [GNPP88, I, 1.15]. The objects of $\square_n^+$ are identified with the sequences $\alpha = (\alpha_0, \alpha_1, \ldots, \alpha_n)$ such that $\alpha_i \in \{0, 1\}$ for $0 \leq i \leq n$. For $n = -1$, we set $\square_{-1}^+ = \{0\}$ and for $n = 0$ we have $\square_0^+ = \{0 \to 1\}$. We denote by $\square_n$ the full subcategory consisting of all objects of $\square_n^+$ except the initial object $(0, \ldots, 0)$. Clearly, the category $\square_n^+$ can be identified with the category of $\square_n$ with an augmentation map to $\{0\}$.

DEFINITION 2.1. A *diagram of schemes* is a functor Φ from a category C^{op} to the category of schemes. A *finite diagram of schemes* is a diagram of schemes such that the aforementioned category C has finitely many objects and morphisms; in this case such a functor will be called a C-*scheme*. A morphism of diagrams of schemes $\Phi : \mathsf{C}^{\text{op}} \to \underline{\text{Sch}}$ to $\Psi : \mathsf{D}^{\text{op}} \to \underline{\text{Sch}}$ is the combined data of a functor $\Gamma : \mathsf{C}^{\text{op}} \to \mathsf{D}^{\text{op}}$ together with a natural transformation of functors $\eta : \Phi \to \Psi \circ \Gamma$.

REMARK 2.2. With these definitions, the class of (finite) diagrams of schemes can be made into a category. Likewise the set of C-schemes can also be made into a category (where the functor $\Gamma : \mathsf{C}^{\text{op}} \to \mathsf{C}^{\text{op}}$ is always chosen to be the identity functor).

REMARK 2.3. Let I be a category. If instead of a functor to the category of reduced schemes, one considers a functor to the category of topological spaces, or the category of categories, one can define I-topological spaces, and I-categories in the obvious way.

If $X_. : I^{\text{op}} \to \underline{\text{Sch}}$ is an I-scheme, and $i \in \text{Ob} \, I$, then X_i will denote the scheme corresponding to i. Likewise if $\phi \in \text{Mor} \, I$ is a morphism $\phi : j \to i$, then X_ϕ will denote the corresponding morphism $X_\phi : X_i \to X_j$. If $f : Y_. \to X_.$ is a morphism of I-schemes, we denote by f_i the induced morphism $Y_i \to X_i$. If $X_.$ is an I-scheme, a closed sub-I-scheme is a morphism of I-schemes $g : Z_. \to X_.$ such that for each $i \in I$, the map $g_i : Z_i \to X_i$ is a closed immersion. We will often suppress the g of the notation if no confusion is likely to arise. More generally, any property of a morphism of schemes (projective, proper, separated, closed immersion, etc...) can be generalized to the notion of a morphism of I-schemes by requiring that for each object i of I, g_i has the desired property (projective, proper, separated, closed immersion, etc...)

DEFINITION 2.4 [GNPP88, I, 2.2]. Suppose that $f : Y_. \to X_.$ is a morphism of I-schemes. Define the *discriminant of f* to be the smallest closed sub-I-scheme

$Z_.$ of $X_.$ such that $f_i : (Y_i - (f_i^{-1}(Z_i))) \to (X_i - Z_i)$ is an isomorphism for all
i.

DEFINITION 2.5 [GNPP88, I, 2.5]. Let $S_.$ be an I-scheme, $f : X_. \to S_.$ a
proper morphism of I-schemes, and $D_.$ the discriminant of f. We say that f is
a *resolution*[6] *of* $S_.$ if $X_.$ is a smooth I-scheme (meaning that each X_i is smooth)
and dim $f_i^{-1}(D_i) < $ dim S_i, for all $i \in$ Ob I.

REMARK 2.6. This is the definition found in [GNPP88]. Note that the maps
are not required to be surjective (of course, the ones one constructs in practice
are usually surjective).

Consider the following example: the map $k[x, y]/(xy) \to k[x]$ which sends
y to 0. We claim that the associated map of schemes is a "resolution" of the
$*$-scheme Spec $k[x, y]/(xy)$. The discriminant is Spec $k[x, y]/(x)$. However,
the preimage is simply the origin on $k[x]$, which has lower dimension than "1".
Resolutions like this one are sometimes convenient to consider.

On the other hand, this definition seems to allow something it perhaps should
not. Choose any variety X of dimension greater than zero and a closed point
$z \in X$. Consider the map $z \to X$ and consider the $*$-scheme X. The discriminant
is all of X. However, the preimage of X is still just a point, which has lower
dimension than X itself, by hypothesis.

In view of these remarks, sometimes it is convenient to assume also that
dim $D_i <$ dim S_i for each $i \in$ Ob I. In the resolutions of I-schemes that we
construct (in particular, in the ones that are used to that prove cubic hyperresolutions exist), this always happens.

Let I be a category. The set of objects of I are given the preorder relation
defined by $i \leq j$ if and only if Hom$_I(i, j)$ is nonempty. We will say that a
category I is ordered if this preorder is a partial order and, for each $i \in$ Ob I,
the only endomorphism of i is the identity [GNPP88, I, C, 1.9]. Note that a
category I is ordered if and only if all isomorphisms and endomorphisms of I
are the identity.

It turns out of that resolutions of I-schemes always exist under reasonable
hypotheses.

THEOREM 2.7 [GNPP88, I, Theorem 2.6]. *Let S be an I-scheme of finite type
over a field k. Suppose that k is a field of characteristic zero and that I is a
finite ordered category. Then there exists a resolution of S.*

In order to construct a resolution $Y_.$ of an I-scheme $X_.$, it might be tempting to
simply resolve each X_i, set Y_i equal to that resolution, and somehow combine

---

[6]A resolution is a distinct notion from a cubic hyperresolution.

this data together. Unfortunately this cannot work, as shown by the example below.

EXAMPLE 2.8. Consider the pinch point singularity,

$$X = \operatorname{Spec} k[x, y, z]/(x^2 y - z^2) = \operatorname{Spec} k[s, t^2, st],$$

and let Z be the closed subscheme defined by the ideal (s, st) (this is the singular set). Let I be the category $\{0 \to 1\}$. Consider the I-scheme defined by $X_0 = X$ and $X_1 = Z$ (with the closed immersion as the map). X_1 is already smooth, and if one resolves X_0, (that is, normalizes it) there is no compatible way to map X_1 (or even another birational model of X_1) to it, since its preimage by normalization will be two-to-one onto $Z \subset X$! The way this problem is resolved is by creating additional components. So to construct a resolution Y, we set $Y_1 = Z = X_1$ (since it was already smooth) and set $Y_0 = \overline{X}_0 \coprod Z$ where $\overline{X}_0$ is the normalization of X_0. The map $Y_1 \to Y_0$ just sends Y_1 (isomorphically) to the new component and the map $Y_0 \to X_0$ is the disjoint union of the normalization and inclusion maps.

One should note that although the theorem proving the existence of resolutions of I-schemes is constructive, [GNPP88], it is often easier in practice to construct an ad-hoc resolution.

Now that we have resolutions of I-schemes, we can discuss cubic hyperresolutions of schemes, in fact, even diagrams of schemes have cubic hyperresolutions! First we will discuss a single iterative step in the process of constructing cubic hyperresolutions. This step is called a 2-resolution.

DEFINITION 2.9 [GNPP88, I, 2.7]. Let S be an I-scheme and $Z_{\bullet}$ a $\square_1^+ \times I$-scheme. We say that $Z_{\bullet}$ is a 2-*resolution* of S if $Z_{\bullet}$ is defined by the following Cartesian square (pullback, or fibred product in the category of (reduced) I-schemes) of morphisms of I-schemes:

$$
\begin{array}{ccc}
Z_{11} & \lhook\joinrel\longrightarrow & Z_{01} \\
\downarrow & & \downarrow f \\
Z_{10} & \lhook\joinrel\longrightarrow & Z_{00}
\end{array}
$$

Here

(i) $Z_{00} = S$,

(ii) Z_{01} is a smooth I-scheme,

(iii)   The horizontal arrows are closed immersions of I-schemes,

(iv) f is a proper I-morphism, and

(v) Z_{10} contains the discriminant of f; in other words, f induces an isomorphism of $(Z_{01})_i - (Z_{11})_i$ over $(Z_{00})_i - (Z_{10})_i$, for all $i \in \operatorname{Ob} I$.

Clearly 2-resolutions always exist under the same hypotheses that resolutions of I-schemes exist: set Z_{01} to be a resolution, Z_{10} to be discriminant (or any appropriate proper closed sub-I-scheme that contains it), and Z_{11} its (reduced) preimage in Z_{01}.

EXAMPLE 2.10. Let $I = \{0\}$ and let S be the I-scheme $\operatorname{Spec} k[t^2, t^3]$. Let $Z_{01} = \mathbb{A}^1 = \operatorname{Spec} k[t]$ and $Z_{01} \to S = Z_{00}$ be the map defined by $k[t^2, t^3] \to k[t]$. The discriminant of that map is the closed subscheme of $S = Z_{00}$ defined by the map $\phi : k[t^2, t^3] \to k$ that sends t^2 and t^3 to zero. Finally we need to define Z_{11}. The usual fibered product in the category of schemes is $k[t]/(t^2)$, but we work in the category of reduced schemes, so instead the fibered product is simply the associated reduced scheme (in this case $\operatorname{Spec} k[t]/(t)$). Thus our 2-resolution is defined by this diagram of rings:

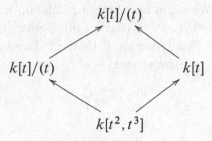

We need one more definition before defining a cubic hyperresolution,

DEFINITION 2.11 [GNPP88, I, 2.11]. Let r be an integer greater than or equal to 1, and let $X_{\cdot}^n$ be a $\square_n^+ \times I$-scheme, for $1 \le n \le r$. Suppose that for all n, $1 \le n \le r$, the $\square_{n-1}^+ \times I$-schemes $X_{00\cdot}^{n+1}$ and $X_{1\cdot}^n$ are equal. Then we define, by induction on r, a $\square_r^+ \times I$-scheme

$$Z_{\cdot} = \operatorname{red}(X_{\cdot}^1, X_{\cdot}^2, \ldots, X_{\cdot}^r)$$

that we call the *reduction* of $(X_{\cdot}^1, \ldots, X_{\cdot}^r)$, in the following way: If $r = 1$, one defines $Z_{\cdot} = X_{\cdot}^1$, if $r = 2$ one defines $Z_{\cdot\cdot} = \operatorname{red}(X_{\cdot}^1, X_{\cdot}^2)$ by

$$Z_{\alpha\beta} = \begin{cases} X_{0\beta}^1 & \text{if } \alpha = (0,0), \\ X_{\alpha\beta}^2 & \text{if } \alpha \in \square_1, \end{cases}$$

for all $\beta \in \square_0^+$, with the obvious morphisms. If $r > 2$, one defines $Z_{\cdot}$ recursively as $\operatorname{red}(\operatorname{red}(X_{\cdot}^1, \ldots, X_{\cdot}^{r-1}), X_{\cdot}^r)$.

Finally we are ready to define cubic hyperresolutions.

DEFINITION 2.12 [GNPP88, I, 2.12]. Let S be an I-scheme. A *cubic hyperresolution augmented over* S is a $\square_r^+ \times I$-scheme $Z_{\cdot}$ such that

$$Z_{\cdot} = \operatorname{red}(X_{\cdot}^1, \ldots, X_{\cdot}^r),$$

where $X_{\bullet}^1$ is a 2-resolution of S, $X_{\bullet}^{n+1}$ is a 2-resolution of X_1^n for $1 \le n < r$, and Z_α is smooth for all $\alpha \in \square_r$.

Now that we have defined cubic hyperresolutions, we should note that they exist under reasonable hypotheses:

THEOREM 2.13 [GNPP88, I, 2.15]. *Let S be an I-scheme. Suppose that k is a field of characteristic zero and that I is a finite (bounded) ordered category. Then there exists $Z_{\bullet}$, a cubic hyperresolution augmented over S such that*

$$\dim Z_\alpha \le \dim S - |\alpha| + 1 \quad \text{for all } \alpha \in \square_r.$$

Below are some examples of cubic hyperresolutions.

EXAMPLE 2.14. Let us begin by computing cubic hyperresolutions of curves so let C be a curve. We begin by taking a resolution $\pi : \overline{C} \to C$ (where $\overline{C}$ is just the normalization). Let P be the set of singular points of C; thus P is the discriminant of π. Finally we let E be the reduced exceptional set of π, therefore we have the Cartesian square

$$\begin{array}{ccc} E & \longrightarrow & \overline{C} \\ \downarrow & & \downarrow{\scriptstyle \pi} \\ P & \longrightarrow & C \end{array}$$

It is clearly already a 2-resolution of C and thus a cubic-hyperresolution of C.

EXAMPLE 2.15. Let us now compute a cubic hyperresolution of a scheme X whose singular locus is itself a smooth scheme, and whose reduced exceptional set of a strong resolution $\pi : \widetilde{X} \to X$ is smooth (for example, any cone over a smooth variety). As in the previous example, let Σ be the singular locus of X and E the reduced exceptional set of π, Then the Cartesian square of reduced schemes

$$\begin{array}{ccc} E & \longrightarrow & \widetilde{X} \\ \downarrow & & \downarrow{\scriptstyle \pi} \\ \Sigma & \longrightarrow & X \end{array}$$

is in fact a 2-resolution of X, just as in the case of curves above.

The obvious algorithm used to construct cubic hyperresolutions does not construct hyperresolutions in the most efficient or convenient way possible. For example, applying the obvious algorithm to the intersection of three coordinate planes gives us the following.

EXAMPLE 2.16. Let $X \cup Y \cup Z$ be the three coordinate planes in $\mathbb{A}^3$. In this example we construct a cubic hyperresolution using the obvious algorithm.

What makes this construction different, is that the dimension is forced to drop when forming the discriminant of a resolution of a diagram of schemes.

Again we begin the algorithm by taking a resolution and the obvious one is $\pi : (X \sqcup Y \sqcup Z) \to (X \cup Y \cup Z)$. The discriminant is $B = (X \cap Y) \cup (X \cap Z) \cup (Y \cap Z)$, the three coordinate axes. The fiber product making the square below Cartesian is simply the exceptional set E shown:

$$E = ((X \cap Y) \cup (X \cap Z)) \sqcup ((Y \cap X) \cup (Y \cap Z)) \sqcup ((Z \cap X) \cup (Z \cap Y)) \longrightarrow X \sqcup Y \sqcup Z$$

$$\downarrow \phi \qquad\qquad\qquad\qquad\qquad\qquad\qquad\qquad\qquad\qquad\qquad \downarrow \pi$$

$$B = (X \cap Y) \cup (X \cap Z) \cup (Y \cap Z) \longrightarrow X \cup Y \cup Z$$

We now need to take a 2-resolution of the $\underline{2}$-scheme $\phi : E \to B$. We take the obvious resolution that simply separates irreducible components. This gives us $\widetilde{E} \to \widetilde{B}$ mapping to $\phi : E \to B$. The discriminant of $\widetilde{E} \to E$ is a set of three points X_0, Y_0 and Z_0 corresponding to the origins in X, Y and Z respectively. The discriminant of the map $\widetilde{B} \to B$ is simply identified as the origin A_0 of our initial scheme $X \cup Y \cup Z$ (recall B is the union of the three axes). The union of that with the images of X_0, Y_0 and Z_0 is again just A_0. The fiber product of the diagram

$$(\widetilde{E} \to \widetilde{B})$$
$$\downarrow$$
$$(\{X_0, Y_0, Z_0\} \to \{A_0\}) \longrightarrow (\phi : E \to B)$$

can be viewed as $\{Q_1, \ldots, Q_6\} \to \{P_1, P_2, P_3\}$ where Q_1 and Q_2 are mapped to P_1 and so on (remember E was the disjoint union of the coordinate axes of X, of Y, and of respectively Z, so $\widetilde{E}$ has six components and thus six origins). Thus we have the diagram

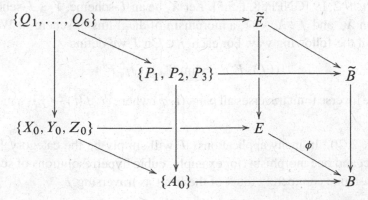

which we can combine with previous diagrams to construct a cubic hyperresolution.

REMARK 2.17. It is possible to find a cubic hyperresolution for the three coordinate planes in $\mathbb{A}^3$ in a different way. Suppose that S is the union of the three coordinate planes $(X, Y, \text{and } Z)$ of $\mathbb{A}^3$. Consider the $\square_2$ or $\square_2^+$ scheme defined by the diagram below, where the dotted arrows are those in $\square_2^+$ but not in $\square_2$.

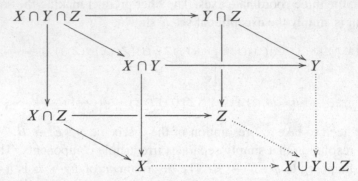

One can verify that this is also a cubic hyperresolution of $X \cup Y \cup Z$.

Now we discuss sheaves on diagrams of schemes, as well as the related notions of push-forward and its right derived functors.

DEFINITION 2.18 [GNPP88, I, 5.3–5.4]. Let $X_\cdot$ be an I-scheme (or even an I-topological space). We define a *sheaf (or pre-sheaf) of abelian groups* $F^\cdot$ on $X_\cdot$ to be the following data:

(i) A sheaf (pre-sheaf) F^i of abelian groups over X_i, for all $i \in \text{Ob } I$, and
(ii) An X_ϕ-morphism of sheaves $F^\phi : F^i \to (X_\phi)_* F^j$ for all morphisms $\phi : i \to j$ of I, required to be compatible in the obvious way.

Given a morphism of diagrams of schemes $f_\cdot : X_\cdot \to Y_\cdot$ one can construct a push-forward functor for sheaves on $X_\cdot$.

DEFINITION 2.19 [GNPP88, I, 5.5]. Let $X_\cdot$ be an I-scheme, $Y_\cdot$ a J-scheme, $F^\cdot$ a sheaf on $X_\cdot$, and $f_\cdot : X_\cdot \to Y_\cdot$ a morphism of diagrams of schemes. We define $(f_\cdot)_* F^\cdot$ in the following way. For each $j \in \text{Ob } J$ we define

$$((f_\cdot)_* F^\cdot)^j = \varprojlim (Y_\phi)_* (f_{i*} F^i)$$

where the inverse limit traverses all pairs (i, ϕ) where $\phi : f(i) \to j$ is a morphism in J^{op}.

REMARK 2.20. In many applications, J will simply be the category $\{0\}$ with one object and one morphism (for example, cubic hyperresolutions of schemes). In that case one can merely think of the limit as traversing I.

REMARK 2.21. One can also define a functor f^*, show that it has a right adjoint and that that adjoint is f_* as defined above [GNPP88, I, 5.5].

DEFINITION 2.22 [GNPP88, I, Section 5]. Let $X_.$ and $Y_.$ be diagrams of topological spaces over I and J respectively, $\Phi : I \to J$ a functor, $f_. : X_. \to Y_.$ a Φ-morphism of topological spaces. If $G^{\cdot}$ is a sheaf over $Y_.$ with values in a complete category C, one denotes by $f_.^* G^{\cdot}$ the sheaf over $X_.$ defined by

$$(f_.^* G^{\cdot})^i = f_i^*(G^{\Phi(i)}),$$

for all $i \in \mathrm{Ob}\, I$. One obtains in this way a functor

$$f_.^* : \mathsf{Sheaves}(Y_., \mathsf{C}) \to \mathsf{Sheaves}(X_., \mathsf{C})$$

Given an I-scheme $X_.$, one can define the category of sheaves of abelian groups $\mathrm{Ab}(X_.)$ on $X_.$ and show that it has enough injectives. Next, one can even define the derived category $D^+(X_., \mathrm{Ab}(X_.))$ by localizing bounded below complexes of sheaves of abelian groups on $X_.$ by the quasi-isomorphisms (those that are quasi-isomorphisms on each $i \in I$). One can also show that $(f_.)_*$ as defined above is left exact so that it has a right derived functor $R(f_.)_*$ [GNPP88, I, 5.8-5.9]. In the case of a cubic hyperresolution of a scheme $f : X_. \to X$,

$$R((f_.)_* F^{\cdot}) = R\varprojlim (R f_{i*} F^i)$$

where the limit traverses the category I of $X_.$.

Final remark. We end our excursion into the world of hyperresolutions here. There are many other things to work out, but we will leave them to the interested reader. Many "obvious" statements need to be proved, but most are relatively straightforward once one gets comfortable using the appropriate language. For those and many more statements, including the full details of the construction of the Du Bois complex and many applications, the reader is encouraged to read [GNPP88].

Acknowledgements

We thank Kevin Tucker, Zsolt Patakfalvi and the referee for reading earlier drafts and making helpful suggestions for improving the presentation.

References

[Amb98] F. Ambro: *The locus of log canonical singularities*, preprint, 1998. arXiv math.AG/9806067

[Bou87] J.-F. Boutot: *Singularités rationnelles et quotients par les groupes réductifs*, Invent. Math. **88** (1987), no. 1, 65–68. MR 88a:14005

[BH93] W. Bruns and J. Herzog: *Cohen-Macaulay rings*, Cambridge Studies in Advanced Mathematics, vol. 39, Cambridge University Press, Cambridge, 1993. MR 95h:13020

[Car85] J. A. Carlson: *Polyhedral resolutions of algebraic varieties*, Trans. Amer. Math. Soc. **292** (1985), no. 2, 595–612. MR 87i:14008

[DB81] P. Du Bois: *Complexe de de Rham filtré d'une variété singulière*, Bull. Soc. Math. France **109** (1981), no. 1, 41–81. MR 82j:14006

[DJ74] P. Dubois and P. Jarraud: *Une propriété de commutation au changement de base des images directes supérieures du faisceau structural*, C. R. Acad. Sci. Paris Sér. A **279** (1974), 745–747. MR0376678 (51 #12853)

[Dur79] A. H. Durfee: *Fifteen characterizations of rational double points and simple critical points*, Enseign. Math. (2) **25** (1979), no. 1-2, 131–163. MR 80m:14003

[Elk78] R. Elkik: *Singularités rationnelles et déformations*, Invent. Math. **47** (1978), no. 2, 139–147. MR501926 (80c:14004)

[Elk81] R. Elkik: *Rationalité des singularités canoniques*, Invent. Math. **64** (1981), no. 1, 1–6. MR621766 (83a:14003)

[Fed83] R. Fedder: *F-purity and rational singularity*, Trans. Amer. Math. Soc. **278** (1983), no. 2, 461–480. MR701505 (84h:13031)

[FW89] R. Fedder and K. Watanabe: *A characterization of F-regularity in terms of F-purity*, Commutative algebra (Berkeley, CA, 1987), Math. Sci. Res. Inst. Publ., vol. 15, Springer, New York, 1989, pp. 227–245. MR 91k:13009

[Fuj08] O. Fujino: *Theory of non-lc ideal sheaves: basic properties*, Kyoto J. Math. **50** (2010), no. 2, 225–245. 2666656

[GKKP10] D. Greb, S. Kebekus, S. J. Kovács, and T. Peternell: *Differential forms on log canonical spaces*, preprint, 2010. arXiv:1003.2913v3

[GT80] S. Greco and C. Traverso: *On seminormal schemes*, Compositio Math. **40** (1980), no. 3, 325–365. MR571055 (81j:14030)

[GNPP88] F. Guillén, V. Navarro Aznar, P. Pascual Gainza, and F. Puerta: *Hyperré-solutions cubiques et descente cohomologique*, Lecture Notes in Mathematics, vol. 1335, Springer-Verlag, Berlin, 1988, Papers from the Seminar on Hodge–Deligne Theory held in Barcelona, 1982. MR 90a:14024

[Har98] N. Hara: *A characterization of rational singularities in terms of injectivity of Frobenius maps*, Amer. J. Math. **120** (1998), no. 5, 981–996. MR 99h:13005

[Har05] N. Hara: *A characteristic p analog of multiplier ideals and applications*, Comm. Algebra **33** (2005), no. 10, 3375–3388. MR 2006f:13006

[HW02] N. Hara and K.-I. Watanabe: *F-regular and F-pure rings vs. log terminal and log canonical singularities*, J. Algebraic Geom. **11** (2002), no. 2, 363–392. MR 2002k:13009

[HY03] N. Hara and K.-I. Yoshida: *A generalization of tight closure and multiplier ideals*, Trans. Amer. Math. Soc. **355** (2003), no. 8, 3143–3174 (electronic). MR 2004i:13003

[Har66] R. Hartshorne: *Residues and duality*, Lecture notes of a seminar on the work of A. Grothendieck, given at Harvard 1963/64. With an appendix by P. Deligne. Lecture Notes in Mathematics, No. 20, Springer-Verlag, Berlin, 1966. MR0222093 (36 #5145)

[Har77] R. Hartshorne: *Algebraic geometry*, Springer-Verlag, New York, 1977, Graduate Texts in Mathematics, No. 52. MR0463157 (57 #3116)

[Hir64] H. Hironaka: *Resolution of singularities of an algebraic variety over a field of characteristic zero. I, II*, Ann. of Math. (2) 79 (1964), 109–203; ibid. (2) **79** (1964), 205–326. MR0199184 (33 #7333)

[HH90] M. Hochster and C. Huneke: *Tight closure, invariant theory, and the Briançon-Skoda theorem*, J. Amer. Math. Soc. **3** (1990), no. 1, 31–116. MR 91g:13010

[HH09] M. Hochster and C. Huneke: *Tight closure in equal characteristic zero*, A preprint of a manuscript, http://www.math.lsa.umich.edu/hochster/dk.ps, 2009.

[HR76] M. Hochster and J. L. Roberts: *The purity of the Frobenius and local cohomology*, Advances in Math. **21** (1976), no. 2, 117–172. MR0417172 (54 #5230)

[Hun96] C. Huneke: *Tight closure and its applications*, CBMS Regional Conference Series in Mathematics, vol. 88, Published for the Conference Board of the Mathematical Sciences, Washington, DC, 1996, With an appendix by Melvin Hochster. MR 96m:13001

[Ish85] S. Ishii: *On isolated Gorenstein singularities*, Math. Ann. **270** (1985), no. 4, 541–554. MR776171 (86j:32024)

[Ish86] S. Ishii: *Small deformations of normal singularities*, Math. Ann. **275** (1986), no. 1, 139–148. MR849059 (87i:14003)

[Ish87] S. Ishii: *Isolated Q-Gorenstein singularities of dimension three*, Complex analytic singularities, Adv. Stud. Pure Math., vol. 8, North-Holland, Amsterdam, 1987, pp. 165–198. MR894292 (89d:32016)

[Kaw07] M. Kawakita: *Inversion of adjunction on log canonicity*, Invent. Math. **167** (2007), no. 1, 129–133. 2264806 (2008a:14025)

[KSB88] J. Kollár and N. I. Shepherd-Barron: *Threefolds and deformations of surface singularities*, Invent. Math. **91** (1988), no. 2, 299–338. MR 88m:14022

[Kol95] J. Kollár: *Shafarevich maps and automorphic forms*, M. B. Porter Lectures, Princeton University Press, Princeton, NJ, 1995. MR 96i:14016

[Kol96] J. Kollár: *Rational curves on algebraic varieties*, Ergebnisse der Mathematik und ihrer Grenzgebiete. 3. Folge. A Series of Modern Surveys in Mathematics, vol. 32, Springer-Verlag, Berlin, 1996. MR 98c:14001

[Kol07] J. Kollár: *Two examples of surfaces with normal crossing singularities*, preprint, 2007. arXiv 0705.0926v2

[Kol08a] J. Kollár: *Hulls and husks*, preprint, 2008. arXiv 0805.0576v2

[Kol08b] J. Kollár: *Semi log resolution*, preprint, 2008. arXiv 0812.3592

[KK10] J. Kollár and S. J. Kovács: *Log canonical singularities are Du Bois*, J. Amer. Math. Soc. **23** (2010), no. 3, 791–813. doi:10.1090/S0894-0347-10-00663-6

[KM98] J. Kollár and S. Mori: *Birational geometry of algebraic varieties*, Cambridge Tracts in Mathematics, vol. 134, Cambridge University Press, Cambridge, 1998, With the collaboration of C. H. Clemens and A. Corti, Translated from the 1998 Japanese original. MR1658959 (2000b:14018)

[Kol92] J. Kollár et. al: *Flips and abundance for algebraic threefolds*, Société Mathématique de France, Paris, 1992, Papers from the Second Summer Seminar on Algebraic Geometry held at the University of Utah, Salt Lake City, Utah, August 1991, Astérisque No. 211 (1992). MR 94f:14013

[Kov99] S. J. Kovács: *Rational, log canonical, Du Bois singularities: on the conjectures of Kollár and Steenbrink*, Compositio Math. **118** (1999), no. 2, 123–133. MR1713307 (2001g:14022)

[Kov00a] S. J. Kovács: *A characterization of rational singularities*, Duke Math. J. **102** (2000), no. 2, 187–191. MR1749436 (2002b:14005)

[Kov00b] S. J. Kovács: *Rational, log canonical, Du Bois singularities. II. Kodaira vanishing and small deformations*, Compositio Math. **121** (2000), no. 3, 297–304. MR1761628 (2001m:14028)

[KSS10] S. J. Kovács, K. Schwede, and K. E. Smith: *The canonical sheaf of Du Bois singularities*, Advances in Mathematics **224** (2010), no. 4, 1618 – 1640.

[Laz04] R. Lazarsfeld: *Positivity in algebraic geometry. II*, Ergebnisse der Mathematik und ihrer Grenzgebiete. 3. Folge. A Series of Modern Surveys in Mathematics [Results in Mathematics and Related Areas. 3rd Series. A Series of Modern Surveys in Mathematics], vol. 49, Springer-Verlag, Berlin, 2004, Positivity for vector bundles, and multiplier ideals. MR2095472 (2005k:14001b)

[MS91] V. B. Mehta and V. Srinivas: *Normal F-pure surface singularities*, J. Algebra **143** (1991), no. 1, 130–143. MR 92j:14044

[MS97] V. B. Mehta and V. Srinivas: *A characterization of rational singularities*, Asian J. Math. **1** (1997), no. 2, 249–271. MR 99e:13009

[PS08] C. A. M. Peters and J. H. M. Steenbrink: *Mixed Hodge structures*, Ergebnisse der Mathematik und ihrer Grenzgebiete. 3. Folge. A Series of Modern Surveys in Mathematics [Results in Mathematics and Related Areas. 3rd Series. A Series of Modern Surveys in Mathematics], vol. 52, Springer-Verlag, Berlin, 2008. MR2393625

[Rei87] M. Reid: *Young person's guide to canonical singularities*, Algebraic geometry, Bowdoin, 1985 (Brunswick, Maine, 1985), Proc. Sympos. Pure Math., vol. 46, Amer. Math. Soc., Providence, RI, 1987, pp. 345–414. MR 89b:14016

[Rei97] M. Reid: *Chapters on algebraic surfaces*, Complex algebraic geometry (Park City, UT, 1993), IAS/Park City Math. Ser., vol. 3, Amer. Math. Soc., Providence, RI, 1997, pp. 3–159. MR1442522 (98d:14049)

[Sai00] M. Saito: *Mixed Hodge complexes on algebraic varieties*, Math. Ann. **316** (2000), no. 2, 283–331. MR1741272 (2002h:14012)

[Sai09] M. Saito: *On the Hodge filtration of Hodge modules*, Mosc. Math. J. **9** (2009), no. 1, 161–191.

[Sch82] P. Schenzel: *Applications of dualizing complexes to Buchsbaum rings*, Adv. in Math. **44** (1982), no. 1, 61–77. MR 83j:13011

[Sch07] K. Schwede: *A simple characterization of Du Bois singularities*, Compos. Math. **143** (2007), no. 4, 813–828. MR2339829

[Sch08] K. Schwede: *Centers of F-purity*, Math. Z. **265** (2010), no. 3, 687–714. MR2644316

[Sch09] K. Schwede: *F-injective singularities are Du Bois*, Amer. J. Math. **131** (2009), no. 2, 445–473. MR2503989

[ST08] K. Schwede and S. Takagi: *Rational singularities associated to pairs*, Michigan Math. J. **57** (2008), 625–658.

[Smi97] K. E. Smith: *F-rational rings have rational singularities*, Amer. J. Math. **119** (1997), no. 1, 159–180. MR 97k:13004

[Smi00] K. E. Smith: *The multiplier ideal is a universal test ideal*, Comm. Algebra **28** (2000), no. 12, 5915–5929, Special issue in honor of Robin Hartshorne. MR 2002d:13008

[Smi01] K. E. Smith: *Tight closure and vanishing theorems*, School on Vanishing Theorems and Effective Results in Algebraic Geometry (Trieste, 2000), ICTP Lect. Notes, vol. 6, Abdus Salam Int. Cent. Theoret. Phys., Trieste, 2001, pp. 149–213. MR1919458 (2003f:13005)

[Ste83] J. H. M. Steenbrink: *Mixed Hodge structures associated with isolated singularities*, Singularities, Part 2 (Arcata, Calif., 1981), Proc. Sympos. Pure Math., vol. 40, Amer. Math. Soc., Providence, RI, 1983, pp. 513–536. MR 85d:32044

[Ste85] J. H. M. Steenbrink: *Vanishing theorems on singular spaces*, Astérisque (1985), no. 130, 330–341, Differential systems and singularities (Luminy, 1983). MR 87j:14026

[Swa80] R. G. Swan: *On seminormality*, J. Algebra **67** (1980), no. 1, 210–229. MR 82d:13006

[Tak04] S. Takagi: *An interpretation of multiplier ideals via tight closure*, J. Algebraic Geom. **13** (2004), no. 2, 393–415. MR 2005c:13002

[Tak08] S. Takagi: *A characteristic p analogue of plt singularities and adjoint ideals*, Math. Z. **259** (2008), no. 2, 321–341. MR 2009b:13004

[TW04] S. Takagi and K.-i. Watanabe: *On F-pure thresholds*, J. Algebra **282** (2004), no. 1, 278–297. MR2097584 (2006a:13010)

[Wat87] K.-i. Watanabe: *On plurigenera of normal isolated singularities. II*, Complex analytic singularities, Adv. Stud. Pure Math., vol. 8, North-Holland, Amsterdam, 1987, pp. 671–685. MR 89h:32029

SÁNDOR J. KOVÁCS
UNIVERSITY OF WASHINGTON
DEPARTMENT OF MATHEMATICS
SEATTLE, WA 98195
UNITED STATES
  kovacs@math.washington.edu

KARL E. SCHWEDE
DEPARTMENT OF MATHEMATICS
THE PENNSYLVANIA STATE UNIVERSITY
UNIVERSITY PARK, PA 16802
UNITED STATES
  kschwede@umich.edu

Topology of Stratified Spaces
MSRI Publications
Volume **58**, 2011

Elliptic genera, real algebraic varieties and quasi-Jacobi forms

ANATOLY LIBGOBER

ABSTRACT. We survey the push-forward formula for elliptic class and various applications obtained in the papers by L. Borisov and the author. We then discuss the ring of quasi-Jacobi forms which allows to characterize the functions which are the elliptic genera of almost complex manifolds and extension of Ochanine elliptic genus to certain singular real algebraic varieties.

Introduction

Interest in the elliptic genus of complex manifolds stems from its appearance in a wide variety of geometric and topological problems. The elliptic genus is an invariant of the complex cobordism class modulo torsion, and hence depends only on the Chern numbers of the manifold. On the other hand, the elliptic genus is a holomorphic function defined on $\mathbb{C} \times \mathbb{H}$, where $\mathbb{H}$ is the upper half plane. In one heuristic approach, the elliptic genus is an index of an operator on the loop space (see [53]) and as such it has counterparts defined for C^∞, oriented or Spin manifolds: these were in fact studied before the complex case [43].

The elliptic genus comes up in the study of the geometry and topology of loop spaces and, more specifically, of the chiral de Rham complex [41]; in the study of invariants of singular algebraic varieties [8] — in particular orbifolds; and, more recently, in the study of Gopakumar–Vafa and Nekrasov conjectures [38; 25]. It is closely related to the fast developing subject of elliptic cohomology [46]. There are various versions of the elliptic genus, including the equivariant, higher elliptic genus obtained by twisting by cohomology classes of the fundamental group, the elliptic genus of pairs and the orbifold elliptic genus. There is inter-

---

The author was partially supported by an NSF grant.

esting connection with singularities of weighted homogeneous polynomials — the so-called Landau–Ginzburg models.

We shall review several recent developments on the elliptic genus; we refer the reader [7] for details on earlier results. Then we shall focus on certain aspects of the elliptic genus: its extension to real singular varieties and its modularity property (or the lack of it). Extensions of the elliptic genus to real singular varieties were suggested by B. Totaro [48]; our approach is based on the push-forward formula for the elliptic class used in [8] to extend elliptic genus from smooth to certain singular complex projective varieties.

In Section 1A we discuss this push-forward formula, which appears as the main technical tool in many applications mentioned later. The rest of Section 1 discusses the relation with other invariants and series of applications based on the material in [9; 8; 6]. It includes a discussion of a relation between elliptic genus and E-function, applications to the McKay correspondence, elliptic genera of non-simply connected manifolds (higher elliptic genera) and generalizations of a formula of R. Dijkgraaf, D. Moore, E. Verlinde, and H. Verlinde. (Other applications in the equivariant context are discussed in R. Waelder's paper in this volume.) The proof of independence of resolutions of the elliptic genus (according to our definition) for certain real algebraic varieties is given later, in Section 3B.

Section 2 deals with modularity properties of the elliptic genus. In the Calabi–Yau cases (of pairs, orbifolds, etc.) the elliptic genus is a weak Jacobi form; see definition below. Also it is important to have a description of the elliptic genus in non-Calabi–Yau situations not just as a function on $\mathbb{C} \times \mathbb{H}$ but as an element of a finite-dimensional algebra of functions. It turns out that in the absence of a Calabi–Yau condition the elliptic genus belong to a very interesting algebra of functions on $\mathbb{C} \times \mathbb{H}$, which we call the algebra of quasi-Jacobi forms, and which is only slightly bigger than the algebra of weak Jacobi forms. This algebra of quasi-Jacobi forms is a counterpart of quasimodular forms [31] and is related to the elliptic genus in the same way as quasimodular forms are related to the Witten genus [55]. The algebra of quasi-Jacobi forms is generated by certain two-variable Eisenstein series, masterfully reviewed by A. Weil in [51], and has many properties parallel to the properties in quasimodular case. A detailed description of the properties of quasi-Jacobi forms appears to be absent in the literature, so we discuss the algebra of such forms in Section 2 — see its introduction. We conclude Section 2 with a discussion of differential operators Rankin–Cohen brackets on the space of Jacobi forms.

Finally in Section 3 we construct an extension of the Ochanine genus to real algebraic varieties with certain class of singularities. This extends results of Totaro [48].

For the readers' convenience we give ample references to prior work on elliptic genus, where more detailed information can be obtained. Section 2, dealing with quasi-Jacobi forms, can be read independently of the rest of the paper.

1. Elliptic genus

1A. Elliptic genus of singular varieties and push-forward formulas. Let X be a projective manifold. We shall use the Chow groups $A_*(X)$ with complex coefficients (see [23]). Let F the ring of functions on $\mathbb{C} \times \mathbb{H}$ where $\mathbb{H}$ is the upper half-plane. The elliptic class of X is an element in $A_*(X) \otimes_{\mathbb{C}} F$ given by

$$\mathcal{ELL}(X) = \prod_i x_i \frac{\theta\left(\frac{x_i}{2\pi i} - z, \tau\right)}{\theta\left(\frac{x_i}{2\pi i}, \tau\right)}[X], \tag{1-1}$$

where

$$\theta(z, \tau) = q^{\frac{1}{8}}(2 \sin \pi z) \prod_{l=1}^{l=\infty}(1 - q^l) \prod_{l=1}^{l=\infty}(1 - q^l e^{2\pi i z})(1 - q^l e^{-2\pi i z}) \tag{1-2}$$

is the Jacobi theta function considered as an element in F with $q = e^{2\pi i \tau}$ [12], the x_i are the Chern roots of the tangent bundle of X, and $[X]$ is the fundamental class of X. The component $Ell(X)$ in $A_0(X) = F$ is the elliptic genus of X.

The components of (1-1) in each degree, evaluated on a class in $A^*(X)$, are linear combinations of symmetric functions in c_i: that is, the Chern classes of X. In particular, $Ell(X)$ depends only on the class of X in the ring $\Omega^U \otimes \mathbb{Q}$ of unitary cobordisms.

The homomorphism $\Omega^U \otimes \mathbb{Q} \to F$ taking X to $Ell(X)$ can be described without reference to theta functions. Let M_{3,A_1} be the class of complex analytic spaces "having only A_1-singularities in codimension three", that is, having only singularities of the following type: the singular set $\mathrm{Sing}\, X$ of $X \in M_{3,A_1}$ is a *manifold* such that $\dim_{\mathbb{C}} \mathrm{Sing}\, X = \dim X - 3$ and for an embedding $X \to Y$ where Y is a manifold and a transversal H to $\mathrm{Sing}\, X$ in Y, the pair

$$(H \cap X, H \cap \mathrm{Sing}\, X)$$

is analytically equivalent to the pair $(\mathbb{C}^4, H_0)$, where H_0 is given by $x^2 + y^2 + z^2 + w^2 = 0$. Each $X \in M_{3,A_1}$ admits two small resolutions $\tilde{X}_1 \to X$ and $\tilde{X}_2 \to X$ in which the exceptional set is a fibration over $\mathrm{Sing}\, X$ with the fiber $\mathbb{P}^1$. One says that the manifolds underlying the resolutions are obtained from each other by a classical flop.

THEOREM 1.1. (*cf.* [47]) *The kernel of the homomorphism $Ell : \Omega^U \otimes \mathbb{Q} \to F$ taking an almost complex manifold X to its elliptic genus $Ell(X)$ is the ideal generated by the classes of differences $\tilde{X}_1 - \tilde{X}_2$ of two small resolutions of a variety in M_{3,A_1}.*

More generally one can fix a class of singular spaces and a type of resolutions and consider the quotient of $\Omega^U \otimes \mathbb{Q}$ by the ideal generated by differences of manifolds underlying resolutions of the same analytic space. The quotient map by this ideal $\Omega^U \otimes \mathbb{Q} \to R$ provides a genus and hence a collection of Chern numbers (linear combination of Chern monomials $c_{i_1} \cdots c_{i_k}[X]$, with $\sum i_s = \dim X$), which can be made explicit via Hirzebruch's procedure with a generating series [28]. These are the Chern numbers which can be defined for the chosen class of singular varieties and chosen class of resolutions. The ideal in Theorem 1.1, it turns out, corresponds to a much larger classes of singular spaces and resolutions. This method of defining Chern classes of singular varieties is an extension of the philosophy underlying a question of Goresky and McPherson [26]: Which Chern numbers can be defined via resolutions independently of the resolution?

DEFINITION 1.2. An analytic space X is called $\mathbb{Q}$-Gorenstein if the divisor D of a meromorphic form $df_1 \wedge \cdots \wedge df_{\dim X}$ is such that for some $n \in \mathbb{Z}$ the divisor nD in locally principal (i.e., K_X is $\mathbb{Q}$-Cartier). In particular, for any codimension-one component E of the exceptional divisor of a map $\pi : \tilde{X} \to X$, the multiplicity $a_E = \mathrm{mult}_E \pi^*(K_X)$ is well defined and a singularity is called log-terminal if there is a resolution π such that $K_{\tilde{X}} = \pi^*(K_X) + \sum a_E E$ and $a_E > -1$. A resolution is called crepant if $a_E = 0$.

THEOREM 1.3. ([8]) *The kernel of the elliptic genus* $\Omega^U \otimes \mathbb{Q} \to F$ *is generated by the differences of* $\tilde{X}_1 - \tilde{X}_2$ *of manifolds underlying crepant resolutions of the singular spaces with* $\mathbb{Q}$-*Gorenstein singularities admitting crepant resolutions.*

The proof of Theorem 1.3 is based on an extension of the elliptic genus $Ell(X)$ of manifolds to the elliptic genus of pairs $Ell(X, D)$, where D is a divisor on X having normal crossings as the only singularities. This is similar to the situation in the study of motivic E-functions of quasiprojective varieties [2; 40]. In fact, other problems such as the the study of McKay correspondence [9] suggest a motivation for looking at triples (X, D, G), where X is a normal variety, G is a finite group acting on X and to introduce the elliptic class $\mathcal{ELL}(X, D, G)$ (see again [9]). More precisely, let $D = \sum a_i D_i$ be a $\mathbb{Q}$-divisor, with the D_i irreducible and $a_i \in \mathbb{Q}$. The pair (X, D) is called Kawamata log-terminal (klt) [35] if $K_X + D$ is $\mathbb{Q}$-Cartier and there is a birational morphism $f : Y \to X$, where Y is smooth and is the union of the proper preimages of components of D, and the components of the exceptional set $E = \bigcup_j E_j$ form a normal crossing divisor such that $K_Y = f^*(K_X + \sum a_i D_i) + \sum \alpha_j E_j$, where $\alpha_j > -1$. (Here K_X, K_Y are the canonical classes of X and Y.) The triple (X, D, G), where X is a nonsingular variety, D is a divisor and G is a finite group of biholomorphic automorphisms is called G-normal [2; 9] if the components of

D form a normal crossings divisor and the isotropy group of any point acts trivially on the components of D containing this point.

DEFINITION 1.4 [9, Definition 3.2]. Let (X, E) be a Kawamata log terminal G-normal pair (in particular, X is smooth and D is a normal crossing divisor) and let $E = -\sum_{k \in \mathcal{K}} \delta_k E_k$. The *orbifold elliptic class* of (X, E, G) is the class in $A_*(X, \mathbb{Q})$ given by

$$\mathcal{ELL}_{\mathrm{orb}}(X, E, G; z, \tau) :=$$

$$\frac{1}{|G|} \sum_{\substack{g,h \\ gh=hg}} \sum_{X^{g,h}} [X^{g,h}] \prod_{\substack{\lambda(g)=0 \\ \lambda(h)=0}} x_\lambda \prod_\lambda \frac{\theta\left(\frac{x_\lambda}{2\pi i} + \lambda(g) - \tau\lambda(h) - z\right)}{\theta\left(\frac{x_\lambda}{2\pi i} + \lambda(g) - \tau\lambda(h)\right)} e^{2\pi i \lambda(h) z}$$

$$\times \prod_k \frac{\theta\left(\frac{e_k}{2\pi i} + \varepsilon_k(g) - \varepsilon_k(h)\tau - (\delta_k+1)z\right)}{\theta\left(\frac{e_k}{2\pi i} + \varepsilon_k(g) - \varepsilon_k(h)\tau - z\right)} \frac{\theta(-z)}{\theta(-(\delta_k+1)z)} e^{2\pi i \delta_k \varepsilon_k(h) z}. \quad (1\text{-}3)$$

where $X^{g,h}$ denotes an irreducible component of the fixed set of the commuting elements g and h and $[X^{g,h}]$ denotes the image of the fundamental class in $A_*(X)$. The restriction of TX to $X^{g,h}$ splits into linearized bundles according to the ([0, 1)-valued) characters λ of $\langle g, h \rangle$, which are sometimes denoted by λ_W, where W is a component of the fixed-point set. Moreover, $e_k = c_1(E_k)$ and ε_k is the character of $\mathcal{O}(E_k)$ restricted to $X^{g,h}$ if E_k contains $X^{g,h}$, and is zero otherwise.

One would like to define the elliptic genus of a Kawamata log-terminal pair (X_0, D_0) as (1-3) calculated for a G-equivariant resolution $(X, E) \to (X_0, D_0)$. Independence of (1-3) of resolution and the proof of (1.3) both depend on the following push-forward formula:

THEOREM 1.5. *Let (X, E) be a Kawamata log-terminal G-normal pair and let Z be a smooth G-equivariant locus in X which is normal crossing to $\mathrm{Supp}\, E$. Let $f : \hat{X} \to X$ denote the blowup of X along Z. Define*

$$\hat{E} = -\sum_k \delta_k \hat{E}_k - \delta \,\mathrm{Exc}\, f,$$

where $\hat{E}_k$ is the proper transform of E_k and δ is determined from $K_{\hat{X}} + \hat{E} = f^*(K_X + E)$. Then $(\hat{X}, \hat{E})$ is a Kawamata log-terminal G-normal pair and

$$f_* \mathcal{ELL}_{\mathrm{orb}}(\hat{X}, \hat{E}, G; z, \tau) = \mathcal{ELL}_{\mathrm{orb}}(X, E, G; z, \tau). \quad (1\text{-}4)$$

Independence from the resolution is a consequence of the weak factorization theorem [1] and of Theorem 1.5; Theorem 1.3 follows since both $\mathcal{ELL}(X_1)$ and $\mathcal{ELL}(X_2)$ coincide with the elliptic genus of the pair $(\tilde{X}, \tilde{D})$, where $\tilde{X}$ is a resolution of X dominating both X_1 and X_2 (here $D = K_{\tilde{X}/X}$; see [8,

Proposition 3.5] and also [52]. For a discussion of the orbifold elliptic genus on orbifolds more general than just global quotients see [20].

1B. Relation to other invariants.

V. Batyrev defined in [2], for a G-normal triple (X, D, G), an E-function $E_{\mathrm{orb}}(X, D, G)$ depending on the Hodge theoretical invariants. (There is also a motivic version; see [2; 40].) Firstly for a quasiprojective algebraic variety W one sets, as in [2, Definition 2.10],

$$E(W, u, v) = (-1)^i \sum_{p,q} \dim Gr_F^p Gr_W^{p+q}(H_c^i(W, \mathbb{C}))u^p v^q, \qquad (1\text{-}5)$$

where F and W are the Hodge and weight filtrations of Deligne's mixed Hodge structure [16; 17]. In particular, $E(W, 1, 1)$ is the topological Euler characteristic of W (with compact support). If W is compact one then obtains Hirzebruch's χ_y-genus [28]:

$$\chi_y(W) = \sum_{i,j}(-1)^q \dim H^q(\Omega_W^p)y^p, \qquad (1\text{-}6)$$

for $v = -1$, $u = y$, and hence the arithmetic genus, signature and so on are special values of (1-5). Secondly, for a G-normal pair as in Definition 1.4 one stratifies $D = \bigcup_{k \in K} D_k$ by strata $D_J^\circ = \bigcap_{j \in J} D_j - \bigcup_{k \in K-J} D_k$, for $J \subset K$ (the intersection being set to X if $J = \varnothing$), and defines

$$E(X, D, G, u, v)$$
$$= \sum_{\substack{\{g\} \\ W \subset X^g}} (uv)^{\sum \varepsilon_{D_i}(g)(\delta_i+1)} \sum_{J \subset \mathcal{K}^g} \prod_{j \in J} \frac{uv-1}{(uv)^{\delta_j+1}-1} E(W \cap D_J^\circ / C(g, J)), \quad (1\text{-}7)$$

where $C(g, J)$ is the subgroup of the centralizer of g leaving $\bigcap_{j \in J} D_j$ invariant.

One shows that for a Kawamata log-terminal pair (X_0, D_0) the E-function $E(X, D, G)$ of a resolution does not depend on the latter but only on X_0, D_0 and G. Hence (1-7) yields an invariant of Kawamata log-terminal G-pairs. The relation with $Ell(X, D, G)$ is the following (Proposition 3.14 of [9]):

$$\lim_{\tau \to i\infty} Ell(X, D, G, z, \tau) = y^{-\frac{1}{2}\dim X} E(X, D, G, y, 1), \qquad (1\text{-}8)$$

where $y = \exp(2\pi i z)$. In particular, in the nonequivariant smooth case the elliptic genus for $q \to 0$ specializes into the Hirzebruch χ_y genus (1-6).

On the other hand, in the nonsingular case, Hirzebruch [29; 30] and Witten [53] defined elliptic genera of complex manifolds which are given by modular forms for the subgroup $\Gamma_0(n)$ on level n in $\mathrm{SL}_2(\mathbb{Z})$, provided the canonical class of the manifold in question is divisible by n.

These genera are of course combinations of Chern numbers, but for $n = 2$ one obtains a combination of Pontryagin classes; i.e., an invariant that depends only on the underlying smooth structure, rather than the (almost) complex structure.

This genus was first introduced by S. Ochanine; see [43] and Section 3. These level-n elliptic genera coincide, up to a dimensional factor, with the specialization $z = (\alpha\tau + b)/n$, for appropriate $\alpha, \beta \in \mathbb{Z}$ specifying particular Hirzebruch level n elliptic genus; see Proposition 3.4 of [6].

1C. Application: The McKay correspondence for the elliptic genus. The classical McKay correspondence is a relation between the representations of the binary dihedral groups $G \subset \mathrm{SU}(2)$ (which are classified according to the root systems of type A_n, D_n, E_6, E_7, E_8) and the irreducible components of the exceptional set of the minimal resolution of $\mathbb{C}^2/G$. In particular, the number of conjugacy classes in G is the same as the number of irreducible components of the minimal resolution. The latter is a special case of the relation between the Euler characteristic $e(\widetilde{X/G})$ of a crepant resolution of the quotient X/G of a complex manifold X by an action of a finite group G and the data of the action on X:

$$e(\widetilde{X/G}) = \sum_{\substack{g,h \\ gh=hg}} e(X^{g,h}). \qquad (1\text{-}9)$$

A refinement of this relation for Hodge numbers and motives is given in [2; 19; 40]. When X is projective one has a refinement in which the Euler characteristic of the manifold in (1-9) is replaced by the elliptic genus of Kawamata log-terminal pairs. More generally, one has the following push-forward formula:

THEOREM 1.6. *Let $(X; D_X)$ be a Kawamata log-terminal pair which is invariant under an effective action of a finite group G on X. Let $\psi \colon X \to X/G$ be the quotient morphism. Let $(X/G; D_{X/G})$ be the quotient pair in the sense that $D_{X/G}$ is the unique divisor on X/G such that $\psi^*(K_{X/G} + D_{X/G}) = K_X + D_X$ (see Definition 2.7 in [9]). Then*

$$\psi_* \mathcal{ELL}_{\mathrm{orb}}(X, D_X, G; z, \tau) = \mathcal{ELL}(X/G, D_{X/G}; z, \tau).$$

In particular, for the components of degree zero one obtains

$$Ell_{\mathrm{orb}}(X, D_X, G, z, \tau) = Ell(X/G, D_{X/G}, z, \tau). \qquad (1\text{-}10)$$

When X is nonsingular and X/G admits a crepant resolution $\widetilde{X/G} \to X/G$, for $q = 0$ one obtains $\chi_y(\widetilde{X/G}) = \chi_y^{\mathrm{orb}}(X, G)$ and hence for $y = 1$ one recovers (1-9).

1D. Application: Higher elliptic genera and K-equivalences. Another application of the push-forward formula in Theorem 1.5 is the invariance of higher elliptic genera under K-equivalences. A question posed in [44], and answered in [3], concerns the higher arithmetic genus $\chi_\alpha(X)$ of a complex manifold X

corresponding to a cohomology class $\alpha \in H^*(\pi_1(X), \mathbb{Q})$ and defined as

$$\int_X T d_X \cup f^*(\alpha), \tag{1-11}$$

where $f : X \to B(\pi_1(X))$ is the classifying map from X to the classifying space of the fundamental group of X. It asks whether the higher arithmetic genus $\chi_\alpha(X)$ is a birational invariant. This question is motivated by Novikov's conjecture: the higher signatures (i.e., the invariant defined for topological manifold X by (1-11) with the Todd class replaced by the L-class) are homotopy invariant [15]. The higher χ_y-genus defined by (1-11) with the Todd class replaced by Hirzebruch's χ_y class [28] comes into the correction terms describing the nonmultiplicativity of χ_y in topologically locally trivial fibrations $\pi : E \to B$ of projective manifolds with nontrivial action of $\pi_1(B)$ on the cohomology of the fibers of π. See [11] for details.

Recall that two manifolds X_1, X_2 are called K-equivalent if there is a smooth manifold $\tilde{X}$ and a diagram

$$\begin{array}{ccc} & \tilde{X} & \\ \phi_1 \swarrow & & \searrow \phi_2 \\ X_1 & & X_2 \end{array} \tag{1-12}$$

in which ϕ_1 and ϕ_2 are birational morphisms and $\phi_1^*(K_{X_1})$ and $\phi_2^*(K_{X_2})$ are linearly equivalent.

The push-forward formula (1.5) leads to:

THEOREM 1.7. *For any $\alpha \in H^*(B\pi, \mathbb{Q})$ the higher elliptic genus*

$$(\mathcal{ELL}(X) \cup f^*(\alpha), [X])$$

is an invariant of K-equivalence. Moreover, if (X, D, G) and $(\hat{X}, \hat{D}, G)$ are G-normal and Kawamata log-terminal and if $\phi : (\hat{X}, \hat{D}) \to (X, D)$ is G-equivariant such that

$$\phi^*(K_X + D) = K_{\hat{X}} + \hat{D}, \tag{1-13}$$

then

$$Ell_\alpha(\hat{X}, \hat{D}, G) = Ell_\alpha(X, D, G).$$

In particular the higher elliptic genera (and hence the higher signatures and $\hat{A}$-genus) are invariant for crepant morphisms. The specialization into the Todd class is birationally invariant (i.e., the invariance condition (1-12) is not needed in the Todd case).

Another consequence is the possibility of defining higher elliptic genera for singular varieties with Kawamata log-terminal singularities and for G-normal pairs (X, D); see [10].

1E. The DMVV formula. The elliptic genus comes into a beautiful product formula for the generating series for the orbifold elliptic genus associated with the action of the symmetric group S_n on products $X \times \cdots \times X$, for which the first case appears in [18], together with a string-theoretical explanation. A general product formula for orbifold elliptic genus of triples is given in [9].

THEOREM 1.8. *Let (X, D) be a Kawamata log-terminal pair. For every $n \geq 0$ consider the quotient of $(X, D)^n$ by the symmetric group S_n, which we will denote by $(X^n / S_n, D^{(n)} / S_n)$. Here we denote by $D^{(n)}$ the sum of pullbacks of D under n canonical projections to X. Then we have*

$$\sum_{n \geq 0} p^n Ell(X^n / S_n, D^{(n)} / S_n; z, \tau) = \prod_{i=1}^{\infty} \prod_{l,m} \frac{1}{(1 - p^i y^l q^m)^{c(mi,l)}}, \quad (1\text{-}14)$$

where the elliptic genus of (X, D) is

$$\sum_{m \geq 0} \sum_{l} c(m, l) y^l q^m$$

and $y = e^{2\pi i z}, q = e^{2\pi i \tau}$.

It is amazing that such a simple-minded construction as the left-hand side of (1-14) leads to the Borcherds lift [4] of Jacobi forms.

1F. Other applications of the elliptic genus. In this section we point out other instances in which the elliptic genus plays a significant role.

The chiral de Rham complex. In [41], the authors construct for a complex manifold X a (bi)-graded sheaf Ω_X^{ch} of vertex operator algebras (with degrees called fermionic charge and conformal weight) with the differential d_{DR}^{ch} having fermionic degree 1 and quasiisomorphic to the de Rham complex of X. An alternative construction using the formal loop space was given in [32]. Each component of fixed conformal weight has a filtration so that graded components are

$$\bigotimes_{n \geq 1} (\Lambda_{-yq^{n-1}} T_X^* \otimes \Lambda_{-y^{-1}q^n} T_X \otimes S_{q^n} T_X^* \otimes S_{q^n} T_X) \quad (1\text{-}15)$$

In particular, it follows that

$$Ell(X, q, y) = y^{-\frac{1}{2} \dim X} \chi(\Omega_X^{ch}) = y^{-\frac{1}{2} \dim X} \text{Supertrace}_{H^*(\Omega_X^{ch})} y^{J[0]} q^{L[0]}, \quad (1\text{-}16)$$

where $J[m], L[n]$ are the operators which are part of the vertex algebra structure. The chiral complex for orbifolds was constructed in [22] and the extension of (1-16) to orbifolds (with discrete torsion) is discussed in [39].

Mirror symmetry. The physics definition of mirror symmetry in terms of conformal field theory suggests that for the elliptic genus, defined as an invariant of a conformal field theory (by an expression similar to the last term in (1-16) — see [54]) one should have for X and its mirror partner $\hat{X}$ the relation

$$Ell(X) = (-1)^{\dim X} Ell(\hat{X}). \qquad (1\text{-}17)$$

This is indeed the case [6, Remark 6.9] for mirror symmetric hypersurfaces in toric varieties in the sense of Batyrev.

Elliptic genus of Landau–Ginzburg models. The physics literature (see [34], for example) also associates to a weighted homogeneous polynomial a conformal field theory (the Landau–Ginzburg model) and in particular the elliptic genus. Moreover it is expected that the orbifoldized Landau–Ginzburg model will coincide with the conformal field theory of the hypersurface corresponding to this weighted homogeneous polynomial. In particular, one expects a certain identity expressing equality of the orbifoldized elliptic genus corresponding to the weighted homogeneous polynomial (or a more general Landau Ginzburg model) and the elliptic genus of the corresponding hypersurface. In [42] the authors construct a vertex operator algebra related by a correspondence of this type to the cohomology of the chiral de Rham complex of the hypersurface in $\mathbb{P}^n$, and obtain in particular the expression for the elliptic genus of a hypersurface as an orbifoldization. In [27] the authors obtain an expression for the one-variable Hirzebruch's genus as an orbifoldization.

Concluding remarks. There are several other interesting issues which should be mentioned in a discussion of the elliptic genus. It plays an important role in work of J. Li, K. F. Liu and J. Zhou [38] in connection with the Gopakumar–Vafa conjecture (see also [25]). The elliptic genus was defined for proper schemes with 1-perfect obstruction theory [21]. In fact one has well defined cobordism classes in Ω^U associated to such objects [14]. In the case of surfaces with normal singularities, one can extend the definition of elliptic genus beyond log-terminal singularities [50]. The elliptic genus is central in the study of elliptic cohomology [46]. Much of the discussion above can be extended to the equivariant context [49]; a survey of this is given in Waelder's paper in this volume.

2. Quasi-Jacobi forms

The Eisenstein series

$$e_k(\tau) = \sum_{\substack{(m,n)\in\mathbb{Z}^2 \\ (m,n)\neq(0,0)}} \frac{1}{(m\tau + n)^k}, \quad \tau \in \mathbb{H},$$

fails to be modular for $k = 2$, but the algebra generated by the functions $e_k(\tau)$, $k \geq 2$, called the algebra of quasimodular forms on $\mathrm{SL}_2(\mathbb{Z})$, has many interesting properties [57]. For example, there is a correspondence between quasimodular forms and real analytic functions on $\mathbb{H}$ which have the same $\mathrm{SL}_2(\mathbb{Z})$ transformation properties as modular forms. Moreover, the algebra of quasimodular forms has a structure of $\mathcal{D}$-module and supports an extension of Rankin–Cohen operations on modular forms.

In this section we show that there is an algebra of functions on $\mathbb{C} \times \mathbb{H}$ closely related to the algebra of Jacobi forms of index zero with similar properties. This algebra is generated by the Eisenstein series $\sum (z + \omega)^{-n}$, the sum being over elements ω of a lattice $W \subset \mathbb{C}$. It has a description in terms of real analytic functions satisfying a functional equation of Jacobi forms and having other properties of quasimodular forms mentioned in the last paragraph. It turns out that the space of functions on $\mathbb{C} \times \mathbb{H}$ generated by elliptic genera of arbitrary (possibly not Calabi–Yau) complex manifolds belong to this algebra of quasi-Jacobi forms.

DEFINITION 2.1. A *weak* (resp. meromorphic) Jacobi form of index $t \in \frac{1}{2}\mathbb{Z}$ and weight k for a finite index subgroup of the Jacobi group $\Gamma_1^J = \mathrm{SL}_2(\mathbb{Z}) \propto \mathbb{Z}^2$ is a holomorphic (resp. meromorphic) function χ on $\mathbb{H} \times \mathbb{C}$ having expansion $\sum c_{n,r} q^n \zeta^r$ in $q = \exp(2\pi\sqrt{-1}\tau)$ with $\mathrm{Im}\,\tau$ sufficiently large and satisfying the functional equations

$$\chi\left(\frac{a\tau + b}{c\tau + d}, \frac{z}{c\tau + d}\right) = (c\tau + d)^k e^{2\pi i t c z^2/(c\tau+d)} \chi(\tau, z),$$

$$\chi(\tau, z + \lambda\tau + \mu) = (-1)^{2t(\lambda+\mu)} e^{-2\pi i t(\lambda^2\tau + 2\lambda z)} \chi(\tau, z)$$

for all elements $\left[\left(\begin{smallmatrix} a & b \\ c & d \end{smallmatrix}\right), 0\right]$ and $\left[\left(\begin{smallmatrix} 1 & 0 \\ 0 & 1 \end{smallmatrix}\right), (a,b)\right]$ in Γ. The algebra of Jacobi forms is the bigraded algebra $J = \bigoplus J_{t,k}$ and the algebra of Jacobi forms of index zero is the subalgebra $J_0 = \bigoplus_k J_{0,k} \subset J$.

For appropriate l a Jacobi form can be expanded in (Fourier) series in $q^{1/l}$, with l depending on Γ. We shall need below the real analytic functions

$$\lambda(z, \tau) = \frac{z - \bar{z}}{\tau - \bar{\tau}} \quad \text{and} \quad \mu(\tau) = \frac{1}{\tau - \bar{\tau}}. \tag{2-1}$$

They have the transformation properties

$$\lambda\left(\frac{z}{c\tau + d}, \frac{a\tau + b}{c\tau + d}\right) = (c\tau + d)\lambda(z, \tau) - 2icz, \tag{2-2}$$

$$\lambda(z + m\tau + n, \tau) = \lambda(z, \tau) + m,$$

$$\mu\left(\frac{a\tau + b}{c\tau + d}\right) = (c\tau + d)^2 \mu(\tau) - 2ic(c\tau + d). \tag{2-3}$$

DEFINITION 2.2. An *Almost meromorphic Jacobi form* of weight k, index zero and depth (s, t) is a (real) meromorphic function in $\mathbb{C}\{q^{1/l}, z\}[z^{-1}, \lambda, \mu]$, with λ, μ given by (2-1), which

(a) satisfies the functional equations (2.1) of Jacobi forms of weight k and index zero, and
(b) which has degree at most s in λ and at most t in μ.

DEFINITION 2.3. A *quasi-Jacobi form* is a constant term of an almost meromorphic Jacobi form of index zero considered as a polynomial in the functions λ, μ; in other words, a meromorphic function f_0 on $\mathbb{H} \times \mathbb{C}$ such that there exist meromorphic functions $f_{i,j}$ such that each $f_0 + \sum f_{i,j} \lambda^i \mu^j$ is an almost meromorphic Jacobi form.

From the algebraic independence of λ, μ over the field of meromorphic functions in q, z one deduces:

PROPOSITION 2.4. *F is a quasi-Jacobi of depth (s, t) if and only if*

$$(c\tau + d)^{-k} f\left(\frac{a\tau + b}{c\tau + d}, \frac{z}{c\tau + d}\right) = \sum_{\substack{i \leq s \\ j \leq t}} S_{i,j}(f)(\tau, z)\left(\frac{cz}{c\tau + d}\right)^i \left(\frac{c}{c\tau + d}\right)^j,$$

$$f(\tau, z + a\tau + b) = \sum_{i \leq s} T_i(f)(\tau, z)a^i.$$

We turn to some basic examples of quasi-Jacobi forms.

DEFINITION 2.5 [51]. Consider the sequence of functions on $\mathbb{H} \times \mathbb{C}$ given by

$$E_n(z, \tau) = \sum_{(a,b)\in\mathbb{Z}^2} \frac{1}{(z + a\tau + b)^n}$$

(These series were used in [24] under the name twisted Eisenstein series.)

The series $E_n(z, \tau)$ converges absolutely for $n \geq 3$ and for $n = 1, 2$ defined via "Eisenstein summation" as

$$\sum_e (\cdot) = \lim_{A \to \infty} \sum_{a=-A}^{a=A} \lim_{B \to \infty} \sum_{b=-B}^{b=B} (\cdot),$$

though we shall omit the subscript e. The series $E_2(z, \tau)$ is related to the Weierstrass function as follows:

$$\wp(z, \tau) = \frac{1}{z^2} + \sum_{\substack{(a,b)\in\mathbb{Z} \\ (a,b)\neq 0}} \frac{1}{(z + a\tau + b)^2} - \frac{1}{(a\tau + b)^2}$$

$$= E_2(z, \tau) - \lim_{z \to 0}\left(E_2(z, \tau) - \frac{1}{z^2}\right).$$

Moreover,

$$e_n = \lim_{z \to 0} \left(E_n(z, \tau) - \frac{1}{z^n} \right) = \sum_{\substack{(a,b) \in \mathbb{Z} \\ (a,b) \neq 0}} \frac{1}{(a\tau + b)^n}$$

is the Eisenstein series, in the notation of [51]. The algebra of functions of $\mathbb{H}$ generated by the Eisenstein series $e_n(\tau)$ for $n \geq 2$ is the algebra of quasimodular forms for $\mathrm{SL}_2(\mathbb{Z})$ [55; 57].

Now we describe the algebra of quasi-Jacobi forms for the Jacobi group Γ_1^J.

PROPOSITION 2.6. *The functions E_n are weak meromorphic Jacobi forms of index zero and weight n for $n \geq 3$. E_1 is a quasi-Jacobi form of index 0 weight 1 and depth $(1, 0)$. $E_2 - e_2$ is a weak Jacobi form of index zero and weight 2 and E_2 is a quasi-Jacobi form of weight 2, index zero and depth $(0, 1)$.*

PROOF. The first part follows from the absolute convergence of the series (2.5) for $n \geq 3$. We have the transformation formulas

$$E_1 \left(a\tau + bc\tau + d, \frac{z}{c\tau + d} \right) = (c\tau + d)E_1(\tau, z) + \frac{\pi i c}{2} z, \tag{2-4}$$

$$E_1(\tau, z + m\tau + n) = E_1(\tau, z) - 2\pi i m, \tag{2-5}$$

$$E_2 \left(\frac{a\tau + b}{c\tau + d}, \frac{z}{c\tau + d} \right) = (c\tau + d)^2 E_2(\tau, z) - \tfrac{1}{2}\pi i c(c\tau + d), \tag{2-6}$$

$$E_2(\tau, z + a\tau + b) = E_2(\tau, z). \tag{2-7}$$

Equalities (2-4) and (2-6) follow from $E_1(\tau, z) = 1/z - \sum e_{2k}(\tau) z^{2k-1}$ and $E_2(\tau, z) = 1/z^2 + \sum_k (2k-1) e_{2k} z^{2k-2}$, respectively; see [51, Chapter 3, (10)]. Equality (2-7) is immediate form the definition of Eisenstein summation, while (2-5) follows from [51]. □

REMARK 2.7. The Eisenstein series $e_k(\tau)$, for $k \geq 4$, belong to the algebra of quasi-Jacobi forms. Indeed, from [51, Chapter IV, (7), (35)] one has

$$E_4 = (E_2 - e_2)^2 - 5e_4; \quad E_3^2 = (E_2 - e_2)^2 - 15e_4(E_2 - e_2) - 35e_4.$$

PROPOSITION 2.8. *The algebra of Jacobi forms (for Γ_1^J) of index zero and weight $t \geq 2$ is generated by $E_2 - e_2, E_3, E_4$.*

A short way to show this is to notice that the ring of such Jacobi forms is isomorphic to the ring of cobordisms of SU-manifolds modulo flops (Section 1A) via an isomorphism sending a complex manifold X of dimension d to $Ell(X) \cdot (\theta'(0)/\theta(z))^d$. This ring of cobordisms in turn is isomorphic to $\mathbb{C}[x_1, x_2, x_3]$, where x_1 is the cobordism class of a K3 surface and x_2, x_3 are the cobordism classes of certain four- and six-manifolds [48]. The graded algebra $\mathbb{C}[E_2 - e_2, E_3, E_4]$ is isomorphic to the same ring of polynomials (Examples 2.14) and the claim follows.

PROPOSITION 2.9. *The algebra of quasi-Jacobi forms is the algebra of functions on $\mathbb{H} \times \mathbb{C}$ generated by the functions $E_n(z, \tau)$ and $e_2(\tau)$.*

PROOF. The coefficient of λ^s for an almost meromorphic Jacobi form $F(\tau, z) = \sum_{i \leq s} f_i \lambda^i$ of depth $(s, 0)$ is a holomorphic Jacobi form of index zero and weight $k - s$; thus, by the previous proposition, it is a polynomial in $E_2 - e_2, E_3, \ldots$. Moreover $f_0 - E_1^s f_s$ is a quasi-Jacobi form of index zero and weight at most $s-1$. Hence, by induction, the ring of quasi-Jacobi forms of index zero and depth $(*, 0)$ can be identified with $\mathbb{C}[E_1, E_2 - e_2, E_3, \ldots]$. Similarly, the coefficient μ^t of an almost meromorphic Jacobi form $F = \sum_{j \leq t} (\sum f_{i,j} \lambda^i) \mu^j$ is an almost meromorphic Jacobi form of depth $(s, 0)$, and $F - (\sum_i f)i, s\lambda^i) E_2^t$ has depth (s', t') with $t' < t$. The claim follows. $\square$

Here is an alternative description of the algebra of quasi-Jacobi forms:

PROPOSITION 2.10. *The algebra of functions generated by the coefficients of the Taylor expansion in x of the function:*

$$\frac{\theta(x + z)\theta'(0)}{\theta(x)\theta(z)} - \left(\frac{1}{x} + \frac{1}{z}\right) = \sum_{i \geq 1} F_i x^i$$

is the algebra of quasi-Jacobi forms for $\mathrm{SL}_2(\mathbb{Z})$.

PROOF. From [45] we have the transformation formulas

$$\theta\left(\frac{a\tau + b}{c\tau + d}, \frac{z}{c\tau + d}\right) = \zeta(c\tau + d)^{1/2} e^{\pi i c z^2/(c\tau + d)} \theta(\tau, z),$$

$$\theta'\left(\frac{a\tau + b}{c\tau + d}, 0\right) = \zeta(c\tau + d)^{3/2} \theta'(\tau, 0),$$

$$\theta(\tau, z + m\tau + n) = (-1)^{m+n} e^{-2\pi i m z - \pi i m^2 \tau} \theta(\tau, z).$$

They imply that the function

$$\Phi(x, z, \tau) = \frac{x\theta(x + z)\theta'(0)}{\theta(x)\theta(z)} \tag{2-8}$$

satisfies the functional equations

$$\Phi\left(\frac{a\tau + b}{c\tau + d}, \frac{x}{c\tau + d}, \frac{z}{c\tau + d}\right) = e^{2\pi i c z x/(c\tau + d)} \Phi(x, z, \tau),$$

$$\Phi(x, z + m\tau + n, \tau) = e^{2\pi i m x} \Phi(x, z, \tau). \tag{2-9}$$

In particular, in the expansion

$$\frac{d^2 \log \Phi}{dx^2} = \sum H_i x^i, \tag{2-10}$$

the left-hand side is invariant under the transformations in (2-9) and the coefficient H_i is a Jacobi form of weight i and index zero for any i. Moreover the coefficients F_i in $\Phi(x, z, \tau) = 1 + \sum F_i(z, \tau)x^i$ are polynomials in F_1 and the H_i. What remains to show is that the E_i determine F_1 and the H_i, for $i \geq 1$, and vice versa.

Recall that E_i has index zero (is invariant with respect to shifts) and weight i. We shall use the expressions

$$\Phi(x, z, \tau) = \frac{x+z}{z} \exp\left(\sum_{k>0} \frac{2}{k!}(x^k + z^k - (x+z)^k)G_k(\tau)\right), \qquad (2\text{-}11)$$

where

$$G_k(\tau) = -\frac{B_k}{2k} + \sum_{l=1}^{\infty} \sum_{d|l}(d^{k-1})q^l; \qquad (2\text{-}12)$$

see [56]. On the other hand, from [51, III.7 (10)] we have

$$E_n(z, \tau) = \frac{1}{z^n} + (-1)^n \sum_{2m \geq n}^{\infty} \binom{2m-1}{n-1})e_{2m}z^{2m-n}, \qquad (2\text{-}13)$$

where

$$e_{2m} = {\sum}'\left(\frac{1}{m\tau+n}\right)^{2m} = \frac{2(2\pi\sqrt{-1})^k}{(k-1)!}G_k \quad \text{for } k = 2m; \qquad (2\text{-}14)$$

see [51, III.7] and [55, p. 220]. We have

$$\frac{d^2 \log \Phi(x, z, \tau)}{dx^2}$$

$$= \sum_{i\geq 1} \frac{(-1)^i i x^{i-1}}{z^{i+1}} + \sum_{i\geq 2} \frac{2}{(i-2)!}(x^{i-2} - (x+z)^{i-2})G_i(\tau). \quad (2\text{-}15)$$

Now, using (2-14) and identities with binomial coefficients, we obtain for the coefficient of x^{l-2} for $l \geq 2$ in the Laurent expansion the value

$$\frac{(-1)^{l-1}(l-1)}{z^l} - \sum_{i\geq 2, i>l} \frac{2}{(i-2)!}\binom{i-2}{l-2}z^{i-l}G_i(\tau)$$

$$= \frac{(-1)^{l-1}(l-1)}{z^{l-1}} - \sum_{\substack{i\geq 2 \\ i>l}} \frac{1}{(2\pi\sqrt{-1})^i}(i-1)\binom{i-2}{l-2}z^{i-l}e_i$$

$$= \frac{(-1)^{l-1}(l-1)}{z^l} - (l-1)\frac{1}{(2\pi\sqrt{-1})^l} \sum_{\substack{i\geq 2 \\ i>l}} \binom{i-1}{l-1}e_i\left(\frac{z}{2\pi\sqrt{-1}}\right)^{i-l} \qquad (2\text{-}16)$$

This yields

$$H_{l-2}(2\pi\sqrt{-1}z,\tau) = (-1)^{l-1}\frac{(l-1)}{(2\pi\sqrt{-1})^l}(E_l - e_l),$$

and the claim follows since formula (15) in [56] yields

$$F_1(z,\tau) = \frac{1}{z} - 2\sum_{r\geq 0}G_{r+1}\frac{z^r}{r!} = \frac{1}{z} - \frac{1}{(2\pi\sqrt{-1})}\sum_{r\geq 0}e_r\left(\frac{z}{2\pi\sqrt{-1}}\right)^r, \quad (2\text{-}17)$$

that is,

$$F_1(2\pi i\sqrt{-1}z,\tau) = \frac{1}{2\pi\sqrt{-1}}E_1(z,\tau) \qquad\qquad \Box$$

REMARK 2.11. The algebra of quasi-Jacobi forms $\mathbb{C}[e_2, E_1, E_2, \dots]$ is closed under differentiation with respect to τ and ∂_z. Indeed, one has

$$2\pi i\frac{\partial E_1}{\partial\tau} = E_3 - E_1 E_2, \qquad\qquad \frac{\partial E_1}{\partial z} = -E_2,$$

$$2\pi i\frac{\partial E_2}{\partial\tau} = 3E_4 - 2E_1 E_3 - E_2^2, \qquad\qquad \frac{\partial E_2}{\partial z} = -2E_3,$$

and hence $\mathbb{C}[\dots, E_i, \dots]$ is a $\mathcal{D}$-module, where $\mathcal{D}$ is the ring of differential operators generated by $\partial/\partial\tau$ and $\partial/\partial z$ over the ring of holomorphic Jacobi group invariant functions on $H\times\mathbb{C}$. As is clear from the discussion, the ring of Eisenstein series $\mathbb{C}[\dots, E_i, \dots]$ has a natural identification with the ring of real valued almost meromorphic Jacobi forms $\mathbb{C}[E_1^*, E_2^*, E_3, \dots]$ on $\mathbb{H}\times\mathbb{C}$ having index zero, where

$$E_1^* = E_1 + 2\pi i\frac{\operatorname{Im}x}{\operatorname{Im}\tau}, \quad E_2^* = E_2 + \frac{1}{\operatorname{Im}\tau}. \qquad (2\text{-}18)$$

THEOREM 2.12. *The algebra of quasi-Jacobi forms of depth* $(k,0)$, $k\geq 0$, *is isomorphic to the algebra of complex unitary cobordisms modulo flops.*

In another direction, the depth of quasi-Jacobi forms allows one to "measure" the deviation of the elliptic genus of a non-Calabi–Yau manifold from being a Jacobi form.

THEOREM 2.13. *Elliptic genera of manifolds of dimension at most d span the subspace of forms of depth $(d,0)$ in the algebra of quasi-Jacobi forms. If a complex manifold satisfies $c_1^k = 0$ and $c_1^{k-1}\neq 0$,* [1] *its elliptic genus is a quasi-Jacobi form of depth $(s,0)$, where $s\leq k-1$.*

---

[1] More generally, k is the smallest among indices i with $c_1^i \in \operatorname{Ann}(c_2, \dots, c_{\dim M})$; an example of such a manifold is an n-manifold having a $(n-k)$-dimensional Calabi–Yau factor.

PROOF. It follows from the proof of Proposition (2.10) that

$$\frac{d^2 \log \Phi}{dx^2} = \sum_{i \geq 2} (-1)^{i-1} \frac{i-1}{(2\pi\sqrt{-1})^i} (E_i - e_i) x^{i-2},$$

which yields

$$\Phi = e^{E_1 x} \prod_i e^{(1/i)(-1)^{i-1}(i-1)/(2\pi\sqrt{-1})^i (E_i - e_i) x^i}. \qquad (2\text{-}19)$$

The Hirzebruch characteristic series is

$$\Phi\left(\frac{x}{2\pi i}\right) \frac{\theta(z)}{\theta'(0)};$$

compare (1-1). Hence, if $c(TX) = \Pi(1 + x_k)$, then

$$Ell(X) = \left(\frac{\theta(z)}{\theta'(0)}\right)^{\dim X} \prod_{i,k} e^{E_1 x_k} e^{(1/i)(-1)^{i-1}(i-1)(E_i - e_i) x_k^i}[X]$$

$$= \left(\frac{\theta(z)}{\theta'(0)}\right)^{\dim X} e^{c_1(X) E_1} \prod_{i,k} e^{(1/i)(-1)^{i-1}(i-1)(E_i - e_i) x_k^i}[X], \quad (2\text{-}20)$$

where $[X]$ is the fundamental class of X. In other words, if $c_1 = 0$, the elliptic class is a polynomial in $E_i - e_i$ with $i \geq 2$, and hence the elliptic genus is a Jacobi form [36]. Moreover of $c_1^k = 0$ the degree of this polynomial is at most k in E_1, and the claim follows. $\qquad \square$

EXAMPLE 2.14. Expression (2-20) can be used to get formulas for the elliptic genus of specific examples in terms of Eisenstein series E_n. For example, for a surface in $\mathbb{P}^3$ having degree d one has

$$\left(E_1^2 (\tfrac{1}{2} d^2 - 4d + 8) d + (E_2 - e_2)(\tfrac{1}{2} d^2 - 2) d\right) \left(\frac{\theta(z)}{\theta'(0)}\right)^2$$

In particular for $d = 1$ one obtains

$$\left(\tfrac{9}{2} E_1^2 - \tfrac{3}{2}(E_2 - e_2)\right) \left(\frac{\theta(z)}{\theta'(0)}\right)^2.$$

One can compare this with the double series that is a special case of the general formula for the elliptic genus of toric varieties in [6]. This leads to a two-variable version of the identity discussed in [6, Remark 5.9]. In fact following [5] one can define the subalgebra of "toric quasi-Jacobi forms" of the algebra of quasi-Jacobi forms, extending the toric quasimodular forms considered in [5]. This issue will be addressed elsewhere.

Next we consider one more similarity between meromorphic Jacobi forms and modular forms: there is a natural noncommutative deformation of the ordinary product of Jacobi forms similar to the deformation of the product modular forms constructed using Rankin–Cohen brackets [57]. In fact we have the following Jacobi counterpart of Rankin–Cohen brackets:

PROPOSITION 2.15. *Let f and g be Jacobi forms of index zero and weights k and l, respectively. Then*

$$[f, g] = k\left(\partial_\tau f - \frac{1}{2\pi i}E_1\partial_z f\right)g - l\left(\partial_\tau g - \frac{1}{2\pi i}E_1(z, \tau)\partial_z g\right)f$$

is a Jacobi form of weight $k + l + 2$. More generally, let

$$D = \partial_\tau - \frac{1}{2\pi i}E_1\partial_z.$$

Then the Cohen–Kuznetsov series (see [57])

$$\tilde{f}_D(z, \tau, X) = \sum_{n=0}^{\infty} \frac{D^n f(z, \tau)X^n}{n!(k)_n},$$

where $(k)_n = k(k + 1)\cdots(k + n - 1)$ is the Pochhammer symbol, satisfies

$$\tilde{f}_D\left(\frac{a\tau + b}{c\tau + d}, \frac{z}{c\tau + d}, \frac{z}{c\tau + d}, \frac{X}{(c\tau + d)^2}\right)$$

$$= (c\tau + d)^k \exp\left(\frac{c}{c\tau + d}\frac{X}{2\pi i}\right)f_D(\tau, z, X),$$

$$\tilde{f}_D(\tau, z + a\tau + b, X) = \tilde{f}_D(\tau, z, X).$$

In particular, the coefficient $[f, g]_n/(k)_n(l)_n$ of X^n in $\tilde{f}_D(\tau, z, -X)\tilde{g}_D(\tau, z, X)$ is a Jacobi form of weight $k + l + 2n$. It is given explicitly in terms of $D^i f$ and $D^j g$ by the same formulas as the classical RC brackets.

PROOF. The main point is that the operator $\partial_\tau - \frac{1}{2\pi i}E_1\partial_z$ has the same deviation from transforming a Jacobi form into another as ∂_τ has on modular forms. Indeed:

$$\left(\partial_\tau - \frac{1}{2\pi i}E_1\partial_z\right)f\left(\frac{a\tau + b}{c\tau + d}, \frac{z}{c\tau + d}\right)$$

$$= \left(kc(c\tau + d)^{k+1}f(\tau, z) + zc(c\tau + d)^{k+1}\partial_z f(\tau, z) + (c\tau + d)^{k+2}\partial_\tau f(\tau, z)\right.$$

$$\left. - \frac{1}{2\pi i}((c\tau + d)E_1(\tau, z) + 2\pi icz)(c\tau + d)^{k+1}\partial_z f\right)$$

$$= (c\tau + d)^{k+2}\left(f(\tau, z) - \frac{1}{2\pi i}E_1 f_z\right) + kc(c\tau + d)^{k+1}f(\tau, z).$$

Moreover,

$$\left(\partial_\tau - \frac{1}{2\pi i}E_1\partial_z\right)f(\tau, z+a\tau+b) = f_\tau + af_z - \frac{1}{2\pi i}(E_1 - 2\pi i a)f_z$$
$$= \left(\partial_\tau - \frac{1}{2\pi i}E_1\partial_z\right)f(\tau, z).$$

The rest of the proof runs as in [57].   □

REMARK 2.16. The brackets introduced in Proposition 2.15 are different from the Rankin–Cohen bracket introduced in [13].

3. Real singular varieties

The Ochanine genus of an oriented differentiable manifold X can be defined using the following series with coefficients in $\mathbb{Q}[\![q]\!]$ as the Hirzebruch characteristic power series (see [37] and references there):

$$Q(x) = \frac{x/2}{\sinh(x/2)}\prod_{n=1}^\infty\left(\frac{(1-q^n)^2}{(1-q^n e^x)(1-q^n e^{-x})}\right)^{(-1)^n} \tag{3-1}$$

As was mentioned in Section 1B, this genus is a specialization of the two-variable elliptic genus (at $z = \frac{1}{2}$). Evaluation of the Ochanine genus of a manifold using (3-1) and viewing the result as function of τ on the upper half-plane (where $q = e^{2\pi i\tau}$) yields a modular form on $\Gamma_0(2) \subset \mathrm{SL}_2(\mathbb{Z})$; see [37].

In this section we discuss elliptic genera for real algebraic varieties. It particular we address Totaro's proposal [48] that "it should be possible to define Ochanine genus for a large class of compact oriented real analytic spaces." In this direction we have:

THEOREM 3.1 [48]. *The quotient of* MSO *by the ideal generated by oriented real flops and complex flops (that is, the ideal generated by* $X' - X$, *where* X' *and* X *are related by a real or complex flop) is*

$$\mathbb{Z}[\delta, 2\gamma, 2\gamma^2, 2\gamma^4],$$

with $\mathbb{CP}^2$ *corresponding to* δ *and* $\mathbb{CP}^4$ *to* $2\gamma + \delta^2$. *This quotient ring is the the image of* MSO$_*$ *under the Ochanine genus.*

In particular the Ochanine genus of a small resolution is independent of its choice for singular spaces having singularities only along nonsingular strata and having in normal directions only singularities which are cones in $\mathbb{R}^4$ or $\mathbb{C}^4$.

Our goal is to find a wider class of singular real algebraic varieties for which the Ochanine genus of a resolution is independent of the choice of the latter.

3A. Real singularities. For the remainder of this paper "real algebraic variety" means an *oriented* quasiprojective variety $X_\mathbb{R}$ over $\mathbb{R}$, $X(\mathbb{R})$ is the set of its $\mathbb{R}$-points with the Euclidean topology, $X_\mathbb{C} = X_\mathbb{R} \times_{\mathrm{Spec}\,\mathbb{R}} \mathrm{Spec}\,\mathbb{C}$ is the complexification and $X(\mathbb{C})$ the analytic space of its complex points. We also assume that $\dim_\mathbb{R} X(\mathbb{R}) = \dim_\mathbb{C} X(\mathbb{C})$.

DEFINITION 3.2. A real algebraic variety $X_\mathbb{R}$ as above is called $\mathbb{Q}$-Gorenstein log-terminal if the analytic space $X(\mathbb{C})$ is $\mathbb{Q}$-Gorenstein log-terminal.

EXAMPLE 3.3. *The affine variety*

$$x_1^2 - x_2^2 + x_3^2 - x_4^2 = 0 \qquad (3\text{-}2)$$

in $\mathbb{R}^4$ is three-dimensional Gorenstein log-terminal and admits a crepant resolution.

Indeed, it is well known that the complexification of the Gorenstein singularity (3-2) admits a small (and hence crepant) resolution having $\mathbb{P}^1$ as its exceptional set.

EXAMPLE 3.4. *The three-dimensional complex cone in $\mathbb{C}^4$ given by $z_1^2 + z_2^2 + z_3^2 + z_4^2 = 0$ considered as a codimension-two subvariety of $\mathbb{R}^8$ is a $\mathbb{Q}$-Gorenstein log-terminal variety over $\mathbb{R}$ and its complexification admits a crepant resolution.*

Indeed, this codimension-two subvariety is a real analytic space which is the intersection of two quadrics in $\mathbb{R}^8$ given by

$$a_1^2 + a_2^2 + a_3^2 + a_4^2 - b_1^2 - b_2^2 - b_3^2 - b_4^2 = 0 = a_1b_1 + a_2b_2 + a_3b_3 + a_4b_4, \quad (3\text{-}3)$$

where $a_i = \mathrm{Re}\,z_i, b_i = \mathrm{Im}\,z_i$. The complexification is the cone over complete intersection of two quadrics in $\mathbb{P}^7$. Moreover, the defining equations of this complete intersection become, after the change of coordinates $x_i = a_i + \sqrt{-1}b_i$, $y_i = a_i - \sqrt{-1}b_i$,

$$x_1^2 + x_2^2 + x_3^2 + x_4^2 = 0 \quad \text{and} \quad y_1^2 + y_2^2 + y_3^2 + y_4^2 = 0. \qquad (3\text{-}4)$$

The singular locus is the union of two disjoint two-dimensional quadrics and the singularity along each is A_1 (i.e., the intersection of the transversal to it in $\mathbb{P}^7$ has an A_1 singularity). To resolve (3-3), one can blow up $\mathbb{C}^8$ at the origin, which results in a $\mathbb{C}$-fibration over the complete intersection (3-4). It can be resolved by small resolutions along two nonsingular components of the singular locus of (3-4). A direct calculation (considering, for example, the order of the pole of the form $dx_2 \wedge dx_3 \wedge dx_4 \wedge dy_2 \wedge dy_3 \wedge dy_4/(x_1y_1)$ along the intersection of the exceptional locus of the blow-up $\tilde{\mathbb{C}}^8$ of $\mathbb{C}^8$ with the proper preimage of (3-4) in $\tilde{\mathbb{C}}^8$) shows that we have a log-terminal resolution of the Gorenstein singularity which is the complexification of (3-3).

3B. The elliptic genus of resolutions of real varieties with $\mathbb{Q}$-Gorenstein log-terminal singularities.

Let X be a real algebraic manifold and let $D = \sum \alpha_k D_k$, with each α_k in $\mathbb{Q}$, be a divisor on the complexification $X_{\mathbb{C}}$ of X (i.e., the D_k are irreducible components of D). Let x_i denote the Chern roots of the tangent bundle of $X_{\mathbb{C}}$ and denote by d_k the classes corresponding to D_k (Section 1).

DEFINITION 3.5. Let X be a real algebraic manifold and D a divisor on the complexification $X_{\mathbb{C}}$ of X. The Ochanine class $\mathcal{ELL}_O(X, D)$ of the pair (X, D) is the specialization

$$\mathcal{ELL}\left(X_{\mathbb{C}}, D, q, z = \tfrac{1}{2}\right)$$

of the two-variable elliptic class of the pair $\mathcal{ELL}(X_{\mathbb{C}}, D, q, z)$ given by

$$\left(\prod_l \frac{\left(\frac{x_l}{2\pi i}\right)\theta\left(\frac{x_l}{2\pi i} - z\right)\theta'(0)}{\theta(-z)\theta\left(\frac{x_l}{2\pi i}\right)}\right) \times \left(\prod_k \frac{\theta\left(\frac{d_k}{2\pi i} - (\alpha_k + 1)z\right)\theta(-z)}{\theta\left(\frac{d_k}{2\pi i} - z\right)\theta(-(\alpha_k + 1)z)}\right). \tag{3-5}$$

The Ochanine elliptic genus of the pair (X, D) as above is

$$Ell(X_{\mathbb{R}}, D) = \sqrt{\mathcal{ELL}\left(X_{\mathbb{C}}, D, q, \tfrac{1}{2}\right)} \cup cl(X(\mathbb{R}))[X(\mathbb{C})]. \tag{3-6}$$

Here $\sqrt{\mathcal{ELL}}$ denotes the class corresponding to the unique series with constant term 1 and having $\mathcal{ELL}$ as its square.

The class of pairs above is the class (1-3) considered in Definition 1.4 in the case where group G is trivial. One can define an orbifold version of this class as well, specializing (1-3) to $z = \tfrac{1}{2}$. See [8] for further discussion of the class $\mathcal{ELL}(X, D)$.

The relation with Ochanine's definition is as follows: if D is the trivial divisor on $X_{\mathbb{C}}$, the result coincides with the genus [43]. More precisely:

LEMMA 3.6. Let $X_{\mathbb{R}}$ be a real algebraic manifold with nonsingular complexification $X_{\mathbb{C}}$. Then

$$Ell(X_{\mathbb{R}}) = \sqrt{\mathcal{ELL}(T_{X(\mathbb{C})})} \cup cl(X(\mathbb{R}))[X(\mathbb{C})].$$

PROOF. Indeed, we have

$$0 \to T_{X(\mathbb{R})} \to T_{X(\mathbb{C})}|_{X(\mathbb{R})} \to T_{X(\mathbb{R})} \to 0, \tag{3-7}$$

with the identification of the normal bundle to $X_{\mathbb{R}}$ with its tangent bundle given by multiplication by $\sqrt{-1}$. Hence $\mathcal{ELL}(X_{\mathbb{R}})^2 = i^*(\mathcal{ELL}X_{\mathbb{C}})$, where $i : X_{\mathbb{R}} \to X_{\mathbb{C}}$ is the canonical embedding. Now the lemma follows from the identification of the characteristic series (3-1) and specialization $z = \tfrac{1}{2}$ of the series in (1-1) (see [6]) and the identification which is just a definition of the class $cl_Z \in$

$H^{\dim_{\mathbb{R}} Y - \dim_{\mathbb{R}} Z}$ of a submanifold Z of a manifold Y: $\mathrm{cl}_Z \cup \alpha[Y] = i^*(\alpha) \cap [Z]$ for any $\alpha \in H^{\dim_{\mathbb{R}} Z}(Y)$. Indeed, we have

$$
\begin{aligned}
Ell(X_{\mathbb{R}}) = \mathcal{E}LL(T_{X(\mathbb{R})})[X(\mathbb{R})] &= \sqrt{\mathcal{E}LL(T_{X(\mathbb{C})})}\big|_{X(\mathbb{R})}[X(\mathbb{R})] \\
&= \sqrt{\mathcal{E}LL(T_{X(\mathbb{C})}) \cup \mathrm{cl}(X(\mathbb{R}))[X(\mathbb{C})]}. \quad \square
\end{aligned}
$$

Our main result in this section is the following:

THEOREM 3.7. *Let* $\pi : (\tilde{X}, \tilde{D}) \to (X, D)$ *be a resolution of singularities of a real algebraic pair with* $\mathbb{Q}$*-Gorenstein log-terminal singularities; i.e.,* $K_{\tilde{X}} + \tilde{D} = \pi^*(K_X + D)$. *Then the elliptic genus of the pair* $(\tilde{X}, \tilde{D})$ *is independent of the resolution. In particular, if a real algebraic variety* X *has a crepant resolution, its elliptic genus is independent of a choice of crepant resolution.*

PROOF. Indeed for a blowup $f : (\tilde{X}, \tilde{D}) \to (X, D)$ we have

$$
f_*\left(\sqrt{\mathcal{E}LL(\tilde{X}, \tilde{D}, q, \tfrac{1}{2})}\right) = \sqrt{\mathcal{E}LL(X, D, q, \tfrac{1}{2})} \tag{3-8}
$$

This is a special case of the push-forward formula (1-4) in theorem 1.5, with G being the trivial group. Hence

$$
\begin{aligned}
\mathcal{E}LL_{\mathbb{O}}(X_{\mathbb{R}}, D) &= \sqrt{\mathcal{E}LL(X_{\mathbb{C}}, D, q, \tfrac{1}{2})} \cup \mathrm{cl}(X_{\mathbb{R}})[X_{\mathbb{C}}] \\
&= \sqrt{\mathcal{E}LL(\tilde{X}_{\mathbb{C}}, \tilde{D}, q, \tfrac{1}{2})} \cup f^*([X_{\mathbb{R}}] \cap [X_{\mathbb{C}}]) = \mathcal{E}LL(\tilde{X}_{\mathbb{R}}, \tilde{D}),
\end{aligned}
$$

as follows from projection formula since $f^*(\mathrm{cl}[X_{\mathbb{R}}]) = [\mathrm{cl}\,\tilde{X}_{\mathbb{R}}]$ and since f_* is the identity on H_0.

For a crepant resolution one has $D = 0$ and hence by Lemma 3.6 the elliptic genus of $X_{\mathbb{R}}$ is the Ochanine genus of the real manifold, which is its crepant resolution. $\square$

REMARK 3.8. Examples 3.3 and 3.4 show that singularities admitting a crepant resolution include real three-dimensional cones and real points of complex three-dimensional cones.

Acknowledgement

The author wants to express his gratitude to Lev Borisov. The material in Section 1 is a survey of joint papers with him, and the results of Section 2 are based on discussions with him several years ago.

References

[1] D. Abramovich, K. Karu, K. Matsuki, J. Włodarsczyk, Torification and factorization of birational maps. J. Amer. Math. Soc., 15 (2002) no. 3, 531–572.

[2] V. Batyrev, Non-archimedean integrals and stringy Euler numbers of log-terminal pairs. J. Eur. Math. Soc. 1 (1999), no. 1, 5–33.

[3] J. Block, S. Weinberger, Higher Todd classes and holomorphic group actions. Pure Appl. Math. Q. 2 (2006), no. 4:2, 1237–1253.

[4] R. Borcherds, Automorphic forms on $O_{s+2,2}(\mathbf{R})^+$ and generalized Kac–Moody algebras. In Proceedings of the International Congress of Mathematicians (Zürich, 1994), 744–752, Birkhäuser, Basel, 1995.

[5] L. Borisov, P. Gunnells, Toric modular forms, Inventiones Math. 144 (2001), no. 2, 297–325.

[6] L. Borisov, A. Libgober, Elliptic genera of toric varieties and applications to mirror symmetry. Invent. Math. 140 (2000), no. 2, 453–485.

[7] L. Borisov, A. Libgober, Elliptic genera of singular varieties, orbifold elliptic genus and chiral de Rham complex. In Mirror symmetry, IV (Montreal, QC, 2000), 325–342, AMS/IP Stud. Adv. Math., 33, Amer. Math. Soc., Providence, RI, 2002.

[8] L. Borisov, A. Libgober, Elliptic genera of singular varieties. Duke Math. J. 116 (2003), no. 2, 319–351.

[9] L. Borisov, A. Libgober, McKay correspondence for elliptic genera. Ann. of Math. (2) 161 (2005), no. 3, 1521–1569.

[10] L. Borisov, A. Libgober, Higher elliptic genera. Math. Res. Lett. 15 (2008), no. 3, 511–520.

[11] S. Cappell, A. Libgober, L. Maxim, J. Shaneson, Hodge genera and characteristic classes of complex algebraic varieties. Electron. Res. Announc. Math. Sci. 15 (2008), 1–7.

[12] K. Chandrasekharan *Elliptic functions*, Grundlehren Math. Wiss. **281**, Springer, Berlin, 1985.

[13] Y. J. Choie, W. Eholzer, Rankin–Cohen operators for Jacobi and Siegel forms. J. Number Theory 68 (1998), no. 2, 160–177.

[14] I. Ciocan-Fontanine, M. Kapranov, Virtual fundamental classes via dg-manifolds, arXiv math/0703214.

[15] J. Davis, *Manifold aspects of the Novikov conjecture*, in Surveys in surgery theory, vol. 1, 195–224, Ann. Math. Studies **145**, Princeton Univ. Press, Princeton, 2000.

[16] P. Deligne, Théorie de Hodge II, Publ. Math. IHES 40 (1971), 5–58.

[17] P. Deligne, Théorie de Hodge III, Publ. Math. IHES 44 (1974), 5–77.

[18] R. Dijkgraaf, D. Moore, E. Verlinde, H. Verlinde, *Elliptic genera of symmetric products and second quantized strings*. Comm. Math. Phys. **185** (1997), no. 1, 197–209.

[19] J. Denef, F. Loeser, Geometry on arc spaces of algebraic varieties. In Proceedings of European Congress of Mathematics, vol. I (Barcelona, 2000), 327–348, Progr. Math. **201**, Birkhäuser, Basel, 2001.

[20] C. Dong, K. F. Liu, X. Ma, On orbifold elliptic genus. In Orbifolds in mathematics and physics (Madison, WI, 2001), 87–105, Contemp. Math. **310**, Amer. Math. Soc., Providence, RI, 2002.

[21] B. Fantechi, L. Göttsche Riemann-Roch theorems and elliptic genus for virtually smooth Schemes. arXiv 0706.0988.

[22] E. Frenkel, M. Szczesny, Chiral de Rham complex and orbifolds. J. Algebraic Geom. 16 (2007), no. 4, 599–624.

[23] W. Fulton, Intersection Theory (second edition), Springer, Berlin, 1998.

[24] M. Gaberdiel, C. Keller, Differential operators for elliptic genera. arXiv 0904.1831.

[25] E. Gasparim, Melissa Liu, The Nekrasov conjecture for toric surfaces, arXiv 0808.0884

[26] M. Goresky, R. MacPherson. Problems and bibliography on intersection homology. In Intersection cohomology, edited by A. Borel et al., 221–233. Birkhäuser, Boston, 1984.

[27] V. Gorbounov,S. Ochanine, Mirror symmetry formulae for the elliptic genus of complete intersections. J. Topol. 1 (2008), no. 2, 429–445.

[28] F. Hirzebruch, Topological methods in algebraic geometry (reprint of the 1978 edition), Classics in Mathematics, Springer, Berlin, 1995.

[29] F. Hirzebruch, Elliptic genera of level N for complex manifolds. In Differential geometric methods in theoretical physics (Como, 1987), edited by K. Bleuer and M. Werner, NATO Adv. Sci. Inst. Ser. C: Math. Phys. Sci. **250**. Dordrecht, Kluwer, 1988.

[30] F. Hirzebruch, T. Berger, R. Jung, Manifolds and modular forms. Aspects of Mathematics **E20**. Vieweg, Braunschweig, 1992.

[31] M. Kaneko, D. Zagier, A generalized Jacobi theta function and quasimodular forms. In The moduli space of curves (Texel Island, 1994), 165–172, Progr. Math. **129**, Birkhäuser, Boston, 1995.

[32] M. Kapranov, E. Vasserot, Vertex algebras and the formal loop space. Publ. Math. IHES 100 (2004), 209–269.

[33] M. Kapranov, E. Vasserot, Formal loops IV: Chiral differential operators, arXiv math/0612371.

[34] T. Kawai, Y. Yamada, and S.-K. Yang, Elliptic genera and $N = 2$ superconformal field theory, Nuclear Phys. B 414 (1994), no. 1–2, 191–212.

[35] Y. Kawamata, K. Matsuda, K. Matsuki, Introduction to the minimal model problem. In Algebraic geometry (Sendai, 1985), 283–360, Adv. Stud. Pure Math. **10**, North-Holland, Amsterdam, 1987.

[36] I. Krichever, Generalized elliptic genera and Baker–Akhiezer functions, Math. Notes, 47 (1990), no. 2, 132–142.

[37] Elliptic curves and modular forms in algebraic topology, edited by P. S. Landweber. Springer, Berlin, 1988.

[38] J. Li, K. F. Liu and J. Zhou. Topological string partition functions and equivariant indices, Asian J. Math. 10 (2006), no. 1, 81–114.

[39] A. Libgober, M. Szczesny, Discrete torsion, orbifold elliptic genera and the chiral de Rham complex. Pure Appl. Math. Q. 2 (2006), no. 4:2, 1217–1236.

[40] E. Looijenga, Motivic measures. In Séminaire Bourbaki, 1999/2000. Astérisque 276 (2002), 267–297.

[41] F. Malikov, V. Schechtman, A. Vaintrob, Chiral de Rham complex. Comm. Math. Phys. 204 (1999), no. 2, 439–473.

[42] F. Malikov, V. Gorbounov. Vertex algebras and the Landau–Ginzburg/Calabi–Yau correspondence. Mosc. Math. J. 4 (2004), no. 3, 729–779.

[43] S. Ochanine, Sur les genres multiplicatifs definis par des intégrales elliptiques. Topology 26 (1987), no. 2, 143–151.

[44] J. Rosenberg, An analogue of the Novikov conjecture in complex algebraic geometry. Trans. Amer. Math. Soc. 360 (2008), no. 1, 383–394.

[45] J. Tannery, J. Molk, Elements de la théorie des fonctions elliptiques. Chelsea, Bronx, NY, 1972.

[46] C. B. Thomas, Elliptic cohomology. Kluwer/Plenum, New York, 1999.

[47] B. Totaro, Chern numbers for singular varieties and elliptic homology. Ann. Math. (2) 151 (2000), no. 2, 757–791.

[48] B. Totaro. The elliptic genus of a singular variety. In Elliptic cohomology, 360–364, London Math. Soc. Lecture Note Ser. 342, Cambridge Univ. Press, Cambridge, 2007.

[49] R. Waelder, Equivariant elliptic genera, Pacific Math. J. 235 (2008), no. 2, 345–277.

[50] R. Waelder, Singular McKay correspondence for normal surfaces. arXiv 0810.3634

[51] A. Weil, Elliptic functions according to Eisenstein and Kronecker. Springer, Berlin, 1999.

[52] C. L. Wang, K-equivalence in birational geometry and characterizations of complex elliptic genera. J. Algebraic Geom. 12 (2003), no. 2, 285–306.

[53] E. Witten, The index of the Dirac operator in loop space. In Elliptic curves and modular forms in algebraic topology (Princeton, 1986), 161–181 edited by P. S. Landweber, Lecture Notes in Math. 1326, Springer, New York, 1988.

[54] E. Witten, Mirror manifolds and topological field theory. In Essays on mirror manifolds. edited by Shing-Tung Yau, 120–158. International Press, Hong Kong, 1992.

[55] D. Zagier, Note on the Landweber-Stong elliptic genus. In Elliptic curves and modular forms in algebraic topology (Princeton, 1986), edited by P. S. Landweber, 216–224, Lecture Notes in Math. **1326**, Springer, New York, 1988.

[56] D. Zagier, Periods of modular forms and Jacobi theta functions. Invent. Math. 104 (1991), no. 3, 449–465.

[57] D. Zagier, Elliptic modular forms and their applications. In The 1-2-3 of modular forms, edited by Kristian Ranestad, 1–103, Universitext, Springer, Berlin, 2008.

ANATOLY LIBGOBER
DEPARTMENT OF MATHEMATICS
UNIVERSITY OF ILLINOIS
CHICAGO, IL 60607
  libgober@math.uic.edu

Topology of Stratified Spaces
MSRI Publications
Volume **58**, 2011

The weight filtration for real algebraic varieties

CLINT MCCRORY AND ADAM PARUSIŃSKI

ABSTRACT. Using the work of Guillén and Navarro Aznar we associate to each real algebraic variety a filtered chain complex, the weight complex, which is well-defined up to a filtered quasi-isomorphism, and induces on Borel–Moore homology with $\mathbb{Z}_2$ coefficients an analog of the weight filtration for complex algebraic varieties.

The weight complex can be represented by a geometrically defined filtration on the complex of semialgebraic chains. To show this we define the weight complex for Nash manifolds and, more generally, for arc-symmetric sets, and we adapt to Nash manifolds the theorem of Mikhalkin that two compact connected smooth manifolds of the same dimension can be connected by a sequence of smooth blowups and blowdowns.

The weight complex is acyclic for smooth blowups and additive for closed inclusions. As a corollary we obtain a new construction of the virtual Betti numbers, which are additive invariants of real algebraic varieties, and we show their invariance by a large class of mappings that includes regular homeomorphisms and Nash diffeomorphisms.

The weight filtration of the homology of a real variety was introduced by Totaro [37]. He used the work of Guillén and Navarro Aznar [15] to show the existence of such a filtration, by analogy with Deligne's weight filtration for complex varieties [10] as generalized by Gillet and Soulé [14]. There is also earlier unpublished work on the real weight filtration by M. Wodzicki, and more recent unpublished work on weight filtrations by Guillén and Navarro Aznar [16].

Totaro's weight filtration for a compact variety is associated to the spectral sequence of a cubical hyperresolution. (For an introduction to cubical hyperresolutions of complex varieties see [34], Chapter 5.) For complex varieties

---

Mathematics Subject Classification: Primary: 14P25. Secondary: 14P10, 14P20.

Research partially supported by a grant from Mathématiques en Pays de la Loire (MATPYL)..

this spectral sequence collapses with rational coefficients, but for real varieties, where it is defined with $\mathbb{Z}_2$ coefficients, the spectral sequence does not collapse in general. We show, again using the work of Guillén and Navarro Aznar, that the weight spectral sequence is itself a natural invariant of a real variety. There is a functor that assigns to each real algebraic variety a filtered chain complex, the *weight complex*, that is unique up to filtered quasi-isomorphism, and functorial for proper regular morphisms. The weight spectral sequence is the spectral sequence associated to this filtered complex, and the weight filtration is the corresponding filtration of Borel–Moore homology with coefficients in $\mathbb{Z}_2$.

Using the theory of Nash constructible functions we give an independent construction of a functorial filtration on the complex of semialgebraic chains in Kurdyka's category of arc-symmetric sets [19; 21], and we show that the filtered complex obtained in this way represents the weight complex of a real algebraic variety. We obtain in particular that the weight complex is invariant under regular rational homeomorphisms of real algebraic sets in the sense of Bochnak, Coste and Roy [5].

The characteristic properties of the weight complex describe how it behaves with respect to generalized blowups (acyclicity) and inclusions of open subvarieties (additivity). The initial term of the weight spectral sequence yields additive invariants for real algebraic varieties, the virtual Betti numbers [24]. Thus we obtain that the virtual Betti numbers are invariants of regular homeomorphisms of real algebraic sets. For real toric varieties, the weight spectral sequence is isomorphic to the toric spectral sequence introduced by Bihan, Franz, McCrory, and van Hamel [4].

In Section 1 we prove the existence and uniqueness of the filtered weight complex of a real algebraic variety. The weight complex is the unique acyclic additive extension to all varieties of the functor that assigns to a nonsingular projective variety the complex of semialgebraic chains with the canonical filtration. To apply the extension theorems of Guillén and Navarro Aznar [15], we work in the category of schemes over $\mathbb{R}$, for which one has resolution of singularities, the Chow–Hironaka Lemma (see [15, (2.1.3)]), and the compactification theorem of Nagata [28]. We obtain the weight complex as a functor of schemes and proper regular morphisms.

In Section 2 we characterize the weight filtration of the semialgebraic chain complex using resolution of singularities. In Section 3 we introduce the Nash constructible filtration of semialgebraic chains, following Pennaneac'h [32], and we show that it gives the weight filtration. A key tool is Mikhalkin's theorem [26] that any two connected closed C^∞ manifolds of the same dimension can be connected by a sequence of blowups and blowdowns. Section 4 we present several applications to real geometry.

In Section 5 we show that for a real toric variety the Nash constructible filtration is the same as the filtration on cellular chains defined by Bihan *et al*. using toric topology.

1. The homological weight filtration

We begin with a brief discussion of the extension theorem of Guillén and Navarro Aznar. Suppose that G is a functor defined for smooth varieties over a field of characteristic zero. The main theorem of [15] gives a criterion for the extension of G to a functor G' defined for all (possibly singular) varieties. This criterion is a relation between the value of G on a smooth variety X and the value of G on the blowup of X along a smooth center. The extension G' satisfies a generalization of this blowup formula for any morphism $f : \tilde{X} \to X$ of varieties that is an isomorphism over the complement of a subvariety Y of X. If one requires an even stronger additivity formula for $G'(X)$ in terms of $G'(Y)$ and $G'(X \setminus Y)$, then one can assume that the original functor G is defined only for smooth projective varieties.

The structure of the target category of the functor G is important in this theory. The prototype is the derived category of chain complexes in an abelian category. That is, the objects are chain complexes, and the set of morphisms between two complexes is expanded to include the inverses of quasi-isomorphisms (morphisms that induce isomorphisms on homology). Guillén and Navarro introduce a generalization of the category of chain complexes called a *descent category*, which has a class of morphisms E that are analogous to quasi-isomorphisms, and a functor **s** from diagrams to objects that is analogous to the total complex of a diagram of chain complexes.

In our application we consider varieties over the field of real numbers, and the target category is the derived category of filtered chain complexes of vector spaces over $\mathbb{Z}_2$. Since this category is closely related to the classical category of chain complexes, it is not hard to check that it is a descent category. Our starting functor G is rather simple: It assigns to a smooth projective variety the complex of semialgebraic chains with the canonical filtration. The blowup formula follows from a short exact sequence (1-3) for the homology groups of a blowup.

Now we turn to a precise statement and proof of Theorem 1.1, which is our main result.

By a *real algebraic variety* we mean a reduced separated scheme of finite type over $\mathbb{R}$. By a *compact* variety we mean a scheme that is complete (proper over $\mathbb{R}$). We adopt the following notation of Guillén and Navarro Aznar [15]. Let $\mathbf{Sch}_c(\mathbb{R})$ be the category of real algebraic varieties and proper regular morphisms, *i. e.* proper morphisms of schemes. By Reg we denote the subcategory of compact nonsingular varieties, and by $\mathbf{V}(\mathbb{R})$ the category of projective nonsingular varieties. A proper morphism or a compactification of varieties will always be understood in the scheme-theoretic sense.

In this paper we are interested in the topology of the set of real points of a real algebraic variety X. Let $\underline{X}$ denote the set of real points of X. The set $\underline{X}$, with its sheaf of regular functions, is a real algebraic variety in the sense of Bochnak, Coste and Roy [5]. For a variety X we denote by $C_*(X)$ the complex of semialgebraic chains of $\underline{X}$ with coefficients in $\mathbb{Z}_2$ and closed supports. The homology of $C_*(X)$ is the Borel–Moore homology of $\underline{X}$ with $\mathbb{Z}_2$ coefficients, and will be denoted by $H_*(X)$.

1A. Filtered complexes. Let $\mathcal{C}$ be the category of bounded complexes of $\mathbb{Z}_2$ vector spaces with increasing bounded filtration,

$$K_* = \cdots \leftarrow K_0 \leftarrow K_1 \leftarrow K_2 \leftarrow \cdots, \quad \cdots \subset F_{p-1} K_* \subset F_p K_* \subset F_{p+1} K_* \subset \cdots.$$

Such a filtered complex defines a spectral sequence $\{E^r, d^r\}$, $r = 1, 2, \ldots$, with

$$E_{p,q}^0 = \frac{F_p K_{p+q}}{F_{p-1} K_{p+q}}, \qquad E_{p,q}^1 = H_{p+q}\left(\frac{F_p K_*}{F_{p-1} K_*}\right),$$

that converges to the homology of K_*,

$$E_{p,q}^\infty = \frac{F_p(H_{p+q} K_*)}{F_{p-1}(H_{p+q} K_*)},$$

where $F_p(H_n K_*) = \text{Image}[H_n(F_p K_*) \to H_n(K_*)]$; see [22, Thm. 3.1]. A *quasi-isomorphism* in $\mathcal{C}$ is a filtered quasi-isomorphism, that is, a morphism of filtered complexes that induces an isomorphism on E^1. Thus a quasi-isomorphism induces an isomorphism of the associated spectral sequences.

Following (1.5.1) in [15], we denote by $\text{Ho}\,\mathcal{C}$ the category $\mathcal{C}$ localized with respect to filtered quasi-isomorphisms.

Every bounded complex K_* has a *canonical filtration* [8] given by

$$F_p^{\text{can}} K_* = \begin{cases} K_q & \text{if } q > -p, \\ \ker \partial_q & \text{if } q = -p, \\ 0 & \text{if } q < -p. \end{cases}$$

We have

$$E_{p,q}^1 = H_{p+q}\left(\frac{F_p^{\text{can}} K_*}{F_{p-1}^{\text{can}} K_*}\right) = \begin{cases} H_{p+q}(K_*) & \text{if } p + q = -p, \\ 0 & \text{otherwise.} \end{cases} \tag{1-1}$$

Thus a quasi-isomorphism of complexes induces a filtered quasi-isomorphism of complexes with canonical filtration.

To certain types of diagrams in $\mathcal{C}$ we can associate an element of $\mathcal{C}$, the *simple filtered complex* of the given diagram. We use notation from [15]. For $n \geq 0$ let

$\square_n^+$ be the partially ordered set of subsets of $\{0, 1, \ldots, n\}$. A *cubical diagram* of type $\square_n^+$ in a category $\mathcal{X}$ is a contravariant functor from $\square_n^+$ to $\mathcal{X}$. If $\mathcal{K}$ is a cubical diagram in $\mathcal{C}$ of type $\square_n^+$, let $K_{*,S}$ be the complex labeled by the subset $S \subset \{0, 1, \ldots n\}$, and let $|S|$ denote the number of elements of S. The simple complex $\mathbf{s}\,\mathcal{K}$ is defined by

$$\mathbf{s}\,\mathcal{K}_k = \bigoplus_{i+|S|-1=k} \mathcal{K}_{i,S}$$

with differentials $\partial : \mathbf{s}\,\mathcal{K}_k \to \mathbf{s}\,\mathcal{K}_{k-1}$ defined as follows. For each S let ∂' : $K_{i,S} \to K_{i-1,S}$ be the differential of $K_{*,S}$. If $T \subset S$ and $|T| = |S| - 1$, let $\partial_{T,S} : K_{*,S} \to K_{*,T}$ be the chain map corresponding to the inclusion of T in S. If $a \in K_{i,S}$, let

$$\partial''(a) = \sum \partial_{T,S}(a),$$

where the sum is over all $T \subset S$ such that $|T| = |S| - 1$, and

$$\partial(a) = \partial'(a) + \partial''(a).$$

The filtration of $\mathbf{s}\,\mathcal{K}$ is given by $F_p\,\mathbf{s}\,\mathcal{K} = \mathbf{s}\,F_p\mathcal{K}$,

$$(F_p\,\mathbf{s}\,\mathcal{K})_k = \bigoplus_{i+|S|-1=k} F_p(\mathcal{K}_{i,S}).$$

The simple complex functor $\mathbf{s}$ is defined for cubical diagrams in the category $\mathcal{C}$, but not for diagrams in the derived category $\mathrm{Ho}\,\mathcal{C}$, since a diagram in $\mathrm{Ho}\,\mathcal{C}$ does not necessarily correspond to a diagram in $\mathcal{C}$. However, for each $n \geq 0$, the functor $\mathbf{s}$ is defined on the derived category of cubical diagrams of type $\square_n^+$. (A quasi-isomorphism in the category of cubical diagrams of type $\square_n^+$ is a morphism of diagrams that is a quasi-isomorphism on each object in the diagram.)

To address this technical problem, Guillén and Navarro Aznar introduce the Φ-*rectification* of a functor with values in a derived category [15, (1.6)], where Φ is the category of finite orderable diagrams [15, (1.1.2)]. A (Φ-)rectification of a functor G with values in a derived category $\mathrm{Ho}\,\mathcal{C}$ is an extension of G to a functor of diagrams, with values in the derived category of diagrams, satisfying certain naturality properties [15, (1.6.5)]. A factorization of G through the category $\mathcal{C}$ determines a canonical rectification of G. One says that G is *rectified* if a rectification of G is given.

1B. The weight complex. To state the next theorem, we only need to consider diagrams in of type $\square_0^+$ or type $\square_1^+$. The inclusion of a closed subvariety $Y \subset X$ is a $\square_0^+$-diagram in $\mathbf{Sch}_c(\mathbb{R})$. An *acyclic square* ([15], (2.1.1)) is a $\square_1^+$-diagram in $\mathbf{Sch}_c(\mathbb{R})$,

$$
\begin{array}{ccc}
\tilde{Y} & \longrightarrow & \tilde{X} \\
\downarrow & & \downarrow{\scriptstyle \pi} \\
Y & \xrightarrow{\ i\ } & X
\end{array}
\tag{1-2}
$$

where i is the inclusion of a closed subvariety, $\tilde{Y} = \pi^{-1}(Y)$, and the restriction of π is an isomorphism $\tilde{X} \setminus \tilde{Y} \to X \setminus Y$. An *elementary acyclic square* is an acyclic square such that X is compact and nonsingular, Y is nonsingular, and π is the blowup of X along Y.

For a real algebraic variety X, let $F^{\mathrm{can}}C_*(X)$ denote the complex $C_*(X)$ of semialgebraic chains with the canonical filtration.

THEOREM 1.1. *The functor*

$$
F^{\mathrm{can}}C_* : \mathbf{V}(\mathbb{R}) \to \mathrm{Ho}\,\mathcal{C}
$$

that associates to a nonsingular projective variety M the semialgebraic chain complex of M with canonical filtration admits an extension to a functor defined for all real algebraic varieties and proper regular morphisms,

$$
\mathcal{W}C_* : \mathbf{Sch}_c(\mathbb{R}) \to \mathrm{Ho}\,\mathcal{C},
$$

such that $\mathcal{W}C_*$ is rectified and has the following properties:

(i) *Acyclicity. For an acyclic square (1-2) the simple filtered complex of the diagram*

$$
\begin{array}{ccc}
\mathcal{W}C_*(\tilde{Y}) & \longrightarrow & \mathcal{W}C_*(\tilde{X}) \\
\downarrow & & \downarrow \\
\mathcal{W}C_*(Y) & \longrightarrow & \mathcal{W}C_*(X)
\end{array}
$$

is acyclic (quasi-isomorphic to the zero complex).

(ii) *Additivity. For a closed inclusion $Y \subset X$, the simple filtered complex of the diagram*

$$
\mathcal{W}C_*(Y) \to \mathcal{W}C_*(X)
$$

is naturally quasi-isomorphic to $\mathcal{W}C_*(X \setminus Y)$.

Such a functor $\mathcal{W}C_*$ is unique up to a unique quasi-isomorphism.

PROOF. This theorem follows from [15], Theorem $(2.2.2)^{op}$. By Proposition $(1.7.5)^{op}$ of [15], the category $\mathcal{C}$, with the class of quasi-isomorphisms and the operation of simple complex s defined above, is a category of homological descent. Since it factors through $\mathcal{C}$, the functor $F^{can}C_*$ is Φ-rectified ([15], (1.6.5), (1.1.2)). Clearly $F^{can}C_*$ is additive for disjoint unions (condition (F1) of [15]). It remains to check condition (F2) for $F^{can}C_*$, that the simple filtered complex associated to an elementary acyclic square is acyclic.

Consider the elementary acyclic square (1-2). Let $\mathcal{K}$ be the simple complex associated to the $\square_1^+$-diagram

$$
\begin{array}{ccc}
F^{can}C_*(\tilde{Y}) & \longrightarrow & F^{can}C_*(\tilde{X}) \\
\downarrow & & \downarrow \\
F^{can}C_*(Y) & \longrightarrow & F^{can}C_*(X)
\end{array}
$$

By definition of the canonical filtration, for each p we have

$$(F_p \, s \, \mathcal{K})_k / (F_{p-1} \, s \, \mathcal{K})_k \neq 0 \quad \text{only for} \quad -p+2 \geq k \geq -p-1,$$

and the complex $(E_{p,*}^0, d^0)$ has the form

$$0 \to \frac{(F_p \, s \, \mathcal{K})_{-p+2}}{(F_{p-1} \, s \, \mathcal{K})_{-p+2}} \to \frac{(F_p \, s \, \mathcal{K})_{-p+1}}{(F_{p-1} \, s \, \mathcal{K})_{-p+1}}$$

$$\to \frac{(F_p \, s \, \mathcal{K})_{-p}}{(F_{p-1} \, s \, \mathcal{K})_{-p}} \to \frac{(F_p \, s \, \mathcal{K})_{-p-1}}{(F_{p-1} \, s \, \mathcal{K})_{-p-1}} \to 0.$$

A computation gives

$$H_{-p+2}(E_{p,*}^0) = 0,$$

$$H_{-p+1}(E_{p,*}^0) = \text{Ker}[H_{-p}(\tilde{Y}) \to H_{-p}(Y) \oplus H_{-p}(\tilde{X})],$$

$$H_{-p}(E_{p,*}^0) =$$

$$\text{Ker}[H_{-p}(Y) \oplus H_{-p}(\tilde{X}) \to H_{-p}(X)]/\text{Im}[H_{-p}(\tilde{Y}) \to H_{-p}(Y) \oplus H_{-p}(\tilde{X})],$$

$$H_{-p-1}(E_{p,*}^0) = H_{-p}(X)/\text{Im}[H_{-p}(Y) \oplus H_{-p}(\tilde{X}) \to H_{-p}(X)].$$

These groups are zero because for all k we have the short exact sequence of an elementary acyclic square,

$$0 \to H_k(\tilde{Y}) \to H_k(Y) \oplus H_k(\tilde{X}) \to H_k(X) \to 0; \qquad (1-3)$$

see [25], proof of Proposition 2.1.   □

REMARK 1.2. This above argument shows that the functor F^{can} is acyclic on any acyclic square (1-2), provided the varieties $X, Y, \tilde{X}, \tilde{Y}$ are nonsingular and compact.

REMARK 1.3. In Section 3 below, we show that the functor $\mathcal{W}C_*$ factors through the category of filtered chain complexes. This explains why $\mathcal{W}C_*$ is rectified.

If X is a real algebraic variety, the *weight complex* of X is the filtered complex $\mathcal{W}C_*(X)$. A stronger version of the uniqueness of $\mathcal{W}C_*$ is given by the following naturality theorem.

THEOREM 1.4. *Let $A_*, B_* : \mathbf{V}(\mathbb{R}) \to \mathcal{C}$ be functors whose localizations $\mathbf{V}(\mathbb{R}) \to \mathrm{Ho}\,\mathcal{C}$ satisfy the disjoint additivity condition* (F1) *and the elementary acyclicity condition* (F2) *of* [15]. *If $\tau : A_* \to B_*$ is a morphism of functors, then the localization of τ extends uniquely to a morphism $\tau' : \mathcal{W}A_* \to \mathcal{W}B_*$.*

PROOF. This follows from $(2.1.5)^{\mathrm{op}}$ and $(2.2.2)^{\mathrm{op}}$ of [15]. □

Thus if $\tau : A_*(M) \to B_*(M)$ is a quasi-isomorphism for all nonsingular projective varieties M, then $\tau' : \mathcal{W}A_*(X) \to \mathcal{W}B_*(X)$ is a quasi-isomorphism for all varieties X.

PROPOSITION 1.5. *For all real algebraic varieties X, the homology of the complex $\mathcal{W}C_*(X)$ is the Borel–Moore homology of X with $\mathbb{Z}_2$ coefficients,*

$$H_n(\mathcal{W}C_*(X)) = H_n(X).$$

PROOF. Let $\mathcal{D}$ be the category of bounded complexes of $\mathbb{Z}_2$ vector spaces. The forgetful functor $\mathcal{C} \to \mathcal{D}$ induces a functor $\varphi : \mathrm{Ho}\,\mathcal{C} \to \mathrm{Ho}\,\mathcal{D}$. To see this, let A'_*, B'_* be filtered complexes, and let $A_* = \varphi(A'_*)$ and $B_* = \varphi(B'_*)$. A quasi-isomorphism $f : A'_* \to B'_*$ induces an isomorphism of the corresponding spectral sequences, which implies that f induces an isomorphism $H_*(A_*) \to H_*(B_*)$; in other words $f : A_* \to B_*$ is a quasi-isomorphism.

Let $C_* : \mathbf{Sch}_c(\mathbb{R}) \to \mathrm{Ho}\,\mathcal{D}$ be the functor that assigns to every real algebraic variety X the complex of semialgebraic chains $C_*(X)$. Then C_* satisfies properties (1) and (2) of Theorem 1.1. Acyclicity of C_* for an acyclic square (1-2) follows from the short exact sequence of chain complexes

$$0 \to C_*(\tilde{Y}) \to C_*(Y) \oplus C_*(\tilde{X}) \to C_*(X) \to 0.$$

The exactness of this sequence follows immediately from the definition of semialgebraic chains. Similarly, additivity of C_* for a closed embedding $Y \to X$ follows from the short exact sequence of chain complexes

$$0 \to C_*(Y) \to C_*(X) \to C_*(X \setminus Y) \to 0.$$

Now consider the functor $\mathcal{W}C_* : \mathbf{Sch}_c(\mathbb{R}) \to \mathrm{Ho}\,\mathcal{C}$ given by Theorem 1.1. The functors $\varphi \circ \mathcal{W}C_*$ and $C_* : \mathbf{Sch}_c(\mathbb{R}) \to \mathrm{Ho}\,\mathcal{D}$ are extensions of $C_* : \mathbf{V}(\mathbb{R}) \to \mathrm{Ho}\,\mathcal{D}$, so by [15] Theorem $(2.2.2)^{\mathrm{op}}$ we have that $\varphi(\mathcal{W}C_*(X))$ is quasi-isomorphic to $C_*(X)$ for all X. Thus $H_*(\mathcal{W}C_*(X)) = H_*(X)$, as desired. □

1C. The weight spectral sequence. If X is a real algebraic variety, the *weight spectral sequence* of X, $\{E^r, d^r\}$, $r = 1, 2, \ldots$, is the spectral sequence of the weight complex $\mathcal{W}C_*(X)$. It is well-defined by Theorem 1.1, and it converges to the homology of X by Proposition 1.5. The associated filtration of the homology of X is the *weight filtration*:

$$0 = \mathcal{W}_{-k-1}H_k(X) \subset \mathcal{W}_{-k}H_k(X) \subset \cdots \subset \mathcal{W}_0 H_k(X) = H_k(X),$$

where $H_k(X)$ is the homology with closed supports (Borel–Moore homology) with coefficients in $\mathbb{Z}_2$. (We show that $\mathcal{W}_{-k-1}H_k(X) = 0$ in Corollary 1.10.) The dual weight filtration on cohomology with compact supports is discussed in [25].

REMARK 1.6. We do not know the relation of the weight filtration of a real algebraic variety X to Deligne's weight filtration [10] on $H_*(X_{\mathbb{C}}; \mathbb{Q})$, the Borel–Moore homology with rational coefficients of the complex points $X_{\mathbb{C}}$. By analogy with Deligne's weight filtration, there should also be a weight filtration on the homology of X with classical compact supports and coefficients in $\mathbb{Z}_2$ (dual to cohomology with closed supports). We plan to study this filtration in subsequent work.

The weight spectral sequence $E^r_{p,q}$ is a second quadrant spectral sequence. (We will show in Corollary 1.10 that if $E^1_{p,q} \neq 0$ then (p, q) lies in the closed triangle with vertices $(0, 0)$, $(0, d)$, $(-d, 2d)$, where $d = \dim X$.) The reindexing

$$p' = 2p + q, \quad q' = -p, \quad r' = r + 1$$

gives a standard first quadrant spectral sequence, with

$$\tilde{E}^2_{p',q'} = E^1_{-q', p'+2q'}.$$

(If $\tilde{E}^2_{p',q'} \neq 0$ then (p', q') lies in the closed triangle with vertices $(0, 0)$, $(d, 0)$, $(0, d)$, where $d = \dim X$.) Note that the total grading is preserved: $p' + q' = p + q$.

The virtual Betti numbers [25] are the Euler characteristics of the rows of $\tilde{E}^2$, that is,

$$\beta_q(X) = \sum_p (-1)^p \dim_{\mathbb{Z}_2} \tilde{E}^2_{p,q}. \tag{1-4}$$

To prove this assertion we will show that the numbers $\beta_q(X)$ defined by (1-4) are additive and equal to the classical Betti numbers for X compact and nonsingular.

For each $q \geq 0$ consider the chain complex defined by the q-th row of the $\tilde{E}^1$ term,

$$C_*(X, q) = (\tilde{E}^1_{*,q}, \tilde{d}^1_{*,q}),$$

where $\tilde{d}^1_{p,q} : \tilde{E}^1_{p,q} \to \tilde{E}^1_{p-1,q}$. This chain complex is well-defined up to quasi-isomorphism, and its Euler characteristic is $\beta_q(X)$.

The additivity of $\mathcal{W}C_*$ implies that if Y is a closed subvariety of X then the chain complex $C_*(X \setminus Y, q)$ is quasi-isomorphic to the mapping cone of the chain map $C_*(Y, q) \to C_*(X, q)$, and hence there is a long exact sequence of homology groups

$$\cdots \to \tilde{E}^2_{p,q}(Y) \to \tilde{E}^2_{p,q}(X) \to \tilde{E}^2_{p,q}(X \setminus Y) \to \tilde{E}^2_{p-1,q}(Y) \cdots.$$

Therefore for each q we have

$$\beta_q(X) = \beta_q(X \setminus Y) + \beta_q(Y).$$

This is the additivity property of the virtual Betti numbers.

REMARK 1.7. Navarro Aznar pointed out to us that $C_*(X, q)$ is actually well-defined up to chain homotopy equivalence. One merely applies [15], Theorem $(2.2.2)^{\mathrm{op}}$, to the functor that assigns to a nonsingular projective variety M the chain complex

$$C_k(M, q) = \begin{cases} H_q(M) & \text{if } k = 0, \\ 0 & \text{if } k \neq 0, \end{cases}$$

in the category of bounded complexes of $\mathbb{Z}_2$ vector spaces localized with respect to chain homotopy equivalences. This striking application of the theorem of Guillén and Navarro Aznar led to our proof of the existence of the weight complex.

We say the weight complex is *pure* if the reindexed weight spectral sequence has $\tilde{E}^2_{p,q} = 0$ for $p \neq 0$. In this case the numbers $\beta_q(X)$ equal the classical Betti numbers of X.

PROPOSITION 1.8. *If X is a compact nonsingular variety, the weight complex $\mathcal{W}C_*(X)$ is pure. In other words, if $k \neq -p$ then*

$$H_k\left(\frac{W_p C_*(X)}{W_{p-1} C_*(X)}\right) = 0.$$

PROOF. For X projective and nonsingular, the filtered complex $\mathcal{W}C_*(X)$ is quasi-isomorphic to $C_*(X)$ with the canonical filtration. The inclusion $\mathbf{V}(\mathbb{R}) \to$ Reg has the extension property in (2.1.10) of [15]; the proof is similar to that in (2.1.11) of the same reference. Therefore by Theorem $(2.1.5)^{\mathrm{op}}$ [15], the functor $F^{\mathrm{can}}C_* : \mathbf{V}(\mathbb{R}) \to \mathrm{Ho}\,\mathcal{C}$ extends to a functor Reg $\to \mathrm{Ho}\,\mathcal{C}$ that is additive for disjoint unions and acyclic, and this extension is unique up to quasi-isomorphism. But $F^{\mathrm{can}}C_* : \mathrm{Reg} \to \mathrm{Ho}\,\mathcal{C}$ is such an extension, since $F^{\mathrm{can}}C_*$ is additive for disjoint unions in Reg and acyclic for acyclic squares in Reg. (Compare the proof of Theorem 1.1 and Remark 1.2.) $\square$

If X is compact, we will show that the reindexed weight spectral sequence $\tilde{E}^r_{p,q}$ is isomorphic to the spectral sequence of a *cubical hyperresolution* of X [15]. (The definition of cubical hyperresolution given in Chapter 5 of [34] is too weak for our purposes; see Example 1.12 below.)

A cubical hyperresolution of X is a special type of $\square_n^+$-diagram with final object X and all other objects compact and nonsingular. Removing X gives a $\square_n$-diagram, which is the same thing as a $\triangle_n$-diagram, *i.e.* a diagram labeled by the simplices contained in the standard n-simplex $\triangle_n$. (Subsets of $\{0, 1, \ldots, n\}$ of cardinality $i + 1$ correspond to i-simplices.)

The spectral sequence of a cubical hyperresolution is the spectral sequence of the filtered complex $(C_*, \hat{F})$, with $C_k = \bigoplus_{i+j=k} C_j X^{(i)}$, where $X^{(i)}$ is the disjoint union of the objects labeled by i-simplices of $\triangle_n$, and the filtration $\hat{F}$ is by skeletons,

$$\hat{F}_p C_k = \bigoplus_{i \le p} C_{k-i} X^{(i)}$$

The resulting first quadrant spectral sequence $\hat{E}^r_{p,q}$ converges to the homology of X, and the associated filtration is the weight filtration defined by Totaro [37].

Let $\partial = \partial' + \partial''$ be the boundary operator of the complex C_*, where $\partial'_i : C_j X^{(i)} \to C_j X^{(i-1)}$ is the simplicial boundary operator, and $\partial''_j : C_j X^{(i)} \to C_{j-1} X^{(i)}$ is $(-1)^i$ times the boundary operator on semialgebraic chains.

PROPOSITION 1.9. *If X is a compact variety, the weight spectral sequence E of X is isomorphic to the spectral sequence $\hat{E}$ of a cubical hyperresolution of X:*

$$E^r_{p,q} \cong \hat{E}^{r+1}_{2p+q,-p}.$$

Thus $\hat{E}^r_{p,q} \cong \tilde{E}^r_{p,q}$, the reindexed weight spectral sequence introduced above.

PROOF. The acyclicity property of the weight complex — condition (1) of Theorem 1.1 — implies that WC_* is acyclic for cubical hyperresolutions (see [15], proof of Theorem (2.1.5)). In other words, if the functor WC_* is applied to a cubical hyperresolution of X, the resulting $\square_n^+$-diagram in $\mathcal{C}$ is acyclic. This says that $WC_*(X)$ is filtered quasi-isomorphic to the total filtered complex of the double complex $WC_{i,j} = WC_j X^{(i)}$. Since the varieties $X^{(i)}$ are compact and nonsingular, this filtered complex is quasi-isomorphic to the total complex $C_k = \bigoplus_{i+j=k} C_j X^{(i)}$ with the canonical filtration,

$$F_p^{\mathrm{can}} C_k = \mathrm{Ker}\, \partial''_{-p} \oplus \bigoplus_{j > -p} C_j X^{(k-j)}.$$

Thus the spectral sequence of this filtered complex is the weight spectral sequence $E^r_{p,q}$.

We now compare the two increasing filtrations F^{can} and $\hat{F}$ on the complex C_*. The weight spectral sequence E is associated to the filtration F^{can}, and the cubical hyperresolution spectral sequence $\hat{E}$ is associated to the filtration $\hat{F}$. We show that $F^{\mathrm{can}} = \mathrm{Dec}\,(\hat{F})$, the *Deligne shift* of $\hat{F}$; for this notion see [8, (1.3.3)] or [34, A.49].

Let $\hat{F}'$ be the filtration

$$\hat{F}'_p C_k = \hat{Z}^1_{p,k-p} = \mathrm{Ker}[\partial : \hat{F}_p C_k \to C_{k-1}/\hat{F}_{p-1} C_{k-1}]$$

and $\hat{E}'$ the associated spectral sequence. By definition of the Deligne shift,

$$\hat{F}'_p C_k = \mathrm{Dec}\,\hat{F}_{p-k} C_k.$$

Now since $\partial = \partial' + \partial''$ it follows that

$$\hat{F}'_p C_k = F^{\mathrm{can}}_{p-k} C_k,$$

and $F^{\mathrm{can}}_{p-k} C_k = F^{\mathrm{can}}_{-q} C_k$, where $p + q = k$. Thus we can identify the spectral sequences

$$(\hat{E}')^{r+1}_{p,q} = E^r_{-q,p+2q} \quad \text{for } r \geq 1.$$

On the other hand, the inclusion $\hat{F}'_p C_k \to \hat{F}_p C_k$ induces an isomorphism of spectral sequences

$$(\hat{E}')^r_{p,q} \cong \hat{E}^r_{p,q} \quad \text{for } r \geq 2. \qquad \square$$

COROLLARY 1.10. *Let X be a real algebraic variety of dimension d, with weight spectral sequence E and weight filtration $\mathcal{W}$. For all p, q, r, if $E^r_{p,q} \neq 0$ then $p \leq 0$ and $-2p \leq q \leq d - p$. Thus for all k we have $\mathcal{W}_{-k-1} H_k(X) = 0$.*

PROOF. For X compact this follows from Proposition 1.9 and the fact that $\hat{E}^r_{p,q} \neq 0$ implies $p \geq 0$ and $0 \leq q \leq d - p$. If U is a noncompact variety, let X be a real algebraic compactification of U, and let $Y = X \setminus U$. We can assume that $\dim Y < d$. The corollary now follows from the additivity property of the weight complex (condition (2) of Theorem 1.1). $\qquad \square$

EXAMPLE 1.11. If X is a compact divisor with normal crossings in a nonsingular variety, a cubical hyperresolution of X is given by the decomposition of X into irreducible components. (The corresponding simplicial diagram associates to an i-simplex the disjoint union of the intersections of $i + 1$ distinct irreducible components of X.) The spectral sequence of such a cubical hyperresolution is the Mayer–Vietoris (or Čech) spectral sequence associated to the decomposition. Example 3.3 of [25] is an algebraic surface X in affine 3-space such that X is the union of three compact nonsingular surfaces with normal crossings and the weight spectral sequence of X does not collapse: $\tilde{E}^2 \neq \tilde{E}^\infty$. The variety $U = \mathbb{R}^3 \setminus X$ is an example of a nonsingular noncompact variety with noncollapsing weight spectral sequence. (The additivity property (2) of Theorem 1.1 can be used to compute the spectral sequence of U.)

EXAMPLE 1.12. For a compact complex variety the Deligne weight filtration can be computed from the skeletal filtration of a simplicial smooth resolution of *cohomological descent* (see [9, (5.3)] or [34, (5.1.3)]). In particular, a rational homology class α has maximal weight if and only if α is in the image of the homology of the zero-skeleton of the resolution.

The following example shows that for real varieties the cohomological descent condition on a resolution is too weak to recover the weight filtration.

We construct a simplicial smooth variety $X_\bullet \to X$ of cohomological descent such that X is compact and the weight filtration of X does not correspond to the skeletal filtration of $X_\bullet$. Let $X = X_0 = S^1$, the unit circle in the complex plane, and let $f : X_0 \to X$ be the double cover $f(z) = z^2$. Let $X_\bullet$ be the Gabrielov–Vorobjov–Zell resolution associated to the map f [13] . Thus

$$X_n = X_0 \times_X X_0 \times_X \cdots (n+1) \cdots \times_X X_0,$$

a compact smooth variety of dimension 1. This resolution is of cohomological descent since the fibers of the geometric realization $|X_\bullet| \to X$ are contractible (see [13] or [34, (5.1.3)]).

Let $\alpha \in H_1(X)$ be the nonzero element ($\mathbb{Z}_2$ coefficients). Now $\alpha \in \mathcal{W}_{-1} H(X)$ since X is compact and nonsingular. Therefore, for every cubical hyperresolution of X, α lies in the image of the homology of the zero-skeleton (*i.e.*, the filtration of α with respect to the spectral sequence $\hat{E}$ is 0). But the filtration of α with respect to the skeletons of the resolution $X_\bullet \to X$ is greater than 0 since $\alpha \notin \mathrm{Im}[f_* : H_1(X_0) \to H_1(X)]$. In fact α has filtration 1 with respect to the skeletons of this resolution.

2. A geometric filtration

We define a functor

$$\mathcal{G}C_* : \mathbf{Sch}_c(\mathbb{R}) \to \mathcal{C}$$

that assigns to each real algebraic variety X the complex $C_*(X)$ of semialgebraic chains of X (with coefficients in $\mathbb{Z}_2$ and closed supports), together with a filtration

$$0 = \mathcal{G}_{-k-1}C_k(X) \subset \mathcal{G}_{-k}C_k(X) \subset \mathcal{G}_{-k+1}C_k(X)$$
$$\subset \cdots \subset \mathcal{G}_0 C_k(X) = C_k(X). \quad (2\text{-}1)$$

We prove in Theorem 2.8 that the functor $\mathcal{G}C_*$ realizes the weight complex functor $\mathcal{W}C_* : \mathbf{Sch}_c(\mathbb{R}) \to \mathrm{Ho}\,\mathcal{C}$ given by Theorem 1.1. Thus the filtration $\mathcal{G}_*$ of chains gives the weight filtration of homology.

2A. Definition of the filtration $\mathcal{G}_*$. The filtration will first be defined for compact varieties. Recall that $\underline{X}$ denotes the set of real points of the real algebraic variety X.

THEOREM 2.1. *There exists a unique filtration* (2-1) *on semialgebraic $\mathbb{Z}_2$-chains of compact real algebraic varieties with the following properties. Let X be a compact real algebraic variety and let $c \in C_k(X)$. Then*

(1) *If $Y \subset X$ is an algebraic subvariety such that $\operatorname{Supp} c \subset \underline{Y}$, then*

$$c \in \mathcal{G}_p C_k(X) \iff c \in \mathcal{G}_p C_k(Y).$$

(2) *Let $\dim X = k$ and let $\pi : \tilde{X} \to X$ be a resolution of X such that there is a normal crossing divisor $D \subset \tilde{X}$ with $\operatorname{Supp} \partial(\pi^{-1}c) \subset \underline{D}$. Then for $p \geq -k$,*

$$c \in \mathcal{G}_p C_k(X) \iff \partial(\pi^{-1}c) \in \mathcal{G}_p C_{k-1}(D).$$

We call a resolution $\pi : \tilde{X} \to X$ *adapted* to $c \in C_k(X)$ if it satisfies condition (2) above. For the definition of the support $\operatorname{Supp} c$ and the pullback $\pi^{-1}c$ see the Appendix.

PROOF. We proceed by induction on k. If $k = 0$ then $0 = \mathcal{G}_{-1}C_0(X) \subset \mathcal{G}_0 C_0(X) = C_0(X)$. In the rest of this subsection we assume the existence and uniqueness of the filtration for chains of dimension $< k$, and we prove the statement for chains of dimension k.

LEMMA 2.2. *Let $X = \bigcup_{i=1}^{s} X_i$ where X_i are subvarieties of X. Then for $m < k$,*

$$c \in \mathcal{G}_p C_m(X) \iff c|_{X_i} \in \mathcal{G}_p C_m(X_i) \text{ for all } i.$$

PROOF. By (1) we may assume that $\dim X = m$ and then that all X_i are distinct of dimension m. Thus an adapted resolution of X is a collection of adapted resolutions of each component of X. ☐

See the Appendix for the definition of the restriction $c|_{X_i}$.

PROPOSITION 2.3. *The filtration $\mathcal{G}_p$ given by Theorem 2.1 is functorial; that is, for a regular morphism $f : X \to Y$ of compact real algebraic varieties, $f_*(\mathcal{G}_p C_m(X)) \subset \mathcal{G}_p C_m(Y)$, for $m < k$.*

PROOF. We prove that if the filtration satisfies the statement of Theorem 2.1 for chains of dimension $< k$ and is functorial on chains of dimension $< k - 1$ then it is functorial on chains of dimension $k - 1$.

Let $c \in C_{k-1}(X)$, and let $f : X \to Y$ be a regular morphism of compact real algebraic varieties. By (1) of Theorem 2.1 we may assume $\dim X = \dim Y =$

$k - 1$ and by Lemma 2.2 that X and Y are irreducible. We may assume that f is dominant; otherwise $f_*c = 0$. Then there exists a commutative diagram

$$
\begin{array}{ccc}
\tilde{X} & \xrightarrow{\tilde{f}} & \tilde{Y} \\
\pi_X \downarrow & & \downarrow \pi_Y \\
X & \xrightarrow{f} & Y
\end{array}
$$

where π_X is a resolution of X adapted to c and π_Y a resolution of Y adapted to f_*c. Then

$$c \in \mathcal{G}_p(X) \iff \partial(\pi_X^{-1}c) \in \mathcal{G}_p(\tilde{X}) \implies \tilde{f}_* \partial(\pi_X^{-1}c) \in \mathcal{G}_p(\tilde{Y}),$$

$$\tilde{f}_* \partial(\pi_X^{-1}c) = \partial \tilde{f}_*(\pi_X^{-1}c) = \partial(\pi_Y^{-1} f_*c),$$

$$\partial(\pi_Y^{-1} f_*c) \in \mathcal{G}_p(\tilde{Y}) \iff f_*c \in \mathcal{G}_p(Y),$$

where the implication in the first line follows from the inductive assumption. $\square$

COROLLARY 2.4. *The boundary operator ∂ preserves the filtration $\mathcal{G}_p$:*

$$\partial \mathcal{G}_p C_m(X)) \subset \mathcal{G}_p C_{m-1}(X) \quad \text{for } m < k.$$

PROOF. Let $\pi : \tilde{X} \to X$ be a resolution of X adapted to c. Let $\tilde{c} = \pi^{-1}c$. Then $c = \pi_* \tilde{c}$ and

$$c \in \mathcal{G}_p \iff \partial \tilde{c} \in \mathcal{G}_p \implies \partial c = \partial \pi_* \tilde{c} = \pi_* \partial \tilde{c} \in \mathcal{G}_p. \qquad \square$$

Let $c \in C_k(X)$, $\dim X = k$. In order to show that condition (2) of Theorem 2.1 is independent of the choice of $\tilde{\pi}$ we need the following lemma.

LEMMA 2.5. *Let X be a nonsingular compact real algebraic variety of dimension k and let $D \subset X$ be a normal crossing divisor. Let $c \in C_k(X)$ satisfy Supp $\partial c \subset D$. Let $\pi : \tilde{X} \to X$ be the blowup of a nonsingular subvariety $C \subset X$ that has normal crossings with D. Then*

$$\partial c \in \mathcal{G}_p C_{k-1}(X) \iff \partial(\pi^{-1}(c)) \in \mathcal{G}_p C_{k-1}(\tilde{X}).$$

PROOF. Let $\tilde{D} = \pi^{-1}(D)$. Then $\tilde{D} = E \cup \bigcup \tilde{D}_i$, where $E = \pi^{-1}(C)$ is the exceptional divisor and $\tilde{D}_i$ denotes the strict transform of D_i. By Lemma 2.2,

$$\partial c \in \mathcal{G}_p C_{k-1}(X) \iff \partial c|_{D_i} \in \mathcal{G}_p C_{k-1}(D_i) \quad \text{for all } i.$$

Let $\partial_i c = \partial c|_{D_i}$. The restriction $\pi_i = \pi|_{\tilde{D}_i} : \tilde{D}_i \to D_i$ is the blowup with smooth center $C \cap D_i$. Hence, by the inductive assumption,

$$\partial(\partial_i c) \in \mathcal{G}_p C_{k-2}(D_i) \iff \partial \pi_i^{-1}(\partial_i c) = \partial\big(\partial(\pi^{-1}(c))|_{\tilde{D}_i}\big) \in \mathcal{G}_p C_{k-2}(\tilde{D}_i)$$

By the inductive assumption of Theorem 2.1,

$$\partial(\partial_i c) \in \mathcal{G}_p C_{k-2}(D_i) \iff \partial_i c \in \mathcal{G}_p C_{k-1}(D_i),$$

and we have similar properties for $\partial(\pi^{-1}(c))|_{\tilde{D}_i}$ and $\partial(\pi^{-1}(c))|_E$.

Thus, to complete the proof it suffices to show that if $\partial\big(\partial(\pi^{-1}(c))|_{\tilde{D}_i}\big)$ lies in $\mathcal{G}_p C_{k-2}(\tilde{D}_i)$ for all i, then $\partial\big(\partial(\pi^{-1}(c))|_E\big) \in \mathcal{G}_p C_{k-2}(E)$. This follows from

$$0 = \partial(\partial \pi^{-1}(c)) = \partial\Big(\sum_i \partial(\pi^{-1}(c))|_{\tilde{D}_i} + \partial(\pi^{-1}(c))|_E\Big). \qquad \square$$

Let $\pi_i : X_i \to X$, $i = 1,2$, be two resolutions of X adapted to c. Then there exists $\sigma : \tilde{X}_1 \to X_1$, the composition of finitely many blowups with smooth centers that have normal crossings with the strict transforms of all exceptional divisors, such that $\pi_1 \circ \sigma$ factors through X_2,

$$
\begin{array}{ccc}
\tilde{X}_1 & \xrightarrow{\ \sigma\ } & X_1 \\
{\scriptstyle \rho}\big\downarrow & & \big\downarrow{\scriptstyle \pi_1} \\
X_2 & \xrightarrow{\ \pi_2\ } & X
\end{array}
$$

By Lemma 2.5,

$$\partial(\pi_1^{-1}(c)) \in \mathcal{G}_p C_{k-1}(X_1) \iff \partial(\sigma^{-1}(\pi_1^{-1}(c))) \in \mathcal{G}_p C_{k-1}(\tilde{X}_1).$$

On the other hand,

$$\rho_* \partial(\sigma^{-1}(\pi_1^{-1}(c))) = \rho_* \partial(\rho^{-1}(\pi_2^{-1}(c))) = \partial(\pi_2^{-1}(c)),$$

and consequently by Proposition 2.3 we have

$$\partial(\pi_1^{-1}(c)) \in \mathcal{G}_p C_{k-1}(X_1) \implies \partial(\pi_2^{-1}(c)) \in \mathcal{G}_p C_{k-1}(X_2).$$

By symmetry, $\partial(\pi_2^{-1}(c)) \in \mathcal{G}_p(X)$ implies $\partial(\pi_1^{-1}(c)) \in \mathcal{G}_p(X)$. This completes the proof of Theorem 2.1. $\qquad \square$

2B. Properties of the filtration $\mathcal{G}_*$.

Let U be a (not necessarily compact) real algebraic variety and let X be a real algebraic compactification of U. We extend the filtration $\mathcal{G}_p$ to U as follows. If $c \in C_*(U)$, let $\bar{c} \in C_*(X)$ be its closure. We define

$$c \in \mathcal{G}_p C_k(U) \iff \bar{c} \in \mathcal{G}_p C_k(X).$$

See the Appendix for the definition of the closure of a chain.

PROPOSITION 2.6. $\mathcal{G}_p C_k(U)$ *is well-defined; that is, for two compactifications X_1 and X_2 of U, we have*

$$c_1 \in \mathcal{G}_p C_k(X_1) \iff c_2 \in \mathcal{G}_p C_k(X_2),$$

where c_i denotes the closure of c in X_i, $i = 1,2$.

PROOF. We may assume that $k = \dim U$. By a standard argument, any two compactifications can be dominated by a third one. Indeed, denote the inclusions by $i_i : U \hookrightarrow X_i$. Then the Zariski closure X of the image of (i_1, i_2) in $X_1 \times X_2$ is a compactification of U.

Thus we may assume that there is a morphism $f : X_2 \to X_1$ that is the identity on U. Then, by functoriality, $c_2 \in \mathcal{G}_p C_k(X_2)$ implies $c_1 = f_*(c_2) \in \mathcal{G}_p C_k(X_1)$. By the Chow–Hironaka lemma there is a resolution $\pi_1 : \tilde{X}_1 \to X_1$, adapted to c_1, that factors through $f : \pi_1 = f \circ g$. Then $c_1 \in \mathcal{G}_p C_k(X_1)$ is equivalent to $\pi_1^{-1}(c_1) \in \mathcal{G}_p C_k(\tilde{X}_1)$; but this implies that $c_2 = g_*(\pi_1^{-1}(c_1)) \in \mathcal{G}_p C_k(X_2)$, as needed. $\qquad\square$

THEOREM 2.7. *The filtration $\mathcal{G}_*$ defines a functor $\mathcal{G}C_* : \mathbf{Sch}_c(\mathbb{R}) \to \mathcal{C}$ with the following properties*:

(1) *For an acyclic square (1-2) the following sequences are exact*:

$$0 \to \mathcal{G}_p C_k(\tilde{Y}) \to \mathcal{G}_p C_k(Y) \oplus \mathcal{G}_p C_k(\tilde{X}) \to \mathcal{G}_p C_k(X) \to 0,$$

$$0 \to \frac{\mathcal{G}_p C_k(\tilde{Y})}{\mathcal{G}_{p-1} C_k(\tilde{Y})} \to \frac{\mathcal{G}_p C_k(Y)}{\mathcal{G}_{p-1} C_k(Y)} \oplus \frac{\mathcal{G}_p C_k(\tilde{X})}{\mathcal{G}_{p-1} C_k(\tilde{X})} \to \frac{\mathcal{G}_p C_k(X)}{\mathcal{G}_{p-1} C_k(X)} \to 0.$$

(2) *For a closed inclusion $Y \subset X$, with $U = X \setminus Y$, the following sequences are exact*:

$$0 \to \mathcal{G}_p C_k(Y) \to \mathcal{G}_p C_k(X) \to \mathcal{G}_p C_k(U) \to 0,$$

$$0 \to \frac{\mathcal{G}_p C_k(Y)}{\mathcal{G}_{p-1} C_k(Y)} \to \frac{\mathcal{G}_p C_k(X)}{\mathcal{G}_{p-1} C_k(X)} \to \frac{\mathcal{G}_p C_k(U)}{\mathcal{G}_{p-1} C_k(U)} \to 0.$$

PROOF. The exactness of the first sequence of (2) follows directly from the definitions (moreover, this sequence splits via $c \mapsto \bar{c}$). The exactness of the second sequence of (2) now follows by a diagram chase. Similarly, the exactness of the first sequence of (1) follows from the definitions, and the exactness of the second sequence of (1) is proved by a diagram chase. $\qquad\square$

For any variety X, the filtration $\mathcal{G}_*$ is contained in the canonical filtration,

$$\mathcal{G}_p C_k(X) \subset F_p^{\mathrm{can}} C_k(X), \tag{2-2}$$

since $\partial_k(\mathcal{G}_{-k} C_k(X)) = 0$. Thus on the category of nonsingular projective varieties we have a morphism of functors

$$\sigma : \mathcal{G}C_* \to F^{\mathrm{can}} C_*.$$

THEOREM 2.8. *For every nonsingular projective real algebraic variety M,*

$$\sigma(M) : \mathcal{G}C_*(M) \to F^{\mathrm{can}} C_*(M)$$

is a filtered quasi-isomorphism. Hence, for every real algebraic variety X the localization of σ induces a quasi-isomorphism $\sigma'(X) : \mathcal{G}C_*(X) \to \mathcal{W}C_*(X)$.

Theorem 2.8 follows from Corollary 3.11 and Corollary 3.12, which will be shown in the next section.

3. The Nash constructible filtration

In this section we introduce the *Nash constructible filtration*

$$0 = \mathcal{N}_{-k-1}C_k(X) \subset \mathcal{N}_{-k}C_k(X) \subset \mathcal{N}_{-k+1}C_k(X)$$
$$\subset \cdots \subset \mathcal{N}_0 C_k(X) = C_k(X) \quad (3\text{-}1)$$

on the semialgebraic chain complex $C_*(X)$ of a real algebraic variety X. We show that this filtration induces a functor

$$\mathcal{N}C_* : \mathbf{Sch}_c(\mathbb{R}) \to \mathcal{C}$$

that realizes the weight complex functor $\mathcal{W}C_* : \mathbf{Sch}_c(\mathbb{R}) \to \mathrm{Ho}\,\mathcal{C}$. In order to prove this assertion in Theorem 3.11, we have to extend $\mathcal{N}C_*$ to a wider category of sets and morphisms. The objects of this category are certain semialgebraic subsets of the set of real points of a real algebraic variety, and they include in particular all connected components of real algebraic subsets of $\mathbb{RP}^n$. The morphisms are certain proper continuous semialgebraic maps between these sets. This extension is crucial for the proof. As a corollary we show that for real algebraic varieties the Nash constructible filtration $\mathcal{N}_*$ coincides with the geometric filtration $\mathcal{G}_*$ of Section 2A. In this way we complete the proof of Theorem 2.8.

For real algebraic varieties, the Nash constructible filtration was first defined in an unpublished paper of H. Pennaneac'h [32], by analogy with the algebraically constructible filtration [31; 33]. Theorem 3.11 implies, in particular, that the Nash constructible filtration of a compact variety is the same as the filtration given by a cubical hyperresolution; this answers affirmatively a question of Pennaneac'h [32, (2.9)].

3A. Nash constructible functions on $\mathbb{RP}^n$ and arc-symmetric sets.
In real algebraic geometry it is common to work with real algebraic subsets of the affine space $\mathbb{R}^n \subset \mathbb{RP}^n$ instead of schemes over $\mathbb{R}$, and with (entire) regular rational mappings as morphisms; see for instance [3] or [5]. Since $\mathbb{RP}^n$ can be embedded in $\mathbb{R}^N$ by a biregular rational map ([3], [5] (3.4.4)), this category also contains algebraic subsets of $\mathbb{RP}^n$.

A *Nash constructible function* on $\mathbb{RP}^n$ is a function $\varphi : \mathbb{RP}^n \to \mathbb{Z}$ such that there exist a finite family of regular rational mappings $f_i : Z_i \to \mathbb{RP}^n$ defined on projective real algebraic sets Z_i, connected components Z_i' of Z_i, and integers m_i, such that for all $x \in \mathbb{RP}^n$,

$$\varphi(x) = \sum_i m_i \chi(f_i^{-1}(x) \cap Z_i'), \qquad (3\text{-}2)$$

where χ is the Euler characteristic. Nash constructible functions were introduced in [24]. Nash constructible functions on $\mathbb{RP}^n$ form a ring.

EXAMPLE 3.1.

(1) If $Y \subset \mathbb{RP}^n$ is Zariski constructible (a finite set-theoretic combination of algebraic subsets), then its characteristic function $\mathbf{1}_Y$ is Nash constructible.

(2) A subset $S \subset \mathbb{RP}^n$ is called *arc-symmetric* if every real analytic arc $\gamma :$ $(a, b) \to \mathbb{RP}^n$ either meets S at isolated points or is entirely included in S. Arc-symmetric sets were first studied by K. Kurdyka in [19]. As shown in [24], a semialgebraic set $S \subset \mathbb{RP}^n$ is arc-symmetric if and only if it is closed in $\mathbb{RP}^n$ and $\mathbf{1}_S$ is Nash constructible. By the existence of arc-symmetric closure [19; 21], for a set $S \subset \mathbb{RP}^n$ the function $\mathbf{1}_S$ is Nash constructible and only if S is a finite set-theoretic combination of semialgebraic arc-symmetric subsets of $\mathbb{RP}^n$. If $\mathbf{1}_S$ is Nash constructible we say that S is an $\mathcal{AS}$ set.

(3) Any connected component of a compact algebraic subset of $\mathbb{RP}^n$ is arc-symmetric. So is any compact real analytic and semialgebraic subset of $\mathbb{RP}^n$.

(4) Every Nash constructible function on $\mathbb{RP}^n$ is in particular *constructible* (constant on strata of a finite semialgebraic stratification of $\mathbb{RP}^n$). Not all constructible functions are Nash constructible. By [24], every constructible function $\varphi : \mathbb{RP}^n \to 2^n \mathbb{Z}$ is Nash constructible.

Nash constructible functions form the smallest family of constructible functions that contains characteristic functions of connected components of compact real algebraic sets, and that is stable under the natural operations inherited from sheaf theory: pullback by regular rational morphisms, pushforward by proper regular rational morphisms, restriction to Zariski open sets, and duality; see [24]. In terms of the *pushforward* (fiberwise integration with respect to the Euler characteristic) the formula (3-2) can be expressed as $\varphi = \sum_i m_i f_{i*} \mathbf{1}_{Z_i'}$. Duality is closely related to the *link operator*, an important tool for studying the topological properties of real algebraic sets. For more on Nash constructible function see [7] and [21].

If $S \subset \mathbb{RP}^n$ is an $\mathcal{AS}$ set (*i.e.* $\mathbf{1}_S$ is Nash constructible), we say that a function on S is *Nash constructible* if it is the restriction of a Nash constructible function on $\mathbb{RP}^n$. In particular, this defines Nash constructible functions on affine real algebraic sets. (In the non-compact case this definition is more restrictive than that of [24].)

3B. Nash constructible functions on real algebraic varieties. Let X be a real algebraic variety and let $\underline{X}$ denote the set of real points on X. We call a function $\varphi : \underline{X} \to \mathbb{Z}$ *Nash constructible* if its restriction to every affine chart is Nash constructible. The following lemma shows that this extends our definition of Nash constructible functions on affine real algebraic sets.

LEMMA 3.2. *If X_1 and X_2 are projective compactifications of the affine real algebraic variety U, then $\varphi : \underline{U} \to \mathbb{Z}$ is the restriction of a Nash constructible function on $\underline{X}_1$ if and only if φ is the restriction of a Nash constructible function on $\underline{X}_2$.*

PROOF. We may suppose that there is a regular projective morphism $f : X_1 \to X_2$ that is an isomorphism on U; cf. the proof of Proposition 2.6. Then the statement follows from the following two properties of Nash constructible functions. If $\varphi_2 : \underline{X}_2 \to \mathbb{Z}$ is Nash constructible, so is its pullback $f^*\varphi_2 = \varphi_2 \circ f : \underline{X}_1 \to \mathbb{Z}$. If $\varphi_1 : \underline{X}_1 \to \mathbb{Z}$ is Nash constructible, so is its pushforward $f_*\varphi_1 : \underline{X}_2 \to \mathbb{Z}$. $\square$

PROPOSITION 3.3. *Let X be a real algebraic variety and let $Y \subset X$ be a closed subvariety. Let $U = X \setminus Y$. Then $\varphi : \underline{X} \to \mathbb{Z}$ is Nash constructible if and only if the restrictions of φ to $\underline{Y}$ and $\underline{U}$ are Nash constructible.*

PROOF. It suffices to check the assertion for X affine; this case is easy. $\square$

THEOREM 3.4. *Let X be a complete real algebraic variety. The function $\varphi : \underline{X} \to \mathbb{Z}$ is Nash constructible if and only if there exist a finite family of regular morphisms $f_i : Z_i \to X$ defined on complete real algebraic varieties Z_i, connected components Z'_i of $\underline{Z}_i$, and integers m_i, such that for all $x \in \underline{X}$,*

$$\varphi = \sum_i m_i \, f_{i*} \mathbf{1}_{Z'_i}. \tag{3-3}$$

PROOF. If X is complete but not projective, then X can be dominated by a birational regular morphism $\pi : \tilde{X} \to X$, with $\tilde{X}$ projective (Chow's Lemma). Let $Y \subset X$, $\dim Y < \dim X$, be a closed subvariety such that π induces an isomorphism $\tilde{X} \setminus \pi^{-1}(Y) \to X \setminus Y$. Then, by Proposition 3.3, $\varphi : \underline{X} \to \mathbb{Z}$ is Nash constructible if and only if $\pi^*\varphi$ and φ restricted to $\underline{Y}$ are Nash constructible.

Let Z be a complete real algebraic variety and let $f : Z \to X$ be a regular morphism. Let Z' be a connected component of $\underline{Z}$. We show that $\varphi = f_*\mathbf{1}_{Z'}$ is Nash constructible. This is obvious if both X and Z are projective. If they are not, we may dominate both X and Z by projective varieties, using Chow's Lemma, and reduce to the projective case by induction on dimension.

Let $\varphi : \underline{X} \to \mathbb{Z}$ be Nash constructible. Suppose first that X is projective. Then $\underline{X} \subset \mathbb{RP}^n$ is a real algebraic set. Let $A \subset \mathbb{RP}^m$ be a real algebraic set and let $f : A \to \underline{X}$ be a regular rational morphism $f = g/h$, where h does not vanish on A, cf. [3]. Then the graph of f is an algebraic subset $\Gamma \subset \mathbb{RP}^n \times \mathbb{RP}^m$

and the set of real points of a projective real variety Z. Let A' be a connected component of A, and Γ' the graph of f restricted to A'. Then $f_*1_{A'} = \pi_*1_{\Gamma'}$, where π denotes the projection on the second factor.

If X is complete but not projective, we again dominate it by a birational regular morphism $\pi : \tilde{X} \to X$, with $\tilde{X}$ projective. Let $\varphi : \underline{X} \to \mathbb{Z}$ be Nash constructible. Then $\tilde{\varphi} = \varphi \circ \pi : \underline{\tilde{X}} \to \mathbb{Z}$ is Nash constructible. Thus, by the case considered above, there are regular morphisms $\tilde{f}_i : \tilde{Z}_i \to \tilde{X}$, and connected components $\tilde{Z}'_i$ such that

$$\tilde{\varphi}(x) = \sum_i m_i \, \tilde{f}_i *1_{\tilde{Z}'_i}.$$

Then $\pi_*\tilde{\varphi} = \sum_i m_i \, \tilde{\pi}_* f_i *1_{\tilde{Z}'_i}$ and differs from φ only on the set of real points of a variety of dimension smaller than $\dim X$. We complete the argument by induction on dimension. $\qquad\square$

If X is a real algebraic variety, we again say that $S \subset \underline{X}$ is an $\mathcal{AS}$ set if 1_S is Nash constructible, and $\varphi : S \to \mathbb{Z}$ is *Nash constructible* if the extension of φ to $\underline{X}$ by zero is a Nash constructible function on $\underline{X}$.

COROLLARY 3.5. *Let X, Y be complete real algebraic varieties and let S be an $\mathcal{AS}$ subset of $\underline{X}$, and T an $\mathcal{AS}$ subset of $\underline{Y}$. Let $\varphi : S \to \mathbb{Z}$ and $\psi : T \to \mathbb{Z}$ be Nash constructible. Let $f : S \to T$ be a map with $\mathcal{AS}$ graph $\Gamma \subset \underline{X} \times \underline{Y}$ and let $\pi_X : X \times Y \to X$ and $\pi_Y : X \times Y \to Y$ denote the standard projections. Then*

$$f_*(\varphi) = (\pi_Y)_*(1_\Gamma \cdot \pi_X^*\varphi) \tag{3-4}$$

and

$$f^*(\psi) = (\pi_X)_*(1_\Gamma \cdot \pi_Y^*\psi) \tag{3-5}$$

are Nash constructible.

3C. Definition of the Nash constructible filtration. Denote by $\mathcal{X}_{\mathcal{AS}}$ the category of locally compact $\mathcal{AS}$ subsets of real algebraic varieties as objects and continuous proper maps with $\mathcal{AS}$ graphs as morphisms.

Let $T \in \mathcal{X}_{\mathcal{AS}}$. We say that $\varphi : T \to \mathbb{Z}$ is *generically Nash constructible on T in dimension k* if φ coincides with a Nash constructible function everywhere on T except on a semialgebraic subset of T of dimension $< k$. We say that φ is *generically Nash constructible on T* if φ is Nash constructible in dimension $d = \dim T$.

Let $c \in C_k(T)$, and let $-k \le p \le 0$. We say that c is *p-Nash constructible*, and write $c \in \mathcal{N}_p C_k(T)$, if there exists $\varphi_{c,p} : T \to 2^{k+p}\mathbb{Z}$, generically Nash constructible in dimension k, such that

$$c = \{x \in T \; ; \; \varphi_{c,p}(x) \notin 2^{k+p+1}\mathbb{Z}\} \text{ up to a set of dimension } < k. \tag{3-6}$$

up to a set of dimension less than k. The choice of $\varphi_{c,p}$ is not unique. Let Z denote the Zariski closure of $\operatorname{Supp} c$. By multiplying $\varphi_{c,p}$ by $\mathbf{1}_Z$, we may always assume that $\operatorname{Supp}\varphi \subset Z$ and hence, in particular, that $\dim \operatorname{Supp}\varphi_{c,p} \leq k$.

We say that $c \in C_k(T)$ is *pure* if $c \in \mathcal{N}_{-k}C_k(T)$. By Theorem 3.9 of [21] and the existence of arc-symmetric closure [19; 21], $c \in C_k(T)$ is pure if and only if $\operatorname{Supp} c$ coincides with an $\mathcal{AS}$ set (up to a set of dimension smaller than k). For T compact this means that c is pure if and only if c can be represented by an arc-symmetric set. By [24], if $\dim T = k$ then every semialgebraically constructible function $\varphi : T \to 2^k \mathbb{Z}$ is Nash constructible. Hence $\mathcal{N}_0 C_k(T) = C_k(T)$.

The boundary operator preserves the Nash constructible filtration:

$$\partial \mathcal{N}_p C_k(T) \subset \mathcal{N}_p C_{k-1}(T).$$

Indeed, if $c \in C_k(T)$ is given by (3-6) and $\dim \operatorname{Supp}\varphi_{c,p} \leq k$, then

$$\partial c = \{x \in Z \; ; \; \varphi_{\partial c,p}(x) \notin 2^{k+p}\mathbb{Z}\}, \tag{3-7}$$

where $\varphi_{\partial c,p}$ equals $\frac{1}{2}\Lambda\varphi_{c,p}$ for k odd and $\frac{1}{2}\Omega\varphi_{c,p}$ for k even [24]. A geometric interpretation of this formula is as follows; see [7]. Let Z be the Zariski closure of $\operatorname{Supp} c$, so $\dim Z = k$ if $c \neq 0$. Let W be an algebraic subset of Z such that $\dim W < k$ and $\varphi_{c,p}$ is locally constant on $Z \setminus W$. At a generic point x of W, we define $\partial_W \varphi_{c,p}(x)$ as the average of the values of $\varphi_{c,p}$ on the local connected components of $Z \setminus W$ at x. It can be shown that $\partial_W \varphi_{c,p}(x)$ is generically Nash constructible in dimension $k - 1$. (For k odd it equals $(\frac{1}{2}\Lambda\varphi_{c,p})|_W$ and for k even it equals $(\frac{1}{2}\Omega\varphi_{c,p})|_W$; see [24].)

We say that a square in $\mathcal{X}_{AS}$

$$
\begin{array}{ccc}
\tilde{S} & \longrightarrow & \tilde{T} \\
\downarrow & & \downarrow{\scriptstyle \pi} \\
S & \xrightarrow{\;i\;} & T
\end{array}
\tag{3-8}
$$

is acyclic if i is a closed inclusion, $\tilde{S} = \pi^{-1}(Y)$ and the restriction of π is a homeomorphism $\tilde{T} \setminus \tilde{S} \to T \setminus S$.

THEOREM 3.6. *The functor* $\mathcal{N}C_* : \mathcal{X}_{AS} \to \mathcal{C}$, *defined on the category* $\mathcal{X}_{AS}$ *of locally compact* $\mathcal{AS}$ *sets and continuous proper maps with* $\mathcal{AS}$ *graphs, satisfies:*

(1) *For an acyclic square* (3-8) *the sequences*

$$0 \to \mathcal{N}_p C_k(\tilde{S}) \to \mathcal{N}_p C_k(S) \oplus \mathcal{N}_p C_k(\tilde{T}) \to \mathcal{N}_p C_k(T) \to 0,$$

$$0 \to \frac{\mathcal{N}_p C_k(\tilde{S})}{\mathcal{N}_{p-1} C_k(\tilde{S})} \to \frac{\mathcal{N}_p C_k(S)}{\mathcal{N}_{p-1} C_k(S)} \oplus \frac{\mathcal{N}_p C_k(\tilde{T})}{\mathcal{N}_{p-1} C_k(\tilde{T})} \to \frac{\mathcal{N}_p C_k(T)}{\mathcal{N}_{p-1} C_k(T)} \to 0,$$

are exact.

(2) *For a closed inclusion $S \subset T$, the restriction to $U = T \setminus S$ induces a morphism of filtered complexes $\mathcal{N}C_*(T) \to \mathcal{N}C_*(U)$, and the sequences*

$$0 \to \mathcal{N}_p C_k(S) \to \mathcal{N}_p C_k(T) \to \mathcal{N}_p C_k(U) \to 0,$$

$$0 \to \frac{\mathcal{N}_p C_k(S)}{\mathcal{N}_{p-1} C_k(S)} \to \frac{\mathcal{N}_p C_k(T)}{\mathcal{N}_{p-1} C_k(T)} \to \frac{\mathcal{N}_p C_k(U)}{\mathcal{N}_{p-1} C_k(U)} \to 0,$$

are exact.

PROOF. We first show that $\mathcal{N}C_*$ is a functor; that is, for a proper morphism $f : T \to S$, $f_* \mathcal{N}_p C_k(T) \subset \mathcal{N}_p C_k(S)$. Let $c \in \mathcal{N}_p C_k(T)$ and let $\varphi = \varphi_{c,p}$ be a Nash constructible function on T satisfying (3-6) (up to a set of dimension $< k$). Then

$$f_* c = \{ y \in S \; ; \; f_*(\psi)(y) \notin 2^{k+p+1} \mathbb{Z} \};$$

that is, $\varphi_{f_* c, p} = f_* \varphi_{c,p}$.

For a closed inclusion $S \subset T$, the restriction to $U = T \setminus S$ of a Nash constructible function on T is Nash constructible. Therefore the restriction defines a morphism $\mathcal{N}C_*(T) \to \mathcal{N}C_*(U)$. The exactness of the first sequence of (2) can be verified easily by direct computation. We note, moreover, that for fixed k the morphism

$$\mathcal{N}_* C_k(T) \to \mathcal{N}_* C_k(U)$$

splits (the splitting does not commute with the boundary), by assigning to $c \in \mathcal{N}_p C_k(U)$ its closure $\bar{c} \in C_k(T)$. Let $\varphi : T \to 2^{k+p} \mathbb{Z}$ be a Nash constructible function such that $\varphi|_{T \setminus S} = \varphi_{c,p}$. Then $\bar{c} = \{ x \in T \; ; \; (\mathbf{1}_T - \mathbf{1}_S) \varphi(x) \notin 2^{k+p+1} \mathbb{Z} \}$ up to a set of dimension $< k$.

The exactness of the second sequence of (2) and the sequences of (1) now follow by standard arguments. (See the proof of Theorem 2.7.) $\qquad\square$

3D. The Nash constructible filtration for Nash manifolds.

A *Nash function* on an open semialgebraic subset U of $\mathbb{R}^N$ is a real analytic semialgebraic function. Nash morphisms and Nash manifolds play an important role in real algebraic geometry. In particular a connected component of compact nonsingular real algebraic subset of $\mathbb{R}^n$ is a Nash submanifold of $\mathbb{R}^N$ in the sense of [5] (2.9.9). Since $\mathbb{R}\mathbb{P}^n$ can be embedded in $\mathbb{R}^N$ by a rational diffeomorphism ([3], [5] (3.4.2)) the connected components of nonsingular projective real algebraic varieties can be considered as Nash submanifolds of affine space. By the Nash Theorem [5, 14.1.8], every compact C^∞ manifold is C^∞-diffeomorphic to a Nash submanifold of an affine space, and moreover such a model is unique up to Nash diffeomorphism [5, Corollary 8.9.7]. In what follows by a *Nash manifold* we mean a compact Nash submanifold of an affine space.

Compact Nash manifolds and the graphs of Nash morphisms on them are $\mathcal{AS}$ sets. If N is a Nash manifold, the Nash constructible filtration is contained in the canonical filtration,

$$\mathcal{N}_p C_k(N) \subset F_p^{\mathrm{can}} C_k(N), \tag{3-9}$$

since $\partial_k(\mathcal{N}_{-k} C_k(N)) = 0$. Thus on the category of Nash manifolds and Nash maps have a morphism of functors

$$\tau : \mathcal{N}C_* \to F^{\mathrm{can}} C_*.$$

THEOREM 3.7. *For every Nash manifold* N,

$$\tau(N) : \mathcal{N}C_*(N) \to F^{\mathrm{can}} C_*(N)$$

is a filtered quasi-isomorphism.

PROOF. We show that for all p and k, $\tau(N)$ induces an isomorphism

$$\tau_* : H_k(\mathcal{N}_p C_*(N)) \cong H_k(F_p^{\mathrm{can}} C_*(N)). \tag{3-10}$$

Then, by the long exact homology sequences of $(\mathcal{N}_p C_*(N), \mathcal{N}_{p-1} C_*(N))$ and $(F_p^{\mathrm{can}} C_*(N), F_{p-1}^{\mathrm{can}} C_*(N))$,

$$\tau_* : H_k\left(\frac{\mathcal{N}_p C_*(N)}{\mathcal{N}_{p-1} C_*(N)} \right) \to H_k\left(\frac{F_p^{\mathrm{can}} C_*(N)}{F_{p-1}^{\mathrm{can}} C_*(N)} \right)$$

is an isomorphism, which shows the claim of the theorem.

We proceed by induction on the dimension of N. We call a Nash morphism $\pi : \tilde{N} \to N$ a *Nash multi-blowup* if π is a composition of blowups along nowhere dense Nash submanifolds.

PROPOSITION 3.8. *Let* N, N' *be compact connected Nash manifolds of the same dimension. Then there exist multi-blowups* $\pi : \tilde{N} \to N, \sigma : \tilde{N}' \to N'$ *such that* $\tilde{N}$ *and* $\tilde{N}'$ *are Nash diffeomorphic.*

PROOF. By a theorem of Mikhalkin (see [26] and Proposition 2.6 in [27]), any two connected closed C^∞ manifolds of the same dimension can be connected by a sequence of C^∞ blowups and and then blowdowns with smooth centers. We show that this C^∞ statement implies an analogous statement in the Nash category.

Let M be a closed C^∞ manifold. By the Nash–Tognoli Theorem there is a nonsingular real algebraic set X, *a fortiori* a Nash manifold, that is C^∞-diffeomorphic to M. Moreover, by approximation by Nash mappings, any two Nash models of M are Nash diffeomorphic; see Corollary 8.9.7 in [5]. Thus in order to show Proposition 3.8 we need only the following lemma.

LEMMA 3.9. *Let $C \subset M$ be a C^∞ submanifold of a closed C^∞ manifold M. Suppose that M is C^∞-diffeomorphic to a Nash manifold N. Then there exists a Nash submanifold $D \subset N$ such that the blowups $Bl(M, C)$ of M along C and $Bl(N, D)$ of N along D are C^∞-diffeomorphic.*

Proof. By the relative version of Nash–Tognoli Theorem proved by Akbulut and King, as well as Benedetti and Tognoli (see for instance Remark 14.1.15 in [5]), there is a nonsingular real algebraic set X and a C^∞ diffeomorphism $\varphi : M \to X$ such that $Y = \varphi(C)$ is a nonsingular algebraic set. Then the blowups $Bl(M, C)$ of M along C and $Bl(X, Y)$ of X along Y are C^∞-diffeomorphic. Moreover, since X and N are C^∞-diffeomorphic, they are Nash diffeomorphic by a Nash diffeomorphism $\psi : X \to N$. Then $Bl(X, Y)$ and $Bl(N, \psi(Y))$ are Nash diffeomorphic. This proves the lemma and the proposition. $\square$

LEMMA 3.10. *Let N be a compact connected Nash manifold and let $\pi : \tilde{N} \to N$ denote the blowup of N along a nowhere dense Nash submanifold Y. Then $\tau(N)$ is a quasi-isomorphism if and only if $\tau(\tilde{N})$ is a quasi-isomorphism.*

PROOF. Let $\tilde{Y} = \pi^{-1}(Y)$ denote the exceptional divisor of π. For each p consider the diagram

$$\to H_{k+1}(\mathcal{N}_p C_*(N)) \to H_k(\mathcal{N}_p C_*(\tilde{Y})) \to H_k(\mathcal{N}_p C_*(Y)) \oplus H_k(\mathcal{N}_p C_*(\tilde{N})) \to$$

$$\downarrow \qquad\qquad\qquad \downarrow \qquad\qquad\qquad \downarrow$$

$$\to H_{k+1}(F_p^{\mathrm{can}} C_*(N)) \to H_k(F_p^{\mathrm{can}} C_*(\tilde{Y})) \to H_k(F_p^{\mathrm{can}} C_*(Y)) \oplus H_k(F^{\mathrm{can}} C_*(\tilde{N})) \to$$

The top row is exact by Theorem 3.6. For all manifolds N and for all p and k, we have

$$H_k(F_p^{\mathrm{can}} C_*(N)) = \begin{cases} H_k(N) & \text{if } k \geq -p, \\ 0 & \text{if } k < -p, \end{cases}$$

so the short exact sequences (1-3) give that the bottom row is exact. The lemma now follows from the inductive assumption and the Five Lemma. $\square$

Consequently it suffices to show that $\tau(N)$ is a quasi-isomorphism for a single connected Nash manifold of each dimension n. We check this assertion for the standard sphere S^n by showing that

$$H_k(\mathcal{N}_p C_*(S^n)) = \begin{cases} H_k(S^n) & \text{if } k = 0 \text{ or } n \text{ and } p \geq -k, \\ 0 & \text{otherwise.} \end{cases}$$

Let $c \in \mathcal{N}_p C_k(S^n)$, $k < n$, be a cycle described as in (3-6) by the Nash constructible function $\varphi_{c,p} : Z \to 2^{k+p}\mathbb{Z}$, where Z is the Zariski closure of Supp c. Then c can be contracted to a point. More precisely, choose $p \in S^n \setminus Z$. Then $S^n \setminus \{p\}$ and $\mathbb{R}^n$ are isomorphic. Define a Nash constructible function

$\Phi : Z \times \mathbb{R} \to 2^{k+p+1}\mathbb{Z}$ by the formula

$$\Phi(x,t) = \begin{cases} 2\varphi_{c,p}(x) & \text{if } t \in [0,1], \\ 0 & \text{otherwise.} \end{cases}$$

Then

$$c \times [0,1] = \{(x,t) \in Z \times \mathbb{R} \; ; \; \Phi(x,t) \notin 2^{k+p+2}\mathbb{Z}\};$$

so $c \times [0,1] \in \mathcal{N}_p C_{k+1}(Z \times \mathbb{R})$. The morphism $f : Z \times \mathbb{R} \to \mathbb{R}^n$, $f(x,t) = tx$, is proper and for $k > 0$

$$\partial f_*(c \times [0,1]) = f_*(\partial c \times [0,1]) = c,$$

which shows that c is a boundary in $\mathcal{N}_p C_*(S^n)$. If $k = 0$ then $\partial f_*(c \times [0,1]) = c - (\deg c)[0]$.

If $c \in \mathcal{N}_p C_n(S^n)$ is a cycle, then c is a cycle in $C_n(S^n)$; that is, either $c = 0$ or $c = [S^n]$. This completes the proof of Theorem 3.7. $\qquad\square$

3E. Consequences for the weight filtration.

COROLLARY 3.11. *For every real algebraic variety X the localization of τ induces a quasi-isomorphism $\tau'(X) : \mathcal{N}C_*(X) \to \mathcal{W}C_*(X)$.*

PROOF. Theorem 3.6 yields that the functor $\mathcal{N}C_* : \mathbf{Sch}_c(\mathbb{R}) \to \mathrm{Ho}\,\mathcal{C}$ satisfies properties (1) and (2) of Theorem 1.1. Hence Theorem 3.7 and Theorem 1.4 give the desired result. $\qquad\square$

COROLLARY 3.12. *Let X be a real algebraic variety. Then for all p and k, $\mathcal{N}_p C_k(X) = \mathcal{G}_p C_k(X)$.*

PROOF. We show that the Nash constructible filtration satisfies properties (1) and (2) of Theorem 2.1. This is obvious for property (1). We show property (2). Let $\tilde{c} = \pi^{-1}(c)$. First we note that

$$c \in \mathcal{N}_p C_k(X) \iff \tilde{c} \in \mathcal{N}_p C_k(\tilde{X}).$$

Indeed, ($\Leftarrow$) follows from functoriality, since $c = \pi_*(\tilde{c})$. If c is given by (3-1) then $\pi^*(\varphi_{c,p})$ is Nash constructible and describes $\tilde{c}$. Thus it suffices to show

$$\tilde{c} \in \mathcal{N}_p C_k(\tilde{X}) \iff \partial\tilde{c} \in \mathcal{N}_p C_{k-1}(\tilde{X})$$

for $p \geq -k$, with the implication ($\Rightarrow$) being obvious. If $p = -k$ then each cycle is arc-symmetric. (Such a cycle is a union of connected components of $\tilde{X}$, since $\tilde{X}$ is nonsingular and compact.) For $p > -k$ suppose, contrary to our claim, that

$$\tilde{c} \in \mathcal{N}_p C_k(\tilde{X}) \setminus \mathcal{N}_{p-1} C_k(\tilde{X}) \quad \text{and} \quad \partial\tilde{c} \in \mathcal{N}_{p-1} C_{k-1}(\tilde{X}).$$

By Corollary 3.11 and Proposition 1.8

$$H_k\left(\frac{\mathcal{N}_p C_*(\tilde{X})}{\mathcal{N}_{p-1}C_*(\tilde{X})}\right) = 0,$$

and $\tilde{c}$ has to be a relative boundary. But $\dim \tilde{X} = k$ and $C_{k+1}(\tilde{X}) = 0$. This completes the proof. $\qquad\square$

4. Applications to real algebraic and analytic geometry

Algebraic subsets of affine space, or more generally Z-open or Z-closed affine or projective sets in the sense of Akbulut and King [3], are $\mathcal{AS}$ sets. So are the graphs of regular rational mappings. Therefore Theorems 3.6 and 3.7 give the following result.

THEOREM 4.1. *The Nash constructible filtration of closed semialgebraic chains defines a functor from the category of affine real algebraic sets and proper regular rational mappings to the category of bounded chain complexes of $\mathbb{Z}_2$ vector spaces with increasing bounded filtration.*

This functor is additive and acyclic; that is, it satisfies properties (1) *and* (2) *of Theorem 3.6; and it induces the weight spectral sequence and the weight filtration on Borel–Moore homology with coefficients in $\mathbb{Z}_2$.*

For compact nonsingular algebraic sets, the reindexed weight spectral sequence is pure: $\tilde{E}^2_{p,q} = 0$ *for* $p > 0$.

For the last claim of the theorem we note that every compact affine real algebraic set that is nonsingular in the sense of [3] and [5] admits a compact nonsingular complexification. Thus the claim follows from Theorem 3.7.

The purity of $\tilde{E}^2$ implies the purity of $\tilde{E}^\infty$: $\tilde{E}^\infty_{p,q} = 0$ for $p > 0$. Consequently every nontrivial homology class of a nonsingular compact affine or projective real algebraic variety can be represented by a semialgebraic arc-symmetric set, a result proved directly in [18] and [21].

REMARK 4.2. Theorem 3.6 and Theorem 3.7 can be used in more general contexts. A compact real analytic semialgebraic subset of a real algebraic variety is an $\mathcal{AS}$ set. A compact semialgebraic set that is the graph of a real analytic map, or more generally the graph of an arc-analytic mapping (*cf.* [21]), is arc-symmetric. In Section 3E we have already used that compact affine Nash manifolds and graphs of Nash morphisms defined on compact Nash manifolds are arc-symmetric.

The weight filtration of homology is an isomorphism invariant but not a homeomorphism invariant; this is discussed in [25] for the dual weight filtration of cohomology.

PROPOSITION 4.3. *Let X and Y be locally compact $\mathcal{AS}$ sets, and let $f : X \to Y$ be a homeomorphism with $\mathcal{AS}$ graph. Then $f_* : \mathcal{N}C_*(X) \to \mathcal{N}C_*(Y)$ is an isomorphism of filtered complexes.*

Consequently, f_* induces an isomorphism of the weight spectral sequences of X and Y and of the weight filtrations of $H_*(X)$ and $H_*(Y)$. Thus the virtual Betti numbers (1-4) *of X and Y are equal.*

PROOF. The first claim follows from the fact that $\mathcal{N}C_* : \mathcal{X}_{AS} \to \mathcal{C}$ is a functor; see the proof of Theorem 3.6. The rest of the proposition then follows from Theorem 3.6 and Theorem 3.7. □

REMARK 4.4. Proposition 4.3 applies, for instance, to regular homeomorphisms such as $f : \mathbb{R} \to \mathbb{R}$, $f(x) = x^3$. The construction of the virtual Betti numbers of [25] was extended to $\mathcal{AS}$ sets by G. Fichou in [11], where their invariance by Nash diffeomorphism was shown. The arguments of [25] and [11] use the weak factorization theorem of [1].

4A. The virtual Poincaré polynomial. Let X be a locally compact $\mathcal{AS}$ set. The virtual Betti numbers give rise to the *virtual Poincaré polynomial*

$$\beta(X) = \sum_i \beta_i(X) t^i. \tag{4-1}$$

For real algebraic varieties the virtual Poincaré polynomial was first introduced in [25]. For $\mathcal{AS}$ sets, not necessarily locally compact, it was defined in [11]. It satisfies the following properties [25; 11]:

(i) *Additivity:* For finite disjoint union $X = \bigsqcup X_i$, we have $\beta(X) = \sum \beta(X_i)$.
(ii) *Multiplicativity:* $\beta(X \times Y) = \beta(X) \cdot \beta(Y)$.
(iii) *Degree:* For $X \neq \varnothing$, $\deg \beta(X) = \dim X$ and the leading coefficient $\beta(X)$ is strictly positive.

(If X is not locally compact we can decompose it into a finite disjoint union of locally compact $\mathcal{AS}$ sets $X = \bigsqcup X_i$ and define $\beta(X) = \sum \beta(X_i)$.)

We say that a function $X \to e(X)$ defined on real algebraic sets is an *invariant* if it an isomorphism invariant, that is $e(X) = e(Y)$ if X and Y are isomorphic (by a biregular rational mapping). We say that e is additive if e takes values in an abelian group and $e(X \setminus Y) = e(X) - e(Y)$ for all $Y \subset X$. We say e is multiplicative if e takes values in a ring and $e(X \times Y) = e(X)e(Y)$ for all X, Y. The following theorem states that the virtual Betti polynomial is a universal additive, or additive and multiplicative, invariant defined on real algebraic sets (or real points of real algebraic varieties in general), among those invariants that do not distinguish Nash diffeomorphic compact nonsingular real algebraic sets.

THEOREM 4.5. *Let e be an additive invariant defined on real algebraic sets. Suppose that for every pair X, Y of Nash diffeomorphic nonsingular compact real algebraic sets we have $e(X) = e(Y)$. Then there exists a unique group homomorphism $h_e : \mathbb{Z}[t] \to G$ such that $e = h_e \circ \beta$. If, moreover, e is multiplicative then h_e is a ring homomorphism.*

PROOF. Define $h(t^n) = e(\mathbb{R}^n)$. We claim that the additive invariant $\varphi(X) = h(\beta(X)) - e(X)$ vanishes for every real algebraic set X. This is the case for $X = \mathbb{R}^n$ since $\beta(\mathbb{R}^n) = t^n$. By additivity, this is also the case for $S^n = \mathbb{R}^n \sqcup pt$. By the existence of an algebraic compactification and resolution of singularities, it suffices to show the claim for compact nonsingular real algebraic sets.

Let X be a compact nonsingular real algebraic set and let $\tilde{X}$ be the blowup of X along a smooth nowhere dense center. Then, using induction on $\dim X$, we see that $\varphi(X) = 0$ if and only if $\varphi(\tilde{X}) = 0$. By the relative version of the Nash–Tognoli Theorem, the same result holds if we have that $\tilde{X}$ is Nash diffeomorphic to the blowup of a nowhere dense Nash submanifold of X. Thus the claim and hence the first statement follows from Mikhalkin's Theorem. □

Following earlier results of Ax and Borel, K. Kurdyka showed in [20] that any regular injective self-morphism $f : X \to X$ of a real algebraic variety is surjective. It was then showed in [29] that an injective continuous self-map $f : X \to X$ of a locally compact $\mathcal{AS}$ set, such that the graph of f is an $\mathcal{AS}$ set, is a homeomorphism. The arguments of both [20] and [29] are topological and use the continuity of f in essential way. The use of additive invariants allows us to handle the non-continuous case.

THEOREM 4.6. *Let X be an $\mathcal{AS}$ set and let $f : X \to X$ be a map with $\mathcal{AS}$ graph. If f is injective then it is surjective.*

PROOF. It suffices to show that there exists a finite decomposition $X = \bigsqcup X_i$ into locally compact $\mathcal{AS}$ sets such that for each i, f restricted to X_i is a homeomorphism onto its image. Then, by Corollary 4.3,

$$\beta\left(X \setminus \bigsqcup_i f(X_i)\right) = \beta(X) - \sum_i \beta(X_i) = 0,$$

and hence, by the degree property, $X \setminus \bigsqcup_i f(X_i) = \varnothing$.

To get the required decomposition first we note that by classical theory there exists a semialgebraic stratification of $X = \bigsqcup S_j$ such that f restricted to each stratum is real analytic. We show that we may choose strata belonging to the class $\mathcal{AS}$. (We do not require the strata to be connected.) By [20] and [29], each semialgebraic subset A of a real algebraic variety V has a minimal $\mathcal{AS}$ closure in V, denoted $\bar{A}^{\mathcal{AS}}$. Moreover if A is $\mathcal{AS}$ then $\dim \bar{A}^{\mathcal{AS}} \setminus A < \dim A$. Therefore, we may take as the first subset of the decomposition the complement in X of the $\mathcal{AS}$ closure of the union of strata S_j of dimension $< \dim X$, and then proceed by induction on dimension.

Let $X = \bigsqcup S_j$ be a stratification with $\mathcal{AS}$ strata and such that f is analytic on each stratum. Then, for each stratum S_j, we apply the above argument to f^{-1} defined on $f(S_j)$. The induced subdivision of $f(S_j)$, and hence of S_j, satisfies the required property.   □

Of course, in general, surjectivity does not apply injectivity for a self-map. Nevertheless we have the following result.

THEOREM 4.7. *Let X be an $\mathcal{AS}$ set and let $f : X \to X$ be a surjective map with $\mathcal{AS}$ graph. Suppose that there exist a finite $\mathcal{AS}$ decomposition $X = \bigsqcup Y_i$ and $\mathcal{AS}$ sets F_i such that for each i, $f^{-1}(Y_i)$ is homeomorphic to $Y_i \times F_i$ by a homeomorphism with $\mathcal{AS}$ graph. Then f is injective.*

PROOF. We have

$$0 = \beta(X) - \beta(f(X)) = \sum \beta(Y_i)(\beta(F_i) - 1).$$

Therefore $\beta(F_i) - 1 = 0$ for each i; otherwise the polynomial on the right-hand side would be nonzero with strictly positive leading coefficient.   □

4B. Application to spaces of orderings. Let V be an irreducible real algebraic subset of $\mathbb{R}^N$. A function $\varphi : V \to \mathbb{Z}$ is called *algebraically constructible* if it satisfies one of the following equivalent properties [24; 30]:

(i) There exist a finite family of proper regular morphisms $f_i : Z_i \to V$, and integers m_i, such that for all $x \in V$,

$$\varphi(x) = \sum_i m_i \chi(f_i^{-1}(x) \cap Z_i).   (4\text{-}2)$$

(ii) There are finitely many polynomials $P_i \in \mathbb{R}[x_1, \ldots, x_N]$ such that for all $x \in V$,

$$\varphi(x) = \sum_i \operatorname{sgn} P_i(x).$$

Let $K = K(V)$ denote the field of rational functions of V. A function $\varphi : V \to \mathbb{Z}$ is generically algebraically constructible if and only if can be identified, up to a set of dimension smaller $\dim V$, with the signature of a quadratic form over K. Denote by $\mathcal{X}$ the real spectrum of K. A (semialgebraically) constructible function on V, up to a set of dimension smaller $\dim V$; can be identified with a continuous function $\varphi : \mathcal{X} \to \mathbb{Z}$; see [5, Chapter 7], [23], and [6]. The representation theorem of Becker and Bröcker gives a fan criterion for recognizing generically algebraically constructible function on V. The following two theorems are due to I. Bonnard.

THEOREM 4.8 [6]. *A constructible function $\varphi : V \to \mathbb{Z}$ is generically algebraically constructible if and only for any finite fan F of $\mathcal{X}$*

$$\sum_{\sigma \in F} \varphi(\sigma) \equiv 0 \mod |F|. \tag{4-3}$$

For the notion of a fan see [5, Chapter 7], [23], and [6]. The number of elements $|F|$ of a finite fan F is always a power of 2. It is known that for every finite fan F of $\mathcal{X}$ there exists a valuation ring B_F of K compatible with F, and on whose residue field the fan F induces exactly one or two distinct orderings. Denote by $\mathcal{F}$ the set of these fans of K for which the residue field induces only one ordering.

THEOREM 4.9 [6]. *A constructible function $\varphi : V \to \mathbb{Z}$ is generically Nash constructible if and only if* (4-3) *holds for every fan $F \in \mathcal{F}$.*

The following question is due to M. Coste and M. A. Marshall [23, Question 2]:

Suppose that a constructible function $\varphi : V \to \mathbb{Z}$ satisfies (4-3) *for every fan F of K with $|F| \leq 2^n$. Does there exists a generically algebraically constructible function $\psi : V \to \mathbb{Z}$ such that for each $x \in V$, $\varphi(x) - \psi(x) \equiv 0 \mod 2^n$?*

We give a positive answer to the Nash constructible analog of this question.

THEOREM 4.10. *Suppose that a constructible function $\varphi : V \to \mathbb{Z}$ satisfies* (4-3) *for every fan $F \in \mathcal{F}$ with $|F| \leq 2^n$. Then there exists a generically Nash constructible function $\psi : V \to \mathbb{Z}$ such that for each $x \in V$, $\varphi(x) - \psi(x) \equiv 0 \mod 2^n$.*

PROOF. We proceed by induction on n and on $k = \dim V$. The case $n = 0$ is trivial.

Suppose $\varphi : V \to \mathbb{Z}$ satisfies (4-3) for every fan $F \in \mathcal{F}$ with $|F| \leq 2^n$, $n \geq 1$. By the inductive assumption, φ is congruent modulo 2^{n-1} to a generically Nash constructible function ψ_{n-1}. By replacing φ by $\varphi - \psi_{n-1}$, we may suppose 2^{n-1} divides φ.

We may also suppose V compact and nonsingular, just choosing a model for $K = K(V)$. Moreover, by resolution of singularities, we may assume that φ is constant in the complement of a normal crossing divisor $D = \bigcup D_i \subset V$.

Let c be given by (3-6) with $\varphi_{c,p} = \varphi$ and $p = n - k - 1$. At a generic point x of D_i define $\partial_{D_i} \varphi(x)$ as the average of the values of φ on the local connected components of $V \setminus D$ at x. Then $\partial c = \sum_i \partial_i c$, where $\partial_i c$ is described by $\partial_{D_i} \varphi$ as in (3-7) (see [7]). Note that the constructible functions $\partial_{D_i} \varphi$ satisfy the inductive assumption for $n - 1$. Hence each $\partial_{D_i} \varphi$ is congruent to a generically Nash constructible function modulo 2^{n-1}. In other words $\partial c \in \mathcal{N}_p C_{k-1}(V)$.

Then by Corollary 3.12 we have $c \in \mathcal{N}_p C_k(V)$, which implies the statement of the theorem. $\qquad\square$

Using Corollary 3.12 we obtain the following result. The original proof was based on the fan criterion (Theorem 4.9).

PROPOSITION 4.11 [7]. *Let $V \subset \mathbb{R}^N$ be compact, irreducible, and nonsingular. Suppose that the constructible function $\varphi : V \to \mathbb{Z}$ is locally constant in the complement of a normal crossing divisor $D = \bigcup D_i \subset V$. Then φ is generically Nash constructible if and only if $\partial_D \varphi$ is generically Nash constructible.*

PROOF. We show only ($\Leftarrow$). Suppose $2^{k+p} | \varphi$ generically, where $k = \dim V$, and let c be given by (3-6) with $\varphi_{c,p} = \varphi$. Then by our assumption $\partial c \in \mathcal{N}_p C_{k-1}(V)$. By Corollary 3.12 we have $c \in \mathcal{N}_p C_k(V)$, which shows that, modulo 2^{k+p+1}, φ coincides with a generically Nash constructible function ψ. Then we apply the same argument to $\varphi - \psi$. $\qquad\square$

REMARK 4.12. We note that Proposition 4.11 implies neither Theorem 4.10 nor Corollary 3.12. Similarly the analog of this proposition proved in [6] does not give an answer to Coste and Marshall's question.

5. The toric filtration

In their investigation of the relation between the homology of the real and complex points of a toric variety [4], Bihan *et al.* define a filtration on the cellular chain complex of a real toric variety. We prove that this filtered complex is quasi-isomorphic to the semialgebraic chain complex with the Nash constructible filtration. Thus the toric filtered chain complex realizes the weight complex, and the real toric spectral sequence of [4] is isomorphic to the weight spectral sequence.

For background on toric varieties see [12]. We use a simplified version of the notation of [4]. Let Δ be a rational fan in $\mathbb{R}^n$, and let X_Δ be the real toric variety defined by Δ. The group $\mathbb{T} = (\mathbb{R}^*)^n$ acts on X_Δ, and the k-dimensional orbits $\mathcal{O}_\sigma$ of this action correspond to the codimension k cones σ of Δ.

The positive part X_Δ^+ of X_Δ is a closed semialgebraic subset of X_Δ, and there is a canonical retraction $r : X_\Delta \to X_\Delta^+$ that can be identified with the orbit map of the action of the finite group $T = (S^0)^n$ on X_Δ, where $S^0 = \{-1, +1\} \subset \mathbb{R}^*$. The T-quotient of the k-dimensional $\mathbb{T}$-orbit $\mathcal{O}_\sigma$ is a semialgebraic k-cell c_σ of X_Δ^+, and $\mathcal{O}_\sigma$ is a disjoint union of k-cells, each of which maps homeomorphically onto c_σ by the quotient map. This decomposition defines a cell structure on X_Δ such that X_Δ^+ is a subcomplex and the quotient map is cellular. Let $C_*(\Delta)$ be the cellular chain complex of X_Δ with coefficients in $\mathbb{Z}_2$. The closures of the cells of X_Δ are not necessarily compact, but they are semialgebraic subsets of X_Δ. Thus we have a chain map

$$\alpha : C_*(\Delta) \to C_*(X_\Delta) \tag{5-1}$$

from cellular chains to semialgebraic chains.

The *toric filtration* of the cellular chain complex $C_*(\Delta)$ is defined as follows [4]. For each $k \geq 0$ we define vector subspaces

$$0 = T_{-k-1}C_k(\Delta) \subset T_{-k}C_k(\Delta) \subset T_{-k+1}C_k(\Delta) \subset \cdots \subset T_0 C_k(\Delta) = C_k(\Delta), \tag{5-2}$$

such that $\partial_k(T_p C_k(\Delta)) \subset T_p C_{k-1}(\Delta)$ for all k and p.

Let σ be a cone of the fan Δ, with codim $\sigma = k$. Let $C_k(\sigma)$ be the subspace of $C_k(\Delta)$ spanned by the k-cells of $\mathcal{O}_\sigma$. Then

$$C_k(\Delta) = \bigoplus_{\text{codim}\,\sigma = k} C_k(\sigma).$$

The orbit $\mathcal{O}_\sigma$ has a distinguished point $x_\sigma \in c_\sigma \subset X_\Delta^+$. Let $T_\sigma = T/T^{x_\sigma}$, where T^{x_σ} is the T-stabilizer of x_σ. We identify the orbit $T \cdot x_\sigma$ with the multiplicative group T_σ. Each k-cell of $\mathcal{O}_\sigma$ contains a unique point of the orbit $T \cdot x_\sigma$. Thus we can make the identification $C_k(\sigma) = C_0(T_\sigma)$, the set of formal sums $\sum_i a_i[g_i]$, where $a_i \in \mathbb{Z}_2$ and $g_i \in T_\sigma$. The multiplication of T_σ defines a multiplication on $C_0(T_\sigma)$, so that $C_0(T_\sigma)$ is just the group algebra of T_σ over $\mathbb{Z}_2$.

Let $\mathcal{I}_\sigma$ be the augmentation ideal of the algebra $C_0(T_\sigma)$, that is,

$$\mathcal{I}_\sigma = \text{Ker}[\varepsilon : C_0(T_\sigma) \to \mathbb{Z}_2] \quad \text{with } \varepsilon \sum_i a_i[g_i] = \sum_i a_i.$$

For $p \leq 0$ we define $T_p C_k(\sigma)$ to be the subspace corresponding to the ideal $(\mathcal{I}_\sigma)^{-p} \subset C_0(T_\sigma)$, and we let

$$T_p C_k(\Delta) = \sum_{\text{codim}\,\sigma = k} T_p C_k(\sigma).$$

If $\sigma < \tau$ in Δ and codim $\tau = $ codim $\sigma - 1$, the geometry of Δ determines a group homomorphism $\varphi_{\tau\sigma} : T_\sigma \to T_\tau$ (see [4]). Let $\partial_{\tau\sigma} : C_k(\sigma) \to C_{k-1}(\tau)$ be the induced algebra homomorphism. We have $\partial_{\tau\sigma}(\mathcal{I}_\sigma) \subset \mathcal{I}_\tau$. The boundary map $\partial_k : C_k(\Delta) \to C_{k-1}(\Delta)$ is given by $\partial_k(\sigma) = \sum_\tau \partial_{\tau\sigma}(\tau)$, and $\partial_k(T_p C_k(\Delta)) \subset T_p C_{k-1}(\Delta)$, so $T_p C_*(\Delta)$ is a subcomplex of $C_*(\Delta)$.

PROPOSITION 5.1. *For all $k \geq 0$ and $p \leq 0$, the chain map α (5-1) takes the toric filtration (5-2) to the Nash filtration (3-1),*

$$\alpha(T_p C_k(\Delta)) \subset \mathcal{N}_p C_k(X_\Delta).$$

PROOF. It suffices to show that for every cone $\sigma \in \Delta$ with $\operatorname{codim} \sigma = k$,

$$\alpha(\mathcal{T}_p C_k(\sigma)) \subset \mathcal{N}_p C_k(\mathcal{O}_\sigma).$$

The variety $\mathcal{O}_\sigma$ is isomorphic to $(\mathbb{R}^*)^k$, the toric variety of the trivial fan $\{0\}$ in $\mathbb{R}^k$, and the action of T_σ on $\mathcal{O}_\sigma$ corresponds to the action of $T_k = \{-1, +1\}^k$ on $(\mathbb{R}^*)^k$. The k-cells of $(\mathbb{R}^*)^k$ are its connected components. Let $\mathcal{I}_k \subset C_0(T_k)$ be the augmentation ideal. Let $q = -p$, so $0 \leq q \leq k$. The vector space $C_0(T_k)$ has dimension 2^k, and for each q the quotient $\mathcal{I}^q / \mathcal{I}^{q+1}$ has dimension $\binom{k}{q}$. A basis for $\mathcal{I}^q / \mathcal{I}^{q+1}$ can be defined as follows. Let $t_1, \ldots, t_k$ be the standard generators of the multiplicative group T_k,

$$t_i = (t_{i1}, \ldots, t_{ik}), \; t_{ij} = \begin{cases} -1 & \text{if } i = j, \\ 1 & \text{if } i \neq j. \end{cases}$$

If $S \subset \{1, \ldots, k\}$, let T_S be the subgroup of T_k generated by $\{t_i \; ; \; i \in S\}$, and define $[T_S] \in C_0(T_k)$ by

$$[T_S] = \sum_{t \in T_S} [t].$$

Then $\{[T_S] \; ; \; |S| = q\}$ is a basis for $\mathcal{I}^q / \mathcal{I}^{q+1}$ (see [4]).

To prove that $\alpha((\mathcal{I}_k)^q) \subset \mathcal{N}_{-q} C_k((\mathbb{R}^*)^k)$ we just need to show that if $|S| = q$ then $\alpha([T_S]) \in \mathcal{N}_{-q} C_k((\mathbb{R}^*)^k)$. Now the chain $\alpha([T_S]) \in C_k((\mathbb{R}^*)^k)$ is represented by the semialgebraic set $A_S \subset (\mathbb{R}^*)^k$,

$$A_S = \{(x_1, \ldots, x_k) \; ; \; x_i > 0, \; i \notin S\},$$

and $\varphi = 2^{k-q} \mathbf{1}_{A_S}$ is Nash constructible. To see this consider the compactification $(\mathbb{P}^1(\mathbb{R}))^k$ of $(\mathbb{R}^*)^k$. We have $\varphi = \tilde{\varphi}|(\mathbb{R}^*)^k$, where $\tilde{\varphi} = f_* \mathbf{1}_{(\mathbb{P}^1(\mathbb{R}))^k}$, with $f : (\mathbb{P}^1(\mathbb{R}))^k \to (\mathbb{P}^1(\mathbb{R}))^k$ defined as follows. If $z = (u : v) \in \mathbb{P}^1(\mathbb{R})$, let $f_1(z) = (u : v)$, and $f_2(z) = (u^2 : v^2)$. Then

$$f(z_1, \ldots, z_k) = (w_1, \ldots, w_k), \; w_i = \begin{cases} f_1(z_i) & \text{if } i \in S, \\ f_2(z_i) & \text{if } i \notin S. \end{cases}$$

This completes the proof. $\square$

LEMMA 5.2. *Let σ be a codimension k cone of Δ, and let*

$$C_i(\sigma) = \begin{cases} C_k(\sigma) & \text{if } i = k, \\ 0 & \text{if } i \neq k. \end{cases}$$

For all $p \leq 0$,

$$\alpha_* : H_*(\mathcal{T}_p C_*(\sigma)) \to H_*(\mathcal{N}_p C_*(\mathcal{O}_\sigma))$$

is an isomorphism.

PROOF. Again we only need to consider the case $\mathcal{O}_\sigma = (\mathbb{R}^*)^k$, where σ is the trivial cone 0 in $\mathbb{R}^n$. Now

$$H_i(C_*(0)) = \begin{cases} C_k(0) & \text{if } i = k, \\ 0 & \text{if } i \neq k, \end{cases}$$

and

$$H_i(C_*((\mathbb{R}^*)^k)) = \begin{cases} \operatorname{Ker} \partial_k & \text{if } i = k, \\ 0 & \text{if } i \neq k, \end{cases}$$

where $\partial_k : C_k((\mathbb{R}^*)^k) \to C_{k-1}((\mathbb{R}^*)^k)$. The vector space $\operatorname{Ker} \partial_k$ has basis the cycles represented by the components of $(\mathbb{R}^*)^k$, and $\alpha : C_k(0) \to C_k((\mathbb{R}^*)^k)$ is a bijection from the cells of $C_k(0)$ to the components of $(\mathbb{R}^*)^k$. Thus $\alpha : C_k(0) \to \operatorname{Ker} \partial_k$ is an isomorphism of vector spaces. Therefore α takes the basis $\{A_S \; ; \; |S| = q\}_{q=0,\dots,k}$ to a basis of $\operatorname{Ker} \partial_k$. The proof of Proposition 5.1 shows that if $|S| \geq q$ then $A_S \in \mathcal{N}_{-q}C_k((\mathbb{R}^*)^k)$. We claim further that if $|S| < q$ then $A_S \notin \mathcal{N}_{-q}C_k((\mathbb{R}^*)^k)$. It follows that $\{A_S \; ; \; |S| \geq q\}$ is a basis for $H_k(\mathcal{N}_{-q}C_*((\mathbb{R}^*)^k))$, and so

$$\alpha_* : H_*(\mathcal{T}_{-q}C_*(0)) \to H_*(\mathcal{N}_{-q}C_*((\mathbb{R}^*)^k))$$

is an isomorphism, as desired.

To prove the claim, it suffices to show that if $\bar{A}_S$ is the closure of A_S in $\mathbb{R}^n$, then $\bar{A}_S \notin \mathcal{N}_{-q}C_k((\mathbb{R}^*)^k)$. We show this by induction on k. The case $k = 1$ is clear: If $\bar{A} = \{x \; ; \; x \geq 0\}$ then $\bar{A} \notin \mathcal{N}_{-1}C_1(\mathbb{R})$ because $\partial \bar{A} \neq 0$. In general $\bar{A}_S = \{(x_1, \dots, x_k) \; ; \; x_i \geq 0, i \notin S\}$. Suppose $\bar{A}_S$ is $(-q)$-Nash constructible for some $q > |S|$. Then there exists $\varphi : \mathbb{R}^k \to 2^{k-q}\mathbb{Z}$ generically Nash constructible in dimension k such that $\bar{A}_S = \{x \in \mathbb{R}^k \; ; \; \varphi(x) \notin 2^{k-q+1}\mathbb{Z}\}$, up to a set of dimension $< k$. Let $j \notin S$, and let $W_j = \{(x_1, \dots, x_k) \; ; \; x_j = 0\} \cong \mathbb{R}^{k-1}$. Then $\partial_{W_j} \varphi : W_j \to 2^{k-q-1}\mathbb{Z}$, and $\bar{A}_S \cap W_j = \{x \in W_j \; ; \; \partial_{W_j}\varphi(x) \notin 2^{k-q}\mathbb{Z}\}$, up to a set of dimension $< k - 1$. Hence $\bar{A}_S \cap W_j \in \mathcal{N}_{-q}C_{k-1}(W_j)$. But

$$\bar{A}_S \cap W_j = \{(x, \dots, x_k) \; ; \; x_j = 0, x_i \geq 0, i \notin S\},$$

and so by the inductive hypothesis $\bar{A}_S \cap W_j \notin \mathcal{N}_{-q}C_{k-1}(W_j)$, which is a contradiction. $\qquad \square$

LEMMA 5.3. *For every toric variety X_Δ and every $p \leq 0$,*

$$\alpha_* : H_*(T_pC_*(\Delta)) \to H_*(\mathcal{N}_pC_*(X_\Delta))$$

is an isomorphism.

PROOF. We show by induction on orbits that the lemma is true for every variety Z that is a union of orbits in the toric variety X_Δ. Let Σ be a subset of Δ, and let $\Sigma' = \Sigma \setminus \{\sigma\}$, where $\sigma \in \Sigma$ is a minimal cone, *i. e.* there is no $\tau \in \Sigma$ with $\tau < \sigma$. Let Z and Z' be the unions of the orbits corresponding to cones in Σ

and Σ', respectively. Then Z' is closed in Z, and $Z \setminus Z' = \mathcal{O}_\sigma$. We have a commutative diagram with exact rows:

$$\cdots \to H_i(\mathcal{T}_p C_*(\Sigma')) \to H_i(\mathcal{T}_p C_*(\Sigma)) \to H_i(\mathcal{T}_p C_*(\sigma)) \to H_{i-1}(\mathcal{T}_p C_*(\Sigma')) \to \cdots$$

$$\downarrow{\scriptstyle\beta_i} \qquad\qquad \downarrow{\scriptstyle\gamma_i} \qquad\qquad \downarrow{\scriptstyle\alpha_i} \qquad\qquad \downarrow{\scriptstyle\beta_{i-1}}$$

$$\cdots \to H_i(\mathcal{N}_p C_*(\Sigma')) \to H_i(\mathcal{N}_p C_*(\Sigma)) \to H_i(\mathcal{N}_p C_*(\sigma)) \to H_{i-1}(\mathcal{N}_p C_*(\Sigma')) \to \cdots$$

By Lemma 5.3 α_i is an isomorphism for all i. By inductive hypothesis β_i is an isomorphism for all i. Therefore γ_i is an isomorphism for all i. $\square$

THEOREM 5.4. *For every toric variety X_Δ and every $p \leq 0$,*

$$\alpha_* : H_* \left(\frac{\mathcal{T}_p C_*(\Delta)}{\mathcal{T}_{p-1} C_*(\Delta)} \right) \to H_* \left(\frac{\mathcal{N}_p C_*(X_\Delta)}{\mathcal{N}_{p-1} C_*(X_\Delta)} \right)$$

is an isomorphism.

PROOF. This follows from Lemma 5.3 and the long exact homology sequences of the pairs $(\mathcal{T}_p C_*(\Delta), \mathcal{T}_{p-1} C_*(\Delta))$ and $(\mathcal{N}_p C_*(X_\Delta), \mathcal{N}_{p-1} C_*(X_\Delta))$. $\square$

Thus for every toric variety X_Δ the toric filtered complex $\mathcal{T}C_*(\Delta)$ is quasi-isomorphic to the Nash constructible filtered complex $\mathcal{N}C_*(X_\Delta)$, and so the toric spectral sequence [4] is isomorphic to the weight spectral sequence.

EXAMPLE 5.5. For toric varieties of dimension at most 4, the toric spectral sequence collapses [4; 35]. V. Hower [17] discovered that the spectral sequence does not collapse for the 6-dimensional projective toric variety associated to the matroid of the Fano plane.

Appendix: Semialgebraic chains

In this appendix we denote by X a locally compact semialgebraic set (*i.e.* a semialgebraic subset of the set of real points of a real algebraic variety) and by $C_*(X)$ the complex of semialgebraic chains of X with closed supports and coefficients in $\mathbb{Z}_2$. The complex $C_*(X)$ has the following geometric description, which is equivalent to the usual definition using a semialgebraic triangulation [5, 11.7].

A *semialgebraic chain c* of X is an equivalence class of closed semialgebraic subsets of X. For $k \geq 0$, let $S_k(X)$ be the $\mathbb{Z}_2$ vector space generated by the closed semialgebraic subsets of X of dimension $\leq k$. Then $C_k(X)$ is the $\mathbb{Z}_2$ vector space obtained as the quotient of $S_k(X)$ by the following relations:

(i) If A and B are closed semialgebraic subsets of X of dimension at most k, then

$$A + B \sim \mathrm{cl}(A \div B),$$

where $A \div B = (A \cup B) \setminus (A \cap B)$ is the symmetric difference of A and B, and cl denotes closure.

(ii) If A is a closed semialgebraic subset of X and dim $A < k$, then $A \sim 0$.

If the chain c is represented by the semialgebraic set A, we write $c = [A]$. If $c \in C_k(X)$, the *support* of c, denoted Supp c, is the smallest closed semialgebraic set representing c. If $c = [A]$ then Supp $c = \{x \in A \; ; \; \dim_x A = k\}$.

The *boundary* operator $\partial_k : C_k(X) \to C_{k-1}(X)$ can be defined using the link operator Λ on constructible functions [24]. If $c \in C_k(X)$ with $c = [A]$, then $\partial_k c = [\partial A]$, where $\partial A = \{x \in A \; ; \; \Lambda 1_A(x) \equiv 1 \pmod 2\}$. The operator ∂_k is well-defined, and $\partial_{k-1} \partial_k = 0$, since $\Lambda \circ \Lambda = 2\Lambda$.

If $f : X \to Y$ is a proper continuous semialgebraic map, the *pushforward* homomorphism $f_* : C_k(X) \to C_k(Y)$ is defined as follows. Let A be a representative of c. Then $f(A) \sim B_1 + \cdots + B_l$, where each closed semialgebraic set B_i has the property that $\#(A \cap f^{-1}(y))$ is constant mod 2 on $B_i \setminus B_i'$ for some closed semialgebraic set $B_i' \subset B_i$ with dim $B_i' < k$. For each i let $n_i \in \mathbb{Z}_2$ be this constant value. Then $f_*(c) = n_1[B_1] + \cdots + n_l[B_l]$.

Alternately, $f_*(c) = [B]$, where $B = \text{cl}\{y \in Y \; ; \; f_* 1_A(y) \equiv 1 \pmod 2\}$, and f_* is pushforward for constructible functions [24]. From this definition it is easy to prove the standard properties $g_* f_* = (gf)_*$ and $\partial_k f_* = f_* \partial_k$.

We use two basic operations on semialgebraic chains: restriction and closure. These operations do not commute with the boundary operator in general.

Let $c \in C_k(X)$ and let $Z \subset X$ be a locally closed semialgebraic subset. If $c = [A]$, we define the *restriction* by $c|_Z = [A \cap Z] \in C_k(Z)$. This operation is well-defined. If U is an open semialgebraic subset of X, then $\partial_k(c|_U) = (\partial_k c)|_U$.

Now let $c \in C_k(Z)$ with $Z \subset X$ locally closed semialgebraic. If $c = [A]$ we define the *closure* by $\bar{c} = [\text{cl}(A)] \in C_k(X)$, where $\text{cl}(A)$ is the closure of A in X. Closure is a well-defined operation on semialgebraic chains.

By means of the restriction and closure operations, we define the pullback of a chain in the following situation, which can be applied to an acyclic square (1-2) of real algebraic varieties. Consider a square of locally closed semialgebraic sets,

$$
\begin{array}{ccc}
\tilde{Y} & \longrightarrow & \tilde{X} \\
\downarrow & & \downarrow \pi \\
Y & \xrightarrow{\; i \;} & X
\end{array}
$$

such that $\pi : \tilde{X} \to X$ is a proper continuous semialgebraic map, i is the inclusion of a closed semialgebraic subset, $\tilde{Y} = \pi^{-1}(Y)$, and the restriction of π is a homeomorphism $\pi' : \tilde{X} \setminus \tilde{Y} \to X \setminus Y$. Let $c \in C_k(X)$. We define the *pullback*

$\pi^{-1}c \in C_k(\tilde{X})$ by the formula

$$\pi^{-1}c = \overline{((\pi')^{-1})_*(c|_{X \setminus Y})}.$$

Pullback does not commute with the boundary operator in general.

Acknowledgement

We thank Michel Coste for comments on a preliminary version of this paper.

References

[1] D. Abramovich, K. Karu, K. Matsuki, J. Włodarczyk, *Torification and factorization of birational maps*, J. Amer. Math. Soc. **29** (2002), 531–572.

[2] S. Akbulut, H. King, *The topology of real algebraic sets*, Enseign. Math. **29** (1983), 221–261.

[3] S. Akbulut, H. King, *Topology of Real Algebraic Sets*, MSRI Publ. **25**, Springer, New York, 1992.

[4] F. Bihan, M. Franz, C. McCrory, J. van Hamel, *Is every toric variety an M-variety?*, Manuscripta Math. **120** (2006), 217–232.

[5] J. Bochnak, M. Coste, M.-F. Roy, *Real Algebraic Geometry*, Springer, New York, 1992.

[6] I. Bonnard, *Un critère pour reconaitre les fonctions algébriquement constructibles*, J. Reine Angew. Math. **526** (2000), 61–88.

[7] I. Bonnard, *Nash constructible functions*, Manuscripta Math. **112** (2003), 55–75.

[8] P. Deligne, *Théorie de Hodge II*, IHES Publ. Math. **40** (1971), 5–58.

[9] P. Deligne, *Théorie de Hodge III*, IHES Publ. Math. **44** (1974), 5–77.

[10] P. Deligne, *Poids dans la cohomologie des variétés algébriques*, Proc. Int. Cong. Math. Vancouver (1974), 79–85.

[11] G. Fichou, *Motivic invariants of arc-symmetric sets and blow-Nash equivalence*, Compositio Math. **141** (2005) 655–688.

[12] W. Fulton, *Introduction to Toric Varieties*, Annals of Math. Studies **131**, Princeton, 1993.

[13] A. Gabrielov, N. Vorobjov, T. Zell, *Betti numbers of semialgebraic and sub-Pfaffian sets*, J. London Math. Soc. (2) **69** (2004), 27–43.

[14] H. Gillet, C. Soulé, *Descent, motives, and K-theory*, J. Reine Angew. Math. **478** (1996), 127–176.

[15] F. Guillén, V. Navarro Aznar, *Un critère d'extension des foncteurs définis sur les schémas lisses*, IHES Publ. Math. **95** (2002), 1–83.

[16] F. Guillén, V. Navarro Aznar, *Cohomological descent and weight filtration* (2003). (Abstract: http://congreso.us.es/rsme-ams/sesionpdf/sesion13.pdf.)

[17] V. Hower, *A counterexample to the maximality of toric varieties*, Proc. Amer. Math. Soc., **136** (2008), 4139–4142.

[18] W. Kucharz, *Homology classes represented by semialgebraic arc-symmetric sets*, Bull. London Math. Soc. **37** (2005), 514–524.

[19] K. Kurdyka, *Ensembles semi-algébriques symétriques par arcs*, Math. Ann. **281** (1988), 445–462.

[20] K. Kurdyka, *Injective endomorphisms of real algebraic sets are surjective*, Math. Ann. **313** no.1 (1999), 69–83

[21] K. Kurdyka, A. Parusiński, *Arc-symmetric sets and arc-analytic mappings*, Panoramas & Syntheses **24**, Soc. Math. France (2007), 33–67.

[22] S. MacLane, *Homology*, Springer, Berlin 1963.

[23] M. A. Marshall, *Open questions in the theory of spaces of orderings*, J. Symbolic Logic **67** (2002), no. 1, 341–352.

[24] C. McCrory, A. Parusiński, *Algebraically constructible functions*, Ann. Sci. Éc. Norm. Sup. **30** (1997), 527–552.

[25] C. McCrory, A. Parusiński, *Virtual Betti numbers of real algebraic varieties*, Comptes Rendus Acad. Sci. Paris, Ser. I, **336** (2003), 763–768. (See also http://arxiv.org/pdf/math.AG/0210374.)

[26] G. Mikhalkin, *Blowup equivalence of smooth closed manifolds*, Topology **36** (1997), 287–299.

[27] G. Mikhalkin, *Birational equivalence for smooth manifolds with boundary*, Algebra i Analiz 11 (1999), no. 5, 152–165. In Russian; translation in St. Petersburg Math. J. 11 (2000), no. 5, 827–836

[28] M. Nagata, *Imbedding of an abstract variety in a complete variety*, J. Math. Kyoto U. **2** (1962), 1–10.

[29] A. Parusiński, *Topology of injective endomorphisms of real algebraic sets*, Math. Ann. **328** (2004), 353–372.

[30] A. Parusiński, Z. Szafraniec, *Algebraically constructible functions and signs of polynomials*, Manuscripta Math. **93** (1997), no. 4, 443–456.

[31] H. Pennaneac'h, *Algebraically constructible chains*, Ann. Inst. Fourier (Grenoble) **51** (2001), no. 4, 939–994,

[32] H. Pennaneac'h, *Nash constructible chains*, preprint Università di Pisa, (2003).

[33] H. Pennaneac'h, *Virtual and non-virtual algebraic Betti numbers*, Adv. Geom. **5** (2005), no. 2, 187–193.

[34] C. Peters, J. Steenbrink, *Mixed Hodge Structures*, Springer, Berlin, 2008.

[35] A. Sine, *Problème de maximalité pour les variétés toriques*, Thèse Doctorale, Université d'Angers 2007.

[36] R. Thom, *Quelques propriétés globales des variétés différentiables*, Comm. Math. Helv. **28** (1954), 17–86.

[37] B. Totaro, *Topology of singular algebraic varieties*, Proc. Int. Cong. Math. Beijing (2002), 533-541.

CLINT MCCRORY
MATHEMATICS DEPARTMENT
UNIVERSITY OF GEORGIA
ATHENS, GA 30602
UNITED STATES
  clint@math.uga.edu

ADAM PARUSIŃSKI
LABORATOIRE J.-A. DIEUDONNÉ
U.M.R. N° 6621 DU C.N.R.S.
UNIVERSITÉ DE NICE - SOPHIA ANTIPOLIS
PARC VALROSE
06108 NICE CEDEX 02
FRANCE
  adam.parusinski@unice.fr

Topology of Stratified Spaces
MSRI Publications
Volume **58**, 2011

On Milnor classes
of complex hypersurfaces

LAURENTIU MAXIM

ABSTRACT. We revisit known results about the Milnor class of a singular
complex hypersurface, and rephrase some of them in a way that allows for
a better comparison with the topological formula of Cappell and Shaneson for
the L-class of such a hypersurface. Our approach is based on Verdier's specialization property for the Chern–MacPherson class, and simple constructible
function calculus.

1. Introduction

It is well-known that for a compact complex hypersurface X with only isolated singularities the sum of the Milnor numbers at the singular points measures
(up to a sign) the difference between the topological Euler characteristic of X
and that of a nonsingular hypersurface linearly equivalent to X, provided such
a hypersurface exists. This led Parusiński to a generalization of the notion of
Milnor number to nonisolated hypersurface singularities [16], which in the case
of isolated singularities reduces to the sum of Milnor numbers at the singular
points.

For a (possibly singular) compact complex hypersurface X, the Euler characteristic $\chi(X)$ equals the degree of the zero-dimensional component of the
Chern–MacPherson homology class $c_*(X)$; see [15]. On the other hand, the
Euler characteristic of a nonsingular hypersurface linearly equivalent to X is
just the degree of the Poincaré dual of the Chern class of the virtual tangent
bundle of X, that is, the degree of the Fulton–Johnson class $c_*^{FJ}(X)$ [10; 11].
Thus, Parusiński's Milnor number equals (up to a sign) the degree of the homology class $c_*^{FJ}(X) - c_*(X)$. It is therefore natural to try to understand the

This work was partially supported by NSF and a PSC-CUNY Research Award.

higher-degree components of this difference class, which usually is called the *Milnor class* of X. The study of the Milnor class also comes up naturally while searching for a Verdier-type Riemann–Roch theorem for the Chern–MacPherson classes (see [20; 22; 23]); indeed, the Milnor class measures the defect of commutativity in a Verdier–Riemann–Roch diagram for MacPherson's Chern class transformation.

While the problem of understanding the Milnor class in terms of invariants of singularities can be formulated in more general contexts (e.g., for local complete intersections, or regular embeddings in arbitrary codimension, see [19; 20]), in this note we restrict ourselves, for simplicity, only to the case of hypersurfaces (i.e., regular embeddings in codimension 1) in complex manifolds. We recall known results about the Milnor class of a singular hypersurface, and rephrase some of these results in a way that, we believe, reflects better the geometry of the singular locus in terms of its stratification. For more comprehensive surveys on Milnor classes, the interested reader is advised to consult [2; 3; 17; 22].

The approach presented in this note is based on a well-known specialization argument [21], and simple calculus of constructible functions as developed in [9]. While this approach is not new (see [18; 19; 20] for similar considerations), the formulation of our main results (Theorem 4.3, Corollary 4.4 and Theorem 4.6) has the advantage of being conceptually very simple, and it allows for a better comparison with the topological formula of Cappell and Shaneson [7; 8] for the L-classes of singular hypersurfaces. Indeed, we also explore a Chern-class analogue of Goresky–MacPherson's homology L-class [12], defined via the constructible function associated to the intersection chain complex of a variety (see [9]). This class, which for a variety X is denoted by $Ic_*(X)$, encodes very detailed information about the geometry of a fixed Whitney stratification of X. In the case of hypersurfaces, we compare this class with the Fulton–Johnson class, and derive a formula for their difference in terms of invariants of the singular locus.

2. Canonical bases for the group of constructible functions

Let X be a topological space with a finite partition $\mathcal{V}$ into a disjoint union of finitely many connected subsets V satisfying the *frontier condition*:

$$W \cap \overline{V} \neq \emptyset \implies W \subset \overline{V}.$$

The main examples of such spaces are complex algebraic or compact complex analytic varieties with a fixed Whitney stratification. Consider on $\mathcal{V}$ the partial order given by

$$W \leq V \iff W \subset \overline{V}.$$

We also write $W < V$ if $W \leq V$ and $W \neq V$.

Let $F_{\mathcal{V}}(X)$ be the abelian group of $\mathcal{V}$-constructible functions on X, that is, functions $\alpha : X \to \mathbb{Z}$ such that $\alpha|_V$ is constant for all $V \in \mathcal{V}$. This is a free abelian group with basis

$$\mathcal{B}_1 := \{ \, 1_V \mid V \in \mathcal{V} \, \},$$

so that any $\alpha \in F_{\mathcal{V}}(X)$ can be written as

$$\alpha = \sum_{V \in \mathcal{V}} \alpha(V) \cdot 1_V. \tag{2-1}$$

In what follows, we will discuss two more canonical bases on $F_{\mathcal{V}}(X)$, see [9] for complete details. First, the collection

$$\mathcal{B}_2 := \{ \, 1_{\overline{V}} \mid V \in \mathcal{V} \, \}$$

is also a basis for $F_{\mathcal{V}}(X)$, since

$$1_{\overline{V}} = \sum_{W \leq V} 1_W$$

and the transition matrix $A = (a_{W,V})$, where $a_{W,V}$ is defined as 1 if $W \leq V$ and 0 otherwise, is upper triangular with respect to $\leq$, with all diagonal entries equal to 1 (so A is invertible). In this basis, a constructible function $\alpha \in F_{\mathcal{V}}(X)$ can be expressed by the identity

$$\alpha = \sum_{V} \alpha(V) \cdot \hat{1}_{\overline{V}}, \tag{2-2}$$

(see [9, Proposition 2.1]), where for each $V \in \mathcal{V}$, we define $\hat{1}_{\overline{V}}$ inductively by the formula

$$\hat{1}_{\overline{V}} = 1_{\overline{V}} - \sum_{W < V} \hat{1}_{\overline{W}}.$$

Note that if there is a stratum $S \in \mathcal{V}$ which is dense in X, i.e., $\overline{S} = X$, so $V \leq S$ for all $V \in \mathcal{V}$, then (2-2) can be rewritten as

$$\alpha = \alpha(S) \cdot 1_X + \sum_{V < S} (\alpha(V) - \alpha(S)) \cdot \hat{1}_{\overline{V}}. \tag{2-3}$$

If moreover $\alpha|_S = 0$, this reduces further to

$$\alpha = \sum_{V < S} \alpha(V) \cdot \hat{1}_{\overline{V}}. \tag{2-4}$$

In order to describe the third basis for the group of constructible functions, assume moreover that X is a topological pseudomanifold with a stratification $\mathcal{V}$ by finitely many oriented strata of *even* dimension. Then, by definition, the strata of $\mathcal{V}$ satisfy the frontier condition, and $\mathcal{V}$ is locally topologically trivial along each stratum V, with fibers the cone on a compact pseudomanifold $L_{V,X}$, the *link* of V in X. Each stratum V, and also its closure $\overline{V}$, get an induced stratification of the same type. Important examples are provided by a complex algebraic (or analytic) Whitney stratification of a reduced complex algebraic (or compact complex analytic) variety.

· For each $V \in \mathcal{V}$, let $IC_{\overline{V}}$ be the intersection cohomology complex [13] associated to the closure of V in X. This is a $\mathcal{V}$-constructible complex of sheaves (i.e., the restrictions of its cohomology sheaves to strata $W < V$ are locally constant), satisfying the normalization property that $IC_{\overline{V}}|_V = \mathbb{Q}_V$ (following Borel's indexing conventions). After extending by zero, we regard all these intersection chain sheaves as complexes on X. Let us fix for each $W \in \mathcal{V}$ a point $w \in W$ with inclusion $i_w : \{w\} \hookrightarrow X$. We now define a constructible function $ic_{\overline{V}} \in F_{\mathcal{V}}(X)$ by taking stalkwise the Euler characteristic for the complex $IC_{\overline{V}}$. That is, for $w \in W < V$ we let

$$ic_{\overline{V}}(w) := \chi(i_w^* IC_{\overline{V}}) = \chi(IH^*(c^\circ L_{W,V})) \overset{\text{def}}{=} I\chi(c^\circ L_{W,V}), \qquad (2\text{-}5)$$

where $c^\circ L_{W,V}$ denotes the open cone on the link $L_{W,V}$ of W in $\overline{V}$, and $I\chi(-)$ stands for the intersection homology Euler characteristic. Moreover,

$$ic_{\overline{V}}|_V = 1_V. \qquad (2\text{-}6)$$

Since clearly $\operatorname{supp}(ic_{\overline{V}}) = \overline{V}$, it is now easy to see that the collection

$$\mathcal{B}_3 := \{ ic_{\overline{V}} \mid V \in \mathcal{V} \}$$

is another distinguished basis of $F_{\mathcal{V}}(X)$. Indeed, by (2-6), the transition matrix to the basis $\{1_V\}$ is upper triangular with respect to $\leq$, with all diagonal entries equal to 1, so it is invertible. The advantage of working with the latter basis is that it carries more information about the geometry of the chosen stratification.

Now assume that X has an open dense stratum $S \in \mathcal{V}$ so that $V \leq S$ for all $V \in \mathcal{V}$, e.g., X is an irreducible reduced complex algebraic (resp., compact complex analytic) variety. For each $V \in \mathcal{V} \setminus \{S\}$ define inductively

$$\widehat{ic}(\overline{V}) := ic_{\overline{V}} - \sum_{W < V} \widehat{ic}(\overline{W}) \cdot I\chi(c^\circ L_{W,V}) \in F_{\mathcal{V}}(X). \qquad (2\text{-}7)$$

Then any $\mathcal{V}$-constructible function $\alpha \in F_{\mathcal{V}}(X)$ can be represented with respect to the basis $\{ic_{\overline{V}} \mid V \in \mathcal{V}\}$ by the following identity (see [9, Theorem 3.1]):

$$\alpha = \alpha(s) \cdot ic_X + \sum_{V < S} \left(\alpha(v) - \alpha(s) \cdot I\chi(c^\circ L_{V,Y}) \right) \cdot \widehat{ic}(\overline{V}). \qquad (2\text{-}8)$$

In the particular case when $\alpha|_S = 0$, i.e., $\operatorname{supp}(\alpha) \subset X \setminus S$, this reduces to the identity

$$\alpha = \sum_{V < S} \alpha(V) \cdot \widehat{ic}(\overline{V}), \qquad (2\text{-}9)$$

which will become very important in the context of computing Milnor classes of singular complex hypersurfaces. Also, if we plug $\alpha = 1_X$ in equation (2-8), we obtain under the assumptions in this paragraph the following comparison formula (also valid if we replace X by the closure of any given stratum of $\mathcal{V}$):

$$1_X = ic_X + \sum_{V < S} \left(1 - I\chi(c^\circ L_{V,Y})\right) \cdot \widehat{ic}(\overline{V}). \qquad (2\text{-}10)$$

3. Chern classes of singular varieties

For the rest of the paper we specialize to the complex algebraic (respectively, compact complex analytic) context, with X a reduced complex algebraic (resp., compact complex analytic) variety. There are several generalizations of the (total) Chern class of complex manifolds to the context of such singular varieties. Among these we mention here the Chern–MacPherson class [15] and the Fulton–Johnson class [10; 11]. Both coincide with the Poincaré dual of the Chern class if the variety is smooth.

3.1. The Chern–MacPherson class. The group $F_c(X)$ of complex algebraically (resp., analytically) constructible functions is defined as the direct limit of groups $F_{\mathcal{V}}(X)$, with respect to the directed system $\{\mathcal{V}\}$ of Whitney stratifications of X. Moreover, there is a functorial pushdown transformation of constructible functions, namely, a proper complex algebraic (resp., analytic) map $f : X \to Y$ induces a group homomorphism

$$f_* : F_c(X) \to F_c(Y),$$

defined by

$$f_*(\alpha)(y) := \chi(\alpha|_{f^{-1}(y)}),$$

for $\chi : F_c(X) \to \mathbb{Z}$ the constructible function which for a closed algebraic (resp., analytic) subspace Z of X is given by

$$\chi(1_Z) := \chi(H^*(Z)) = \chi(Z).$$

In particular, for such a closed subset $Z \subset X$ we have that

$$f_*(1_Z)(y) = \chi(Z \cap f^{-1}(y)).$$

The fact that the pushdown f_* is well-defined requires a stratification of the morphism f (see [15]).

The Chern class transformation of MacPherson [15] is the group homomorphism

$$c_* : F_c(X) \to H_{2*}^{BM}(X; \mathbb{Z}),$$

which commutes with proper pushdowns, and is uniquely characterized by this property together with the normalization axiom asserting that

$$c_*(1_X) = c^*(TX) \cap [X]$$

if X is a complex algebraic (resp., analytic) manifold. Here $c^*(TX)$ is the Chern cohomology class of the tangent bundle TX. Also $H_{2*}^{BM}(-)$ stands for the even-dimensional Borel–Moore homology. The Chern–MacPherson class of X is then defined as

$$c_*(X) := c_*(1_X) \in H_{2*}^{BM}(X; \mathbb{Z}).$$

If X is compact, the degree of $c_*(X)$ is just $\chi(X)$, the topological Euler characteristic of X. Similarly, we set

$$Ic_*(X) := c_*(ic_X),$$

which is another possible extension of Chern classes of manifolds to the singular setting. Of course, if X is smooth then $c_*(X) = Ic_*(X)$, but in general they differ for singular varieties, their difference being a measure of the singular locus, which, moreover, is computable in terms of the geometry of the stratification. Indeed, by applying c_* to the identity (2-10), we obtain the following comparison formula:

$$c_*(X) - Ic_*(X) = \sum_{V < S} \left(1 - I\chi(c^\circ L_{V,Y})\right) \cdot \widehat{Ic}_*(\overline{V}). \tag{3-1}$$

If X is compact, the degree of $Ic_*(X)$ is just $I\chi(X)$, the intersection homology Euler characteristic of X.

3.2. The Fulton–Johnson class. Assume that X is a local complete intersection embedded in a complex manifold M with inclusion

$$X \overset{i}{\hookrightarrow} M.$$

If $N_X M$ denotes the normal cone of X in M, then the *virtual tangent bundle* of X, that is,

$$T_{\mathrm{vir}} X := [i^* TM - N_X M] \in K^0(X), \tag{3-2}$$

is a well-defined element in the Grothendieck group of vector bundles on X (e.g., see [11][Ex.4.2.6]), so one can associate to the pair (M, X) an *intrinsic* homology class, $c_*^{FJ}(X) \in H_{2*}^{BM}(X; \mathbb{Z})$, called the Fulton–Johnson class and defined as follows (see [10; 11]):

$$c_*^{FJ}(X) := c^*(T_{\mathrm{vir}} X) \cap [X]. \tag{3-3}$$

Of course, if X is also smooth, then $T_{\text{vir}}X$ coincides with the (class of the) usual tangent bundle of X, and $c_*^{FJ}(X)$ is in this case just the Poincaré dual of $c^*(TX)$.

4. Milnor classes of hypersurfaces

This section is devoted to comparing the two notions of Chern classes mentioned in the previous section. For simplicity, we restrict to the case when X is a hypersurface in a complex manifold M. As already mentioned, the Chern–MacPherson class and the Fulton–Johnson class coincide if X is smooth. However, they differ in the singular case. For example, if X has only isolated singularities, the difference is (up to a sign) the sum of the Milnor numbers attached to the singular points. For this reason, the difference $c_*^{FJ}(X) - c_*(X)$ is usually called the *Milnor class* of X, and is denoted by $\mathcal{M}(X)$.[1] The Milnor class is a homology class supported on the singular locus of X, and it has been recently studied by many authors using quite different methods, e.g., see [1; 2; 3; 4; 5; 6; 18; 17; 19; 20; 22]. For example, it was computed in [18] (see also [17; 22]) as a weighted sum in the Chern–MacPherson classes of closures of singular strata of X, the weights depending only on the normal information to the strata. The approach we follow here is that of [19; 20], and relies only on the simple calculus of constructible functions, as outlined in Section 2, together with a well-known specialization argument due to Verdier [21].

Assume in what follows that X is a reduced complex analytic hypersurface, which is globally defined as the zero-set of a holomorphic function $f : M \to \mathbb{D}$ with a critical value at $0 \in \mathbb{D}$, for M a compact complex manifold and $\mathbb{D}$ the open unit disc about $0 \in \mathbb{C}$. For each point $x \in X$, we have a corresponding Milnor fibration with fiber

$$M_{f,x} := B_\delta(x) \cap f^{-1}(t)$$

for appropriate choices of $0 < |t| \ll \delta \ll 1$.

Denote by L the trivial line bundle on M, obtained by pulling back by f the tangent bundle of $\mathbb{C}$. Then the virtual tangent bundle of X can be identified with

$$T_{\text{vir}}X = [TM|_X - L|_X]. \tag{4-1}$$

For each $t \neq 0$ small enough, each fiber $X_t := f^{-1}(t)$ is a compact complex manifold. Moreover, by compactness, given a regular neighborhood $\mathcal{U}$ of X in M, there is a sufficiently small t so that $X_t \subset \mathcal{U}$. Denote by i_t the corresponding

---

[1] The definition of the Milnor class usually includes a sign, but for simplicity we choose to ignore it here.

inclusion map. Also, let $r : \mathcal{U} \to X$ be the obvious deformation retract. *Verdier's specialization map* in homology is then defined as the composition

$$\psi_H = r_* \circ i_{t_*} : H_*(X_t) \to H_*(X). \tag{4-2}$$

There is also a specialization map defined on the level of constructible functions [21],

$$\psi_{CF} : F_c(M) \to F_c(X), \tag{4-3}$$

which is just the constructible function version of Deligne's nearby cycle functor for constructible complexes of sheaves. This is defined by the formula

$$\psi_{CF}(\alpha)(x) = \chi(\alpha \cdot 1_{M_{f,x}}). \tag{4-4}$$

In particular,

$$\psi_{CF}(1_M) = \mu_X \in F_c(X), \tag{4-5}$$

where $\mu_X : X \to \mathbb{Z}$ is the constructible function defined by the rule:

$$\mu_X(x) := \chi(M_{f,x}), \tag{4-6}$$

for all $x \in X$. This definition justifies the analogy with the nearby cycle functor defined on the level of constructible complexes of sheaves.

Verdier's specialization property for the Chern–MacPherson classes [21] asserts that for any $\alpha \in F_c(M)$ we have:

$$\psi_H c_*(\alpha|_{X_t}) = c_*(\psi_{CF}(\alpha)). \tag{4-7}$$

In particular, by letting $\alpha = 1_M$ and using (4-5), we have that

$$\psi_H c_*(X_t) = c_*(\mu_X). \tag{4-8}$$

We can now state the following easy (known) consequence:

PROPOSITION 4.1.

$$\mathcal{M}(X) = c_*(\tilde{\mu}_X), \tag{4-9}$$

where $\tilde{\mu}_X \in F_c(X)$ is the constructible function supported on the singular locus of X, whose value at $x \in X$ is defined by the Euler characteristic of the reduced cohomology of the corresponding Milnor fiber, i.e.,

$$\tilde{\mu}_X(x) := \chi(\tilde{H}^*(M_{f,x})). \tag{4-10}$$

PROOF. First note that, since X_t is smooth,

$$\psi_H c_*(X_t) = \psi_H c_*^{FJ}(X_t) = c_*^{FJ}(X), \tag{4-11}$$

where the last equality follows from the fact that the homology specialization map ψ_H carries (the dual of) the Chern classes of $TM|_{X_t}$ and $L|_{X_t}$ into (the dual of) the Chern classes of $TM|_X$ and $L|_X$, respectively [21].

On the other hand,

$$c_*(X) = c_*(\mu_X) - c_*(\tilde{\mu}_X), \tag{4-12}$$

so the desired identity follows by combining (4-8) and (4-11). $\qquad \square$

REMARK 4.2. Note that $\tilde{\mu}_X$ is the constructible function analogue of Deligne's vanishing cycle functor defined on constructible sheaves. Indeed,

$$\tilde{\mu}_X = \phi_{CF}(1_M), \tag{4-13}$$

where $\phi_{CF} := \psi_{CF} - i^*$, for $i^* : F_c(M) \to F_c(X)$ the pullback (restriction) of constructible functions defined by $i^*(\alpha) := \alpha \circ i$.

We are now ready to prove the main result of this note:

THEOREM 4.3. *Let M be a compact complex manifold, and X a reduced hypersurface defined by the zero-set of a holomorphic function $f : M \to \mathbb{D}$ with a critical value at the origin. Fix a Whitney stratification $\mathcal{V}$ on X, and for each stratum $V \in \mathcal{V}$ fix a point $v \in V$ with corresponding Milnor fiber $M_{f,v}$. Then the Milnor class of X, i.e., the class*

$$\mathcal{M}(X) := c_*^{FJ}(X) - c_*(X) \in H_*(X),$$

can be computed by the formula

$$\mathcal{M}(X) = \sum_{\substack{V \in \mathcal{V} \\ V \subset \mathrm{Sing}(X)}} \chi(\tilde{H}^*(M_{f,v})) \cdot \big(c_*(\overline{V}) - c_*(\overline{V} \setminus V)\big)$$

$$= \sum_{\substack{V \in \mathcal{V} \\ V \subset \mathrm{Sing}(X)}} \chi(\tilde{H}^*(M_{f,v})) \cdot \hat{c}_*(\overline{V}),$$

where for a stratum $V \in \mathcal{V}$ we let $\hat{c}_*(\overline{V})$ be defined inductively as

$$\hat{c}_*(\overline{V}) := c_*(\overline{V}) - \sum_{W < V} \hat{c}_*(\overline{W}).$$

If, moreover, X is irreducible and we let S denote the dense open stratum in X, then:

$$\mathcal{M}(X) = \sum_{V < S} \chi(\tilde{H}^*(M_{f,v})) \cdot \widehat{Ic}_*(\overline{V}), \tag{4-14}$$

where for each $V \in \mathcal{V}$, $\widehat{Ic}_*(\overline{V})$ is defined inductively by

$$\widehat{Ic}_*(\overline{V}) := Ic_*(\overline{V}) - \sum_{W < V} I\chi(c^\circ L_{W,V}) \cdot \widehat{Ic}_*(\overline{W}),$$

for $L_{W,V}$ the link of W in V.[2]

PROOF. Recall that by Proposition 4.1 we have:

$$\mathcal{M}(X) = c_*(\tilde{\mu}_X). \qquad (4\text{-}15)$$

Moreover, the function $\tilde{\mu}_X : X \to \mathbb{Z}$ is constructible with respect to the Whitney stratification $\mathcal{V}$. Therefore, as in (2-1) and (2-2), we can write

$$\tilde{\mu}_X = \sum_{V \in \mathcal{V}} \tilde{\mu}_X(v) \cdot 1_V = \sum_{V \in \mathcal{V}} \tilde{\mu}_X(v) \cdot (1_{\overline{V}} - 1_{\overline{V} \setminus V}) = \sum_{V \in \mathcal{V}} \tilde{\mu}_X(v) \cdot \hat{1}_{\overline{V}}.$$

Since smooth points have contractible Milnor fibers, only strata contained in the singular locus of X contribute to the above sums. The first part of the theorem follows from (4-15) by applying the Chern–MacPherson transformation c_* to the last two of the above equalities.

If X is irreducible with dense open stratum S, then as in (2-9) we can write

$$\tilde{\mu}_X = \sum_{V < S} \tilde{\mu}_X(v) \cdot \hat{i} c(\overline{V}).$$

By applying c_*, we obtain the desired identity (4-14) from (4-15). $\qquad \square$

As a consequence, the Chern–MacPherson class and the Fulton–Johnson class coincide in dimensions greater than the dimension of the singular locus. And it can be seen from any of the above formulae that if X has only isolated singularities, the Milnor class is (up to a sign) just the sum of the Milnor numbers at the singular points.

By combining (3-1) and (4-14) we also obtain a comparison formula for the Fulton–Johnson class $c_*^{FJ}(X)$ and the Chern class $Ic_*(X)$ defined via the intersection cohomology chain sheaf.

COROLLARY 4.4. *If X as above is a reduced irreducible hypersurface with dense open stratum S, then*

$$\mathcal{IM}(X) := c_*^{FJ}(X) - Ic_*(X) = \sum_{V < S} \widehat{Ic}_*(\overline{V}) \cdot \big(\chi(M_{f,v}) - I\chi(c^\circ L_{V,X})\big).$$

$$(4\text{-}16)$$

Note that by constructible function calculus, we have that

$$\mathcal{IM}(X) = c_*(\widetilde{I\mu}_X), \qquad (4\text{-}17)$$

for $\widetilde{I\mu}_X : X \to \mathbb{Z}$ the $\mathcal{V}$-constructible function whose value at $v \in V$ is given by

$$\widetilde{I\mu}_X(v) = \chi(M_{f,v}) - I\chi(c^\circ L_{V,X}). \qquad (4\text{-}18)$$

---

[2]By the functoriality of c_*, we can regard all classes $c_*(\overline{V})$, $\hat{c}_*(\overline{V})$ and $\widehat{Ic}_*(\overline{V})$ associated to a stratum $V \in \mathcal{V}$ as homology classes in $H_*(X)$. This is the reason why we apply the Chern–MacPherson transformation c_* only to closed subvarieties of X.

By its definition, $\widetilde{I\mu}_X$ is supported on the singular locus of X, so (2-9) can be used directly to prove (4-16).

REMARK 4.5. Our formula (4-16) should be compared to the topological formula of Cappell and Shaneson [7; 8] for the Goresky–MacPherson L-class [12] of an irreducible reduced complex hypersurface $X \subset M$ as above, namely,

$$L_*(T_{\mathrm{vir}}X) - L_*(X) = \sum_{V < S} L_*(\overline{V}) \cdot \sigma(\mathrm{lk}(V)), \qquad (4\text{-}19)$$

where $\sigma(\mathrm{lk}(V)) \in \mathbb{Z}$ is a certain signature invariant associated to the link pair of the stratum V in (M, X). Here $L_*(T_{\mathrm{vir}}X) := L^*(T_{\mathrm{vir}}X) \cap [X]$, with L^* the L-polynomial of Hirzebruch [14] defined in terms of the power series $x/\tanh(x)$. The comparison is motivated by the fact that the L-class of a singular variety X is a topological invariant associated to the intersection cohomology complex of the variety. We should point out that the Cappell–Shaneson formula holds in much greater generality, namely for real codimension two PL embeddings with even codimension strata, and its proof relies on powerful algebraic cobordism decompositions of self-dual sheaves. However, we believe that in the context of complex algebraic/analytic geometry, a simpler proof could be given by using a specialization argument similar to the one presented here.

More generally, assume that $i : X \hookrightarrow M$ is a regular embedding in codimension one of complex algebraic (resp., compact complex analytic) spaces with M smooth. Then X is locally defined in M by one equation $\{f = 0\}$, and the specialization map $\psi_{CF} : F_c(M) \to F_c(X)$ is still well-defined, as it is independent of the chosen local equation for X. In particular, we still have that $\psi_{CF}(1_M) = \mu$, whose value at a point $x \in X$ is given by the Euler characteristic of a local Milnor fiber at x. In other words, if $\{f = 0\}$ is a defining equation for X near x, then

$$\tilde{\mu}_X(x) := \chi(\tilde{H}^*(M_{f,x})), \qquad (4\text{-}20)$$

for $M_{f,x}$ the corresponding Milnor fiber. Then arguments similar to those used in this section apply to this more general situation, and yield the following result (see [20, Corollary 0.2] for equation (4-21) below):

THEOREM 4.6. *Let $i : X \hookrightarrow M$ be a regular embedding in codimension one of complex algebraic (resp., compact complex analytic) spaces with M smooth. Then,*

$$\mathcal{M}(X) = c^*(N_X M)^{-1} \cap c_*(\tilde{\mu}_X), \qquad (4\text{-}21)$$

with $\tilde{\mu}$ the constructible function supported on the singular locus of X, whose value at a point $x \in X$ is given by the Euler characteristic of the reduced cohomology of a local Milnor fiber at x. So, if we assume X irreducible with dense *open stratum S, then in the notations of Theorem 4.3 we get*

$$\mathcal{M}(X) = \sum_{V<S} c^*(N_X M)^{-1} \cap \left(c_*(\overline{V}) - c_*(\overline{V} \setminus V)\right) \cdot \tilde{\mu}_X(v)$$

$$= \sum_{V<S} c^*(N_X M)^{-1} \cap \hat{c}_*(\overline{V}) \cdot \tilde{\mu}_X(v)$$

$$= \sum_{V<S} c^*(N_X M)^{-1} \cap \widehat{Ic}_*(\overline{V}) \cdot \tilde{\mu}_X(v).$$

Similar considerations apply to $\mathcal{IM}(X)$. (Again, by functoriality, we regard all classes defined on the closure of a given stratum as homology classes in X.)

We conclude this note by recalling some functoriality results for the Milnor class of hypersurfaces (see [20; 24] for complete details). More precisely, we are concerned with the behavior of the Milnor class under a proper pushdown. Similar results were obtained in [9] for the Chern–MacPherson classes $c_*(-)$ and $Ic_*(-)$, respectively.

Let us consider the cartesian diagram

$$\begin{array}{ccc} \tilde{X} & \xrightarrow{\ j\ } & \tilde{M} \\ {\scriptstyle f}\downarrow & & \downarrow{\scriptstyle \pi} \\ X & \xrightarrow{\ i\ } & M \end{array}$$

with M and $\tilde{M}$ compact analytic manifolds, and $\pi : \tilde{M} \to M$ a proper morphism. Also assume that i and j are regular closed embeddings of (local) codimension one, with M irreducible. Then it's easy to see that $N_{\tilde{X}}\tilde{M} \simeq f^*(N_X M)$. Therefore, by (4-21) and the projection formula, one has

$$f_*\mathcal{M}(\tilde{X}) = f_*(c^*(N_{\tilde{X}}\tilde{M})^{-1} \cap c_*(\tilde{\mu}_{\tilde{X}}))$$

$$= f_*(f^*c^*(N_X M)^{-1} \cap c_*(\tilde{\mu}_{\tilde{X}}))$$

$$= c^*(N_X M)^{-1} \cap f_*c_*(\tilde{\mu}_{\tilde{X}}).$$

Next, by the functoriality of c_* and the definition of $\tilde{\mu}_{\tilde{X}}$ in (4-13) we obtain

$$f_*c_*(\tilde{\mu}_{\tilde{X}}) = c_* f_*(\tilde{\mu}_{\tilde{X}}) = c_* f_*\phi_{CF}(1_{\tilde{M}}) = c_*\phi_{CF}(\pi_*(1_{\tilde{M}})),$$

where the last identity follows by proper base change. Assume now that π (hence also f) is an *Euler morphism*, i.e., the Euler characteristics of all its fibers are the same (e.g., π is smooth), and denote this value by χ_f. Then $\pi_*(1_{\tilde{M}}) = \chi_f \cdot 1_M$, and it follows in this case that

$$f_*\mathcal{M}(\tilde{X}) = \chi_f \cdot \mathcal{M}(X). \tag{4-22}$$

But in the case of a general morphism we have that

$$f_*\mathcal{M}(\tilde{X}) = \chi_f \cdot \mathcal{M}(X) + c^*(N_X M)^{-1} \cap c_*\phi_{CF}(\alpha), \qquad (4\text{-}23)$$

for $\alpha := \pi_*(1_{\tilde{M}}) - \chi_f \cdot 1_M$, with χ_f the Euler characteristic of the generic fiber of π. Note that α is supported on the critical locus of the morphism π.

To this end, we note that the above considerations can also be used to study the push-forward of the class $\mathcal{IM}(\tilde{X})$ in the case when $\tilde{X}$ is pure-dimensional and X is irreducible and reduced. Let us choose a stratification $\mathcal{V}$ on X with dense open stratum S, so that $f_*(1_{\tilde{X}}), f_*(ic_{\tilde{X}}) \in F_{\mathcal{V}}(X)$ (e.g., choose $\tilde{\mathcal{V}}$ and $\mathcal{V}$ complex Whitney stratifications on $\tilde{X}$ and X, respectively, so that f is a stratified submersion, and $1_X, ic_{\tilde{X}} \in F_{\tilde{\mathcal{V}}}(X)$). Then, since

$$\mathcal{IM}(\tilde{X}) = \mathcal{M}(\tilde{X}) + (c_*(\tilde{X}) - Ic_*(\tilde{X})),$$

a formula for $f_*\mathcal{IM}(\tilde{X})$ can be derived by using (4-23), together with the formulae from [9, Propositions 3.4 and 3.6] for the push-forward of the Chern classes $c_*(\tilde{X})$ and $Ic_*(\tilde{X})$, respectively. We leave the details as an exercise for the interested reader. We only want to point out that for an Euler morphism (with smooth generic fiber), we obtain

$$f_*\mathcal{IM}(\tilde{X})$$
$$= \chi_f \cdot \mathcal{IM}(X) + \sum_{V < S} \left(\chi_f I\chi(c^\circ L_{V,X}) - I\chi(f^{-1}(c^\circ L_{V,X})) \right) \cdot \widehat{ic}(\overline{V}), \quad (4\text{-}24)$$

where χ_f is the Euler characteristic of the generic fiber F of f (and π).

Acknowledgements

I am grateful to Jörg Schürmann for many inspiring conversations on this subject. I also thank Sylvain Cappell, Anatoly Libgober and Julius Shaneson for constant encouragement and advice.

References

[1] Aluffi, P., *Chern classes for singular hypersurfaces*, Trans. Amer. Math. Soc. **351** (1999), no. 10, 3989–4026.

[2] Aluffi, P., *Characteristic classes of singular varieties*, in Topics in cohomological studies of algebraic varieties, 1–32, Birkhäuser, Basel, 2005.

[3] Brasselet, J.-P., *From Chern classes to Milnor classes–a history of characteristic classes for singular varieties*, in Singularities (Sapporo 1998), 31–52, Adv. Stud. Pure Math., 29, Kinokuniya, Tokyo, 2000.

[4] Brasselet, J.-P., Lehmann, D., Seade, J., Suwa, T., *Milnor numbers and classes of local complete intersections*, Proc. Japan Acad. Ser. A Math. Sci. **75** (1999), no. 10, 179–183.

[5] Brasselet, J.-P., Lehmann, D., Seade, J., Suwa, T., *Milnor classes of local complete intersections*, Trans. Amer. Math. Soc. **354** (2001), 1351–1371.

[6] Brasselet, J.-P., Seade, J., Suwa, T., *An explicit cycle representing the Fulton–Johnson class, I*, Séminaire & Congrès **10** (2005), p. 21–38.

[7] Cappell, S., Shaneson, J., *Characteristic classes, singular embeddings, and intersection homology*, Proc. Natl. Acad. Sci. USA, Vol **84** (1991), 3954–3956.

[8] Cappell, S., Shaneson, J., *Singular spaces, characteristic classes, and intersection homology*, Ann. of Math. **134** (1991), 325–374.

[9] Cappell, S. E., Maxim, L. G., Shaneson, J. L., *Euler characteristics of algebraic varieties*, Comm. Pure Appl. Math. **61** (2008), no. 3, 409–421.

[10] Fulton, W., Johnson, K., *Canonical classes on singular varieties*, Manuscripta Math **32** (1980), 381–389.

[11] Fulton, W., *Intersection theory*, Second edition. Springer, Berlin, 1998.

[12] Goresky, M., MacPherson, R., *Intersection homology theory*, Topology **19** (1980), no. 2, 135–162.

[13] Goresky, M., MacPherson, R., *Intersection homology theory, II*, Inv. Math. **72** (1983), no. 2, 77–129.

[14] Hirzebruch, F., *Topological methods in algebraic geometry*, Springer, New York, 1966.

[15] MacPherson, R., *Chern classes for singular algebraic varieties*, Ann. of Math. (2) **100** (1974), 423–432.

[16] Parusiński, A., *A generalization of the Milnor number*, Math. Ann. **281** (1988), 247–254.

[17] Parusiński, A., *Characteristic classes of singular varieties*, in Singularity theory and its applications, 347–367, Adv. Stud. Pure Math., 43, Math. Soc. Japan, Tokyo, 2006.

[18] Parusiński, A., Pragacz, P., *Characteristic classes of hypersurfaces and characteristic cycles*, J. Algebraic Geom. **10** (2001), no. 1, 63–79.

[19] Schürmann, J., *Lectures on characteristic classes of constructible functions*, in Topics in cohomological studies of algebraic varieties, 175–201, Birkhäuser, Basel, 2005.

[20] Schürmann, J., *A generalized Verdier-type Riemann–Roch theorem for Chern–Schwartz–MacPherson classes*, arXiv:math/0202175.

[21] Verdier, J.-L., *Spécialisation des classes de Chern*, Astérisque **82–83**, 149–159 (1981).

[22] Yokura, S., *On characteristic classes of complete intersections*, in Algebraic geometry: Hirzebruch 70 (Warsaw, 1998), 349–369, Contemp. Math., **241**, Amer. Math. Soc., Providence, RI, 1999.

[23] Yokura, S., *On a Verdier-type Riemann–Roch for Chern–Schwartz–MacPherson class*, Topology and its Applications 94 (1999), no. 1–3, 315–327.

[24] Yokura, S., *An application of bivariant theory to Milnor classes*, Topology and its Applications 115 (2001), 43–61.

LAURENTIU MAXIM
LAURENTIU MAXIM
DEPARTMENT OF MATHEMATICS
UNIVERSITY OF WISCONSIN-MADISON
480 LINCOLN DRIVE
MADISON, WI 53706-1388
UNITED STATES
maxim@math.wisc.edu

Topology of Stratified Spaces
MSRI Publications
Volume **58**, 2011

An introduction to intersection homology with general perversity functions

GREG FRIEDMAN

ABSTRACT. We provide an expository survey of the different notions of perversity in intersection homology and how different perversities require different definitions of intersection homology theory itself. We trace the key ideas from the introduction of intersection homology by Goresky and MacPherson through to the recent and ongoing work of the author and others.

CONTENTS

1. Introduction

When Goresky and MacPherson first introduced intersection homology [32], they required its perversity parameters to satisfy a fairly rigid set of constraints. Their perversities were functions on the codimensions of strata, $\bar{p} : \mathbb{Z}^{\geq 2} \to \mathbb{Z}$,

2000 Mathematics Subject Classification: Primary: 55N33, 57N80; Secondary: 55N45, 55N30, 57P10.
Keywords: intersection homology, perversity, pseudomanifold, Poincaré duality, Deligne sheaf, intersection pairing.

satisfying

$$\bar{p}(2) = 0 \text{ and } \bar{p}(k) \leq \bar{p}(k+1) \leq \bar{p}(k) + 1.$$

These strict requirements were necessary for Goresky and MacPherson to achieve their initial goals for intersection homology: that the intersection homology groups $I^{\bar{p}}H_*(X)$ should satisfy a generalized form of Poincaré duality for stratified pseudomanifolds and that they should be topological invariants, i.e., they should be independent of the choice of stratification of X.

In the ensuing years, perversity parameters have evolved as the applications of intersection homology have evolved, and in many cases the basic definitions of intersection homology itself have had to evolve as well. Today, there are important results that utilize the most general possible notion of a perversity as a function

$$\bar{p} : \{\text{components of singular strata of a stratified pseudomanifold}\} \to \mathbb{Z}.$$

In this setting, one usually loses topological invariance of intersection homology (though this should be seen not as a loss but as an opportunity to study stratification data), but duality results remain, at least if one chooses the right generalizations of intersection homology. Complicating this choice is the fact that there are a variety of approaches to intersection homology to begin with, even using Goresky and MacPherson's perversities. These include (at the least) the original simplicial chain definition [32]; Goresky and MacPherson's Deligne sheaves [33; 6]; King's singular chain intersection homology [32]; Cheeger's L^2 cohomology and L^2 Hodge theory [16]; perverse differential forms on Thom–Mather stratified spaces (and, later, on unfoldable spaces [7]), first published by Brylinski [8] but attributed to Goresky and MacPherson; and the theory of perverse sheaves [4]. Work to find the "correct" versions of these theories when general perversities are allowed has been performed by the author, using stratified coefficients for simplicial and singular intersection chains [26]; by Saralegi, using "relative" intersection homology and perverse differential forms in [54]; and by the author, generalizing the Deligne sheaf in [22]. Special cases of non-Goresky–MacPherson perversities in the L^2 Hodge theory setting have also been considered by Hausel, Hunsicker, and Mazzeo [37]; Hunsicker and Mazzeo [39]; and Hunsicker [38]. And arbitrary perversities have been available from the start in the theory of perverse sheaves!

This paper is intended to serve as something of a guidebook to the different notions of perversities and as an introduction to some new and exciting work in this area. Each stage of development of the idea of perversities was accompanied by a flurry of re-examinings of what it means to have an intersection homology theory and what spaces such a theory can handle as input, and each such re-examining had to happen within one or more of the contexts listed above. In many cases, the outcome of this re-examination led to a modification or expansion of the basic definitions. This has resulted in a, quite justified, parade of papers consumed with working through all the technical details. However, technicalities often have the unintended effect of obscuring the few key main ideas. Our goal then is to present these key ideas and their consequences in an expository fashion, referring the reader to the relevant papers for further technical developments and results. We hope that such a survey will provide something of an introduction to and overview of the recent and ongoing work of the author, but we also hope to provide a readable (and hopefully accurate!) historical account of this particular chain of ideas and an overview of the work of the many researchers who have contributed to it. We additionally hope that such an overview might constitute a suitable introduction for those wishing to learn about the basics of intersection homology and as preparation for those wishing to pursue the many intriguing new applications that general perversities bring to the theory.

This exposition is not meant to provide a comprehensive historical account but merely to cover one particular line of development. We will focus primarily on the approaches to intersection homology by simplicial and singular chains and by sheaf theory. We will touch only tangentially upon perverse differential forms when we consider Saralegi's work in Section 10; we advise the reader to consult [54] for the state of the art, as well as references to prior work, in this area. Also, we will not discuss L^2-cohomology. This is a very active field of research, as is well-demonstrated elsewhere in this volume [30], but the study of L^2-cohomology and L^2 Hodge theories that yield intersection homology with general perversities remains under development. The reader should consult the papers cited above for the work that has been done so far. We will briefly discuss perverse sheaves in Section 8.2, but the reader should consult [4] or any of the variety of fine surveys on perverse sheaves that have appeared since for more details.

We will not go into many of the myriad results and applications of intersection homology theory, especially those beyond topology proper in analysis, algebraic geometry, and representation theory. For broader references on intersection homology, the reader might start with [6; 42; 2]. These are also excellent sources for the material we will be assuming regarding sheaf theory and derived categories and functors.

We proceed roughly in historical order as follows: Section 2 provides the original Goresky–MacPherson definitions of PL pseudomanifolds and PL chain intersection homology. We also begin to look closely at the cone formula for intersection homology, which will have an important role to play throughout. In Section 3, we discuss the reasons for the original Goresky–MacPherson condi-

tions on perversities and examine some consequences, and we introduce King's
singular intersection chains. In Section 4, we turn to the sheaf-theoretic definition of intersection homology and introduce the Deligne sheaf. We discuss
the intersection homology version of Poincaré duality, then we look at our
first example of an intersection homology result that utilizes a non-Goresky–
MacPherson perversity, the Cappell–Shaneson superduality theorem.

In Section 5, we discuss *subperversities* and *superperversities*. Here we first
observe the schism that occurs between chain-theoretic and sheaf-theoretic intersection homology when perversities do not satisfy the Goresky–MacPherson
conditions. Section 6 introduces *stratified coefficients*, which were developed
by the author in order to correct the chain version of intersection homology for
it to conform with the Deligne sheaf version.

In Section 7, we discuss the further evolution of the chain theory to the most
general possible perversities and the ensuing results and applications. Section 8
contains the further generalization of the Deligne sheaf to general perversities,
as well as a brief discussion of perverse sheaves and how general perversity
intersection homology arises in that setting. Some indications of recent work
and work-in-progress with these general perversities is provided in Section 9.

Finally, Sections 10 and 11 discuss some alternative approaches to intersection homology with general perversities. In Section 10, we discuss Saralegi's
"relative intersection chains", which are equivalent to the author's stratified coefficients when both are defined. In Section 11, we present the work of Habegger
and Saper from [35]. This work encompasses another option to correcting the
schism presented in Section 5 by providing a sheaf theory that agrees with King's
singular chains, rather than the other way around; however, the Habegger–Saper
theory remains rather restrictive with respect to acceptable perversities.

2. The original definition of intersection homology

We begin by recalling the original definition of intersection homology as
given by Goresky and MacPherson in [32]. We must start with the spaces that
intersection homology is intended to study.

2.1. Piecewise linear stratified pseudomanifolds. The spaces considered by
Goresky and MacPherson in [32] were *piecewise linear (PL) stratified pseudomanifolds*. An n-dimensional PL stratified pseudomanifold X is a piecewise
linear space (meaning it is endowed with a compatible family of triangulations)
that also possesses a filtration by closed PL subspaces (the stratification)

$$X = X^n \supset X^{n-2} \supset X^{n-3} \supset \cdots \supset X^1 \supset X^0 \supset X^{-1} = \varnothing$$

satisfying the following properties:

(a) $X - X^{n-2}$ is dense in X,
(b) for each $k \geq 2$, $X^{n-k} - X^{n-k-1}$ is either empty or is an $n - k$ dimensional PL manifold,
(c) if $x \in X^{n-k} - X^{n-k-1}$, then x has a *distinguished neighborhood* N that is PL homeomorphic to $\mathbb{R}^{n-k} \times cL$, where cL is the open cone on a compact $k - 1$ dimensional stratified PL pseudomanifold L. Also, the stratification of L must be compatible with the stratification of X.

A PL stratified pseudomanifold X is oriented (or orientable) if $X - X^{n-2}$ has the same property.

A few aspects of this definition deserve comment. Firstly, the definition is inductive: to define an n-dimensional PL stratified pseudomanifold, we must already know what a $k - 1$ dimensional PL stratified pseudomanifold is for $k - 1 < n$. The base case occurs for $n = 0$; a 0-pseudomanifold is a discrete set of points. Secondly, there is a gap from n to $n - 2$ in the filtration indices. This is more-or-less intended to avoid issues of pseudomanifolds with boundary, although there are now established ways of dealing with these issues that we will return to below in Section 5.

The sets X^i are called *skeleta*, and we can verify from condition (b) that each has dimension $\leq i$ as a PL complex. The sets $X_i := X^i - X^{i-1}$ are traditionally called *strata*, though it will be more useful for us to use this term for the connected components of X_i, and we will favor this latter usage rather than speaking of "stratum components."[1] The strata of $X^n - X^{n-2}$ are called *regular strata*, and the other strata are called *singular strata*. The space L is called the link of x or of the stratum containing x. For a PL stratified pseudomanifold L is uniquely determined up to PL homeomorphism by the stratum containing x. The cone cL obtains a natural stratification from that of L: $(cL)^0$ is the cone point and for $i > 0$, $(cL)^i = L^{i-1} \times (0, 1) \subset cL$, where we think of cL as

$$\frac{L \times [0, 1)}{(x, 0) \sim (y, 0)}.$$

The compatibility condition of item (c) of the definition means that the PL homeomorphism should take $X^i \cap N$ to $\mathbb{R}^{n-k} \times (cL)^{i-(n-k)}$.

Roughly, the definition tells us the following. An n-dimensional PL stratified pseudomanifold X is mostly the n-manifold $X - X^{n-2}$, which is dense in X. (In much of the literature, X^{n-2} is also referred to as Σ, the singular locus of X.) The rest of X is made up of manifolds of various dimensions, and these must fit together nicely, in the sense that each point in each stratum should have

---

[1] It is perhaps worth noting here that the notation we employ throughout mostly will be consistent with the author's own work, though not necessarily with all historical sources.

a neighborhood that is a trivial fiber bundle, whose fibers are cones on lower-dimensional stratified spaces.

We should note that examples of such spaces are copious. Any complex analytic or algebraic variety can be given such a structure (see [6, Section IV]), as can certain quotient spaces of manifolds by group actions. PL pseudomanifolds occur classically as spaces that can be obtained from a pile of n-simplices by gluing in such a way that each $n - 1$ face of an n-simplex is glued to exactly one $n - 1$ face of one other n-simplex. (Another classical condition is that we should be able to move from any simplex to any other, passing only through interiors of $n - 1$ faces. This translates to say that $X - X^{n-2}$ is path connected, but we will not concern ourselves with this condition.) Other simple examples arise by taking open cones on manifolds (naturally, given the definition), by suspending manifolds (or by repeated suspensions), by gluing manifolds and pseudomanifolds together in allowable ways, etc. One can construct many useful examples by such procedures as "start with this manifold, suspend it, cross that with a circle, suspend again, ..." For more detailed examples, the reader might consult [6; 2; 44].

More general notions of stratified spaces have co-evolved with the various approaches to intersection homology, mostly by dropping or weakening requirements. We shall attempt to indicate this evolution as we progress.

2.2. Perversities. Besides the spaces on which one is to define intersection homology, the other input is the perversity parameter. In the original Goresky–MacPherson definition, a perversity $\bar{p}$ is a function from the integers ≥ 2 to the non-negative integers satisfying the following properties:

(a) $\bar{p}(2) = 0$.
(b) $\bar{p}(k) \leq \bar{p}(k + 1) \leq \bar{p}(k) + 1$.

These conditions say that a perversity is something like a sub-step function. It starts at 0, and then each time the input increases by one, the output either stays the same or increases by one. Some of the most commonly used perversities include the zero perversity $\bar{0}(k) = 0$, the top perversity $\bar{t}(k) = k - 2$, the lower-middle perversity $\bar{m}(k) = \lfloor \frac{k-2}{2} \rfloor$, and the upper middle perversity $\bar{n}(k) = \lfloor \frac{k-1}{2} \rfloor$.

The idea of the perversity is that the input number k represents the codimension of a stratum $X_{n-k} = X^{n-k} - X^{n-k-1}$ of an n-dimensional PL stratified pseudomanifold, while the output will control the extent to which the PL chains in our homology computations will be permitted to interact with these strata.

The reason for the arcane restrictions on $\bar{p}$ will be made clear below in Section 3. We will call any perversity satisfying conditions (a) and (b) a *Goresky–MacPherson perversity*, or a *GM perversity*.

2.3. Intersection homology.

At last, we are ready to discuss intersection homology.

Let X be an n-dimensional PL stratified pseudomanifold, and let $C_*^T(X)$ denote the simplicial chain complex of X with respect to the triangulation T. The PL chain complex $C_*(X)$ is defined to be $\varinjlim_T C_*^T(X)$, where the limit is taken with respect to the directed set of compatible triangulations. This PL chain complex is utilized by Goresky and MacPherson in [32] (see also [6]), and it is useful in a variety of other contexts (see [46], for instance). However, it turns out that this is somewhat technical overkill for the basic definition of intersection homology, as what follows can also be performed in $C_*^T(X)$, assuming T is sufficiently refined with respect to the the stratification of X (for example, pick any T, take two barycentric subdivisions, and you're set to go — see [45]).

We now define the perversity $\bar{p}$ intersection chain complex $I^{\bar{p}}C_*(X) \subset C_*(X)$. We say that a PL j-simplex σ is $\bar{p}$-allowable provided

$$\dim(\sigma \cap X_{n-k}) \leq j - k + \bar{p}(k)$$

for all $k \geq 2$. We say that a PL i-chain $\xi \in C_i(X)$ is $\bar{p}$-allowable if each i-simplex occurring with nonzero coefficient in ξ is $\bar{p}$-allowable and if each $i - 1$ simplex occurring with nonzero coefficient in $\partial\xi$ is $\bar{p}$-allowable. Notice that the simplices in ξ must satisfy the simplex allowability condition with $j = i$ while the simplices of $\partial\xi$ must satisfy the condition with $j = i - 1$.

Then $I^{\bar{p}}C_*(X)$ is defined to be the complex of allowable chains. It follows immediately from the definition that this is indeed a chain complex. The intersection homology groups are $I^{\bar{p}}H_*(X) = H_*(I^{\bar{p}}C_*(X))$.

Some remarks are in order.

REMARK 2.1. The allowability condition at first seems rather mysterious. However, the condition $\dim(\sigma \cap X_{n-k}) \leq j - k$ would be precisely the requirement that σ and X_{n-k} intersect in general position if X_{n-k} were a submanifold of X. Thus introducing a perversity can be seen as allowing deviation from general position to a degree determined by the perversity. This seems to be the origin of the nomenclature.

REMARK 2.2. It is a key observation that if ξ is an i-chain, then it is not every $i - 1$ face of every i-simplex of ξ that must be checked for its allowability, but only those that survive in $\partial\xi$. Boundary pieces that cancel out do not need to be checked for allowability. This seemingly minor point accounts for many subtle phenomena, including the next remark.

REMARK 2.3. Intersection homology with coefficients $I^{\bar{p}}H_*(X; G)$ can be defined readily enough beginning with $C_*(X; G)$ instead of $C_*(X)$. However, $I^{\bar{p}}C_*(X; G)$ is generally NOT the same as $I^{\bar{p}}C_*(X) \otimes G$. This is precisely due to the boundary cancellation behavior: extra boundary cancellation in chains may occur when G is a group with torsion, leading to allowable chains in $I^{\bar{p}}C_*(X; G)$ that do not come from any G-linear combinations of allowable chains in $I^{\bar{p}}C_*(X; \mathbb{Z})$. For more details on this issue, including many examples, the reader might consult [29].

REMARK 2.4. In [32], Goresky and MacPherson stated the allowability condition in terms of skeleta, not strata. In other words, they define a j-simplex to be allowable if

$$\dim(\sigma \cap X^{n-k}) \leq j - k + \bar{p}(k)$$

for all $k \geq 2$. However, it is not difficult to check that the two conditions are equivalent for the perversities we are presently considering. When we move on to more general perversities, below, it becomes necessary to state the condition in terms of strata rather than in terms of skeleta.

2.4. Cones. It turns out that understanding cones plays a crucial role in almost all else in intersection homology theory, which perhaps should not be too surprising, as pseudomanifolds are all locally products of cones with euclidean space. Most of the deepest proofs concerning intersection homology can be reduced in some way to what happens in these distinguished neighborhoods. The euclidean part turns out not to cause too much trouble, but cones possess interesting and important behavior.

So let L be a compact $k - 1$ dimensional PL stratified pseudomanifold, and let cL be the open cone on L. Checking allowability of a j-simplex σ with respect to the cone vertex $\{v\} = (cL)^0$ is a simple matter, since the dimension of $\sigma \cap \{v\}$ can be at most 0. Thus σ can allowably intersect v if and only if $0 \leq j - k + \bar{p}(k)$, i.e., if $j \geq k - \bar{p}(k)$. Now, suppose ξ is an allowable i-cycle in L. We can form the chain $\bar{c}\xi \in I^{\bar{p}}C_{i+1}(cL)$ by taking the cone on each simplex in the chain (by extending each simplex linearly to the cone point). We can check using the above computation (and a little more work that we'll suppress) that $\bar{c}\xi$ is allowable if $i + 1 \geq k - \bar{p}(k)$, and thus $\xi = \partial \bar{c}\xi$ is a boundary; see [6, Chapters I and II]. Similar, though slightly more complicated, computations show that any allowable cycle in cL is a boundary. Thus $I^{\bar{p}}H_i(cL) = 0$ if $i \geq k - 1 - \bar{p}(k)$. On the other hand, if $i < k - 1 - \bar{p}(k)$, then no i-chain ξ can intersect v nor can any chain of which it might be a boundary. Thus ξ is left to its own devices in $cL - v$, i.e., $I^{\bar{p}}H_i(cL) = I^{\bar{p}}H_i(cL - v) \cong I^{\bar{p}}H_i(L \times (0, 1))$. It turns out that intersection homology satisfies the Künneth theorem when one factor is euclidean space and we take the obvious product stratification (see [6, Chapter I]), or alternatively we can use the invariance of intersection homology under stratum-preserving homotopy equivalences (see [23]), and so in this range $I^{\bar{p}}H_i(cL) \cong I^{\bar{p}}H_i(L)$.

Altogether then, we have

$$I^{\bar{p}} H_i(cL^{k-1}) \cong \begin{cases} 0 & \text{if } i \geq k-1-\bar{p}(k), \\ I^{\bar{p}} H_i(L) & \text{if } i < k-1-\bar{p}(k). \end{cases} \tag{2-1}$$

We will return to this formula many times.

3. Goresky–MacPherson perversities

The reasons for the original Goresky–MacPherson conditions on perversities, as enumerated in Section 2.2, are far from obvious. Ultimately, they come down to the two initially most important properties of intersection homology: its topological invariance and its Poincaré duality.

The topological invariance property of traditional intersection homology says that when $\bar{p}$ is a Goresky–MacPherson perversity and X is a stratified pseudomanifold (PL or topological, as we'll get to soon) then $I^{\bar{p}} H_*(X)$ depends only on X and not on the choice of stratification (among those allowed by the definition). This is somewhat surprising considering how the intersection chain complex depends on the strata.

The desire for $I^{\bar{p}} H_*(X)$ to be a topological invariant leads fairly quickly to the condition that we should not allow $\bar{p}(k)$ to be negative. This will be more evident once we get to the sheaf-theoretic formulation of intersection homology, but for now, consider the cone formula (2-1) for cL^{k-1}, and suppose $\bar{p}(k) < 0$. Then we can check that no allowable PL chain may intersect v. Thus we see that the intersection homology of cL is the same as if we removed the cone point altogether. A little more work (see [22, Corollary 2.5]) leads more generally to the conclusion that if $\bar{p}(k) < 0$, then $I^{\bar{p}} H_*(X) \cong I^{\bar{p}} H_*(X - X_k)$. This would violate the topological invariance since, for example, topological invariance tells us that if M^n is a manifold then $I^{\bar{p}} H_*(M) \cong H_*(M)$, no matter how we stratify it[2]. But if we now allow, say, a locally-flat PL submanifold N^{n-k} and stratify by $M \supset N$, then if $\bar{p}(k) < 0$ we would have $H_*(M) \cong I^{\bar{p}} H_*(M) \cong I^{\bar{p}} H_*(M - N) \cong H_*(M - N)$. This presents a clear violation of topological invariance.

The second Goresky–MacPherson condition, that $\bar{p}(k) \leq \bar{p}(k+1) \leq \bar{p}(k)+1$, also derives from topological invariance considerations. The following example is provided by King [41, p. 155]. We first note that, letting SX denote the suspension of X, we have $cSX \cong \mathbb{R} \times cX$ (ignoring the stratifications). This is not hard to see topologically (recall that cX is the *open* cone on X). But now if we assume X is $k-1$ dimensional and that we take the obvious stratifications

---

[2]Note that one choice of stratification is the trivial one containing a single regular stratum, in which case it is clear from the definition that $I^{\bar{p}} H_*(M) \cong H_*(M)$.

of $\mathbb{R} \times cX$ (assuming some initial stratification on X), then

$$I^{\bar{p}} H_i(\mathbb{R} \times cX) \cong \begin{cases} 0 & \text{if } i \geq k-1-\bar{p}(k), \\ I^{\bar{p}} H_i(X) & \text{if } i < k-1-\bar{p}(k). \end{cases} \tag{3-1}$$

This follows from the cone formula (2-1) together with the intersection homology Künneth theorem, for which one term is unstratified [41] (or stratum-preserving homotopy equivalence [23]).

But now it also follows by an easy argument, using (2-1) and the Mayer–Vietoris sequence, that

$$I^{\bar{p}} H_i(SX) \cong \begin{cases} I^{\bar{p}} H_{i-1}(X) & \text{if } i > k-1-\bar{p}(k), \\ 0 & \text{if } i = k-1-\bar{p}(k), \\ I^{\bar{p}} H_i(X) & \text{if } i < k-1-\bar{p}(k), \end{cases} \tag{3-2}$$

and, since SX has dimension k,

$$I^{\bar{p}} H_i(cSX) \cong \begin{cases} 0 & \text{if } i \geq k-\bar{p}(k+1), \\ I^{\bar{p}} H_i(SX) & \text{if } i < k-\bar{p}(k+1). \end{cases} \tag{3-3}$$

So, $I^{\bar{p}} H_i(\mathbb{R} \times cX)$ is 0 for $i \geq k-1-\bar{p}(k)$, while $I^{\bar{p}} H_i(cSX)$ must be 0 for $i \geq k-\bar{p}(k+1)$ and also for $i = k-1-\bar{p}(k)$ even if $k-1-\bar{p}(k) < k-\bar{p}(k+1)$. Also, it is not hard to come up with examples in which the terms that are not forced to be zero are, in fact, nonzero. If $k-1-\bar{p}(k) \geq k-\bar{p}(k+1)$ (i.e., $1 + \bar{p}(k) \leq \bar{p}(k+1)$), so that the special case $i = k-1-\bar{p}(k)$ is already in the zero range for $I^{\bar{p}} H_*(cSX)$, then topological invariance would require $k-1-\bar{p}(k) = k-\bar{p}(k+1)$, i.e., $\bar{p}(k+1) = \bar{p}(k)+1$. So if we want topological invariance, $\bar{p}(k+1)$ cannot be greater than $\bar{p}(k)+1$.

On the other hand, if $k-1-\bar{p}(k) < k-\bar{p}(k+1)$, the 0 at $I^{\bar{p}} H_{k-1-\bar{p}(k)}(cSX)$ forced by the suspension formula drops below the truncation dimension cutoff at $k-\bar{p}(k+1)$ that arises from the cone formula. If $k-1-\bar{p}(k) = k-1-\bar{p}(k+1)$ (i.e., $\bar{p}(k) = \bar{p}(k+1)$), no contradiction occurs. But if

$$k-1-\bar{p}(k) < k-1-\bar{p}(k+1)$$

(i.e., $\bar{p}(k+1) < \bar{p}(k)$), then $I^{\bar{p}} H_{k-1-\bar{p}(k+1)}(cSX)$ could be nonzero, which means, via the formula for $I^{\bar{p}} H_*(\mathbb{R} \times cX)$, that we must have $k-1-\bar{p}(k+1) < k-1-\bar{p}(k)$ (i.e., $\bar{p}(k+1) > \bar{p}(k)$), yielding a contradiction.

Hence the only viable possibilities for topological invariance are $\bar{p}(k+1) = \bar{p}(k)$ or $\bar{p}(k+1) = \bar{p}(k)+1$.

It turns out that both possibilities work out. Goresky and MacPherson [33] showed using sheaf theory that any perversity satisfying the two Goresky–MacPherson conditions yields a topologically invariant intersection homology theory. King [41] later gave a non-sheaf proof that holds even when $\bar{p}(2) > 0$.

Why, then, did Goresky and MacPherson limit consideration to perversities for which $\bar{p}(2) = 0$? For one thing, they were primarily concerned with the Poincaré duality theorem for intersection homology, which states that if X is a compact oriented n-dimensional PL stratified pseudomanifold, then there is a nondegenerate pairing

$$I^{\bar{p}} H_i(X;\mathbb{Q}) \otimes I^{\bar{q}} H_{n-i}(X;\mathbb{Q}) \to \mathbb{Q}$$

if $\bar{p}$ and $\bar{q}$ satisfy the Goresky–MacPherson conditions *and* $\bar{p} + \bar{q} = \bar{t}$, or, in other words, $\bar{p}(k) + \bar{q}(k) = k - 2$. If we were to try to allow $\bar{p}(2) > 0$, then we would have to have $\bar{q}(2) < 0$, and we have already seen that this causes trouble with topological invariance. So if we want both duality and invariance, we must have $\bar{p}(2) = \bar{q}(2) = 0$. Without this condition we might possibly have one or the other, but not both. In fact, King's invariance results for $\bar{p}(2) > 0$ implies that duality cannot hold in general when we pair a perversity with $\bar{p}(2) > 0$ with one with $\bar{q}(2) < 0$, at least not without modifying the definition of intersection homology, which we do below.

But there is another interesting reason that Goresky and MacPherson did not obtain King's invariance result for $\bar{p}(2) > 0$. When intersection homology was first introduced in [32], Goresky and MacPherson were unable initially to prove topological invariance. They eventually succeeded by reformulating intersection homology in terms of sheaf theory. But, as it turns out, when $\bar{p}(2) \neq 0$ the original sheaf theory version of intersection homology does not agree with the chain version of intersection homology we have been discussing and for which King proved topological invariance. Furthermore, the sheaf version is not a topological invariant when $\bar{p}(2) > 0$ (some examples can be found in [24]). Due to the powerful tools that sheaf theory brings to intersection homology, the sheaf theoretic point of view has largely overshadowed the chain theory. However, this discrepancy between sheaf theory and chain theory for non-GM perversities turns out to be very interesting in its own right, as we shall see.

3.1. Some consequences of the Goresky–MacPherson conditions. The Goresky–MacPherson perversity conditions have a variety of interesting consequences beyond turning out to be the right conditions to yield both topological invariance and Poincaré duality.

Recall that the allowability condition for an i-simplex σ is that $\dim(\sigma \cap X_k) \leq i - k + \bar{p}(k)$. The GM perversity conditions ensure that $\bar{p}(k) \leq k - 2$, and so for any perversity we must have $i - k + \bar{p}(k) \leq i - 2$. Thus no i-simplex in an allowable chain can intersect any singular stratum in the interiors of its i or its $i - 1$ faces. One simple consequence of this is that no 0- or 1-simplices may intersect X^{n-2}, and so $I^{\bar{p}} H_0(X) \cong H_0(X - X^{n-2})$.

Another consequence is the following fantastic idea, also due to Goresky and MacPherson. Suppose we have a local coefficient system of groups (i.e., a locally constant sheaf) defined on $X - X^{n-2}$, even perhaps one that cannot be extended to all of X. If one looks back at early treatments of homology with local coefficient systems, for example in Steenrod [56], it is sufficient to assign a coefficient group to each simplex of a triangulation (we can think of the group as being located at the barycenter of the simplex) and then to assign to each boundary face map a homomorphism between the group on the simplex and the group on the boundary face. This turns out to be sufficient to define homology with coefficients — what happens on lower dimensional faces does not matter (roughly, everything on lower faces cancels out because we still have $\partial^2 = 0$). Since the intersection i-chains with the GM perversities have the barycenters of their simplices and of their top $i - 1$ faces outside of X^{n-2}, a local coefficient system $\mathcal{G}$ on X^{n-2} is sufficient to define the intersection chain complex $I^{\bar{p}}C_*(X; \mathcal{G})$ and the resulting homology groups. For more details on this construction, see, [26], for example.

Of course now the stratification does matter to some extent since it determines where the coefficient system is defined. However, see [6, Section V.4] for a discussion of stratifications adapted to a given coefficient system defined on an open dense set of X of codimension ≥ 2.

One powerful application of this local coefficient version of intersection homology occurs in [12], in which Cappell and Shaneson study singular knots by considering the knots in their ambient spaces as stratified spaces. They employ a local coefficient system that wraps around the knot to mimic the covering space arguments of classical knot theory. This work also contains one of the first useful applications of intersection homology with non-GM perversities. In order to explain this work, though, we first need to discuss the sheaf formulation of intersection homology, which we pick up in Section 4.

3.2. Singular chain intersection homology. Before moving on to discuss the sheaf-theoretic formulation of intersection homology, we jump ahead in the chronology a bit to King's introduction of singular chain intersection homology in [41]. As one would expect, singular chains are a bit more flexible than PL chains (pun somewhat intended), and the singular intersection chain complex can be defined on any filtered space $X \supset X^{n-1} \supset X^{n-2} \supset \cdots$, with no further restrictions. In fact, the "dimension" indices of the skeleta X^k need no longer have a geometric meaning. These spaces include both PL stratified pseudomanifolds and *topological stratified pseudomanifolds*, the definition of which is the same as of PL pseudomanifolds but with all requirements of piecewise linearity dropped. We also extend the previous definition now to allow an $n-1$ skeleton, and we must extend perversities accordingly to be functions $\bar{p} : \mathbb{Z}^{\geq 1} \to \mathbb{Z}$. King

defines *loose perversities*, which are arbitrary functions of this type. We will
return to these more general perversities in greater detail as we go on.

To define the singular intersection chain complex, which we will denote
$I^{\bar{p}}S_*(X)$, we can no longer use dimension of intersection as a criterion (especially if the index of a skeleton no longer has a dimensional meaning). Instead, the natural generalization of the allowability condition is that a singular
i-simplex $\sigma : \Delta^i \to X$ is allowable if

$$\sigma^{-1}(X_{n-k}) \subset \{i - k + \bar{p}(k) \text{ skeleton of } \Delta^i\}.$$

Once allowability has been defined for simplices, allowability of chains is defined as in the PL case, and we obtain the chain complex $I^{\bar{p}}S_*(X)$ and the
homology groups $I^{\bar{p}}H_*(X)$.

If X is a PL stratified pseudomanifold, the notation $I^{\bar{p}}H_*(X)$ for singular
chain intersection homology causes no confusion; as King observes, the PL and
singular intersection homology theories agree on such spaces. Also as for PL
chains, and by essentially the same arguments, if X has no codimension one
stratum and $\bar{p}$ is a GM perversity, singular intersection homology can take local
coefficients on $X - X^{n-2}$.

From here on, when we refer to chain-theoretic intersection homology, we
will mean both the singular version (in any context) and the PL version (on PL
spaces).

4. Sheaf-theoretic intersection homology

Although intersection homology was developed originally utilizing PL chain
complexes, this approach was soon largely supplanted by the techniques of sheaf
theory. Sheaf theory was brought to bear by Goresky and MacPherson in [33],
originally as a means to demonstrate the topological invariance (stratification
independence) of intersection homology with GM perversities; this was before
King's proof of this fact using singular chains. However, it quickly became
evident that sheaf theory brought many powerful tools along with it, including a
Verdier duality approach to the Poincaré duality problem on pseudomanifolds.
Furthermore, the sheaf theory was able to accommodate topological pseudomanifolds. This sheaf-theoretic perspective has largely dominated intersection
homology theory ever since.

The Deligne sheaf.   We recall that if X^n is a stratified topological pseudomanifold[3], then a primary object of interest is the so-called *Deligne sheaf*. For
notation, we let $U_k = X - X^{n-k}$ for $k \geq 2$, and we let $i_k : U_k \hookrightarrow U_{k+1}$ denote

---

[3]For the moment, we again make the historical assumption that there are no codimension one strata.

the inclusion. Suppose that $\bar{p}$ is a GM perversity. We have seen that intersection homology should allow a local system of coefficients defined only on $X - X^{n-2}$; let $\mathcal{G}$ be such a local system. The Deligne sheaf complex $\mathcal{P}^*$ (or, more precisely $\mathcal{P}^*_{\bar{p},\mathcal{G}}$) is defined by an inductive process. It is[4]

$$\mathcal{P}^* = \tau_{\leq \bar{p}(n)} Ri_{n*} \ldots \tau_{\leq \bar{p}(2)} Ri_{2*}(\mathcal{G} \otimes \mathcal{O}),$$

where $\mathcal{O}$ is the orientation sheaf on $X - X^{n-2}$, Ri_{k*} is the right derived functor of the pushforward functor i_{k*}, and $\tau_{\leq m}$ is the sheaf complex truncation functor that takes the sheaf complex $\mathcal{S}^*$ to $\tau_{\leq m}\mathcal{S}^*$ defined by

$$(\tau_{\leq m}\mathcal{S}^*)^i = \begin{cases} 0 & \text{if } i > m, \\ \ker(d_i) & \text{if } i = m, \\ \mathcal{S}^i & \text{if } i < m. \end{cases}$$

Here d_i is the differential of the sheaf complex. Recall that $\tau_{\leq m}\mathcal{S}^*$ is quasi-isomorphic to $\mathcal{S}^*$ in degrees $\leq m$ and is quasi-isomorphic to 0 in higher degrees.

REMARK 4.1. Actually, the orientation sheaf $\mathcal{O}$ is not usually included here as part of the definition of $\mathcal{P}^*$, or it would be only if we were discussing $\mathcal{P}^*_{\bar{p},\mathcal{G}\otimes\mathcal{O}}$. However, it seems best to include this here so as to eliminate having to continually mess with orientation sheaves when discussing the equivalence of sheaf and chain theoretic intersection homology, which, without this convention, would read that $\mathbb{H}^*(X;\mathcal{P}^*_{\bar{p},\mathcal{G}\otimes\mathcal{O}}) \cong I^{\bar{p}}H_{n-*}(X;\mathcal{G})$; see below. Putting $\mathcal{O}$ into the definition of $\mathcal{P}^*$ as we have done here allows us to leave this nuisance tacit in what follows.

The connection between the Deligne sheaf complex (also called simply the "Deligne sheaf") and intersection homology is that it can be shown that, on an n-dimensional PL pseudomanifold, $\mathcal{P}^*$ is quasi-isomorphic to the sheaf $U \to I^{\bar{p}}C^{\infty}_{n-*}(U;\mathcal{G})$. Here the ∞ indicates that we are now working with Borel–Moore PL chain complexes, in which chains may contain an infinite number of simplices with nonzero coefficients, so long as the collection of such simplices in any chain is locally-finite. This is by contrast to the PL chain complex discussed above for which each chain can contain only finitely many simplices with nonzero coefficient. This sheaf of intersection chains is also soft, and it follows via sheaf theory that the hypercohomology of the Deligne sheaf is isomorphic to the Borel–Moore intersection homology

$$\mathbb{H}^*(X;\mathcal{P}^*) \cong I^{\bar{p}}H^{\infty}_{n-*}(X;\mathcal{G}).$$

---

[4]There are several other indexing conventions. For example, it is common to shift this complex so that the coefficients $\mathcal{G}$ live in degree $-n$ and the truncations become $\tau_{\leq \bar{p}(k)-n}$. There are other conventions that make the cohomologically nontrivial degrees of the complex symmetric about 0 when n is even. We will stick with the convention that $\mathcal{G}$ lives in degree 0 throughout. For details on other conventions, see [33], for example.

It is also possible to recover the intersection homology we introduced initially by using compact supports:

$$\mathbb{H}_c^*(X; \mathcal{P}^*) \cong I^{\bar{p}} H_{n-*}^c(X; \mathcal{G}).$$

Now that we have introduced Borel–Moore chains, we will use "c" to indicate the more familiar compact (finite number of simplices) supports. If the results we discuss hold in both contexts (in particular if X is compact) we will forgo either decoration. More background and details on all of this can be found in [33; 6].

It was shown later, in [26], that a similar connection exists between the Deligne sheaf and singular chain intersection homology on *topological* pseudo-manifolds. Continuing to assume GM perversities, one can also define a sheaf via the sheafification of the presheaf of *singular chains*[5]

$$U \to I^{\bar{p}} S_{n-*}(X, X - \bar{U}; \mathcal{G}).$$

This sheaf turns out to be homotopically fine, and it is again quasi-isomorphic to the Deligne sheaf. Thus, once again, we have

$$\mathbb{H}_c^*(X; \mathcal{P}^*) \cong I^{\bar{p}} H_{n-*}^c(X; \mathcal{G}) \quad \text{and} \quad \mathbb{H}^*(X; \mathcal{P}^*) \cong I^{\bar{p}} H_{n-*}^\infty(X; \mathcal{G}),$$

which is the homology of the chain complex $I^{\bar{p}} S_*^\infty(X; \mathcal{G})$ consisting of chains that can involve an infinite, though locally-finite, number of simplices with nonzero coefficient.

The Goresky–MacPherson proof of topological invariance follows by showing that the Deligne sheaf is uniquely defined up to quasi-isomorphism via a set of axioms that do not depend on the stratification of the space. This proof is given in [33]. However, we would here like to focus attention on what the Deligne sheaf accomplishes locally, particularly in mind of the maxim that a sheaf theory (and sheaf cohomology) is a machine for assembling local information into global. So let's look at the local cohomology (i.e., the stalk cohomology) of the sheaf $\mathcal{P}^*$ at $x \in X_{n-k}$. This is $\mathcal{H}^*(\mathcal{P}^*)_x = H^*(\mathcal{P}_x^*) \cong \varinjlim_{x \in U} \mathbb{H}^*(U; \mathcal{P}^*) \cong \varinjlim_{x \in U} I^{\bar{p}} H_{n-*}^\infty(U; \mathcal{G})$, and we may assume that the limit is taken over the cofinal system of distinguished neighborhoods $N \cong \mathbb{R}^{n-k} \times cL^{k-1}$ containing x. It is not hard to see that $\mathcal{P}^*$ at $x \in X_{n-k}$ depends only on the stages of the iterative Deligne construction up through $\tau_{\leq \bar{p}(k)} Ri_{k*}$ (at least so long as we assume that $\bar{p}$ is nondecreasing[6], as it will be for a GM perversity). Then it follows immediately from the definition of τ that $\mathcal{H}^*(\mathcal{P}^*)_x = 0$ for $* > \bar{p}(k)$. On the other hand, the pushforward construction,

---

[5] Since X is locally compact, we may use either c or ∞ to obtain the same sheaf.

[6] If $\bar{p}$ ever decreases, say at k, then the truncation $\tau_{\leq \bar{p}(k)}$ might kill local cohomology in other strata of lower codimension.

together with a Künneth computation and an appropriate induction step (see [6, Theorem V.2.5]), shows that for $* \leq \bar{p}(k)$ we have

$$
\begin{aligned}
\mathcal{H}^*(\mathcal{P}^*)_x &\cong \mathbb{H}^*(N - N \cap X^{n-k}; \mathcal{P}^*) \\
&\cong \mathbb{H}^*(\mathbb{R}^{n-k} \times (cL - v); \mathcal{P}^*) \\
&\cong \mathbb{H}^*(\mathbb{R}^{n-k+1} \times L; \mathcal{P}^*) \\
&\cong \mathbb{H}^*(L; \mathcal{P}^*|_L).
\end{aligned}
$$

It can also be shown that $\mathcal{P}^*|_L$ is quasi-isomorphic to the Deligne sheaf on L, so $\mathbb{H}^*(L; \mathcal{P}^*|_L) \cong I^{\bar{p}} H_{k-1-*}(L)$.

For future reference, we record the formula

$$
\mathcal{H}^i(\mathcal{P}^*)_x \cong \begin{cases} 0 & \text{if } i > \bar{p}(k), \\ \mathbb{H}^i(L; \mathcal{P}^*) & \text{if } i \leq \bar{p}(k), \end{cases} \tag{4-1}
$$

for $x \in X_{n-k}$ and L the link of x. Once one accounts for the shift in indexing between intersection homology and Deligne sheaf hypercohomology and for the fact that we are now working with Borel–Moore chains, these computations work out to be equivalent to the cone formula (2-1). In fact,

$$
\begin{aligned}
H^*(\mathcal{P}_x^*) &\cong I^{\bar{p}} H_{n-*}^\infty(\mathbb{R}^{n-k} \times cL; \mathcal{G}) \\
&\cong I^{\bar{p}} H_{k-*}^\infty(cL; \mathcal{G}) \qquad \text{(by the Künneth theorem)} \\
&\cong I^{\bar{p}} H_{k-*}(cL, L \times (0, 1); \mathcal{G}),
\end{aligned}
$$

and the cone formula (2-1) translates directly, via the long exact sequence of the pair $(cL, L \times (0, 1))$, to this being 0 for $* > \bar{p}(k)$ and $I^{\bar{p}} H_{k-1-*}(L; \mathcal{G})$ otherwise.

So the Deligne sheaf recovers the local cone formula, and one would be hard pressed to find a more direct or natural way to "sheafify" the local cone condition than the Deligne sheaf construction. This reinforces our notion that the cone formula is really at the heart of intersection homology. In fact, the axiomatic characterization of the Deligne sheaf alluded to above is strongly based upon the sheaf version of the cone formula. There are several equivalent sets of characterizing axioms. The first, $AX1_{\bar{p}, \mathcal{G}}$, is satisfied by a sheaf complex $\mathcal{S}^*$ if

(a) $\mathcal{S}^*$ is bounded and $\mathcal{S}^* = 0$ for $i < 0$,
(b) $\mathcal{S}^*|_{X - X^{n-2}} \cong \mathcal{G} \otimes \mathcal{O}$,[7]
(c) for $x \in X_{n-k}$, $H^i(\mathcal{S}_x^*) = 0$ if $i > \bar{p}(k)$, and
(d) for each inclusion $i_k : U_k \to U_{k+1}$, the "attaching map" α_k given my the composition of natural morphisms $\mathcal{S}^*|_{U_{k+1}} \to i_{k*} i_k^* \mathcal{S}^* \to R i_{k*} i_k^* \mathcal{S}^*$ is a quasi-isomorphism in degrees $\leq \bar{p}(k)$.

---

[7]See Remark 4.1 on page 190.

These axioms should technically be thought of as applying in the derived category of sheaves on X, in which case all equalities and isomorphisms should be thought of as quasi-isomorphisms of sheaf complexes. The first axiom acts as something of a normalization and ensures that S^* lives in the bounded derived category. The second axiom fixes the coefficients on $X - X^{n-2}$. The third and fourth axioms are equivalent to the cone formula (4-1); see [6, Sections V.1 and V.2]. In fact, it is again not difficult to see that the Deligne sheaf construction is designed precisely to satisfy these axioms. It turns out that these axioms completely characterize a sheaf up to quasi-isomorphism (see [6, Section V.2]), and in fact it is by showing that the sheafification of $U \to I^{\bar{p}} S_*(X, X - \bar{U}; \mathcal{G})$ satisfies these axioms that one makes the connection between the sheaf of singular intersection chains and the Deligne sheaf.

Goresky and MacPherson [33] proved the stratification independence of intersection homology by showing that the axioms $AX1$ are equivalent to other sets of axioms, including one that does not depend on the stratification of X. See [6; 33] for more details.

4.1. Duality. It would take us too far afield to engage in a thorough discussion of how sheaf theory and, in particular, Verdier duality lead to proofs of the intersection homology version of Poincaré duality. However, we sketch some of the main ideas, highlighting the role that the perversity functions play in the theory. For complete accounts, we refer the reader to the excellent expository sources [6; 2].

The key to sheaf-theoretic duality is the Verdier dualizing function $\mathcal{D}$. Very roughly, $\mathcal{D}$ functions as a fancy sheaf-theoretic version of the functor $\text{Hom}(\cdot, R)$. In fact, $\mathcal{D}$ takes a sheaf complex S^* to a sheaf complex $Hom^*(S^*, \mathbb{D}_X^*)$, where $\mathbb{D}_X^*$ is the Verdier dualizing sheaf on the space X. In reasonable situations, the dualizing sheaf $\mathbb{D}_X^*$ is quasi-isomorphic (after reindexing) to the sheaf of singular chains on X; see [6, Section V.7.2.]. For us, the most important property of the functor $\mathcal{D}$ is that it satisfies a version of the universal coefficient theorem. In particular, if S^* is a sheaf complex over the Dedekind domain R, then for any open $U \subset X$,

$$\mathbb{H}^i(U; \mathcal{D}S^*) \cong \text{Hom}(\mathbb{H}_c^{-i}(U; S^*); R) \oplus \text{Ext}(\mathbb{H}_c^{-i+1}(U; S^*); R).$$

The key, now, to proving a duality statement in intersection homology is to show that if X is orientable over a ground field F and $\bar{p}$ and $\bar{q}$ are dual perversities, meaning $\bar{p}(k) + \bar{q}(k) = k - 2$ for all $k \geq 2$, then $\mathcal{D}\mathcal{P}_{\bar{p}}^*[-n]$ is quasi-isomorphic to $\mathcal{P}_{\bar{q}}^*$. Here $[-n]$ is the degree shift by $-n$ degrees, i.e., $(S^*[-n])^i = S^{i-n}$, and this shift is applied to $\mathcal{D}\mathcal{P}^*$ (it is not a shifted $\mathcal{P}^*$ being dualized). It then follows from the universal coefficient theorem with field coefficients F

that

$$I^{\bar{q}} H^{\infty}_{n-i}(X; F) \cong \text{Hom}(I^{\bar{p}} H^c_i(X; F), F),$$

which is intersection homology Poincaré duality for pseudomanifolds.

To show that $\mathcal{D}\mathcal{P}^*_{\bar{p}}[-n]$ is quasi-isomorphic to $\mathcal{P}^*_{\bar{q}}$, it suffices to show that $\mathcal{D}\mathcal{P}^*_{\bar{p}}[-n]$ satisfies the axioms $AX1_{\bar{q}}$. Again, we will not go into full detail, but we remark the following main ideas, referring the reader to the axioms $AX1$ outlined above:

(a) On $X - X^{n-2}$, $\mathcal{D}\mathcal{P}^*_{\bar{p}}[-n]$ restricts to the dual of the coefficient system

$$\mathcal{P}^*_{\bar{p}}|_{X-X^{n-2}},$$

which is again a local coefficient system. If X is orientable and the coefficient system is trivial, then so is its dual.

(b) Recall that the third and fourth axioms for the Deligne sheaf concern what happens at a point x in the stratum X_{n-k}. To compute $H^*(\mathcal{S}^*_x)$, we may compute $\varinjlim_{x \in U} \mathbb{H}^*(U; \mathcal{S}^*)$. In particular, if we let each U be a distinguished neighborhood $U \cong \mathbb{R}^{n-k} \times cL$ of x and apply the universal coefficient theorem, we obtain

$$\begin{aligned}
H^i(U; \mathcal{D}\mathcal{P}^*_{\bar{p}}[-n]_x) &\cong \varinjlim_{x \in U} \mathbb{H}^i(\mathcal{D}\mathcal{P}^*_{\bar{p}}[-n]) \qquad\qquad (4\text{-}2) \\
&\cong \varinjlim_{x \in U} \mathbb{H}^{i-n}(U; \mathcal{D}\mathcal{P}^*_{\bar{p}}) \\
&\cong \varinjlim_{x \in U} \text{Hom}(\mathbb{H}^{n-i}_c(U; \mathcal{P}^*_{\bar{p}}), F) \\
&\cong \varinjlim_{x \in U} \text{Hom}(I^{\bar{p}} H^c_i(\mathbb{R}^{n-k} \times cL; F), F) \\
&\cong \varinjlim_{x \in U} \text{Hom}(I^{\bar{p}} H^c_i(cL; F), F).
\end{aligned}$$

The last equality is from the Künneth theorem with compact supports. From the cone formula, we know that this will vanish if $i \geq k - 1 - \bar{p}(k)$, i.e., if $i > k - 2 - \bar{p}(k) = \bar{q}(k)$. This is the third item of $AX1_{\bar{q}}$.

(c) The fourth item of $AX1_{\bar{q}}$ is only slightly more difficult, but the basic idea is the same. By the computations (4-2), $H^i(\mathcal{D}\mathcal{P}^*_{\bar{p}}[-n]_x)$ comes down to computing $I^{\bar{p}} H^c_i(cL; F)$, which we know is isomorphic to $I^{\bar{p}} H^c_i(L; F)$ when $i < k - 1 - \bar{p}(k)$, i.e., $i \leq \bar{q}(k)$. It is then an easy argument to show that in fact the attaching map condition of $AX1_{\bar{q}}$ holds in this range.

(d) The first axiom also follows from these computations; one checks that the vanishing of $H^i(\bar{P}^*_{\bar{p},x})$ for $i < 0$ and for $i > \bar{p}(k)$ for $x \in X_{n-k}$ is sufficient to imply that $H^i(\mathcal{D}\mathcal{P}^*_{\bar{p}}[-n]_x)$ also vanishes for $i < 0$ or i sufficiently large.

We see quite clearly from these arguments precisely why the dual perversity condition $\bar{p}(k) + \bar{q}(k) = k - 2$ is necessary in order for duality to hold.

A more general duality statement, valid over principal ideal domains, was provided by Goresky and Siegel in [34]. However, there is an added requirement that the space X be *locally* $(\bar{p}, R)$-*torsion free*. This means that for each $x \in X_{n-k}$, $I^{\bar{p}} H^{c}_{k-2-\bar{p}(k)}(L_x)$ is R-torsion free, where L_x is the link of x in X. The necessity of this condition is that when working over a principal ideal domain R, the Ext terms of the universal coefficient theorem for Verdier duals must be taken into account. If these link intersection homology groups had torsion, there would be a possibly nonzero Ext term in the computation (4-2) when $i = \bar{q}(k) + 1$, due to the degree shift in the Ext term of the universal coefficient theorem. This would prevent the proof that $\mathcal{D}\mathcal{P}^{*}_{\bar{p}}[-n]$ satisfies $AX1_{\bar{q}}$, so this possibility is eliminated by hypothesis. With these assumption, there result duality pairings analogous to those that occur for manifolds using ordinary homology with $\mathbb{Z}$ coefficients. In particular, one obtains a nondegenerate intersection pairing on homology mod torsion and a nondegenerate torsion linking pairing on torsion subgroups. See [34] and [22] for more details.

This circle of ideas is critical in leading to the need for superperversities in the Cappell–Shaneson superduality theorem, which we shall now discuss.

4.2. Cappell–Shaneson superduality.

The first serious application (of which the author is aware) of a non-GM perversity in sheaf theoretic intersection homology occurs in Cappell and Shaneson's [12], where they develop a generalization of the Blanchfield duality pairing of knot theory to study L-classes of certain codimension 2 subpseudomanifolds of manifolds. Their pairing is a perfect Hermitian pairing between the perversity $\bar{p}$ intersection homology $\mathbb{H}^{*}(X; \mathcal{P}^{*}_{\bar{p}, \mathcal{G}})$ (with $\bar{p}$ a GM perversity) and $\mathbb{H}^{n-1-*}(X; \mathcal{P}^{*}_{\bar{q}, \mathcal{G}*})$, where $\mathcal{G}^*$ is a Hermitian dual system to $\mathcal{G}$ and $\bar{q}$ satisfies $\bar{p}(k) + \bar{q}(k) = k - 1$. This assures that $\bar{q}$ satisfies the GM perversity condition $\bar{q}(k) \leq \bar{q}(k+1) \leq \bar{q}(k)+1$, but it also forces $\bar{q}(2) = 1$. In [26], we referred to such perversities as *superperversities*, though this term was later expanded by the author to include larger classes of perversities $\bar{q}$ for which $\bar{q}(k)$ may be greater than $\bar{t}(k) = k - 2$ for some k.

Cappell and Shaneson worked with the sheaf version of intersection homology throughout. Notice that the Deligne sheaf remains perfectly well-defined despite $\bar{q}$ being a non-GM perversity; the truncation process just starts at a higher degree. Let us sketch how these more general perversities come into play in the Cappell–Shaneson theory.

The Cappell–Shaneson superduality theorem holds in topological settings that generalize those in which one studies the Blanchfield pairing of Alexander modules in knot theory; see [12] for more details. The Alexander modules are the homology groups of infinite cyclic covers of knot complements, and one of the key features of these modules is that they are torsion modules over the principal ideal domain $\mathbb{Q}[t, t^{-1}]$. In fact, the Alexander polynomials are just the products of the torsion coefficients of these modules. Similarly, the Cappell–Shaneson intersection homology groups $\mathbb{H}^*(X; \mathcal{P}^*_{\bar{p}, \mathcal{G}})$ are torsion modules over $\mathbb{Q}[t, t^{-1}]$ (in fact, $\mathcal{G}$ is a coefficient system with stalks equal to $\mathbb{Q}[t, t^{-1}]$ and with monodromy action determined by the linking number of a closed path with the singular locus in X). Now, what happens if we try to recreate the Poincaré duality argument from Section 4.1 in this context? For one thing, the dual of the coefficient system over $X - X^{n-2}$ becomes the dual system $\mathcal{G}^*$. More importantly, all of the Hom terms in the universal coefficient theorem for Verdier duality vanish, because all modules are torsion, but the Ext terms remain. From here, it is possible to finish the argument, replacing all Homs with Exts, but there is one critical difference. Thanks to the degree shift in Ext terms in the universal coefficient theorem, at a point $x \in X_{n-k}$, $H^i(\mathcal{D}\mathcal{P}^*_{\bar{p}, \mathcal{G}}[-n]_x)$ vanishes not for $i > k - 2 - \bar{p}(k)$ but for $i > k - 1 - \bar{p}(k)$, while the attaching isomorphism holds for $i \le k - 1 - \bar{p}(k)$. It follows that $\mathcal{D}\mathcal{P}^*_{\bar{p}, \mathcal{G}}[-n]_x$ is quasi-isomorphic to $\mathcal{P}^*_{\bar{q}, \mathcal{G}^*}$, but now $\bar{q}$ must satisfy $\bar{p}(k) + \bar{q}(k) = k - 1$.

The final duality statement that arises has the form

$$I^{\bar{p}} H_i(X; \mathcal{G})^* \cong \text{Ext}(I^{\bar{q}} H_{n-i-1}(X; \mathcal{G}), \mathbb{Q}[t, t^{-1}])$$
$$\cong \text{Hom}(I^{\bar{q}} H_{n-i-1}(X; \mathcal{G}); \mathbb{Q}(t, t^{-1})/\mathbb{Q}[t, t^{-1}]),$$

where $\bar{p}(k) + \bar{q}(k) = k - 1$, X is compact and orientable, and the last isomorphism is from routine homological algebra. We refer the reader to [12] for the remaining technical details.

Note that this is somewhat related to our brief discussion of the Goresky–Siegel duality theorem. In that theorem, a special condition was added to ensure the vanishing of the extra Ext term. In the Cappell–Shaneson duality theorem, the extra Ext term is accounted for by the change in perversity requirements, but it is important that all Hom terms vanish, otherwise there would still be a mismatch between the degrees in which the Hom terms survive truncation and the degrees in which the Ext terms survive truncation. It might be an enlightening exercise for the reader to work through the details.

While the Cappell–Shaneson superduality theorem generalizes the Blanchfield pairing in knot theory, the author has identified an intersection homology generalization of the Farber–Levine $\mathbb{Z}$-torsion pairing in knot theory [21]. In this case, the duality statement involves Ext^2 terms and requires perversities satisfying the duality condition $\bar{p}(k) + \bar{q}(k) = k$.

5. Subperversities and superperversities

We have already noted that King considered singular chain intersection homology for perversities satisfying $\bar{p}(2) > 0$, and, more generally, he defined in

[41] a *loose* perversity to be an arbitrary function from $\{2, 3, \ldots\}$ to $\mathbb{Z}$. It is not hard to see that the PL and singular chain definitions of intersection homology (with constant coefficients) go through perfectly well with loose perversities, though we have seen that we would expect to forfeit topological invariance (and perhaps Poincaré duality) with such choices. On the sheaf side, Cappell and Shaneson [12] used a perversity with $\bar{p}(2) > 0$ in their superduality theorem. Somewhat surprisingly, however, once we have broken into the realm of non-GM perversities, the sheaf and chain theoretic versions of intersection homology no longer necessarily agree.

A very basic example comes by taking $\bar{p}(k) < 0$ for some k; we will call such a perversity a *subperversity*. In the Deligne sheaf construction, a subperversity will truncate everything away and wind up with the trivial sheaf complex, whose hypercohomology groups are all 0. In the chain construction, however, we have only made it more difficult for a chain to be allowable with respect to the kth stratum. In fact, it is shown in [22, Corollary 2.5] that the condition $\bar{p}(k) < 0$ is homologically equivalent to declaring that allowable chains cannot intersect the kth stratum at all. So, for example, if $\bar{p}(k) < 0$ for all k, then $I^{\bar{p}} H_*^c(X) \cong H_*^c(X - X^{n-2})$.

The discrepancy between sheaf theoretic and chain theoretic intersection homology also occurs when perversities exceed the top perversity $\bar{t}(k) = k - 2$ for some k; we call such perversities *superperversities*. To see what the issue is, let us return once again to the cone formula, which we have seen plays the defining local (and hence global) role in intersection homology. So long as $\bar{p}$ is non-decreasing (and non-negative), the arguments of the preceding section again yield the sheaf-theoretic cone formula (4-1) from the Deligne construction. However, the cone formula can fail in the chain version of superperverse intersection homology.

To understand why, suppose L is a compact $k - 1$ pseudomanifold, so that $(cL)_k = v$, the cone point. Recall from Section 2.4 that the cone formula comes by considering cones on allowable cycles and checking whether or not they are allowable with respect to v. In the dimensions where such cones are allowable, this kills the homology. In the dimensions where the cones are not allowable, we also cannot have any cycles intersecting the cone vertex, and the intersection homology reduces to $I^{\bar{p}} H_i^c(cL - v) \cong I^{\bar{p}} H_i^c(L \times \mathbb{R}) \cong I^{\bar{p}} H_i^c(L)$, the first isomorphism because $cL - v$ is homeomorphic to $L \times \mathbb{R}$ and the second using the Künneth theorem with the unstratified $\mathbb{R}$ (see [41]) or stratum-preserving homotopy equivalence (see [23]). These arguments hold in both the PL and singular chain settings. However, there is a subtle point these arguments overlook when perversities exceed $\bar{t}$.

If $\bar{p}$ is a GM perversity, then $\bar{p}(k) \leq k-2$ and so $k-1-\bar{p}(k) > 0$ and the cone formula guarantees that $I^{\bar{p}}H_0^c(cL)$ is always isomorphic to $I^{\bar{p}}H_0^c(L)$. In fact, we have already observed, in Section 3.1, that 0- and 1-simplices cannot intersect the singular strata. Now suppose that $\bar{p}(k) = k-1$. Then extending the cone formula should predict that $I^{\bar{p}}H_0^c(cL) = 0$. But, in these dimensions, the argument breaks down. For if x is a point in $cL - (cL)^{k-2}$ representing a cycle in $I^{\bar{p}}S_0^c(cL)$, then $\bar{c}x$ is a 1-simplex, and a quick perversity computation shows that it is now an allowable 1-simplex. However, it is not allowable as a chain since $\partial(\bar{c}x)$ has two 0-simplices, one supported at the cone vertex. This cone vertex is not allowable. The difference between this case and the prior ones is that when $i > 0$ the boundary of a cone on an i-cycle is (up to sign) that i-cycle. But when $i = 0$, there is a new boundary component. In the previous computations, this was not an issue because the 1-simplex would not have been allowable either. But now this ruins the cone formula.

In general, a careful computation shows that if L is a compact $k-1$ filtered space and $\bar{p}$ is any loose perversity, then the singular intersection homology cone formula becomes [41]

$$I^{\bar{p}}H_i^c(cL) \cong \begin{cases} 0 & \text{if } i \geq k-1-\bar{p}(k) \text{ and } i \neq 0, \\ \mathbb{Z} & \text{if } i = 0 \text{ and } \bar{p}(k) \geq k-1, \\ I^{\bar{p}}H_i(L) & \text{if } i < k-1-\bar{p}(k). \end{cases} \tag{5-1}$$

Which is the right cone formula? So when we allow superperversities with $\bar{p}(k) > \bar{t}(k) = k-2$, the cone formula (2-1) no longer holds for singular intersection homology, and there is a disagreement with the sheaf theory, for which the sheaf version (4-1) of (2-1) always holds by the construction of the Deligne sheaf (at least so long as $\bar{p}$ is non-decreasing). What, then, is the "correct" version of intersection homology for superperversities (and even more general perversities)? Sheaf theoretic intersection homology allows the use of tools such as Verdier duality, and the superperverse sheaf intersection homology plays a key role in the Cappell–Shaneson superduality theorem. On the other hand, singular intersection homology is well-defined on more general spaces and allows much more easily for homotopy arguments, such as those used in [41; 23; 25; 27].

In [35], Habegger and Saper created a sheaf theoretic generalization of King's singular chain intersection homology provided $\bar{p}(k) \leq \bar{p}(k+1) \leq \bar{p}(k)+1$ and $\bar{p}(2) \geq 0$. This theory satisfies a version of Poincaré duality but is somewhat complicated. We will return to this below in Section 11.

Alternatively, a modification of the chain theory whose homology agrees with the hypercohomology of the Deligne sheaf even for superperversities (up to the appropriate reindexing) was introduced independently by the author in [26] and by Saralegi in [54]. This chain theory has the satisfying property of maintaining the cone formula (2-1) for completely general perversities, even those that are not necessarily non-decreasing, while yielding the usual intersection homology groups for GM perversities. Recently, the author has found also a generalization of the Deligne sheaf construction that yields sheaf complexes whose hyperco-homology groups agree with the homology groups of this chain theory and are the usual ones for GM perversities. Of course these groups generally will not be independent of the stratification, but they do possess Poincaré duality for pseudomanifolds. Thus this theory seems to be a reasonable candidate for the most general possible intersection homology theory. We will describe this theory and its characteristics in the following sections.

Superperversities and codimension one strata. It is a remarkable point of interest that the perversity issues we have been discussing provide some additional insight into why codimension one strata needed to be left out of the definition of stratified pseudomanifolds used by Goresky and MacPherson (though I do not know if it was clear that this was the issue at the time). On the one hand, if we assume that X has a codimension one stratum and let $\bar{p}(1) = 0$, then $\bar{p}(1)$ is greater than $\bar{t}(1)$, which we would expect to be $1 - 2 = -1$, and so we run into the trouble with the cone formula described earlier in this section. On the other hand, if we let $\bar{q}(1) = \bar{t}(1) = -1$, then we run into the trouble with negative perversities described prior to that. In this latter case, the Deligne sheaf is always trivial, yielding only trivial sheaf intersection homology, so there can be no non-trivial Poincaré duality via the sheaf route (note that $\bar{p}(1) = 0$ and $\bar{q}(1) = -1$ are dual perversities at $k = 1$, so any consideration of duality involving the one perversity would necessarily involve the other). Similarly, there is no duality in the chain version since, for example, if $X^n \supset X^{n-1}$ is $S^1 \supset$ pt then easy computations shows that $I^{\bar{q}} H_1(X) \cong H_1^c(S^1 - \mathrm{pt}) = 0$, while $I^{\bar{p}} H_0(X) \cong \mathbb{Z}$. Note that the first computation shows that we have also voided the stratification independence of intersection homology.

One of the nice benefits of our (and Saralegi's) "correction" to chain-theoretic intersection homology is that it allows one to include codimension one strata and still obtain Poincaré duality results. In general, though, the stratification independence does need to be sacrificed. One might argue that this is the preferred trade-off, since one might wish to use duality as a tool to study spaces *together with* their stratifications.

6. "Correcting" the definition of intersection chains

As we observed in the previous section, if $\bar{p}$ is a superperversity (i.e., $\bar{p}(k) > k - 2$ for some k), then the Deligne sheaf version of intersection homology and the chain version of intersection homology need no longer agree. Modifications of the chain theory to correct this anomaly were introduced by the author in [26] and by Saralegi in [54], and these have turned out to provide a platform for the extension of other useful properties of intersection homology, including Poincaré duality. These modifications turn out to be equivalent, as proven in [28]. We first present the author's version, which is slightly more general in that it allows for the use of local coefficient systems on $X - X^{n-1}$.

As we saw in Section 5, the discrepancy between the sheaf cone formula and the chain cone formulas arises because the boundary of a 1-chain that is the cone on a 0-chain has a 0-simplex at the cone point. So to fix the cone formula, it is necessary to find a way to make the extra 0 simplex go away. This is precisely what both the author's and Saralegi's corrections do, though how they do it is described in different ways.

The author's idea, motivated by the fact that Goresky–MacPherson perversity intersection chains need only have their coefficients well-defined on $X - X^{n-2}$, was to extend the coefficients $\mathcal{G}$ on $X - X^{n-1}$ (now allowing codimension one strata) to a *stratified coefficient system* by including a "zero coefficient system" on X^{n-1}. Together these are denote $\mathcal{G}_0$. Then a coefficient on a singular simplex $\sigma : \Delta^i \to X$ is defined by a lift of $\sigma|_{\sigma^{-1}(X-X^{n-1})}$ to the bundle $\mathcal{G}$ on $X - X^{n-1}$ and by a "lift" of $\sigma|_{\sigma^{-1}(X^{n-1})}$ to the 0 coefficient system over X^{n-1}. Boundary faces then inherit their coefficients from the simplices they are boundaries of by restriction. A simplex has coefficient 0 if its coefficient lift is to the zero section over all of Δ^i. In the PL setting, coefficients of PL simplices are defined similarly. In principle, there is no reason the coefficient system over X^{n-1} must be trivial, and one could extend this definition by allowing different coefficient systems on all the strata of X; however, this idea has yet to be investigated.

With this coefficient system $\mathcal{G}_0$, the intersection chain complex $I^{\bar{p}} S_*(X; \mathcal{G}_0)$ is defined exactly as it is with ordinary coefficients — allowability of simplices is determined by the same formula, and chains are allowable if each simplex with a nonzero coefficient in the chain is allowable. So what has changed? The subtle difference is that if a simplex that is in the boundary of a chain has support in X^{n-1}, then that boundary simplex must now have coefficient 0, since that is the only possible coefficient for simplices in X^{n-1}; thus such boundary simplices vanish and need not be tested for allowability. This simple idea turns out to be enough to fix the cone formula.

Indeed, let us reconsider the example of a point x in $cL - (cL)^{k-2}$, together with a coefficient lift to $\mathcal{G}$, representing a cycle in $I^{\bar{p}} S_0^c(cL; \mathcal{G}_0)$, where $\bar{p}(k) = k - 1$. As before, $\bar{c}x$ is a 1-simplex, and it is allowable. Previously, $\bar{c}x$ was not, however, allowable *as a chain* since the component of $\partial(\bar{c}x)$ in the cone vertex was not allowable. However, if we consider the boundary of $\bar{c}x$ in $I^{\bar{p}} S_0^c(cL; \mathcal{G}_0)$, then the simplex at the cone point vanishes because it must have a zero coefficient there. Thus allowability is not violated by $\bar{c}x$; it is now an allowable *chain*.

A slightly more detailed computation (see [26]) shows that, in fact,

$$I^{\bar{p}} H_i^c (cL^{k-1}; \mathcal{G}_0) \cong \begin{cases} 0 & \text{if } i \geq k-1-\bar{p}(k), \\ I^{\bar{p}} H_i^c (L; \mathcal{G}_0) & \text{if } i < k-1-\bar{p}(k), \end{cases} \qquad (6\text{-}1)$$

i.e., we recover the cone formula, even if $\bar{p}(k) > k - 2$.

Another pleasant feature of $I^{\bar{p}} H_*(X; \mathcal{G}_0)$ is that if $\bar{p}$ does happen to be a GM perversity and X has no codimension one strata, then $I^{\bar{p}} H_*(X; \mathcal{G}_0) \cong I^{\bar{p}} H_*(X; \mathcal{G})$, the usual intersection homology. In fact, this follows from our discussion in Section 3.1, where we noted that if $\bar{p}$ is a GM perversity then no allowable i-simplices intersect X^{n-2} in either the interiors of their i faces or the interiors of their $i - 1$ faces. Thus no boundary simplices can lie entirely in X^{n-2} and canceling of boundary simplices due to the stratified coefficient system does not occur. Thus $I^{\bar{p}} H_*(X; \mathcal{G}_0)$ legitimately extends the original Goresky–MacPherson theory. Furthermore, working with this "corrected" cone formula, one can show that the resulting intersection homology groups $I^{\bar{p}} H_*^\infty (X; \mathcal{G}_0)$ agree on topological stratified pseudomanifolds (modulo the usual reindexing issues) with the Deligne sheaf hypercohomology groups (and similarly with compact supports), assuming that $\bar{p}(2) \geq 0$ and that $\bar{p}$ is non-decreasing. This was proven in [26] under the assumption that $\bar{p}(2) = 0$ or 1 and that $\bar{p}(k) \leq \bar{p}(k+1) \leq \bar{p}(k)+1$, but the more general case follows from [22].

Thus, in summary, $I^{\bar{p}} H_*(X; \mathcal{G}_0)$ satisfies the cone formula, generalizes intersection homology with GM perversities, admits codimension one strata, and agrees with the Deligne sheaf for the superperversities we have considered up to this point. It turns out that stratified coefficients also permit useful results for even more general contexts.

REMARK 6.1. A similar idea for modifying the definition of intersection homology for non-GM perversities occurs in the unpublished notes of MacPherson [44]. There, only locally-finite chains in $X - X^{n-1}$ are considered, but their closures in X are used to determine allowability.

7. General perversities

We have now seen that stratified coefficients $\mathcal{G}_0$ allow us to recover the cone formula (6-1) both when $\bar{p}$ is a GM perversity and when it is a non-decreasing superperversity. How far can we push this? The answer turns out to be "quite far!" In fact, the cone formula will hold if $\bar{p}$ is completely arbitrary. Recall that we have defined a stratum of X to be a connected component of any $X_k =$

$X^k - X^{k-1}$. For a stratified pseudomanifold, possibly with codimension one strata, we define a *general perversity* $\bar{p}$ on X to be a function

$$\bar{p} : \{\text{singular strata of X}\} \to \mathbb{Z}.$$

Then a singular simplex $\sigma : \Delta^i \to X$ is $\bar{p}$-allowable if

$$\sigma^{-1}(Z) \subset \{i - \mathrm{codim}(Z) + \bar{p}(Z) \text{ skeleton of } \Delta^i\}$$

for each singular stratum Z of X. Even in this generality, the cone formula (6-1) holds for $I^{\bar{p}} H^c_*(cL^{k-1}; \mathcal{G}_0)$, replacing $\bar{p}(k)$ with $\bar{p}(v)$, where v is the cone vertex.

Such general perversities were considered in [44], following their appearance in the realm of perverse sheaves (see [4] and Section 8.2, below), and they appear in the work of Saralegi on intersection differential forms [53; 54]. They also play an important role in the intersection homology Künneth theorem of [28], which utilizes "biperversities" in which the set $X_k \times Y_l \subset X \times Y$ is given a perversity value depending on $\bar{p}(k)$ and $\bar{q}(l)$ for two perversities $\bar{p}, \bar{q}$ on X and Y, respectively; see Section 9.

In this section, we discuss some of the basic results on intersection homology with general perversities, most of which generalize the known theorems for GM perversities. We continue, for the most part, with the chain theory point of view. In Section 8, we will return to sheaf theory and discuss sheaf-theoretic techniques for handling general perversities.

REMARK 7.1. One thing that we can continue to avoid in defining general perversities is assigning perversity values to regular strata (those in $X - X^{n-1}$) and including this as part of the data to check for allowability. The reason is as follows: If Z is a regular stratum, the allowability conditions for a singular i-simplex σ would include the condition that $\sigma^{-1}(Z)$ lie in the $i + \bar{p}(Z)$ skeleton of Δ^i. If $\bar{p}(Z) \geq 0$, then this is true of any singular i-simplex, and if $\bar{p}(Z) < 0$, then this would imply that the singular simplex must not intersect Z at all, since X^{n-1} is a closed subset of X. Thus there are essentially only two possibilities. The case $\bar{p}(Z) \geq 0$ is the default that we work with already (without explicitly checking the condition that would always be satisfied on regular strata). On the other hand, the case $\bar{p}(Z) < -1$ is something of a degeneration. If $\bar{p}(Z) < 0$ for all regular strata, then all singular chains must be supported in X^{n-1} and so $I^{\bar{p}} S_*(X; \mathcal{G}_0) = 0$. If there are only some regular strata such that $\bar{p}(Z) < 0$, then, letting X^+ denote the pseudomanifold that is the closure of the union of the regular strata Z of X such that $\bar{p}(Z) \geq 0$, we have $I^{\bar{p}} H_*(X; \mathcal{G}_0) \cong I^{\bar{p}} H_*(X^+; \mathcal{G}_0|_{X^+})$. We could have simply studied intersection homology on X^+ in the first place, so we get nothing new. Thus it is reasonable to concern ourselves only with singular strata in defining allowability of simplices.

This being said, there are occasional situations where it is useful in technical formulae to assume that $\bar{p}(Z)$ is defined for all strata. This comes up, for example, in [28], where we define perversities on product strata $Z_1 \times Z_2 \subset X_1 \times X_2$ using formulas such as $Q_{\{\bar{p},\bar{q}\}}(Z_1 \times Z_2) = \bar{p}(Z_1) + \bar{q}(Z_2)$ for perversities $\bar{p}, \bar{q}$. Here $Z_1 \times Z_2$ may be a singular stratum, for example, even if Z_1 is regular but Z_2 is singular. The formula has the desired consequence in [28] by setting $\bar{p}(Z_1) = 0$ for Z_1 regular, and this avoids having to write out several cases.

Efficient perversities. It turns out that such generality contains a bit of overkill. In [22], we define a general perversity $\bar{p}$ to be *efficient* if

$$-1 \le \bar{p}(Z) \le \mathrm{codim}(Z) - 1$$

for each singular stratum $Z \subset X$. Given a general $\bar{p}$, we define its *efficientization* $\check{p}$ as

$$\check{p}(Z) = \begin{cases} \mathrm{codim}(Z) - 1 & \text{if } \bar{p}(Z) \ge \mathrm{codim}(Z) - 1, \\ \bar{p}(Z) & \text{if } 0 \le \bar{p}(Z) \le \mathrm{codim}(Z) - 2, \\ -1 & \text{if } \bar{p}(Z) \le -1. \end{cases}$$

It is shown in [22, Section 2] that $I^{\bar{p}} H_*(X; \mathcal{G}_0) \cong I^{\check{p}} H_*(X; \mathcal{G}_0)$. Thus it is always sufficient to restrict attention to the efficient perversities.

Efficient perversities and interiors of simplices. Efficient perversities have a nice feature that makes them technically better behaved than the more general perversities. If $\bar{p}$ is a perversity for which $\bar{p}(Z) \ge \mathrm{codim}(Z)$ for some singular stratum Z, then any i-simplex σ will be $\bar{p}$-allowable with respect to Z. In particular, Z will be allowed to intersect the image under σ of the interior of Δ^i. As such, $\sigma^{-1}(X - X^{n-1})$ could potentially have an infinite number of connected components, and a coefficient of σ might lift each component to a different branch of $\mathcal{G}$, even if $\mathcal{G}$ is a constant system. This could potentially lead to some pathologies, especially when considering intersection chains from the sheaf point of view. However, if $\bar{p}$ is efficient, then for a $\bar{p}$-allowable σ we must have $\sigma^{-1}(X - X^{n-1})$ within the $i - 1$ skeleton of Δ^i. Hence assigning a coefficient lift value to one point of the interior of Δ^i determines the coefficient value at all points (on $\sigma^{-1}(X - X^{n-1})$) by the unique extension of the lift and on $\sigma^{-1}(X^{n-1})$, where it is 0). This is technically much simpler and makes the complex of chains in some sense smaller.

In [28], the complex $I^{\bar{p}} S_*(X; \mathcal{G}_0)$ was defined with the assumption that this "unique coefficient" property holds, meaning that a coefficient should be determined by its lift at a single point. However, as noted in [28, Appendix], even for inefficient perversities, this does not change the intersection homology. So we are free to assume all perversities are efficient, without loss of any information (at least at the level of quasi-isomorphism), and this provides a reasonable way to avoid the issue entirely.

7.1. Properties of intersection homology with general perversities and stratified coefficients.
One major property that we lose in working with general perversities and stratified coefficients is independence of stratification. However, most of the other basic properties of intersection homology survive, including Poincaré duality, some of them in even a stronger form than GM perversities allow.

Basic properties. Suppose X^n is a topological stratified pseudomanifold, possibly with codimension one strata, let $\mathcal{G}$ be a coefficient system on $X - X^{n-1}$, and let $\bar{p}$ be a general perversity. What properties does $I^{\bar{p}} H_*(X; \mathcal{G}_0)$ possess?

For one thing, the most basic properties of intersection homology remain intact. It is invariant under stratum-preserving homotopy equivalences, and it possesses an excision property, long exact sequences of the pair, and Mayer–Vietoris sequences. The Künneth theorem when one term is an unstratified manifold M holds true (i.e., $I^{\bar{p}} S_*^c(X \times M; (\mathcal{G} \times \mathcal{G}')_0)$ is quasi-isomorphic to $I^{\bar{p}} S_*^c(X; \mathcal{G}_0) \otimes S_*^c(M; \mathcal{G}'_0))$. There are versions of this intersection homology with compact supports and with closed supports. And $U \to I^{\bar{p}} S_*(X, X - \bar{U}; \mathcal{G}_0)$ sheafifies to a homotopically fine sheaf whose hypercohomology groups recover the intersection homology groups, up to reindexing. It is also possible to work with PL chains on PL pseudomanifolds. For more details, see [26; 22].

Duality. Let us now discuss Poincaré duality in our present context.

THEOREM 7.2 (POINCARÉ DUALITY). *If F is a field[8], X is an F-oriented n-dimensional stratified pseudomanifold, and $\bar{p} + \bar{q} = \bar{t}$ (meaning that $\bar{p}(Z) + \bar{q}(Z) = \mathrm{codim}(Z) - 2$ for all singular strata Z), then*

$$I^{\bar{p}} H_i^{\infty}(X; F_0) \cong \mathrm{Hom}(I^{\bar{q}} H_{n-i}^c(X; F_0), F).$$

For compact orientable PL pseudomanifolds without codimension one strata and with GM perversities, this was initially proven in [32] via a combinatorial argument; a proof extending to the topological setting using the axiomatics of the Deligne sheaf and Verdier duality was obtained in [33]. This Verdier duality proof was extended to the current setting in [22] using a generalization of the Deligne sheaf that we will discuss in the following section. It also follows from the theory of perverse sheaves [4]. Recent work of the author and Jim McClure in [31] shows that intersection homology Poincaré duality can be proven using a cap product with an intersection homology orientation class by analogy to

---

[8]Recall that even in the Goresky–MacPherson setting, duality only holds, in general, with field coefficients.

the usual proof of Poincaré duality on manifolds (see, [36], for example). A slightly more restrictive statement (without proof) of duality for general perversities appears in the unpublished lecture notes of MacPherson [44] as far back as 1990.

As is the case for classical intersection homology, more general duality statements hold. These can involve local-coefficient systems, non-orientable pseudomanifolds, and, if X is *locally* $(\bar{p}, R)$-*torsion free* for the principal ideal domain R, then there are torsion linking and mod torsion intersection dualities over R. For complete details, see [22].

Pseudomanifolds with boundary and Lefschetz duality. General perversities and stratified coefficients can also be used to give an easy proof of a Lefschetz version of the duality pairing, one for which X is a pseudomanifold with boundary:

DEFINITION 7.3. An n-dimensional *stratified pseudomanifold with boundary* is a pair $(X, \partial X)$ such that $X - \partial X$ is an n-dimensional stratified pseudomanifold and the *boundary* ∂X is an $n - 1$ dimensional stratified pseudomanifold possessing a neighborhood in X that is stratified homeomorphic to $\partial X \times [0, 1)$, where $[0, 1)$ is unstratified and $\partial X \times [0, 1)$ is given the product stratification.

REMARK 7.4. A pseudomanifold may have codimension one strata that are not part of a boundary, even if they would be considered part of a boundary otherwise. For example, let M be a manifold with boundary ∂M (in the usual sense). If we consider M to be unstratified, then ∂M is the boundary of M. However, if we stratify M by the stratification $M \supset \partial M$, then ∂M is *not* a boundary of M as a stratified pseudomanifold, and in this case M is a stratified pseudomanifold without boundary.

We can now state a Lefschetz duality theorem for intersection homology of pseudomanifolds with boundary.

THEOREM 7.5 (LEFSCHETZ DUALITY). *If F is a field, X is a compact F-oriented n-dimensional stratified pseudomanifold, and $\bar{p} + \bar{q} = \bar{t}$ (meaning that $\bar{p}(Z) + \bar{q}(Z) = \mathrm{codim}(Z) - 2$ for all singular strata Z), then*

$$I^{\bar{p}} H_i(X; F_0) \cong \mathrm{Hom}(I^{\bar{q}} H_{n-i}(X, \partial X; F_0), F).$$

This duality also can be extended to include local-coefficient systems, noncompact or non-orientable pseudomanifolds, and, if X is *locally* $(\bar{p}, R)$-*torsion free* for the principal ideal domain R, then there are torsion linking and mod torsion intersection dualities over R.

In fact, in the setting of intersection homology with general perversities, this Lefschetz duality follows easily from Poincaré duality. To see this, let

$\hat{X} = X \cup_{\partial X} \bar{c} \partial X$, the space obtained by adjoining to X a cone on the boundary (or, equivalently, pinching the boundary to a point). Let v denote the vertex of the cone point. Let $\bar{p}_-$, $\bar{q}_+$ be the dual perversities on $\hat{X}$ such that $\bar{p}_-(Z) = \bar{p}(Z)$ and $\bar{q}_+(Z) = \bar{q}(Z)$ for each stratum Z of X, $\bar{p}_-(v) = -2$, and $\bar{q}_+(v) = n$. Poincaré duality gives a duality isomorphism between $I^{\bar{p}_-} H_*(\hat{X})$ and $I^{\bar{q}_+} H_*(\hat{X})$. But now we simply observe that $I^{\bar{p}_-} H_*(\hat{X}) \cong I^{\bar{p}_-} H_*(\hat{X} - v) \cong I^{\bar{p}} H_*(X)$, because the perversity condition at v ensures that no singular simplex may intersect v. On the other hand, since $I^{\bar{q}_+} H_*(c\partial X) = 0$ by the cone formula, $I^{\bar{q}_+} H_*(\hat{X}) \cong I^{\bar{q}_+} H_*(\hat{X}, \bar{c}\partial X)$ by the long exact sequence of the pair, but $I^{\bar{q}_+} H_*(\hat{X}, \bar{c}\partial X) \cong I^{\bar{q}_+} H_*(X, \partial X) \cong I^{\bar{q}} H_*(X, \partial X)$ by excision.

Notice that general perversities are used in this argument even if $\bar{p}$ and $\bar{q}$ are GM perversities.

PL intersection pairings. As in the classical PL manifold situation, the duality isomorphism of intersection homology arises out of a more general pairing of chains. In [32], Goresky and MacPherson defined the intersection pairing of PL intersection chains in a PL pseudomanifold as a generalization of the classical manifold intersection pairing. For manifolds, the intersection pairing is dual to the cup product pairing in cohomology. Given a ring R and GM perversities $\bar{p}, \bar{q}, \bar{r}$ such that $\bar{p} + \bar{q} \le \bar{r}$, Goresky and MacPherson constructed an intersection pairing

$$I^{\bar{p}} H_i^c(X; R) \otimes I^{\bar{q}} H_j^c(X; R) \to I^{\bar{r}} H_{i+j-n}^c(X; R).$$

This pairing arises by pushing cycles into a stratified version of general position due to McCrory [47] and then taking chain-theoretic intersections.

The Goresky–MacPherson pairing is limited in that a $\bar{p}$-allowable chain and a $\bar{q}$-allowable chain can be intersected only if there is a GM perversity $\bar{r}$ such that $\bar{p} + \bar{q} \le \bar{r}$. In particular, we must have $\bar{p} + \bar{q} \le \bar{t}$. This is more than simply a failure of the intersection of the chains to be allowable with respect to a GM perversity — if $\bar{p} + \bar{q} \not\le \bar{t}$, there are even technical difficulties with defining the intersection product in the first place. See [22, Section 5] for an in depth discussion of the details.

If we work with stratified coefficients, however, the problems mentioned in the preceding paragraphs can be circumvented, and we obtain pairings

$$I^{\bar{p}} H_i(X; R_0) \otimes I^{\bar{q}} H_j(X; R_0) \to I^{\bar{r}} H_{i+j-n}(X; R_0)$$

for *any* general perversities such that $\bar{p} + \bar{q} \le \bar{r}$.

Goresky and MacPherson extended their intersection pairing to topological pseudomanifolds using sheaf theory [33]. This can also be done for general perversities and stratified coefficients, but first we must revisit the Deligne sheaf construction. We do so in the next section.

A new approach to the intersection pairing via intersection cohomology cup products is presently being pursued by the author and McClure in [31].

Further applications. Some further applications of general perversity intersection homology will be discussed below in Section 9.

8. Back to sheaf theory

8.1. A generalization of the Deligne construction.
Intersection chains with stratified coefficients were introduced to provide a chain theory whose homology agrees with the hypercohomology of the Deligne sheaf when $\bar{p}$ is a superperversity, in particular when $\bar{p}(2) > 0$ or when X has codimension one strata. However, when $\bar{p}$ is a general perversity, our new chain formulation no long agrees with the Deligne construction. For one thing, we know that if $\bar{p}$ is ever negative, the Deligne sheaf is trivial. The classical Deligne construction also has no mechanism for handling perversities that assign different values to strata of the same codimension, and, even if we restrict to less general perversities, any decrease in perversity value at a later stage of the Deligne process will truncate away what might have been vital information coming from an earlier stage. Thus, we need a generalization of the Deligne process that incorporates general perversities and stratified coefficients. One method was provided by the author in [22], and we describe this now.

The first step is to modify the truncation functor to be a bit more picky. Rather than truncating a sheaf complex in the same degree at all stalks, we truncate more locally. This new truncation functor is a further generalization of the "truncation over a closed subset" functor presented in [33, Section 1.14] and attributed to Deligne; that functor is used in [33, Section 9] to study extensions of Verdier duality pairings in the context of intersection homology with GM perversities. Our construction is also related to the "intermediate extension" functor in the theory of perverse sheaves; we will discuss this in the next subsection.

DEFINITION 8.1. Let $\mathcal{A}^*$ be a sheaf complex on X, and let $\mathfrak{F}$ be a locally-finite collection of subsets of X. Let $|\mathfrak{F}| = \cup_{V \in \mathfrak{F}} V$. Let P be a function $\mathfrak{F} \to \mathbb{Z}$. Define the presheaf $T^{\mathfrak{F}}_{\leq P} \mathcal{A}^*$ as follows. If U is an open set of X, let

$$T^{\mathfrak{F}}_{\leq P} \mathcal{A}^*(U) = \begin{cases} \Gamma(U; \mathcal{A}^*) & \text{if } U \cap |\mathfrak{F}| = \varnothing, \\ \Gamma(U; \tau_{\leq \inf\{P(V)|V \in \mathfrak{F}, U \cap V \neq \varnothing\}} \mathcal{A}^*) & \text{if } U \cap |\mathfrak{F}| \neq \varnothing. \end{cases}$$

Restriction is well-defined because if $m < n$ then there is a natural inclusion $\tau_{\leq m} \mathcal{A}^* \hookrightarrow \tau_{\leq n} \mathcal{A}^*$.

Let the *generalized truncation sheaf* $\tau^{\mathfrak{F}}_{\leq P} \mathcal{A}^*$ be the sheafification of $T^{\mathfrak{F}}_{\leq P} \mathcal{A}^*$.

For maps $f : \mathcal{A}^* \to \mathcal{B}^*$ of sheaf complexes over X, we can define $\tau^{\mathfrak{F}}_{\leq P} f$ in the obvious way. In fact, $T^{\mathfrak{F}}_{\leq P} f$ is well-defined by applying the ordinary truncation functors on the appropriate subsets, and we obtain $\tau^{\mathfrak{F}}_{\leq P} f$ again by passing to limits in the sheafification process.

Using this truncation, we can modify the Deligne sheaf.

DEFINITION 8.2. Let X be an n-dimensional stratified pseudomanifold, possibly with codimension one strata, let $\bar{p}$ be a general perversity, let $\mathcal{G}$ be a coefficient system on $X - X^{n-1}$, and let $\mathcal{O}$ be the orientation sheaf on $X - X^{n-1}$. Let X_k stand also for the set of strata of dimension k. Then we define the *generalized Deligne sheaf* as[9]

$$\mathcal{Q}^*_{\bar{p},\mathcal{G}} = \tau^{X_0}_{\leq \bar{p}} R i_{n*} \ldots \tau^{X_{n-1}}_{\leq \bar{p}} R i_{1*}(\mathcal{G} \otimes \mathcal{O}).$$

If $\bar{p}$ is a GM perversity, then it is not hard to show directly that $\mathcal{Q}^*_{\bar{p},\mathcal{G}}$ is quasi-isomorphic to the usual Deligne sheaf $\mathcal{P}^*_{\bar{p},\mathcal{G}}$. Furthermore, it is shown in [22] that $\mathcal{Q}^*_{\bar{p},\mathcal{G}}$ is quasi-isomorphic to the sheaf generated by the presheaf

$$U \to I^{\bar{p}} S_{n-*}(X, X - \bar{U}; \mathcal{G}_0),$$

and so $\mathbb{H}^*(\mathcal{Q}^*_{\bar{p},\mathcal{G}}) \cong I^{\bar{p}} H^{\infty}_{n-*}(X; \mathcal{G}_0)$ and similarly for compact supports. It is also true, generalizing the Goresky–MacPherson case, that if $\bar{p} + \bar{q} = \bar{t}$, then $\mathcal{Q}^*_{\bar{p}}$ and $\mathcal{Q}^*_{\bar{q}}$ are appropriately Verdier dual, leading to the expected Poincaré and Lefschetz duality theorems. Furthermore, for any general perversities such that $\bar{p} + \bar{q} \leq \bar{r}$, there are sheaf pairings $\mathcal{Q}^*_{\bar{p}} \otimes \mathcal{Q}^*_{\bar{q}} \to \mathcal{Q}^*_{\bar{r}}$ that generalize the PL intersection pairing. If $\bar{p} + \bar{q} \leq \bar{t}$, there is also a pairing $\mathcal{Q}^*_{\bar{p}} \otimes \mathcal{Q}^*_{\bar{q}} \to \mathbb{D}^*_X[-n]$, where $\mathbb{D}^*_X[-n]$ is the shifted Verdier dualizing complex on X. See [22] for the precise statements of these results.

8.2. Perverse sheaves.

The theory of perverse sheaves provided, as far back as the early 1980s, a context for the treatment of general perversities. To quote Banagl's introduction to [2, Chapter 7]:

> In discussing the proof of the Kazhdan–Lusztig conjecture, Beilinson, Bernstein and Deligne discovered that the essential image of the category of regular holonomic $\mathcal{D}$-modules under the Riemann–Hilbert correspondence gives a natural abelian subcategory of the nonabelian bounded constructible derived category [of sheaves] on a smooth complex algebraic variety. An intrinsic characterization of this abelian subcategory was obtained by Deligne (based on discussions with Beilinson, Bernstein, and MacPherson), and independently by Kashiwara. It was then realized that one still gets an abelian subcategory if the axioms of the characterization

---

[9] This definition differs from that in [22] by the orientation sheaf $\mathcal{O}$ — see Remark 4.1 on page 190. For consistency, we also change notation slightly to include $\mathcal{G}$ as a subscript rather than as an argument.

are modified to accommodate an arbitrary perversity function, with the original axioms corresponding to the middle perversity. The objects of these abelian categories were termed *perverse sheaves*...

Thus, the phrase "perverse sheaves" refers to certain subcategories, indexed by various kinds of perversity functions, of the derived category of bounded constructible sheaf complexes on a space X. The general theory of perverse sheaves can handle general perversities, though the middle perversities are far-and-away those most commonly encountered in the literature (and, unfortunately, many expositions restrict themselves solely to this case). The remarkable thing about these categories of perverse sheaves is that they are abelian, which the derived category is not (it is only "triangulated").[10] The Deligne sheaf complexes on the various strata of X (and with appropriate coefficients systems) turn out to be the simple objects of these subcategories.

The construction of perverse sheaves is largely axiomatic, grounded in a number of quite general categorical structures. It would take us well too far afield to provide all the details. Rather, we provide an extremely rough sketch of the ideas and refer the reader to the following excellent sources: [4], [40, Chapter X], [2, Chapter 7], [5], and [20, Chapter 5]. For a more historical account, the reader should see [43].

The starting point for any discussion of perverse sheaves is the notion of T-*structures*. Very roughly, a T-structure on a triangulated category D is a pair of subcategories $(D^{\leq 0}, D^{\geq 0})$ that are complementary, in the sense that for any S in D, there is a distinguished triangle

$$S_1 \to S \to S_2,$$

with $S_1 \in D^{\leq 0}$ and S_2 in $D^{\geq 0}$. Of course there are a number of axioms that must be satisfied and that we will not discuss here. The notation reflects the canonical T-structure that occurs on the derived category of sheaves on a space X: $D^{\leq 0}(X)$ is defined to be those sheaf complexes $\mathcal{S}^*$ such that $\mathcal{H}^j(\mathcal{S}^*) = 0$ for $j > 0$, and $D^{\geq 0}(X)$ is defined to be those sheaf complexes $\mathcal{S}^*$ such that $\mathcal{H}^j(\mathcal{S}^*) = 0$ for $j < 0$. Here $\mathcal{H}^*(\mathcal{S}^*)$ denotes the derived cohomology sheaf of the sheaf complex $\mathcal{S}^*$, such that $\mathcal{H}^*(\mathcal{S}^*)_x = H^*(\mathcal{S}_x^*)$.

The *heart* (or *core*) of a T-structure is the intersection $D^{\leq 0} \cap D^{\geq 0}$. It is always an abelian category. In our canonical example, the heart consists of the sheaf complexes with nonvanishing cohomology only in degree 0. In this case, the heart is equivalent to the abelian category of sheaves on X. Already from

---

[10]There is an old joke in the literature that perverse sheaves are neither perverse nor sheaves. The first claim reflects the fact that perverse sheaves form abelian categories, which are much less "perverse" than triangulated categories. The second reflects simply the fact that perverse sheaves are actually complexes of sheaves.

this example, we see how truncation might play a role in providing perverse sheaves — in fact, for the sheaf complex $\mathcal{S}^*$, the distinguished triangle in this example is provided by

$$\tau_{\leq 0}\mathcal{S}^* \to \mathcal{S}^* \to \tau_{\geq 0}\mathcal{S}^*.$$

Furthermore, this example can be modified easily by shifting the truncation degree from 0 to any other integer k. This T-structure is denoted by

$$(D^{\leq k}(X), D^{\geq k}(X)).$$

The next important fact about T-structures is that if X is a space, U is an open subspace, $F = X - U$, and T-structures satisfying sufficient axioms on the derived categories of sheaves on U and F are given, they can be "glued" to provide a T-structure on the derived category of sheaves on X. The idea the reader should have in mind now is that of gluing together sheaves truncated at a certain dimension on U and at another dimension on F. This then starts to look a bit like the Deligne process. In fact, let P be a perversity[11] on the two stratum space $X \supset F$, and let $(D^{\leq P(U)}(U), D^{\geq P(U)}(U))$ and $(D^{\leq P(F)}(F), D^{\geq P(F)}(F))$ be T-structures on U and F. Then these T-structures can be glued to form a T-structure on X, denoted by $({}^P D^{\leq 0}, {}^P D^{\geq 0})$.

It turns out that the subcategories ${}^P D^{\leq 0}$ and ${}^P D^{\geq 0}$ can be described quite explicitly. If $i : U \hookrightarrow X$ and $j : F \hookrightarrow X$ are the inclusions, then

$${}^P D^{\leq 0} = \left\{ \mathcal{S}^* \in D^+(X) \,\middle|\, \begin{array}{l} \mathcal{H}^k(i^*\mathcal{S}^*) = 0 \text{ for } k > P(U) \\ \mathcal{H}^k(j^*\mathcal{S}^*) = 0 \text{ for } k > P(F) \end{array} \right\},$$

$${}^P D^{\geq 0} = \left\{ \mathcal{S}^* \in D^+(X) \,\middle|\, \begin{array}{l} \mathcal{H}^k(i^*\mathcal{S}^*) = 0 \text{ for } k < P(U) \\ \mathcal{H}^k(j^!\mathcal{S}^*) = 0 \text{ for } k < P(F) \end{array} \right\}.$$

If $\mathcal{S}^*$ is in the heart of this T-structure, we say it is P-perverse.

More generally, if X is a space with a variety of singular strata Z and P is a perversity on the stratification of X, then it is possible to glue T-structures inductively to obtain the category of P-perverse sheaves. If $j_Z : Z \hookrightarrow X$ are the inclusions, then the P-perverse sheaves are those which satisfy $\mathcal{H}^k(j_Z^*\mathcal{S}^*) = 0$ for $k > P(Z)$ and $\mathcal{H}^k(j_Z^!\mathcal{S}^*) = 0$ for $k < P(Z)$.

These two conditions turn out to be remarkably close to the conditions for $\mathcal{S}^*$ to satisfy the Deligne sheaf axioms $AX1$. In fact, the condition $\mathcal{H}^k(j_Z^*\mathcal{S}^*) = 0$ for $k > P(Z)$ is precisely the third axiom. The condition $\mathcal{H}^k(j_Z^!\mathcal{S}^*) = 0$ for

---

[11]The reason we use P here for a perversity, departing from both our own notation, above, and from the notation in most sources on perverse sheaves (in particular [4]) is that when we use perverse sheaf theory, below, to recover intersection homology, there will be a discrepancy between the perversity P for perverse sheaves and the perversity $\bar{p}$ for the Deligne sheaf.

$k < P(Z)$ implies that the local attaching map is an isomorphism up to degree $P(Z) - 2$; see [6, page 87]. Notice that this is a less strict requirement than that for the Deligne sheaf. Thus, Deligne sheaves are perverse sheaves, but not necessarily vice versa.

The machinery developed in [4] also contains a method for creating sheaf complexes that satisfy the intersection homology axioms $AX1$, though again it is more of an axiomatic construction than the concrete construction provided in Section 8.1. Let $U \subset X$ be an open subset of X that is a union of strata, let $i : U \hookrightarrow X$ be the inclusion, and let $\mathcal{S}^*$ be a P-perverse sheaf on U. Then there is defined in [4] the "intermediate extension functor" $i_{!*}$ such that $i_{!*}\mathcal{S}^*$ is the unique extension in the category of P-perverse sheaves of $\mathcal{S}^*$ to X (meaning that the restriction of $i_{!*}\mathcal{S}^*$ to U is quasi-isomorphic to $\mathcal{S}^*$) such that for each stratum $Z \subset X - U$ and inclusion $j : Z \hookrightarrow X$, we have $\mathcal{H}^k(j^*i_{!*}\mathcal{S}^*) = 0$ for $k \geq P(Z)$ and $\mathcal{H}^k(j^!i_{!*}\mathcal{S}^*) = 0$ for $k \leq P(Z)$. We refer the reader to [4, Section 1.4] or [20, Section 5.2] for the precise definition of the functor $i_{!*}$.

In particular, suppose we let $U = X - X^{n-1}$, that $\mathcal{S}^*$ is just the local system $\mathcal{G}$, and that $\bar{p}$ is a general perversity on X. The sheaf $\mathcal{G}$ is certainly P-perverse on U with respect to the perversity $P(U) = 0$. Now let $P(Z) = \bar{p}(Z) + 1$. It follows that for each singular stratum inclusion $j : Z \hookrightarrow X$, we have $\mathcal{H}^k(j^*i_{!*}\mathcal{G}) = 0$ for $k > \bar{p}(Z)$ and $\mathcal{H}^k(j^!i_{!*}\mathcal{G}) = 0$ for $k \leq \bar{p}(Z) + 1$. In the presence of the first condition, the second condition is equivalent to the attaching map being an isomorphism up through degree $\bar{p}(Z)$; see [6, page 87]. But, according to the axioms $AX1$, these conditions are satisfied by the perversity $\bar{p}$ Deligne sheaf, which is also easily seen to be P-perverse. Thus, since $i_{!*}\mathcal{G}$ is the unique extension of $\mathcal{G}$ with these properties, $i_{!*}\mathcal{G}$ is none other than the Deligne sheaf (up to quasi-isomorphism)! Thus we can think of the Deligne process provided in Section 8.1 as a means to provide a concrete realization of $i_{!*}\mathcal{G}$.

9. Recent and future applications of general perversities

Beyond extending the results of intersection homology with GM perversities, working with general perversities makes possible new results that do not exist in "classical" intersection homology theory. For example, we saw in Sections 7 and 8 that general perversities permit the definition of PL or sheaf-theoretic intersection pairings with no restrictions on the perversities of the intersection homology classes being intersected. In this section, we review some other recent and forthcoming results made possible by intersection homology with general perversities.

Künneth theorems and cup products. In [28], general perversities were used to provide a very general Künneth theorem for intersection homology. Some special cases had been known previously. King [41] showed that for any loose perversity $I^{\bar{p}} H_*^c(M \times X) \cong H_*(C_*^c(M) \otimes I^{\bar{p}} C_*^c(X))$ when X is a pseudomanifold, M is an unstratified manifold, and $(M \times X)^i = M \times X^i$. Special cases of this result were proven earlier by Cheeger [16], Goresky and MacPherson [32; 33], and Siegel [55]. In [18], Cohen, Goresky, and Ji provided counterexamples to the existence of a general Künneth theorem for a single perversity and showed that $I^{\bar{p}} H_*^c(X \times Y; R) \cong H_*(I^{\bar{p}} C_*^c(X; R) \otimes I^{\bar{p}} C_*^c(Y; R))$ for pseudomanifolds X and Y and a principal ideal domain R provided either that

(a) $\bar{p}(a) + \bar{p}(b) \leq \bar{p}(a+b) \leq \bar{p}(a) + \bar{p}(b) + 1$ for all a and b, *or* that
(b) $\bar{p}(a) + \bar{p}(b) \leq \bar{p}(a+b) \leq \bar{p}(a) + \bar{p}(b) + 2$ for all a and b and either X or Y is locally $(\bar{p}, R)$-torsion free.

The idea of [28] was to ask a broader question: for what perversities on $X \times Y$ is the intersection chain complex quasi-isomorphic to the product $I^{\bar{p}} C_*^c(X; R_0) \otimes I^{\bar{q}} C_*^c(X; R_0)$? This question encompasses the Cohen–Goresky–Ji Künneth theorem and the possibility of both GM and non-GM perversities $\bar{p}, \bar{q}$. However, in order to avoid the fairly complicated conditions on a single perversity found by Cohen, Goresky, and Ji, it is reasonable to consider general perversities on $X \times Y$ that assign to a singular stratum $Z_1 \times Z_2$ a value depending on $\bar{p}(Z_1)$ and $\bar{q}(Z_2)$. Somewhat surprisingly, there turn out to be many perversities on $X \times Y$ that provide the desired quasi-isomorphism. The main result of [28] is the following theorem. The statement is reworded here to account for the most general case (see [28, Theorem 3.2, Remark 3.4, Theorem 5.2]), while the statement in [28] is worded to avoid overburdening the reader too much with details of stratified coefficients, which play a minimal role that paper.

THEOREM 9.1. *If R is a principal ideal domain and $\bar{p}$ and $\bar{q}$ are general perversities, then $I^Q H_*^c(X \times Y; R_0) \cong H_*(I^{\bar{p}} C_*^c(X; R_0) \otimes I^{\bar{q}} C_*^c(Y; R_0))$ if the following conditions hold*:

(a) $Q(Z_1 \times Z_2) = \bar{p}(Z_1)$ *if Z_2 is a regular stratum of Y and $Q(Z_1 \times Z_2) = \bar{q}(Z_2)$ if Z_1 is a regular stratum of X.*
(b) *For each pair $Z_1 \times Z_2$ such that Z_1 and Z_2 are each singular strata, either*

(i) $Q(Z_1 \times Z_2) = \bar{p}(Z_1) + \bar{q}(Z_2)$, *or*
(ii) $Q(Z_1 \times Z_2) = \bar{p}(Z_1) + \bar{q}(Z_2) + 1$, *or*
(iii) $Q(Z_1 \times Z_2) = \bar{p}(Z_1) + \bar{q}(Z_2) + 2$ *and the torsion product*

$$I^{\bar{p}} H_{\operatorname{codim}(Z_1)-2-\bar{p}(Z_1)}(L_1; R_0) * I^{\bar{q}} H_{\operatorname{codim}(Z_2)-2-\bar{q}(Z_2)}(L_2; R_0)$$

is zero, where L_1, L_2 are the links of Z_1, Z_2 in X, Y, respectively, and codim *refers to codimension in X or Y, as appropriate.*

Furthermore, if these conditions are not satisfied, then $I^Q H^c_*(X \times Y; R_0)$ will not equal $H_*(I^{\bar{p}} C^c_*(X; R_0) \otimes I^{\bar{q}} C^c_*(Y; R_0))$ in general.

Of course the torsion condition in (iii) will be satisfied automatically if R is a field or if X or Y is locally $(\bar{p}, R)$- or $(\bar{q}, R)$-torsion free. Note also that it is not required that a consistent choice among the above options be made across all products of singular strata — for each such $Z_1 \times Z_2$ one can choose independently which perversity to use from among options (i), (ii), or, assuming the hypothesis, (iii). The theorem can also be generalized further to include stratified local coefficient systems on X or Y; we leave the details to the reader.

This Künneth theorem has opened the way toward other results in intersection homology, including the formulation by the author and Jim McClure of an intersection cohomology cup product over field coefficients that they expect to be dual to the Goresky–MacPherson intersection pairing. There does not seem to have been much past research done on or with intersection cohomology in the sense of the homology groups of cochains $I_{\bar{p}} C^*(X; R_0) = \text{Hom}(I^{\bar{p}} C^c_*(X; R_0); R)$. One important reason would seem to be the prior lack of availability of a geometric cup product. A cup product using the Alexander–Whitney map is unavailable in intersection homology since it does not preserve the admissibility conditions for intersection chains — in other words, breaking chains into "front p-faces and back q-faces" (see [49, Section 48]) might destroy allowability of simplices. However, there is another classical approach to the cup product that can be adapted to intersection cohomology, provided one has an appropriate Künneth theorem. For *ordinary homology*, this alternative approach is to define a diagonal map (with field coefficients) as the composite

$$H^c_*(X) \to H^c_*(X \times X) \overset{\cong}{\leftarrow} H^c_*(X) \otimes H^c_*(X),$$

where the first map is induced by the geometric diagonal inclusion map and the second is the Eilenberg–Zilber shuffle product, which is an isomorphism by the ordinary Künneth theorem with field coefficients (note that the shuffle product should have better geometric properties than the Alexander–Whitney map because it is really just Cartesian product). The appropriate Hom dual of this composition yields the cup product. This process suggests doing something similar in intersection homology with field coefficients, and indeed the Künneth theorem of [28] provides the necessary right-hand quasi-isomorphism in a diagram of the form

$$I^{\bar{s}} H^c_*(X; F_0) \to I^Q H^c_*(X \times X; F_0) \overset{\cong}{\leftarrow} I^{\bar{p}} H^c_*(X; F_0) \otimes I^{\bar{q}} H^c_*(X; F_0).$$

When $\bar{p} + \bar{q} \geq \bar{t} + \bar{s}$, there results a cup product

$$I_{\bar{p}} H^*(X; F_0) \otimes I_{\bar{q}} H^*(X; F_0) \to I_{\bar{s}} H^*(X; F_0).$$

The intersection Künneth theorem also allows for a cap product of the form

$$I_{\bar{p}}H^i(X; F_0) \otimes I^{\bar{s}} H_j^c(X; F_0) \to I^{\bar{q}} H_{j-i}^c(X; F_0)$$

for any field F and any perversities satisfying $\bar{p} + \bar{q} \geq \bar{t} + \bar{s}$. This makes possible a Poincaré duality theorem for intersection (co)homology given by cap products with a fundamental class in $I^{\bar{0}} H_n(X; F_0)$. For further details and applications, the reader is urged to consult [31].

Perverse signatures. Right from its beginnings, there has been much interest and activity in using intersection homology to define signature (index) invariants and bordism theories under which these signatures are preserved. Signatures first appeared in intersection homology in [32] associated to the symmetric intersection pairings on $I^{\bar{m}} H_{2n}(X^{4n}; \mathbb{Q})$ for spaces X with only strata of even codimension, such as complex algebraic varieties. The condition on strata of even codimension ensures that $I^{\bar{m}} H_{2n}(X^{4n}; \mathbb{Q}) \cong I^{\bar{n}} H_{2n}(X^{4n}; \mathbb{Q})$ so that this group is self-dual under the intersection pairing. These ideas were extended by Siegel [55] to the broader class of Witt spaces, which also satisfy $I^{\bar{m}} H_{2n}(X^{4n}; \mathbb{Q}) \cong I^{\bar{n}} H_{2n}(X^{4n}; \mathbb{Q})$. In addition, Siegel developed a bordism theory of Witt spaces, which he used to construct a geometric model for ko-homology at odd primes. Further far reaching generalizations of these signatures have been studied by, among others and in various combinations, Banagl, Cappell, Libgober, Maxim, Shaneson, and Weinberger [1; 3; 10; 11; 9].

Signatures on singular spaces have also been studied analytically via L^2-cohomology and L^2 Hodge theory, which are closely related to intersection homology. Such signatures may relate to duality in string theory, such as through Sen's conjecture on the dimension of spaces of self-dual harmonic forms on monopole moduli spaces. Results in these areas and closely related topics include those of Müller [48]; Dai [19]; Cheeger and Dai [17]; Hausel, Hunsicker, and Mazzeo [37; 39; 38]; Saper [51; 50]; Saper and Stern [52]; and Carron [13; 15; 14]; and work on analytic symmetric signatures is currently being pursued by Albin, Leichtmann, Mazzeo and Piazza. Much more on analytic approaches to invariants of singular spaces can be found in the other papers in the present volume [30].

A different kind of signature invariant that can be defined using non-GM perversities appears in this analytic setting in the works of Hausel, Hunsicker, and Mazzeo [37; 39; 38], in which they demonstrate that groups of L^2 harmonic forms on a manifold with fibered boundary can be identified with cohomology spaces associated to the intersection cohomology groups of varying perversities for a canonical compactification X of the manifold. These *perverse signatures* are the signatures of the nondegenerate intersection pairings on $\text{im}(I^{\bar{p}} H_{2n}(X^{4n}) \to I^{\bar{q}} H_{2n}(X^{4n}, \partial X^{4n}))$, when $\bar{p} \leq \bar{q}$. The signature for Witt spaces mentioned above is a special case in which $\bar{p} = \bar{q} = \bar{m} = \bar{n}$ and $\partial X = \varnothing$. If X is the compactification of the interior of a compact manifold with boundary $(M, \partial M)$ and $\bar{p}(Z) < 0$ and $\bar{q}(Z) \geq \operatorname{codim}(Z) - 1$ for all singular Z, then $I^{\bar{p}} H_*(X) \cong H_*(M)$, $I^{\bar{q}} H_*(X) \cong H_*(M, \partial M)$, and in this case the perverse signature is the classical signature associated to a manifold with boundary.

Using the Lefschetz duality results of general perversity intersection homology described above, Hunsicker and the author are currently undertaking a topological study of the perverse signatures, including research on how Novikov additivity and Wall non-additivity extend to these settings.

10. Saralegi's relative intersection chains

Independently of the author's introduction of stratified coefficients, Saralegi [54] discovered another way, in the case of a constant coefficient system, to obtain an intersection chain complex that satisfies the cone formula (2-1) for general perversities. In [54], he used this chain complex to prove a general perversity version of the de Rham theorem on unfoldable pseudomanifolds. These spaces are a particular type of pseudomanifold on which it is possible to define a differential form version of intersection cohomology over the real numbers. This de Rham intersection cohomology appeared in a paper by Brylinski [8], though he credits Goresky and MacPherson with the idea. Brylinski showed that for GM perversities and on a Thom–Mather stratified space, de Rham intersection cohomology is Hom dual to intersection homology with real coefficients. Working on more general "unfoldable spaces," Brasselet, Hector, and Saralegi later proved a de Rham theorem in [7], showing that this result can be obtained by integration of forms on intersection chains, and this was extended to more general perversities by Saralegi in [53]. However, [53] contains an error in the case of perversities $\bar{p}$ satisfying $\bar{p}(Z) > \operatorname{codim}(Z) - 2$ or $\bar{p}(Z) < 0$ for some singular stratum Z. This error can be traced directly to the failure of the cone formula for non-GM perversities. Saralegi introduced his *relative intersection chains*[12] in [54] specifically to correct this error.

The rough idea of Saralegi's relative chains is precisely the same as the author's motivation for introducing stratified coefficients: when a perversity on a stratum Z is too high (greater than $\operatorname{codim}(Z) - 2$), it is necessary to kill chains living in that stratum in order to preserve the cone formula. The idea of stratified coefficients is to redefine the coefficient system so that such chains are killed by virtue of their coefficients being trivial. The idea of relative chains is instead

---

[12]These should not be confused with relative intersection chains in the sense $I^{\bar{p}} C_*(X, A) \cong I^{\bar{p}} C_*(X) / I^{\bar{p}} C_*(A)$.

to form a quotient group so that the chains living in such strata are killed in the quotient.

More precisely, let $A^{\bar{p}} C_i(X)$ be the group generated by the $\bar{p}$-allowable i-simplices of X (notice that there is no requirement that the boundary of an element of $A^{\bar{p}} C_i(X)$ be allowable), and let $X_{\bar{t}-\bar{p}}$ be the closure of the union of the singular strata Z of X such that $\bar{p}(Z) > \text{codim}(Z) - 2$. Let $A^{\bar{p}} C_i(X_{\bar{t}-\bar{p}})$ be the group generated by the $\bar{p}$ allowable i-simplices with support in $X_{\bar{t}-\bar{p}}$. Then Saralegi's relative intersection chain complex is defined to be

$$
S^{\bar{p}} C_*^c(X, X_{\bar{t}-\bar{p}})
$$
$$
= \frac{\left(A^{\bar{p}} C_*(X) + A^{\bar{p}+1} C_*(X_{\bar{t}-\bar{p}})\right) \cap \partial^{-1}\left(A^{\bar{p}} C_{*-1}(X) + A^{\bar{p}+1} C_{*-1}(X_{\bar{t}-\bar{p}})\right)}{A^{\bar{p}+1} C_*(X_{\bar{t}-\bar{p}}) \cap \partial^{-1} A^{\bar{p}+1} C_{*-1}(X_{\bar{t}-\bar{p}})}.
$$

Roughly speaking, this complex consists of $\bar{p}$-allowable chains in X and slightly more allowable chains $((\bar{p}+1)$-allowable) in $X_{\bar{t}-\bar{p}}$ whose boundaries are also either $\bar{p}$-allowable in X or $\bar{p}+1$ allowable in $X_{\bar{t}-\bar{p}}$, but then we quotient out by those chains supported in $X_{\bar{t}-\bar{p}}$. This quotient step is akin to the stratified coefficient idea of setting simplices supported in X^{n-1} to 0. In fact, there is no harm in extending Saralegi's definition by replacing $X_{\bar{t}-\bar{p}}$ by all of X^{n-1}, since the perversity conditions already guarantee that no simplex of $A^{\bar{p}} C_i(X)$ nor the boundary of any such simplex can have support in those singular strata not in $X_{\bar{t}-\bar{p}}$. In addition, there is also nothing special about the choice $\bar{p}+1$ for allowability of chains in $X_{\bar{t}-\bar{p}}$: the idea is to throw in enough singular chains supported in the singular strata so that the boundaries of any chains in $A^{\bar{p}} C_i(X)$ will also be in "the numerator" (for example, the inallowable 0-simplex in $\partial(\bar{c}x)$ that lives at the cone vertex in our example in Section 5), but then to kill any such extra chains by taking the quotient. In other words, it would be equivalent to define Saralegi's relative intersection chain complex as

$$
\frac{\left(A^{\bar{p}} C_*(X) + S_*(X^{n-1})\right) \cap \partial^{-1}\left(A^{\bar{p}} C_{*-1}(X) + S_{*-1}(X^{n-1})\right)}{S_*(X^{n-1})},
$$

where $S_*(X)$ is the ordinary singular chain complex.

We refer the reader to [28, Appendix A] for a proof[13] that $S^{\bar{p}} C_*(X, X_{\bar{t}-\bar{p}}; G)$ and $I^{\bar{p}} S_*(X; G_0)$ are chain isomorphic, and so, in particular, they yield the same intersection homology groups. It is not clear that there is a well-defined version of $S^{\bar{p}} C_*(X, X_{\bar{t}-\bar{p}})$ with coefficients in a local system $\mathcal{G}$ defined only on

---

[13]The proof in [28] uses a slightly different definition of intersection chains with stratified coefficients than the one given here. However, for any general perversity, the intersection chains with stratified coefficients there are quasi-isomorphic to the ones discussed here, and they are isomorphic for any efficient perversity. See [28, Appendix A].

$X - X^{n-1}$, and so stratified coefficients may be a slightly broader concept. There may also be some technical advantages in sheaf theory to avoiding quotient groups.

11. Habegger and Saper's codimension $\geq c$ intersection homology theory

Finally, we discuss briefly the work of Habegger and Saper [35], in which they introduce what they call *codimension $\geq c$ intersection homology*. This is the sheafification of King's loose perversity intersection homology. In a sense, this is the opposite approach to that of stratified coefficients: stratified coefficients were introduced to provide a chain theory that agrees with the Deligne sheaf construction for superperversities, while codimension $\geq c$ intersection homology provides a Deligne-type sheaf construction whose hypercohomology yields King's intersection homology groups. Habegger and Saper work with perversities $\bar{p} : \mathbb{Z}^{\geq 2} \to \mathbb{Z}$ such that $\bar{p}(k) \leq \bar{p}(k+1) \leq \bar{p}(k)+1$ and[14] $\bar{p}(2) \geq 0$, and they work on cs-sets, which generalize pseudomanifolds (see [41; 35]). In fact, King showed in [41] that intersection homology is independent of the stratification in this setting.

The paper [35] involves many technicalities in order to obtain the most general possible results. We will attempt to simplify the discussion greatly in order to convey what seems to be the primary stream of ideas. However, we urge the reader to consult [35] for the correct details.

Given a perversity $\bar{p}$, the "codimension $\geq c$" in the name of the theory comes from considering

$$c_{\bar{p}} = \min(\{k \in \mathbb{Z}^+ \mid \bar{p}(k) \leq k - 2\} \cup \{\infty\}).$$

In other words, $c_{\bar{p}}$ (or simply c when the perversity is understood) is the first codimension for which $\bar{p}$ takes the values of a GM perversity. Since the condition $\bar{p}(k) \leq \bar{p}(k+1) \leq \bar{p}(k) + 1$ ensures that $\bar{p}$ will be in the Goresky–MacPherson range of values for all $k \geq c$, the number c serves as somewhat of a phase transition. At points in strata of codimension $\geq c$, the cone formula (2-1) holds locally for King's singular intersection chains (i.e., we can use the cone formula to compute the local intersection homology groups in a distinguished neighborhood). For strata of codimension $< c$, the perversity $\bar{p}$ is in the "super" range, and the cone formula fails, as observed in Section 5. So, the idea of Habegger and Saper, building on the Goresky–MacPherson–Deligne axiomatic approach to intersection homology (see Section 4, above) was to find a way to axiomatize a sheaf construction that upholds the cone formula as the Deligne

---

[14]Technically, they allow $\bar{p}(2) < 0$, but in this case their theory is trivial; see [35, Corollary 4.8].

sheaf does for GM perversities, but only on strata of codimension $\geq c$. This idea is successful, though somewhat complicated because the coefficients now must live on $X - X^c$ and must include the sheafification on this subspace of $U \to I^{\bar{p}}S_*(U;\mathcal{G})$.

In slightly more detail (though still leaving out many technicalities), for a fixed $\bar{p}$, let $U_c = X - X^{c_{\bar{p}}}$. Then a codimension c coefficient system $\mathcal{E}^*$ is basically a sheaf on U_c that satisfies the axiomatic properties of the sheafification of $U \to I^{\bar{p}}S_*(U;\mathcal{G})$ there with respect to some stratification of U_c. These axiomatic conditions are a modification of the axioms $AX2$ (see [6, Section V.4]), which, for a GM perversity, are equivalent to the axioms $AX1$ discussed above in Section 4. We will not pursue the axioms $AX2$ in detail here, but we note that the Habegger–Saper modification occurs by requiring certain vanishing conditions to hold only in certain degrees depending on c. This takes into account the failure of the cone formula to vanish in the expected degrees (see Section 5). Then Habegger and Saper define a sheaf complex $\mathcal{P}^*_{\bar{p},\mathcal{E}^*}$ by extending $\mathcal{E}^*$ from U_c to the rest of X by the Deligne process from this point.

Among other results in their paper, Habegger and Saper show that the hypercohomology of their sheaf complex agrees (up to reindexing and with an appropriate choice of coefficients) with the intersection homology of King on PL pseudomanifolds, that this version of intersection homology is a topological invariant, and that there is a duality theorem. To state their duality theorem, let $\bar{q}(k) = k - 2 - \bar{p}(k)$, and let $\bar{q}'(k) = \max(\bar{q}(k), 0) + c_{\bar{p}} - 2$. Then, with coefficients in a field, the Verdier dual $\mathcal{D}_X \mathcal{P}^*_{\bar{p},\mathcal{E}^*}$ is quasi-isomorphic to $\mathcal{P}^*_{\bar{q}',\mathcal{D}_{U_c}(E^*)}[c_p - 2 + n]$. Roughly speaking, and ignoring the shifting of perversities and indices, which is done for technical reasons, this says that if $\bar{p} + \bar{q} = \bar{t}$ and we dualize the sheaf of intersection chains "by hand" on U_c from $\mathcal{E}^*$ to $\mathcal{D}_{U_c}\mathcal{E}^*$, then further extensions by the Deligne process, using perversity $\bar{p}$ for $\mathcal{E}^*$ and perversity $\bar{q}$ for $\mathcal{D}_{U_c}\mathcal{E}^*$, will maintain that duality. If $\bar{p}$ is a GM perversity and X is a pseudomanifold with no codimension one strata, this recovers the duality results of Goresky and MacPherson. Unfortunately, for more general perversities, there does not seem to be an obvious way to translate this duality back into the language of chain complexes, due to the complexity of the dual coefficient system $\mathcal{D}_{U_c}\mathcal{E}^*$ that appears on U_c.

One additional note should be made concerning the duality results in [35]. As mentioned above, Habegger and Saper work on cs-sets. These are more general than pseudomanifolds, primarily in that $X - X^{n-1}$ need not be dense and there is no inductive assumption that the links be pseudomanifolds. These are the spaces on which King demonstrated his stratification independence results in [41]. Thus these results are more general than those we have been discussing on pseudomanifolds, at least as far as the space X is concerned. However, as far as the author can tell, in one sense these duality results are not quite as much more general as they at first appear, as least when considering strata of X that are not in the closure of $X - X^{n-1}$. In particular, if Z is such a stratum and it lies in U_c, then the duality results on it are tautological — induced by the "by hand" dualization of the codimension c coefficient system. But if Z is not in U_c, then the pushforwards of the Deligne process cannot reach it, and $\mathcal{P}^*|_Z = 0$. So at the sheaf level the truly interesting piece of the duality still occurs in the closure of $X - X^{n-1}$. It would be interesting to understand how the choice of coefficient system and "by hand" duality on these "extraneous" strata in U_c (the strata not in the closure of $X - X^{n-1}$) influence the hypercohomology groups and the duality there. We also note that the closure of the union of the regular strata of a cs-set may still not be a pseudomanifold, due to the lack of condition on the links. It would be interesting to explore just how much more general such spaces are and the extent to which the other results we have discussed extend to them.

We refer the reader again to [35] for the further results that can be found there, including results on the intersection pairing and Zeeman's filtration.

Acknowledgments

I sincerely thank my co-organizers and co-editors (Eugénie Hunsicker, Anatoly Libgober, and Laurentiu Maxim) for making an MSRI workshop and this accompanying volume possible. I thank my collaborators Jim McClure and Eugénie Hunsicker as the impetus and encouragement for some of the work that is discussed here and for their comments on earlier drafts of this paper. And I thank an anonymous referee for a number of helpful suggestions that have greatly improved this exposition.

References

[1] Markus Banagl, *The signature of partially defined local coefficient systems*, J. Knot Theory Ramifications 17 (2008), no. 12, 1455–1481.

[2] ———, *Topological invariants of stratified spaces*, Springer Monographs in Mathematics, Springer, New York, 2006.

[3] Markus Banagl, Sylvain Cappell, and Julius Shaneson, *Computing twisted signatures and L-classes of stratified spaces*, Math. Ann. **326** (2003), 589–623.

[4] A. A. Beilinson, J. Bernstein, and P. Deligne, *Faisceaux pervers*, Astérisque **100** (1982), 5–171.

[5] Armand Borel, *Introduction to middle intersection cohomology and perverse sheaves*, Algebraic groups and their generalizations: classical methods (University

Park, PA, 1991), Proc. Sympos. Pure Math., vol. 56, Amer. Math. Soc., Providence, RI, 1994, pp. 25–52.

[6] A. Borel et. al., *Intersection cohomology*, Progress in Mathematics, vol. 50, Birkhäuser, Boston, 1984.

[7] J. P. Brasselet, G. Hector, and M. Saralegi, *Theéorème de deRham pour les variétés stratifiées*, Ann. Global Anal. Geom. **9** (1991), 211–243.

[8] Jean-Luc Brylinski, *Equivariant intersection cohomology*, Contemp. Math. **139** (1992), 5–32.

[9] S. E. Cappell, A. Libgober, L. G. Maxim, and J. L. Shaneson, *Hodge genera of algebraic varieties, II*, preprint, see math.AG/0702380.

[10] Sylvain Cappell, Julius Shaneson, and Shmuel Weinberger, *Classes topologiques caractéristiques pour les actions de groupes sur les espaces singuliers*, C. R. Acad. Sci. Paris Sér. I Math. **313** (1991), no. 5, 293–295.

[11] Sylvain E. Cappell, Laurentiu G. Maxim, and Julius L. Shaneson, *Hodge genera of algebraic varieties, I*, Comm. Pure Appl. Math. **61** (2008), no. 3, 422–449.

[12] Sylvain E. Cappell and Julius L. Shaneson, *Singular spaces, characteristic classes, and intersection homology*, Annals of Mathematics **134** (1991), 325–374.

[13] G. Carron, L^2-*cohomology of manifolds with flat ends*, Geom. Funct. Anal. **13** (2003), no. 2, 366–395.

[14] Gilles Carron, *Cohomologie L^2 des variétés QALE*, preprint, 2005.

[15] _____, *Cohomologie L^2 et parabolicité*, J. Geom. Anal. **15** (2005), no. 3, 391–404.

[16] J. Cheeger, *On the Hodge theorey of Riemannian pseudomanifolds*, Geometry of the Laplace Operator (Proc. Sympos. Pure Math., Univ. Hawaii, Honolulu, Hawaii, 1979) (Providence, RI), vol. 36, Amer. Math. Soc., 1980, pp. 91–146.

[17] Jeff Cheeger and Xianzhe Dai, L^2-*cohomology of spaces with nonisolated conical singularities and nonmultiplicativity of the signature*. Riemannian topology and geometric structures on manifolds, 1–24, Progr. Math., 271, Birkhäuser, Boston, 2009.

[18] Daniel C. Cohen, Mark Goresky, and Lizhen Ji, *On the Künneth formula for intersection cohomology*, Trans. Amer. Math. Soc. **333** (1992), 63–69.

[19] Xianzhe Dai, *Adiabatic limits, nonmultiplicativity of signature, and Leray spectral sequence*, J. Amer. Math. Soc. **4** (1991), no. 2, 265–321.

[20] Alexandru Dimca, *Sheaves in topology*, Springer, Berlin, 2004.

[21] Greg Friedman, *Farber–Levine pairings in intersection homology*, in preparation.

[22] _____, *Intersection homology with general perversities*, Geom. Dedicata **148** (2010), 103–135.

[23] _____, *Stratified fibrations and the intersection homology of the regular neighborhoods of bottom strata*, Topology Appl. **134** (2003), 69–109.

[24] _____, *Superperverse intersection cohomology: stratification (in)dependence*, Math. Z. **252** (2006), 49–70.

[25] _____, *Intersection homology of stratified fibrations and neighborhoods*, Adv. Math. **215** (2007), no. 1, 24–65.

[26] _____, *Singular chain intersection homology for traditional and super-perversities*, Trans. Amer. Math. Soc. **359** (2007), 1977–2019.

[27] _____, *Intersection homology and Poincaré duality on homotopically stratified spaces*, Geom. Topol. **13** (2009), 2163–2204.

[28] Greg Friedman, *Intersection homology Künneth theorems*, Math. Ann. **343** (2009), no. 2, 371–395.

[29] _____, *Intersection homology with field coefficients: K-Witt spaces and K-Witt bordism*, Comm. Pure Appl. Math. **62** (2009), 1265–1292.

[30] Greg Friedman, Eugénie Hunsicker, Anatoly Libgober, and Laurentiu Maxim, *Proceedings of the Workshop on the Topology of Stratified Spaces at MSRI, September 8-12, 2008*, in preparation.

[31] Greg Friedman and James McClure, *Cup and cap product in intersection (co)homology*, in preparation.

[32] Mark Goresky and Robert MacPherson, *Intersection homology theory*, Topology **19** (1980), 135–162.

[33] _____, *Intersection homology, II*, Invent. Math. **72** (1983), 77–129.

[34] Mark Goresky and Paul Siegel, *Linking pairings on singular spaces*, Comment. Math. Helvetici **58** (1983), 96–110.

[35] Nathan Habegger and Leslie Saper, *Intersection cohomology of cs-spaces and Zeeman's filtration*, Invent. Math. **105** (1991), 247–272.

[36] Allen Hatcher, *Algebraic topology*, Cambridge University Press, Cambridge, 2002.

[37] Tamás Hausel, Eugénie Hunsicker, and Rafe Mazzeo, *Hodge cohomology of gravitational instantons*, Duke Math. J. **122** (2004), no. 3, 485–548.

[38] Eugénie Hunsicker, *Hodge and signature theorems for a family of manifolds with fibre bundle boundary*, Geom. Topol. **11** (2007), 1581–1622.

[39] Eugénie Hunsicker and Rafe Mazzeo, *Harmonic forms on manifolds with edges*, Int. Math. Res. Not. (2005), no. 52, 3229–3272.

[40] Masaki Kashiwara and Pierre Schapira, *Sheaves on manifolds*, Grundlehren der Mathematischen Wissenschaften, vol. 292, Springer, Berlin, 1994.

[41] Henry C. King, *Topological invariance of intersection homology without sheaves*, Topology Appl. **20** (1985), 149–160.

[42] Frances Kirwan and Jonathan Woolf, *An introduction to intersection homology theory. second edition*, Chapman & Hall/CRC, Boca Raton, FL, 2006.

[43] Steven L. Kleiman, *The development of intersection homology theory*, Pure Appl. Math. Q. **3** (2007), 225–282.

[44] Robert MacPherson, *Intersection homology and perverse sheaves*, Unpublished Colloquium Lectures.

[45] R. MacPherson and K. Vilonen, *Elementary construction of perverse sheaves*, Invent. Math. **84** (1986), 403–435.

[46] J. E. McClure, *On the chain-level intersection pairing for PL manifolds*, Geom. Topol. **10** (2006), 1391–1424.

[47] Clint McCrory, *Stratified general position*, Algebraic and geometric topology (Proc. Sympos., Univ. California, Santa Barbara, Calif. 1977) (Berlin), Lecture Notes in Math., vol. 664, Springer, 1978, pp. 142–146.

[48] Werner Müller, L^2-*index theory, eta invariants and values of L-functions*, Geometric and topological invariants of elliptic operators (Brunswick, ME, 1988), Contemp. Math., vol. 105, Amer. Math. Soc., Providence, RI, 1990, pp. 145–189.

[49] James R. Munkres, *Elements of algebraic topology*, Addison-Wesley, Reading, MA, 1984.

[50] Leslie Saper, L^2-*cohomology of locally symmetric spaces, I*, Pure Appl. Math. Q. **1** (2005), no. 4, part 3, 889–937.

[51] ———, $\mathcal{L}$-*modules and the conjecture of Rapoport and Goresky-MacPherson*, Automorphic forms. I. Astérisque (2005), no. 298, 319–334.

[52] Leslie Saper and Mark Stern, L_2-*cohomology of arithmetic varieties*, Ann. of Math. (2) **132** (1990), no. 1, 1–69.

[53] Martin Saralegi, *Homological properties of stratified spaces, I*, Illinois Journal of Mathematics **38** (1994), 47–70.

[54] Martintxo Saralegi-Aranguren, *de Rham intersection cohomology for general perversities*, Illinois J. Math. **49** (2005), no. 3, 737–758.

[55] P. H. Siegel, *Witt spaces: a geometric cycle theory for KO-homology at odd primes*, American J. Math. **110** (1934), 571–92.

[56] Norman Steenrod, *The topology of fibre bundles*, Princeton University Press, Princeton, NJ, 1951.

GREG FRIEDMAN
DEPARTMENT OF MATHEMATICS
TEXAS CHRISTIAN UNIVERSITY
BOX 298900
FORT WORTH, TX 76129
g.friedman@tcu.edu

Topology of Stratified Spaces
MSRI Publications
Volume **58**, 2011

The signature of singular spaces and its refinements to generalized homology theories

MARKUS BANAGL

ABSTRACT. These notes are based on an expository lecture that I gave at the workshop "Topology of Stratified Spaces" at MSRI Berkeley in September 2008. We will first explain the definition of a bordism invariant signature for a singular space, proceeding along a progression from less singular to more and more singular spaces, starting out from spaces that have no odd codimensional strata and, after having discussed Goresky–Siegel spaces and Witt spaces, ending up with general (non-Witt) stratified spaces. We will moreover discuss various refinements of the signature to orientation classes in suitable bordism theories based on singular cycles. For instance, we will indicate how one may define a symmetric $\mathbb{L}^{\bullet}$-homology orientation for Goresky–Siegel spaces or a Sullivan orientation for those non-Witt spaces that still possess generalized Poincaré duality. These classes can be thought of as refining the L-class of a singular space. Along the way, we will also see how to compute twisted versions of the signature and L-class.

CONTENTS

The author was in part supported by a research grant of the Deutsche Forschungsgemeinschaft. He thanks Andrew Ranicki for suggestions and clarifying comments.

223

1. Introduction

Let M be a closed smooth n-dimensional manifold. The Hirzebruch L-classes $L^i(M) \in H^{4i}(M;\mathbb{Q})$ of its tangent bundle are powerful tools in the classification of such M, particularly in the high dimensional situation where $n \geq 5$. To make this plausible, we observe first that the $L^i(M)$, with the exception of the top class $L^{n/4}(M)$ if n is divisible by 4, are not generally homotopy invariants of M, and are therefore capable of distinguishing manifolds in a given homotopy type, contrary to the ability of homology and other homotopy invariants. For example, there exist infinitely many manifolds $M_i, i = 1, 2, \ldots$ in the homotopy type of $S^2 \times S^4$, distinguished by the first Pontrjagin class of their tangent bundle $p_1(TM_i) \in H^4(S^2 \times S^4) = \mathbb{Z}$, namely $p_1(TM_i) = Ki$, K a fixed nonzero integer. The first L-class L^1 is proportional to the first Pontrjagin class p_1, in fact they are related by the formula $L^1 = \frac{1}{3}p_1$.

Suppose that M^n, $n \geq 5$, is simply connected, as in the example. The classification of manifolds breaks up into two very different tasks: Classify Poincaré complexes up to homotopy equivalence and, given a Poincaré complex, determine all manifolds homotopy equivalent to it.

In dimension 3, one has a relatively complete answer to the former problem. One can associate purely algebraic data to a Poincaré complex such that two such complexes are homotopy equivalent if, and only if, their algebraic data are isomorphic, see the classification result in [Hen77]. Furthermore, every given algebraic data is realizable as the data of a Poincaré complex; see [Tur90]. In higher dimensions, the problem becomes harder. While one can still associate classifying data to a Poincaré complex, this data is not purely algebraic anymore, though at least in dimension 4, one can endow Poincaré duality chain complexes with an additional structure that allows for classification, [BB08].

The latter problem is the realm of surgery theory. Elements of the *structure set* $S(M)$ of M are represented by homotopy equivalences $N \to M$, where N is another closed smooth manifold, necessarily simply connected, since M is. Two such homotopy equivalences represent the same element of $S(M)$ if there is a diffeomorphism between the domains that commutes with the homotopy equivalences. The goal of surgery theory is to compute $S(M)$. The central tool provided by the theory is the surgery exact sequence

$$L_{n+1} \xrightarrow{\ \cdot\ } S(M) \xrightarrow{\ \eta\ } N(M) \longrightarrow L_n,$$

an exact sequence of pointed sets. The L_n are the 4-periodic simply connected surgery obstruction groups, $L_n = \mathbb{Z}, 0, \mathbb{Z}/2, 0$ for $n \equiv 0, 1, 2, 3 \mod 4$. The term $N(M)$ is the *normal invariant set*, investigated by Sullivan. It is a generalized cohomology theory and a Pontrjagin–Thom type construction yields $N(M) \cong [M, G/O]$, where $[M, G/O]$ denotes homotopy classes of maps from

M into a certain universal space G/O, which does not depend on M. Since $[M, G/O]$ is a cohomology theory, it is particularly important to know its coefficients $\pi_*(G/O)$. While the torsion is complicated, one has modulo torsion

$$\pi_i(G/O) \otimes \mathbb{Q} = \begin{cases} \mathbb{Q}, & i = 4j, \\ 0, & \text{otherwise.} \end{cases}$$

One obtains an isomorphism

$$[M, G/O] \otimes \mathbb{Q} \cong \bigoplus_{j \geq 0} H^{4j}(M; \mathbb{Q}).$$

The group L_{n+1} acts on $S(M)$ so that the point-inverses of η are the orbits of the action, i.e. for all $f, h \in S(M)$ one has $\eta(f) = \eta(h)$ if, and only if, there is a $g \in L_{n+1}$ which moves f to h, $g \cdot f = h$.

Suppose our manifold M is even dimensional. Then L_{n+1} vanishes and thus $\eta(f) = \eta(h)$ implies $f = g \cdot f = h$, so that η is injective. In particular, we obtain an injection

$$S(M) \otimes \mathbb{Q} \hookrightarrow N(M) \otimes \mathbb{Q}.$$

Composing this with $N(M) \otimes \mathbb{Q} \cong \bigoplus H^{4j}(M; \mathbb{Q})$, we obtain an injective map

$$S(M) \otimes \mathbb{Q} \overset{L}{\hookrightarrow} \bigoplus H^{4j}(M; \mathbb{Q}).$$

This map sends a homotopy equivalence $h : N \to M$ to the cohomology class $L^*(h)$ uniquely determined by $h^*(L^*(M) + L^*(h)) = L^*(N)$. Thus M is determined, up to finite ambiguity, by *its homotopy type and its L-classes*. This demonstrates impressively the power of the L-classes as a tool to classify manifolds.

The L-classes are closely related to the signature invariant, and indeed the classes can be defined, following Thom [Tho58], by the signatures of submanifolds, as we shall now outline. The link between the L-classes and the signature is the Hirzebruch signature theorem. It asserts that the evaluation of the top L-class $L^j(M) \in H^n(M; \mathbb{Q})$ of an $n = 4j$-dimensional oriented manifold M on the fundamental class of M equals the signature $\sigma(M)$ of M. Once we know this, we can define $L^*(M)$ as follows. A theorem of Serre states that the Hurewicz map is an isomorphism

$$\pi^k(M) \otimes \mathbb{Q} \cong H^k(M; \mathbb{Q})$$

in the range $n < 2k-1$, where $\pi^*(M)$ denotes the cohomotopy sets of M, whose elements are homotopy classes of maps from M to spheres. Thus, in this range, we may think of a rational cohomology class as a (smooth) map $f : M \to S^k$. The preimage $f^{-1}(p)$ of a regular value $p \in S^k$ is a submanifold and has a signature $\sigma(f^{-1}(p))$. Use the bordism invariance of the signature to conclude that this signature depends only on the homotopy class of f. Assigning $\sigma(f^{-1}(p))$ to the homotopy class of f yields a map $H^k(M;\mathbb{Q}) \to \mathbb{Q}$, that is, a *homology* class $L_k(M) \in H_k(M;\mathbb{Q})$. By Poincaré duality, this class can be dualized back into cohomology, where it agrees with the Hirzebruch classes $L^*(M)$. Note that all you need for this procedure is transversality for maps to spheres in order to get suitable subspaces and a bordism invariant signature defined on these subspaces. Thus, whenever these ingredients are present for a singular space X, we will obtain an L-class $L_*(X) \in H_*(X;\mathbb{Q})$ in the rational homology of X. (This class cannot necessarily be dualized back into cohomology, due to the lack of classical Poincaré duality for singular X.) Therefore, we only need to discuss which classes of singular spaces have a bordism invariant signature. The required transversality results are available for Whitney stratified spaces, for example. The notion of a Whitney stratified space incorporates smoothness in a particularly amenable way into the world of singular spaces. A Whitney stratification of a space X consists of a (locally finite) partition of X into locally closed smooth manifolds of various dimensions, called the *pure strata*. If one stratum intersects the closure of another one, then it must already be completely contained in it. Connected components S of strata have tubular neighborhoods T_S that possess locally trivial projections $\pi_S : T_S \to S$ whose fiber $\pi_S^{-1}(p)$, $p \in S$, is the cone on a compact space $L(p)$ (also Whitney stratified), called the *link* of S at p. It follows that every point p has a neighborhood homeomorphic to $\mathbb{R}^{\dim S} \times \text{cone } L(p)$. Real and complex algebraic varieties possess a natural Whitney stratification, as do orbit spaces of smooth group actions. The pseudomanifold condition means that the singular strata have codimension at least two and the complement of the singular set (the *top stratum*) is dense in X. The figure eight space, for instance, can be Whitney stratified but is not a pseudomanifold. The pinched 2-torus is a Whitney stratifiable pseudomanifold. If we attach a whisker to the pinched 2-torus, then it loses its pseudomanifold property, while retaining its Whitney stratifiability. By [Gor78], a Whitney stratified pseudomanifold X can be triangulated so that the Whitney stratification defines a PL stratification of X.

Inspired by the success of L-classes in manifold theory sketched above, one would like to have L-classes for stratified pseudomanifolds as well. In [CW91], see also [Wei94], Cappell and Weinberger indicate the following result, analogous to the manifold classification result sketched above. Suppose X is a stratified pseudomanifold that has no strata of odd dimension. Assume that all strata S have dimension at least 5, and that all fundamental groups in sight are trivial, that is, all strata are simply connected and all links are simply connected. (A pseudomanifold whose links are all simply connected is called *supernormal*. This is compatible with the notion of a *normal* pseudomanifold, meaning that all links are connected.) Then differences of L-classes give an injection

$$S(X) \otimes \mathbb{Q} \hookrightarrow \bigoplus_{S \subset X} \bigoplus_j H_j(\overline{S}; \mathbb{Q}),$$

where S ranges over the strata of X, $\overline{S}$ denotes the closure of S in X, and $S(X)$ is an appropriately[1] defined structure set for X. This would suggest that L-classes are as powerful in classifying stratified spaces as in classifying manifolds. Since, as we have seen, the definition of L-classes is intimately related to, and can be given in terms of, the signature, we shall primarily investigate the possibility of defining a bordism invariant signature for an oriented stratified pseudomanifold X.

2. Pseudomanifolds without odd codimensional strata

In order to define a signature, one needs an intersection form. But singular spaces do not possess Poincaré duality, in particular no intersection form, in ordinary homology. The solution is to change to a different kind of homology. Motivated by a question of D. Sullivan [Sul70], Goresky and MacPherson define (in [GM80] for PL pseudomanifolds and in [GM83] for topological pseudomanifolds) a collection of groups $IH_*^{\bar{p}}(X)$, called *intersection homology groups of X*, depending on a multi-index $\bar{p}$, called a *perversity*. For these groups, a Poincaré–Lefschetz-type intersection theory can be defined, and a generalized form of Poincaré duality holds, but only between groups with "complementary perversities." More precisely, with $\bar{t}(k) = k - 2$ denoting the top perversity, there are intersection pairings

$$IH_i^{\bar{p}}(X) \otimes IH_j^{\bar{q}}(X) \longrightarrow \mathbb{Z} \tag{2-1}$$

for an oriented closed pseudomanifold X, $\bar{p} + \bar{q} = \bar{t}$ and $i + j = \dim X$, which are nondegenerate when tensored with the rationals. Jeff Cheeger discovered, working independently of Goresky and MacPherson and not being aware of their intersection homology, that Poincaré duality on triangulated pseudomanifolds equipped with a suitable (locally conical) Riemannian metric on the top stratum, can be recovered by using the complex of L^2 differential forms on the top stratum, see [Che80], [Che79] and [Che83]. The connection between his and the work of Goresky and MacPherson was pointed out by Sullivan in 1976. For an introduction to intersection homology see [B+84], [KW06] or [Ban07]. A third method, introduced in [Ban09] and implemented there for pseudomanifolds

---

[1] In [CW91], the structure sets $S(X)$ are defined as the homotopy groups of the homotopy fiber of the assembly map $X \wedge \mathbb{L}_{\bullet}(\mathbb{Z})_0 \to \mathbb{L}_{\bullet}(\mathbb{Z}\pi_1(X))$, constructed in [Ran79]. This can be defined for any space, but under the stated assumptions on X, [CW91] interprets $S(X)$ geometrically in terms of classical structure sets of the strata of X.

with isolated singularities and two-strata spaces with untwisted link bundle, associates to a singular pseudomanifold X an *intersection space* $I^{\bar{p}}X$, whose *ordinary* rational homology has a nondegenerate intersection pairing

$$\widetilde{H}_i(I^{\bar{p}}X;\mathbb{Q}) \otimes \widetilde{H}_j(I^{\bar{q}}X;\mathbb{Q}) \longrightarrow \mathbb{Q}.$$

This theory is not isomorphic, albeit related, to intersection homology. It solves a problem posed in string theory, related to the presence of massless D-branes in the course of conifold transitions.

In sheaf-theoretic language, the groups $IH_*^{\bar{p}}(X)$ are given as the hypercohomology groups of a sheaf complex $\mathbf{IC}_{\bar{p}}^{\bullet}(X)$ over X. If we view this complex as an object of the derived category (that is, we invert quasi-isomorphisms), then $\mathbf{IC}_{\bar{p}}^{\bullet}(X)$ is characterized by certain stalk/costalk vanishing conditions. The rationalization of the above intersection pairing (2-1) is induced on hypercohomology by a duality isomorphism $\mathcal{D}\mathbf{IC}_{\bar{p}}^{\bullet}(X;\mathbb{Q})[n] \cong \mathbf{IC}_{\bar{q}}^{\bullet}(X;\mathbb{Q})$ in the derived category, where $\mathcal{D}$ denotes the Verdier dualizing functor. This means roughly that one does not just have a global chain equivalence to the dual (intersection) chain complex, but a chain equivalence on every open set.

Let X^n be an oriented closed topological stratified pseudomanifold which has only even dimensional strata. A wide class of examples is given by complex algebraic varieties. In this case, the intersection pairing (2-1) allows us to define a signature $\sigma(X)$ by using the two complementary middle perversities $\bar{m}$ and $\bar{n}$:

k	2	3	4	5	6	7	8	9	$\cdots$
$\bar{m}(k)$	0	0	1	1	2	2	3	3	$\cdots$
$\bar{n}(k)$	0	1	1	2	2	3	3	4	$\cdots$

Since $\bar{m}(k) = \bar{n}(k)$ for even values of k, and only these values are relevant for our present X, we have $IH_{n/2}^{\bar{m}}(X) = IH_{n/2}^{\bar{n}}(X)$. Therefore, the pairing (2-1) becomes

$$IH_{n/2}^{\bar{m}}(X;\mathbb{Q}) \otimes IH_{n/2}^{\bar{m}}(X;\mathbb{Q}) \longrightarrow \mathbb{Q}$$

(symmetric if $n/2$ is even), that is, defines a quadratic form on the vector space $IH_{n/2}^{\bar{m}}(X;\mathbb{Q})$. Let $\sigma(X)$ be the signature of this quadratic form. Goresky and MacPherson show that this is a bordism invariant for bordisms that have only strata of even codimension. Since $\mathbf{IC}_{\bar{m}}^{\bullet}(X) = \mathbf{IC}_{\bar{n}}^{\bullet}(X)$, the intersection pairing is induced by a self-duality isomorphism $\mathcal{D}\mathbf{IC}_{\bar{m}}^{\bullet}(X;\mathbb{Q})[n] \cong \mathbf{IC}_{\bar{m}}^{\bullet}(X;\mathbb{Q})$. This is an example of a *self-dual sheaf*.

3. Witt spaces

To form a bordism theory based on pseudomanifold cycles, one could consider bordism based on all (say topological, or PL) closed pseudomanifolds,

$$\Omega_*^{\text{all pseudomfds}}(Y) = \left\{ [X \xrightarrow{f} Y] \mid X \text{ a pseudomanifold} \right\},$$

where the admissible bordisms consist of compact pseudomanifolds with collared boundary, without further restrictions. Now it is immediately clear that the associated coefficient groups vanish, $\Omega_*^{\text{all pseudomfds}}(\text{pt}) = 0$, $* > 0$, since any pseudomanifold X is the boundary of the cone on X, which is an admissible bordism. Thus this naive definition does not lead to an interesting and useful new theory, and we conclude that a subclass of pseudomanifolds has to be selected to define such theories. Given the results on middle perversity intersection homology presented so far, our next approach would be to select the class of all closed pseudomanifolds with only even codimensional strata,

$$\Omega_*^{\text{ev}}(Y) = \left\{ [X \xrightarrow{f} Y] \mid X \text{ has only even codim strata} \right\}$$

(and the same condition is imposed on all admissible bordisms). While we do know that the signature is well-defined on $\Omega_*^{\text{ev}}(\text{pt})$, this is however still not a good theory as this definition leads to a large number of geometrically insignificant generators. Many operations (such as coning or refinement of the stratification) do introduce strata of odd codimension, so we need to allow some strata of this kind, but so as not to destroy Poincaré duality. In [Sie83], Paul Siegel introduced a class of oriented stratified PL pseudomanifolds called *Witt spaces*, by imposing the condition that $IH^{\bar{m}}_{\text{middle}}(\text{Link}(x); \mathbb{Q}) = 0$ for all points x in odd codimensional strata of X. The suspension $X^7 = \Sigma\mathbb{CP}^3$ has two singular points which form a stratum of odd codimension 7. The link is $\mathbb{CP}^3$ with middle homology $H_3(\mathbb{CP}^3) = 0$. Hence X^7 is a Witt space. The suspension $X^3 = \Sigma T^2$ has two singular points which form a stratum of odd codimension 3. The space X^3 is not Witt, since the middle Betti number of the link T^2 is 2. In sheaf-theoretic language, a pseudomanifold X is Witt if and only if the canonical morphism $\mathbf{IC}_{\bar{m}}^\bullet(X; \mathbb{Q}) \to \mathbf{IC}_{\bar{n}}^\bullet(X; \mathbb{Q})$ is an isomorphism (in the derived category). Thus, $\mathbf{IC}_{\bar{m}}^\bullet(X; \mathbb{Q})$ is self-dual on a Witt space, and if X is compact, we have a nonsingular intersection pairing $IH_i^{\bar{m}}(X; \mathbb{Q}) \otimes IH_{n-i}^{\bar{m}}(X; \mathbb{Q}) \to \mathbb{Q}$. Let $\Omega_*^{\text{Witt}}(Y)$ denote Witt space bordism, that is, bordism of closed oriented Witt spaces X mapping continuously into Y. Admissible bordisms are compact pseudomanifolds with collared boundary that satisfy the Witt condition, together with a map into Y.

When is a Witt space X^n a boundary? Suppose the dimension n is odd. Then $X = \partial Y$ with $Y = \text{cone } X$. The cone Y is a Witt space, since the cone-point is a stratum of even codimension in Y. This shows that $\Omega_{2k+1}^{\text{Witt}}(\text{pt}) = 0$ for all k. In particular, the de Rham invariant does not survive in $\Omega_*^{\text{Witt}}(\text{pt})$.

Let $W(\mathbb{Q})$ denote the Witt group of the rationals. Its structure is known and given by

$$W(\mathbb{Q}) \cong W(\mathbb{Z}) \oplus \bigoplus_{p \text{ prime}} W(\mathbb{Z}/p),$$

where $W(\mathbb{Z}) \cong \mathbb{Z}$ via the signature, $W(\mathbb{Z}/2) \cong \mathbb{Z}/2$, and for $p \neq 2$, $W(\mathbb{Z}/p) \cong \mathbb{Z}/4$ or $\mathbb{Z}/2 \oplus \mathbb{Z}/2$. Sending a Witt space X^{4k} to its intersection form on $IH_{2k}^{\bar{m}}(X;\mathbb{Q})$ defines a bordism-invariant element $w(X) \in W(\mathbb{Q})$. Siegel shows that the induced map $w : \Omega_{4k}^{\text{Witt}}(\text{pt}) \to W(\mathbb{Q})$ is an isomorphism for $k > 0$. In dimension zero we get $\Omega_0^{\text{Witt}} = \mathbb{Z}$. If X has dimension congruent 2 modulo 4, then X bounds a Witt space by singular surgery on a symplectic basis for the antisymmetric intersection form. Thus $\Omega_n^{\text{Witt}} = 0$ for n not congruent 0 mod 4. Since $W(\mathbb{Q})$ is just another name for the L-group $L^{4k}(\mathbb{Q})$ and $L^n(\mathbb{Q}) = 0$ for n not a multiple of 4, we can summarize Siegel's result succinctly as saying that $\Omega_*^{\text{Witt}}(\text{pt}) \cong L^*(\mathbb{Q})$ in positive degrees. By the Brown representability theorem, Witt space bordism theory is given by a spectrum MWITT, which is in fact an MSO module spectrum, see [Cur92]. (Regard a manifold as a Witt space with one stratum.) By [TW79], any MSO module spectrum becomes a product of Eilenberg–Mac Lane spectra after localizing at 2. Thus,

$$\text{MWITT}_{(2)} \simeq K(\mathbb{Z}_{(2)}, 0) \times \prod_{j>0} K(L^j(\mathbb{Q})_{(2)}, j)$$

and we conclude that

$$\Omega_n^{\text{Witt}}(Y)_{(2)} \cong H_n(Y;\mathbb{Z})_{(2)} \oplus \bigoplus_{j>0} H_{n-j}(Y; L^j(\mathbb{Q})_{(2)}).$$

(As $\mathbb{Z}_{(2)}$ is flat over $\mathbb{Z}$, we have $S_*(X)_{(2)} = (S_{(2)})_*(X)$ for any spectrum S.) Let us focus on the odd-primary situation. Regard $\mathbb{Z}[\frac{1}{2}, t]$ as a graded ring with $\deg(t) = 4$. Let $\Omega_*^{\text{SO}}(Y)$ denote bordism of smooth oriented manifolds. Considering the signature as a map $\sigma : \Omega_*^{\text{SO}}(\text{pt}) \to \mathbb{Z}[\frac{1}{2}, t]$, $[M^{4k}] \mapsto \sigma(M)t^k$, makes $\mathbb{Z}[\frac{1}{2}, t]$ into an $\Omega_*^{\text{SO}}(\text{pt})$-module and we can form the homology theory

$$\Omega_*^{\text{SO}}(Y) \otimes_{\Omega_*^{\text{SO}}(\text{pt})} \mathbb{Z}[\tfrac{1}{2}, t].$$

On a point, this is

$$\Omega_*^{\text{SO}}(\text{pt}) \otimes_{\Omega_*^{\text{SO}}(\text{pt})} \mathbb{Z}[\tfrac{1}{2}, t] \cong \mathbb{Z}[\tfrac{1}{2}, t],$$

the isomorphism being given by $[M^{4l}] \otimes at^k \mapsto a\sigma(M^{4l})t^{k+l}$. Let $ko_*(Y)$ denote connective KO homology, regarded as a $\mathbb{Z}$-graded, not $\mathbb{Z}/4$-graded, theory.

It is given by a spectrum bo whose homotopy groups vanish in negative degrees and are given by

$$ko_*(\text{pt}) = \pi_*(bo) = \mathbb{Z} \oplus \Sigma^1 \mathbb{Z}/2 \oplus \Sigma^2 \mathbb{Z}/2 \oplus \Sigma^4 \mathbb{Z} \oplus \Sigma^8 \mathbb{Z} \oplus \Sigma^9 \mathbb{Z}/2 \oplus \cdots,$$

repeating with 8-fold periodicity in nonnegative degrees. Inverting 2 kills the torsion in degrees 1 and 2 mod 8 so that $ko_*(\text{pt}) \otimes_{\mathbb{Z}} \mathbb{Z}[\frac{1}{2}] \cong \mathbb{Z}[\frac{1}{2}, t]$. In his MIT-notes [Sul05], Sullivan constructs a natural Conner–Floyd-type isomorphism of homology theories

$$\Omega_*^{\text{SO}}(Y) \otimes_{\Omega_*^{\text{SO}}(\text{pt})} \mathbb{Z}[\tfrac{1}{2}, t] \xrightarrow{\cong} ko_*(Y) \otimes_{\mathbb{Z}} \mathbb{Z}[\tfrac{1}{2}].$$

Siegel [Sie83] shows that Witt spaces provide a geometric description of connective KO homology at odd primes: He constructs a natural isomorphism of homology theories

$$\Omega_*^{\text{Witt}}(Y) \otimes_{\mathbb{Z}} \mathbb{Z}[\tfrac{1}{2}] \xrightarrow{\cong} ko_*(Y) \otimes_{\mathbb{Z}} \mathbb{Z}[\tfrac{1}{2}] \tag{3-1}$$

(which we shall return to later). It reduces to the signature homomorphism on coefficients, i.e. an element $[X^{4k}] \otimes a \in \Omega_{4k}^{\text{Witt}}(\text{pt}) \otimes_{\mathbb{Z}} \mathbb{Z}[\frac{1}{2}]$ maps to $a\sigma(X)t^k \in ko_*(\text{pt}) \otimes_{\mathbb{Z}} \mathbb{Z}[\frac{1}{2}] = \mathbb{Z}[\frac{1}{2}, t]$. This is an isomorphism, since inverting 2 kills the torsion components of the invariant $w(X)$, $W(\mathbb{Q}) \otimes \mathbb{Z}[\frac{1}{2}] \cong W(\mathbb{Z}) \otimes \mathbb{Z}[\frac{1}{2}] \cong \mathbb{Z}[\frac{1}{2}]$. Now $\Omega_*^{\text{SO}}(Y) \otimes_{\Omega_*^{\text{SO}}(\text{pt})} \mathbb{Z}[\frac{1}{2}, t]$ being a quotient of $\Omega_*^{\text{SO}}(Y) \otimes_{\mathbb{Z}} \mathbb{Z}[\frac{1}{2}, t]$, yields a natural surjection

$$\Omega_*^{\text{SO}}(Y) \otimes_{\mathbb{Z}} \mathbb{Z}[\tfrac{1}{2}, t] \twoheadrightarrow \Omega_*^{\text{SO}}(Y) \otimes_{\Omega_*^{\text{SO}}(\text{pt})} \mathbb{Z}[\tfrac{1}{2}, t].$$

Let us consider the diagram of natural transformations

$$\begin{array}{ccc} \Omega_*^{\text{SO}}(Y) \otimes_{\Omega_*^{\text{SO}}(\text{pt})} \mathbb{Z}[\tfrac{1}{2}, t] & \xrightarrow{\cong} & ko_*(Y) \otimes_{\mathbb{Z}} \mathbb{Z}[\tfrac{1}{2}] \\ \uparrow & & \uparrow{\cong} \\ \Omega_*^{\text{SO}}(Y) \otimes_{\mathbb{Z}} \mathbb{Z}[\tfrac{1}{2}, t] & \longrightarrow & \Omega_*^{\text{Witt}}(Y) \otimes_{\mathbb{Z}} \mathbb{Z}[\tfrac{1}{2}], \end{array}$$

where the lower horizontal arrow maps an element $[M \xrightarrow{f} Y] \otimes at^k$ to $[M \times \mathbb{CP}^{2k} \xrightarrow{f\pi_1} Y] \otimes a$. On a point, this arrow thus maps an element $[M^{4l}] \otimes at^k$ to $[M^{4l} \times \mathbb{CP}^{2k}] \otimes a$. Mapping an element $[M^{4l}] \otimes at^k$ clockwise yields $a\sigma(M)t^{k+l} \in ko_*(\text{pt}) \otimes_{\mathbb{Z}} \mathbb{Z}[\frac{1}{2}] \cong \mathbb{Z}[\frac{1}{2}, t]$. Mapping the same element counterclockwise gives

$$a\sigma(M^{4l} \times \mathbb{CP}^{2k})t^{(4l+4k)/4} = a\sigma(M)t^{k+l}$$

also. The diagram commutes and shows that away from 2, the canonical map from manifold bordism to Witt bordism is a surjection. This is a key observation of [BCS03] and frequently allows bordism invariant calculations for Witt spaces to be pulled back to calculations on smooth manifolds. This principle may be viewed as a topological counterpart of resolution of singularities in complex algebraic and analytic geometry (though one should point out that there are complex 2-dimensional singular projective toric varieties $X(\Delta)$ such that no nonzero multiple of $X(\Delta)$ is bordant to any toric resolution of $X(\Delta)$). It is applied in [BCS03] to prove that the twisted signature $\sigma(X; \mathcal{S})$ of a closed oriented Whitney stratified Witt space X^n of even dimension with coefficients in a (Poincaré-) local system $\mathcal{S}$ on X can be computed as the product of the (untwisted) L-class of X and a modified Chern character of the K-theory signature $[\mathcal{S}]_K$ of $\mathcal{S}$,

$$\sigma(X; \mathcal{S}) = \langle \widetilde{ch}[\mathcal{S}]_K, L_*(X) \rangle \in \mathbb{Z}$$

where $L_*(X) \in H_*(X; \mathbb{Q})$ is the total L-class of X. The higher components of the product $\widetilde{ch}[\mathcal{S}]_K \cap L_*(X)$ in fact compute the rest of the twisted L-class $L_*(X; \mathcal{S})$. Such twisted classes come up naturally if one wants to understand the pushforward under a stratified map of characteristic classes of the domain, see [CS91] and [Ban06c].

4. IP spaces: integral duality

Witt spaces satisfy generalized Poincaré duality rationally. Is there a class of pseudomanifolds whose members satisfy Poincaré duality integrally? This requires restrictions more severe than those imposed on Witt spaces. An *intersection homology Poincaré space* ("IP space"), introduced in [GS83], is an oriented stratified PL pseudomanifold such that the middle perversity, middle dimensional intersection homology of even dimensional links vanishes and the torsion subgroup of the middle perversity, lower middle dimensional intersection homology of odd dimensional links vanishes. This condition characterizes spaces for which the integral intersection chain sheaf $\mathbf{IC}^\bullet_{\bar{m}}(X; \mathbb{Z})$ is self-dual. Goresky and Siegel show that for such spaces X^n there are nonsingular pairings

$$IH_i^{\bar{m}}(X)/\operatorname{Tors} \otimes IH_{n-i}^{\bar{m}}(X)/\operatorname{Tors} \longrightarrow \mathbb{Z}$$

and

$$\operatorname{Tors} IH_i^{\bar{m}}(X) \otimes \operatorname{Tors} IH_{n-i-1}^{\bar{m}}(X) \longrightarrow \mathbb{Q}/\mathbb{Z}.$$

Let $\Omega_*^{IP}(\text{pt})$ denote the bordism groups of IP spaces. The signature $\sigma(X)$ of the above intersection pairing is a bordism invariant and induces a homomorphism $\Omega_{4k}^{IP}(\text{pt}) \to \mathbb{Z}$. If $\dim X = n = 4k + 1$, then the number mod 2 of $\mathbb{Z}/2$-summands in $\operatorname{Tors} IH_{2k}^{\bar{m}}(X)$ is a bordism invariant, the *de Rham invariant* $\mathrm{dR}(X) \in \mathbb{Z}/2$ of X. It induces a homomorphism $\mathrm{dR} : \Omega_{4k+1}^{IP}(\text{pt}) \to \mathbb{Z}/2$. Pardon shows in [Par90] that these maps are both isomorphisms for $k \geq 1$ and that all other groups

$\Omega_n^{IP}(\text{pt}) = 0$, $n > 0$. In summary, one obtains

$$\Omega_n^{IP}(\text{pt}) = \begin{cases} \mathbb{Z}, & n \equiv 0(4), \\ \mathbb{Z}/2, & n \geq 5, n \equiv 1(4), \\ 0 & \text{otherwise.} \end{cases}$$

Let $L^*(\mathbb{Z}G)$ denote the symmetric L-groups, as defined by Ranicki, of the group ring $\mathbb{Z}G$ of a group G. For the trivial group $G = e$, these are the homotopy groups $\pi_*(\mathbb{L}^{\bullet}) = \mathbb{L}_*^{\bullet}(\text{pt})$ of the symmetric L-spectrum $\mathbb{L}^{\bullet}$ and are given by

$$L^n(\mathbb{Z}e) = \begin{cases} \mathbb{Z}, & n \equiv 0(4), \\ \mathbb{Z}/2, & n \equiv 1(4), \\ 0 & \text{otherwise.} \end{cases}$$

We notice that this is extremely close to the IP bordism groups, the only difference being a $\mathbb{Z}/2$ in dimension 1. A comparison of their respective coefficient groups thus leads us to expect that the difference between the generalized homology theory $\Omega_*^{IP}(Y)$ given by mapping IP pseudomanifolds X continuously into a space Y and symmetric $\mathbb{L}^{\bullet}$-homology $\mathbb{L}_*^{\bullet}(Y)$ is very small. Indeed, according to [Epp07], there exists a map $\phi : \text{MIP} \to \mathbb{L}^{\bullet}$, where MIP is the spectrum giving rise to IP bordism theory, whose homotopy cofiber is an Eilenberg–Mac Lane spectrum $K(\mathbb{Z}/2, 1)$. The map is obtained by using a description of $\mathbb{L}^{\bullet}$ as a simplicial Ω-spectrum, whose k-th space has its n-simplices given by homotopy classes of $(n-k)$-dimensional n-ads of symmetric algebraic Poincaré complexes (pairs). Similarly, MIP can be described as a simplicial Ω-spectrum, whose k-th space has its n-simplices given by $(n-k)$-dimensional n-ads of compact IP pseudomanifolds. Given these simplicial models, one has to map n-ads of IP spaces to n-ads of symmetric Poincaré complexes. On a suitable incarnation of the middle perversity integral intersection chain sheaf on a compact IP space, a Poincaré symmetric structure can be constructed by copying Goresky's symmetric construction of [Gor84]. Taking global sections and resolving by finitely generated projectives (observing that the cohomology of the section complex is finitely generated by compactness), one obtains a symmetric algebraic Poincaré complex. This assignment can also be done for pairs and behaves well under gluing. The symmetric structure is uniquely determined by its restriction to the top stratum. On the top stratum, which is a manifold, the construction agrees sheaf-theoretically with the construction used classically for manifolds, see e.g. [Bre97]. In particular, if we start with a smooth oriented closed manifold and view it as an IP space with one stratum, then the top stratum is the *entire* space and the constructed symmetric structure agrees with Ranicki's symmetric structure. Modelling MSO and MSTOP as simplicial Ω-spectra consisting of n-ads of smooth oriented manifolds and n-ads of topological oriented manifolds, respectively, we thus see that the diagram

$$\begin{array}{ccc} \text{MSO} & \longrightarrow & \text{MIP} \\ \downarrow & & \downarrow \phi \\ \text{MSTOP} & \longrightarrow & \mathbb{L}^{\bullet} \end{array}$$

homotopy commutes, where the "symmetric signature map" $\text{MSTOP} \to \mathbb{L}^{\bullet}$ of ring spectra has been constructed by Ranicki. (Technically, IP spaces are PL pseudomanifolds, so to obtain the canonical map $\text{MSO} \to \text{MIP}$, it is necessary to find a canonical PL structure on a given smooth manifold. This is possible by J. II. C. Whitchead's triangulation results of [Whi40], where it is shown that every smooth manifold admits a compatible triangulation as a PL manifold and this PL manifold is unique to within a PL homeomorphism; see also [WJ66].) It follows that

$$\begin{array}{ccc} \Omega_*^{\text{SO}}(Y) & \longrightarrow & \Omega_*^{\text{IP}}(Y) \\ \downarrow & & \downarrow \phi_*(Y) \\ \Omega_*^{\text{STOP}}(Y) & \longrightarrow & \mathbb{L}_*^{\bullet}(Y) \end{array} \qquad (4\text{-}1)$$

commutes. Let us verify the commutativity for $Y = \text{pt}$ by hand. If $* = 4k + 2$ or $4k + 3$, then $\mathbb{L}_*^{\bullet}(\text{pt}) = L^*(\mathbb{Z}e) = 0$, so the two transformations agree in these dimensions. Commutativity in dimension 1 follows from $\Omega_1^{\text{SO}}(\text{pt}) = 0$. For $* = 4k + 1, k > 0$, the homotopy cofiber sequence of spectra

$$\text{MIP} \xrightarrow{\phi} \mathbb{L}^{\bullet} \longrightarrow K(\mathbb{Z}/_2, 1)$$

induces on homotopy groups an exact sequence and hence an isomorphism

$$\pi_{4k+1}(\text{MIP}) \xrightarrow{\cong} \pi_{4k+1}(\mathbb{L}^{\bullet}).$$

But both of these groups are $\mathbb{Z}/_2$, whence the isomorphism is the identity map. Thus if M^{4k+1} is a smooth oriented manifold, then $[M^{4k+1}] \in \Omega_{4k+1}^{\text{IP}}(\text{pt})$ maps under ϕ to the de Rham invariant $\text{dR}(M) \in L^{4k+1}(\mathbb{Z}e) = \mathbb{Z}/_2$. Hence the two transformations agree on a point in dimensions $4k + 1$. Again using the exact sequence of homotopy groups determined by the above cofibration sequence, ϕ induces isomorphisms $\pi_{4k}(\text{MIP}) \xrightarrow{\cong} \pi_{4k}(\mathbb{L}^{\bullet})$. Both of these groups are $\mathbb{Z}$, so this isomorphism is ± 1. Consequently, a smooth oriented manifold M^{4k}, defining an element $[M^{4k}] \in \Omega_{4k}^{\text{IP}}(\text{pt})$, maps under ϕ to $\pm \sigma(M) \in L^{4k}(\mathbb{Z}e) = \mathbb{Z}$, and it is $+\sigma(M)$ when the signs in the two symmetric structures are correctly matched.

For an n-dimensional Poincaré space which is either a topological manifold or a combinatorial homology manifold (i.e. a polyhedron whose links of simplices are homology spheres), Ranicki defines a canonical $\mathbb{L}^{\bullet}$-*orientation* $[M]_{\mathbb{L}} \in \mathbb{L}_n^{\bullet}(M)$, see [Ran92]. Its image under the assembly map

$$\mathbb{L}_n^{\bullet}(M) \xrightarrow{A} L^n(\mathbb{Z}\pi_1(M))$$

is the *symmetric signature* $\sigma^*(M)$, which is a homotopy invariant. The class $[M]_{\mathbb{L}}$ itself is a topological invariant. The geometric meaning of the $\mathbb{L}^{\bullet}$-orientation class is that its existence for a geometric Poincaré complex X^n, $n \geq 5$, assembling to the symmetric signature (which any Poincaré complex possesses), implies up to 2-torsion that X is homotopy equivalent to a compact topological manifold. (More precisely, X is homotopy equivalent to a compact manifold if it has an $\mathbb{L}^{\bullet}$-orientation class, which assembles to the *visible* symmetric signature of X.) Cap product with $[M]_{\mathbb{L}}$ induces an $\mathbb{L}^{\bullet}$-homology Poincaré duality isomorphism $(\mathbb{L}^{\bullet})^i(M) \xrightarrow{\cong} \mathbb{L}_{n-i}^{\bullet}(M)$. Rationally, $[M]_{\mathbb{L}}$ is given by the homology L-class of M,

$$[M]_{\mathbb{L}} \otimes 1 = L_*(M) \in \mathbb{L}_n^{\bullet}(M) \otimes \mathbb{Q} \cong \bigoplus_{j \geq 0} H_{n-4j}(M; \mathbb{Q}).$$

Thus, we may view $[M]_{\mathbb{L}}$ as an integral refinement of the L-class of M. Another integral refinement of the L-class is the signature homology orientation class $[M]_{\text{Sig}} \in \text{Sig}_n(M)$, to be defined below. The identity $A[M]_{\mathbb{L}} = \sigma^*(M)$ may then be interpreted as a non-simply connected generalization of the Hirzebruch signature formula. The localization of $[M]_{\mathbb{L}}$ at odd primes is the Sullivan orientation $\Delta(M) \in KO_n(M) \otimes \mathbb{Z}[\frac{1}{2}]$, which we shall return to later. Under the map $\Omega_n^{\text{STOP}}(M) \to \mathbb{L}_n^{\bullet}(M)$, $[M]_{\mathbb{L}}$ is the image of the identity map $[M \xrightarrow{\text{id}} M] \in \Omega_n^{\text{STOP}}(M)$.

We shall now apply ϕ in defining an $\mathbb{L}^{\bullet}$-orientation $[X]_{\mathbb{L}} \in \mathbb{L}_n^{\bullet}(X)$ for an oriented closed n-dimensional IP pseudomanifold X. (For Witt spaces, an $\mathbb{L}^{\bullet}$-orientation and a symmetric signature has been defined in [CSW91].) The identity map $X \to X$ defines an orientation class $[X]_{\text{IP}} \in \Omega_n^{\text{IP}}(X)$.

DEFINITION 4.1. The $\mathbb{L}^{\bullet}$-orientation $[X]_{\mathbb{L}} \in \mathbb{L}_n^{\bullet}(X)$ of an oriented closed n-dimensional IP pseudomanifold X is defined to be the image of $[X]_{\text{IP}} \in \Omega_n^{\text{IP}}(X)$ under the map

$$\Omega_n^{\text{IP}}(X) \xrightarrow{\phi_*(X)} \mathbb{L}_n^{\bullet}(X).$$

If $X = M^n$ is a smooth oriented manifold, then the identity map $M \to M$ defines an orientation class $[M]_{\text{SO}} \in \Omega_n^{\text{SO}}(M)$, which maps to $[M]_{\mathbb{L}}$ under the map

$$\Omega_n^{\text{SO}}(M) \longrightarrow \Omega_n^{\text{STOP}}(M) \longrightarrow \mathbb{L}_n^{\bullet}(M).$$

Thus, the above definition of $[X]_{\mathbb{L}}$ for an IP pseudomanifold X is compatible with manifold theory in view of the commutativity of diagram (4-1). Applying Ranicki's assembly map, it is then straightforward to define the symmetric signature of an IP pseudomanifold.

DEFINITION 4.2. The symmetric signature $\sigma^*(X) \in L^n(\mathbb{Z}\pi_1(X))$ of an oriented closed n-dimensional IP pseudomanifold X is defined to be the image of $[X]_{\mathbb{L}}$ under the assembly map

$$\mathbb{L}_n^{\bullet}(X) \xrightarrow{A} L^n(\mathbb{Z}\pi_1(X)).$$

This then agrees with the definition of the Mishchenko–Ranicki symmetric signature $\sigma^*(M)$ of a manifold $X = M$ because $A[M]_{\mathbb{L}} = \sigma^*(M)$.

5. Non-Witt spaces

All pseudomanifolds previously considered had to satisfy a vanishing condition for the middle dimensional intersection homology of the links of odd codimensional strata. Can a bordism invariant signature be defined for an even larger class of spaces? As pointed out above, taking the cone on a pseudomanifold immediately proves the futility of such an attempt on the full class of all pseudomanifolds. What, then, are the obstructions for an oriented pseudomanifold to possess Poincaré duality compatible with intersection homology?

Let $\mathcal{LK}$ be a collection of closed oriented pseudomanifolds. We might envision forming a bordism group $\Omega_n^{\mathcal{LK}}$, whose elements are represented by closed oriented n-dimensional stratified pseudomanifolds whose links are all homeomorphic to (finite disjoint unions of) elements of $\mathcal{LK}$. Two spaces X and X' represent the same bordism class, $[X] = [X']$, if there exists an $(n + 1)$-dimensional oriented compact pseudomanifold-with-boundary Y^{n+1} such that all links of the interior of Y are in $\mathcal{LK}$ and $\partial Y \cong X \sqcup -X'$ under an orientation-preserving homeomorphism. (The boundary is, as always, to be collared in a stratum-preserving way.) If, for instance,

$$\mathcal{LK} = \{S^1, S^2, S^3, \ldots\},$$

then $\Omega_*^{\mathcal{LK}}$ is bordism of manifolds. If

$$\mathcal{LK} = \text{Odd} \cup \{L^{2l} \mid IH_l^{\bar{m}}(L; \mathbb{Q}) = 0\},$$

where Odd is the collection of all odd dimensional oriented closed pseudomanifolds, then $\Omega_*^{\mathcal{LK}} = \Omega_*^{\text{Witt}}$. The question is: Which other spaces can one throw into this $\mathcal{LK}$, yielding an enlarged collection $\mathcal{LK}' \supset \mathcal{LK}$, such that one can still define a bordism invariant signature $\sigma : \Omega_*^{\mathcal{LK}'} \longrightarrow \mathbb{Z}$ so that the diagram

commutes, where $\Omega_*^{\text{Witt}} \longrightarrow \Omega_*^{\mathcal{LK}'}$ is the canonical map induced by the inclusion $\mathcal{LK} \subset \mathcal{LK}'$? Note that $\sigma(L) = 0$ for every $L \in \mathcal{LK}$. Suppose we took an $\mathcal{LK}'$ that contains a manifold P with $\sigma(P) \neq 0$, e.g. $P = \mathbb{CP}^{2k}$. Then $[P] = 0 \in \Omega_*^{\mathcal{LK}'}$, since P is the boundary of the cone on P, and the cone on P is an admissible bordism in $\Omega_*^{\mathcal{LK}'}$, as the link of the cone-point is P and $P \in \mathcal{LK}'$. Thus, in the above diagram,

$$[P] \overset{\sigma}{\longmapsto} \sigma(P) \neq 0.$$

This argument shows that the desired diagonal arrow cannot exist for any collection $\mathcal{LK}'$ that contains any manifolds with nonzero signature. Thus we are naturally led to consider only links with zero signature, that is, links whose intersection form on middle dimensional homology possesses a Lagrangian subspace. As you move along a stratum of odd codimension, these Lagrangian subspaces should fit together, forming a subsheaf of the middle dimensional cohomology sheaf $\mathbf{H}$ associated to the link-bundle over the stratum. (Actually, no bundle neighborhood structure is required to do this.) So a natural language in which to phrase and solve the problem is sheaf theory.

From the sheaf-theoretic vantage point, the statement that a space X^n does not satisfy the Witt condition means precisely that the canonical morphism $\mathbf{IC}_{\bar{m}}^\bullet(X) \to \mathbf{IC}_{\bar{n}}^\bullet(X)$ from lower to upper middle perversity is not an isomorphism in the derived category. (We are using sheaves of real vector spaces now and shall not indicate this further in our notation.) Thus there is no way to introduce a quadratic form whose signature one could take, using intersection chain sheaves. But one may ask how close to such sheaves one might get by using self-dual sheaves on X. In [Ban02], we define a full subcategory $SD(X)$ of the derived category on X, whose objects $\mathbf{S}^\bullet$ satisfy all the axioms that $\mathbf{IC}_{\bar{n}}^\bullet(X)$ satisfies, with the exception of the last axiom, the costalk vanishing axiom. This axiom is replaced with the requirement that $\mathbf{S}^\bullet$ be self-dual, that is, there is an isomorphism $\mathcal{D}\mathbf{S}^\bullet[n] \cong \mathbf{S}^\bullet$, just as there is for $\mathbf{IC}_{\bar{m}}^\bullet$ on a Witt space. Naturally, this category may be empty, depending on the geometry of X. So we need to develop a structure theorem for $SD(X)$, and this is done in [Ban02]. It turns out that every such object $\mathbf{S}^\bullet$ interpolates between $\mathbf{IC}_{\bar{m}}^\bullet$ and $\mathbf{IC}_{\bar{n}}^\bullet$, i.e. possesses a factorization $IC_{\bar{m}}^{\bullet} \to S^{\bullet} \to IC_{\bar{n}}^{\bullet}$ of the canonical morphism. The two morphisms of the factorization are dual to each other. Note that in the basic two strata case, the mapping cone of the canonical morphism is the middle cohomology sheaf $\mathbf{H}$ of the link-bundle. We prove that the mapping cone of $IC_{\bar{m}}^{\bullet} \to S^{\bullet}$, restricted to the stratum of odd codimension, is a Lagrangian subsheaf of $\mathbf{H}$, so that the circle to the above geometric ideas closes. The main result of [Ban02] is an equivalence of categories between $SD(X)$ and a fibered product of categories of Lagrangian structures, one such category for each stratum of odd codimension. This then is a kind of Postnikov system for $SD(X)$, encoding both the obstruction theory and the constructive technology to manufacture objects in $SD(X)$.

Suppose X is such that $SD(X)$ is not empty. An object $\mathbf{S}^{\bullet}$ in $SD(X)$ defines a signature $\sigma(\mathbf{S}^{\bullet}) \in \mathbb{Z}$ by taking the signature of the quadratic form that the self-duality isomorphism $\mathcal{D}\mathbf{S}^{\bullet}[n] \cong \mathbf{S}^{\bullet}$ induces on the middle dimensional hypercohomology group of $\mathbf{S}^{\bullet}$. Since restricting a self-dual sheaf to a transverse (to the stratification) subvariety again yields a self-dual sheaf on the subvariety, we get a signature for all transverse subvarieties and thus an L-class $L_*(\mathbf{S}^{\bullet}) \in H_*(X; \mathbb{Q})$, using maps to spheres and Serre's theorem as indicated in the beginning. We prove in [Ban06b] that $L_*(\mathbf{S}^{\bullet})$, in particular $\sigma(\mathbf{S}^{\bullet}) = L_0(\mathbf{S}^{\bullet})$, is independent of the choice of $\mathbf{S}^{\bullet}$ in $SD(X)$. Consequently, a non-Witt space has a well-defined L-class $L_*(X)$ and signature $\sigma(X)$, provided $SD(X)$ is not empty.

Let $\mathrm{Sig}_n(\mathrm{pt})$ be the bordism group of pairs $(X, \mathbf{S}^{\bullet})$, where X is a closed oriented topological or PL n-dimensional pseudomanifold and $\mathbf{S}^{\bullet}$ is an object of $SD(X)$. Admissible bordisms are oriented compact pseudomanifolds-with-boundary Y^{n+1}, whose interior $\mathrm{int}\, Y$ is covered with an object of $SD(\mathrm{int}\, Y)$ which pushes to the given sheaf complexes on the boundary. These groups have been introduced in [Ban02] under the name Ω_*^{SD}. Let us compute these groups. The signature $(X, \mathbf{S}^{\bullet}) \mapsto \sigma(\mathbf{S}^{\bullet})$ is a bordism invariant and hence induces a map $\sigma : \mathrm{Sig}_{4k}(\mathrm{pt}) \to \mathbb{Z}$. This map is onto, since e.g. $(\mathbb{CP}^{2k}, \mathbb{R}_{\mathbb{CP}^{2k}}[4k])$ (and disjoint copies of it) is in $\mathrm{Sig}_{4k}(\mathrm{pt})$. However, contrary for example to Witt bordism, σ is also injective: Suppose $\sigma(X, \mathbf{S}^{\bullet}) = 0$. Let Y^{4k+1} be the closed cone on X. Define a self-dual sheaf on the interior of the punctured cone by pulling back $\mathbf{S}^{\bullet}$ from X under the projection from the interior of the punctured cone, $X \times (0, 1)$, to X. According to the Postnikov system of Lagrangian structures for $SD(\mathrm{int}\, Y)$, the self-dual sheaf on the interior of the punctured cone will have a self-dual extension in $SD(\mathrm{int}\, Y)$ if, and only if, there exists a Lagrangian structure at the cone-point (which has odd codimension $4k + 1$ in Y). That Lagrangian structure exists because $\sigma(X, \mathbf{S}^{\bullet}) = 0$. Let $\mathbf{T}^{\bullet} \in SD(\mathrm{int}\, Y)$ be any self-dual extension given by a choice of Lagrangian structure. Then $\partial(Y, \mathbf{T}^{\bullet}) = (X, \mathbf{S}^{\bullet})$ and thus $[(X, \mathbf{S}^{\bullet})] = 0$ in $\mathrm{Sig}_{4k}(\mathrm{pt})$. Clearly, $\mathrm{Sig}_n(\mathrm{pt}) = 0$ for $n \not\equiv 0(4)$

because an anti-symmetric form always has a Lagrangian subspace and the cone on an odd dimensional space is even dimensional, so in these cases there are no extension problems at the cone point — just perform a one-step Goresky–MacPherson–Deligne extension. In summary then, one has

$$\mathrm{Sig}_n(\mathrm{pt}) \cong \begin{cases} \mathbb{Z}, & n \equiv 0(4), \\ 0, & n \not\equiv 0(4). \end{cases}$$

(Note that in particular the de Rham invariant has been disabled and the signature is a complete invariant for these bordism groups.) Minatta [Min04], [Min06] takes this as his starting point and constructs a bordism theory $\mathrm{Sig}_*(-)$, called *signature homology*, whose coefficients are the above groups $\mathrm{Sig}_*(\mathrm{pt})$. Elements of $\mathrm{Sig}_n(Y)$ are represented by pairs $(X, \mathbf{S}^\bullet)$ as above together with a continuous map $X \to Y$. For a detailed proof that $\mathrm{Sig}_*(-)$ is a generalized homology theory when PL pseudomanifolds are used, consult the appendix of [Ban06a]. Signature homology is represented by an MSO module spectrum MSIG, which is also a ring spectrum. Regarding a smooth manifold as a pseudomanifold with one stratum covered by the constant sheaf of rank 1 concentrated in one dimension defines a natural transformation of homology theories $\Omega_*^{\mathrm{SO}}(-) \to \mathrm{Sig}_*(-)$. Thus, MSIG is 2-integrally a product of Eilenberg–Mac Lane spectra,

$$\mathrm{MSIG}_{(2)} \simeq \prod_{j \geq 0} K(\mathbb{Z}_{(2)}, 4j).$$

As for the odd-primary situation, the isomorphism $\mathrm{Sig}_*(\mathrm{pt}) \otimes_{\mathbb{Z}} \mathbb{Z}[\frac{1}{2}] \to \mathbb{Z}[\frac{1}{2}, t]$ given by $[(X^{4k}, \mathbf{S}^\bullet)] \otimes a \mapsto a\sigma(\mathbf{S}^\bullet)t^k$, determines an identification

$$\Omega_*^{\mathrm{SO}}(Y) \otimes_{\Omega_*^{\mathrm{SO}}(\mathrm{pt})} \mathrm{Sig}_*(\mathrm{pt}) \otimes_{\mathbb{Z}} \mathbb{Z}[\tfrac{1}{2}] \xrightarrow{\cong} \Omega_*^{\mathrm{SO}}(Y) \otimes_{\Omega_*^{\mathrm{SO}}(\mathrm{pt})} \mathbb{Z}[\tfrac{1}{2}, t].$$

A natural isomorphism of homology theories

$$\Omega_*^{\mathrm{SO}}(Y) \otimes_{\Omega_*^{\mathrm{SO}}(\mathrm{pt})} \mathrm{Sig}_*(\mathrm{pt}) \otimes_{\mathbb{Z}} \mathbb{Z}[\tfrac{1}{2}] \xrightarrow{\cong} \mathrm{Sig}_*(Y) \otimes_{\mathbb{Z}} \mathbb{Z}[\tfrac{1}{2}]$$

is induced by sending $[M \xrightarrow{f} Y] \otimes [(X, \mathbf{S}^\bullet)]$ to $[(M \times X, \mathbf{P}^\bullet, M \times X \to M \xrightarrow{f} Y)]$, where $\mathbf{P}^\bullet$ is the pullback sheaf of $\mathbf{S}^\bullet$ under the second-factor projection. Composing, we obtain a natural isomorphism

$$\Omega_*^{\mathrm{SO}}(-) \otimes_{\Omega_*^{\mathrm{SO}}(\mathrm{pt})} \mathbb{Z}[\tfrac{1}{2}, t] \xrightarrow{\cong} \mathrm{Sig}_*(-) \otimes_{\mathbb{Z}} \mathbb{Z}[\tfrac{1}{2}],$$

describing signature homology at odd primes in terms of manifold bordism.

Again, it follows in particular that the natural map

$$\Omega_*^{\mathrm{SO}}(Y) \otimes_{\mathbb{Z}} \mathbb{Z}[\tfrac{1}{2}, t] \to \mathrm{Sig}_*(Y) \otimes_{\mathbb{Z}} \mathbb{Z}[\tfrac{1}{2}]$$

is a surjection, which frequently allows one to reduce bordism invariant calculations on non-Witt spaces to the manifold case. We observed this in [Ban06a] and apply it there to establish a multiplicative characteristic class formula for the twisted signature and L-class of non-Witt spaces. Let X^n be a closed oriented Whitney stratified pseudomanifold and let $\mathcal{S}$ be a nondegenerate symmetric local system on X. If $SD(X)$ is not empty, that is, X possesses Lagrangian structures along its strata of odd codimension so that $L_*(X) \in H_*(X; \mathbb{Q})$ is defined, then

$$L_*(X; \mathcal{S}) = \widetilde{\mathrm{ch}}[\mathcal{S}]_K \cap L_*(X).$$

For the special case of the twisted signature $\sigma(X; \mathcal{S}) = L_0(X; \mathcal{S})$, one has therefore

$$\sigma(X; \mathcal{S}) = \langle \widetilde{\mathrm{ch}}[\mathcal{S}]_K, L(X) \rangle.$$

We shall apply the preceding ideas in defining a *Sullivan orientation* $\Delta(X) \in ko_*(X) \otimes \mathbb{Z}[\frac{1}{2}]$ for a pseudomanifold X that possesses generalized Poincaré duality (that is, its self-dual perverse category $SD(X)$ is not empty), but need not satisfy the Witt condition. In [Sul05], Sullivan defined for an oriented rational PL homology manifold M an orientation class $\Delta(M) \in ko_*(M) \otimes \mathbb{Z}[\frac{1}{2}]$, whose Pontrjagin character is the L-class $L_*(M)$. For a Witt space X^n, a Sullivan class $\Delta(X) \in ko_*(X) \otimes \mathbb{Z}[\frac{1}{2}]$ was constructed by Siegel [Sie83], using the intersection homology signature of a Witt space and transversality to produce the requisite Sullivan periodicity squares that represent elements of $KO^{4k}(N, \partial N) \otimes \mathbb{Z}[\frac{1}{2}]$, where N is a regular neighborhood of a codimension $4k$ PL-embedding of X in a high dimensional Euclidean space. An element in $ko^{4k}(N, \partial N) \otimes \mathbb{Z}[\frac{1}{2}]$ corresponds to a unique element in $ko_n(X) \otimes \mathbb{Z}[\frac{1}{2}]$ by Alexander duality. Siegel's isomorphism (3-1) is then given by the Hurewicz-type map

$$\Omega_*^{\mathrm{Witt}}(Y) \otimes \mathbb{Z}[\tfrac{1}{2}] \longrightarrow ko_*(Y) \otimes \mathbb{Z}[\tfrac{1}{2}]$$
$$[X \xrightarrow{f} Y] \otimes 1 \mapsto f_* \Delta(X),$$

where $f_* : ko_*(X) \otimes \mathbb{Z}[\frac{1}{2}] \to ko_*(Y) \otimes \mathbb{Z}[\frac{1}{2}]$. In particular, the transformation (3-1) maps the Witt orientation class $[X]_{\mathrm{Witt}} \in \Omega_n^{\mathrm{Witt}}(X)$, given by the identity map $f = \mathrm{id}_X : X \to X$, to $\Delta(X)$.

REMARK 5.1. In [CSW91], there is indicated an extension to continuous actions of a finite group G on a Witt space X. If the action satisfies a weak condition on the fixed point sets, then there is a homeomorphism invariant class $\Delta^G(X)$ in the equivariant KO-homology of X away from 2, which is the Atiyah–Singer G-signature invariant for smooth actions on smooth manifolds.

Let P be a compact polyhedron. Using Balmer's 4-periodic Witt groups of triangulated categories with duality, Woolf [Woo08] defines groups $W_*^c(P)$, called constructible Witt groups of P because the underlying triangulated categories are the derived categories of sheaf complexes that are constructible with respect to the simplicial stratifications of admissible triangulations of P. (The duality is given by Verdier duality.) Elements of $W_n^c(P)$ are represented by symmetric self-dual isomorphisms $d : \mathbf{S}^\bullet \to (\mathcal{D}\mathbf{S}^\bullet)[n]$. The periodicity isomorphism $W_n^c(P) \cong W_{n+4}^c(P)$ is induced by shifting such a d twice:

$$d[2] : \mathbf{S}^\bullet[2] \longrightarrow (\mathcal{D}\mathbf{S}^\bullet)[n][2] = (\mathcal{D}\mathbf{S}^\bullet)[-2][n+4] = \mathcal{D}(\mathbf{S}^\bullet[2])[n+4].$$

(Shifting only once does not yield a correct symmetric isomorphism with respect to the duality fixed for $W_*^c(P)$.) Woolf shows that for commutative regular Noetherian rings R of finite Krull dimension in which 2 is invertible, for example $R = \mathbb{Q}$, the assignment $P \mapsto W_*^c(P)$ is a generalized homology theory on compact polyhedra and continuous maps. Let K be a simplicial complex triangulating P. Relating both Ranicki's (R, K)-modules on the one hand and constructible sheaves on the other hand to combinatorial sheaves on K, Woolf obtains a natural transformation

$$\mathbb{L}^\bullet(R)_*(K) \longrightarrow W_*^c(|K|)$$

$(|K| = P)$, which he shows to be an isomorphism when every finitely generated R-module can be resolved by a finite complex of finitely generated free R-modules. Again, this applies to $R = \mathbb{Q}$. Given a map $f : X^n \to P$ from a compact oriented Witt space X^n into P, the pushforward $Rf_*(d)$ of the symmetric self-duality isomorphism $d : \mathbf{IC}_{\bar{m}}^\bullet(X) \cong \mathcal{D}\mathbf{IC}_{\bar{m}}^\bullet(X)[n]$ defines an element $[Rf_*(d)] \in W_n^c(P)$. This induces a natural map

$$\Omega_n^{\text{Witt}}(P) \longrightarrow W_n^c(P),$$

which is an isomorphism when $n > \dim P$. Given any $n \geq 0$, we can iterate the 4-periodicity until $n + 4* > \dim P$ and obtain

$$W_n^c(P) \cong W_{n+4}^c(P) \cong \cdots \cong W_{n+4k}^c(P) \cong \Omega_{n+4k}^{\text{Witt}}(P),$$

where $n + 4k > \dim P$. Thus, as Woolf points out, the Witt class of any symmetric self-dual sheaf on P is given, after a suitable even number of shifts, by the pushforward of an intersection chain sheaf on some Witt space. This viewpoint also allows for the interpretation of L-classes as homology operations $W_*^c(-) \to H_*(-)$ or $\Omega_*^{\text{Witt}}(-) \to H_*(-)$. Other characteristic classes arising in complex algebraic geometry can be interpreted through natural transformations as well. MacPherson's Chern class of a variety can be defined as the image $c_*^M(1_X)$ of the function 1_X under a natural transformation $c_*^M : F(-) \to H_*(-)$, where $F(X)$ is the abelian group of constructible functions on X. The Baum–Fulton–MacPherson Todd class can be defined as the image $\text{td}_*^{\text{BMF}}(\mathcal{O}_X)$ of $\mathcal{O}_X$

under a natural transformation

$$\mathrm{td}_*^{\mathrm{BMF}} : G_0(-) \to H_*(-) \otimes \mathbb{Q},$$

where $G_0(X)$ is the Grothendieck group of coherent sheaves on X. In [BSY], Brasselet, Schürmann and Yokura realized two important facts: First, there exists a source $K_0(\mathcal{VAR}/X)$ which possesses natural transformations to all three domains of the characteristic class transformations mentioned. That is, there exist natural transformations

$$K_0(\mathcal{VAR}/-) \tag{5-1}$$

$$F(-) \qquad G_0(-) \qquad \Omega^Y(-),$$

where $\Omega^Y(X)$ is the abelian group of Youssin's bordism classes of self-dual constructible sheaf complexes on X. That source $K_0(\mathcal{VAR}/X)$ is the free abelian group generated by algebraic morphisms $f : V \to X$ modulo the relation

$$[V \xrightarrow{f} X] = [V - Z \xrightarrow{f|} X] + [Z \xrightarrow{f|} X]$$

for every closed subvariety $Z \subset V$. Second, there exists a unique natural transformation, the *motivic characteristic class transformation*,

$$T_{y*} : K_0(\mathcal{VAR}/X) \longrightarrow H_*(X) \otimes \mathbb{Q}[y]$$

such that

$$T_{y*}[\mathrm{id}_X] = T_y(TX) \cap [X]$$

for nonsingular X, where $T_y(TX)$ is Hirzebruch's generalized Todd class of the tangent bundle TX of X. Characteristic classes for singular varieties are of course obtained by taking $T_{y*}[\mathrm{id}_X]$. Under the above three transformations (5-1), $[\mathrm{id}_X]$ is mapped to 1_X, $[\mathcal{O}_X]$, and $[\mathbb{Q}_X[2\dim X]]$ (when X is nonsingular), respectively. Following these three transformations with c_*^{M}, $\mathrm{td}_*^{\mathrm{BMF}}$, and the L-class transformation

$$\Omega^Y(-) \to H_*(-) \otimes \mathbb{Q},$$

one obtains T_{y*} for $y = -1, 0, 1$, respectively. This, then, is an attractive unification of Chern-, Todd- and L-classes of singular complex algebraic varieties, see also Yokura's paper in this volume, as well as [SY07].

The natural transformation

$$\Omega_*^{\mathrm{Witt}}(-) \otimes \mathbb{Z}[\tfrac{1}{2}] \longrightarrow \mathrm{Sig}_*(-) \otimes \mathbb{Z}[\tfrac{1}{2}],$$

given by covering a Witt space X with the middle perversity intersection chain sheaf $\mathbf{S}^\bullet = \mathbf{IC}_{\bar{m}}^\bullet(X)$, which is an object of $SD(X)$, is an isomorphism because on a point, it is given by the signature

$$\Omega_{4k}^{\mathrm{Witt}}(\mathrm{pt}) \otimes \mathbb{Z}[\tfrac{1}{2}] \cong L^{4k}(\mathbb{Q}) \otimes \mathbb{Z}[\tfrac{1}{2}] \cong \mathbb{Z}[\tfrac{1}{2}] \cong \mathrm{Sig}_{4k}(\mathrm{pt}) \otimes \mathbb{Z}[\tfrac{1}{2}]$$

(the infinitely generated torsion of $L^{4k}(\mathbb{Q})$ is killed by inverting 2), and

$$\Omega_j^{\mathrm{Witt}}(\mathrm{pt}) = 0 = \mathrm{Sig}_j(\mathrm{pt})$$

for j not divisible by 4. Inverting this isomorphism and composing with Siegel's isomorphism (3-1), we obtain a natural isomorphism of homology theories

$$D : \mathrm{Sig}_*(-) \otimes \mathbb{Z}[\tfrac{1}{2}] \xrightarrow{\cong} ko_*(-) \otimes \mathbb{Z}[\tfrac{1}{2}].$$

Let X^n be a closed pseudomanifold, not necessarily a Witt space, but still supporting self-duality, i.e. $SD(X)$ is not empty. Choose a sheaf $\mathbf{S}^\bullet \in SD(X)$. Then the pair $(X, \mathbf{S}^\bullet)$, together with the identity map $X \to X$, defines an element $[X]_{\mathrm{Sig}} \in \mathrm{Sig}_n(X)$.

DEFINITION 5.2. The *signature homology orientation class* of an n-dimensional closed pseudomanifold X with $SD(X) \neq \varnothing$, but not necessarily a Witt space, is the element $[X]_{\mathrm{Sig}} \in \mathrm{Sig}_n(X)$.

PROPOSITION 5.3. The orientation class $[X]_{\mathrm{Sig}}$ is well-defined, that is, independent of the choice of sheaf $\mathbf{S}^\bullet \in SD(X)$.

PROOF. Let $\mathbf{T}^\bullet \in SD(X)$ be another choice. In [Ban06b], a bordism $(Y, \mathbf{U}^\bullet)$, $\mathbf{U}^\bullet \in SD(\mathrm{int}\, Y)$, is constructed between $(X, \mathbf{S}^\bullet)$ and $(X, \mathbf{T}^\bullet)$. Topologically, Y is a cylinder $Y \cong X \times I$, but equipped with a nonstandard stratification, of course. The identity map $X \to X$ thus extends over this bordism by taking

$$Y \to X, \quad (x, t) \mapsto x. \qquad \square$$

DEFINITION 5.4. The *Sullivan orientation* of an n-dimensional closed pseudo-manifold X with $SD(X) \neq \varnothing$, but not necessarily a Witt space, is defined as

$$\Delta(X) = D([X]_{\mathrm{Sig}} \otimes 1) \in ko_*(X) \otimes \mathbb{Z}[\tfrac{1}{2}].$$

Let us compare signature homology and $\mathbb{L}^\bullet$-homology away from 2, at 2, and rationally, following [Epp07] and drawing on work of Taylor and Williams, [TW79]. For a spectrum S, let $S_{(\mathrm{odd})}$ denote its localization at odd primes. We have observed above that

$$\mathrm{MSIG}_{(2)} \simeq \prod_{j \geq 0} K(\mathbb{Z}_{(2)}, 4j)$$

and, according to [Epp07] and [Min04],

$$\mathrm{MSIG}_{(\mathrm{odd})} \simeq bo_{(\mathrm{odd})}.$$

Rationally, we have the decomposition

$$\mathrm{MSIG} \otimes \mathbb{Q} \simeq \prod_{j \geq 0} K(\mathbb{Q}, 4j).$$

Thus MSIG fits into a localization pullback square

$$
\begin{array}{ccc}
\mathrm{MSIG} & \xrightarrow{\mathrm{loc}_{(\mathrm{odd})}} & bo_{(\mathrm{odd})} \\
{\scriptstyle \mathrm{loc}_{(2)}} \downarrow & & \downarrow \\
\prod K(\mathbb{Z}_{(2)}, 4j) & \xrightarrow{\lambda} & \prod K(\mathbb{Q}, 4j).
\end{array}
$$

The symmetric L-spectrum $\mathbb{L}^{\bullet}$ is an MSO module spectrum, so it is 2-integrally a product of Eilenberg–Mac Lane spectra,

$$\mathbb{L}^{\bullet}_{(2)} \simeq \prod_{j \geq 0} K(\mathbb{Z}_{(2)}, 4j) \times K(\mathbb{Z}/_2, 4j+1).$$

Comparing this to $\mathrm{MSIG}_{(2)}$, we thus see the de Rham invariants coming in. Away from 2, $\mathbb{L}^{\bullet}$ coincides with bo,

$$\mathbb{L}^{\bullet}_{(\mathrm{odd})} \simeq bo_{(\mathrm{odd})},$$

as does MSIG. Rationally, $\mathbb{L}^{\bullet}$ is again

$$\mathbb{L}^{\bullet} \otimes \mathbb{Q} \simeq \prod_{j \geq 0} K(\mathbb{Q}, 4j).$$

Thus $\mathbb{L}^{\bullet}$ fits into a localization pullback square

$$
\begin{array}{ccc}
\mathbb{L}^{\bullet} & \xrightarrow{\mathrm{loc}_{(\mathrm{odd})}} & bo_{(\mathrm{odd})} \\
{\scriptstyle \mathrm{loc}_{(2)}} \downarrow & & \downarrow \\
\prod K(\mathbb{Z}_{(2)}, 4j) \times K(\mathbb{Z}/_2, 4j+1) & \xrightarrow{\lambda'} & \prod K(\mathbb{Q}, 4j).
\end{array}
$$

The map λ factors as

$$\prod_{j \geq 0} K(\mathbb{Z}_{(2)}, 4j) \xhookrightarrow{\iota} \prod_{j \geq 0} K(\mathbb{Z}_{(2)}, 4j) \times K(\mathbb{Z}/_2, 4j+1) \xrightarrow{\lambda'} \prod_{j \geq 0} K(\mathbb{Q}, 4j),$$

where ι is the obvious inclusion, not touching the 2-torsion nontrivially. Hence, by the universal property of a pullback, we get a map μ from signature homology to $\mathbb{L}^{\bullet}$-homology,

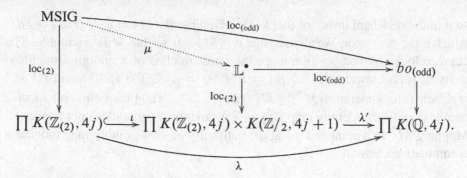

On the other hand, λ' factors as

$$\prod_{j\geq 0} K(\mathbb{Z}_{(2)}, 4j) \times K(\mathbb{Z}/_2, 4j+1) \xrightarrow{\text{proj}} \prod_{j\geq 0} K(\mathbb{Z}_{(2)}, 4j) \xrightarrow{\lambda} \prod_{j\geq 0} K(\mathbb{Q}, 4j),$$

where proj is the obvious projection. Again using the universal property of a pullback, we obtain a map $\nu : \mathbb{L}^{\bullet} \to \text{MSIG}$. The map μ is a homotopy splitting for ν, $\nu\mu \simeq \text{id}$, since proj $\circ \iota = \text{id}$. It follows that via μ, signature homology is a direct summand in symmetric $\mathbb{L}^{\bullet}$-homology. We should like to point out that the diagram

$$\begin{array}{ccc} \text{MSO} & \longrightarrow & \text{MSIG} \\ \downarrow & & \downarrow{\mu} \\ \text{MSTOP} & \longrightarrow & \mathbb{L}^{\bullet} \end{array}$$

does *not* commute. This is essentially due to the fact that the de Rham invariant is lost in Sig_*, but is still captured in $\mathbb{L}^{\bullet}_*$. In more detail, consider the induced diagram on π_5,

$$\begin{array}{ccc} \Omega_5^{\text{SO}}(\text{pt}) & \longrightarrow & \text{Sig}_5(\text{pt}) = 0 \\ \downarrow & & \downarrow{\mu} \\ \Omega_5^{\text{STOP}}(\text{pt}) & \longrightarrow & \mathbb{L}^{\bullet}_5(\text{pt}). \end{array}$$

The clockwise composition in the diagram is zero, but the counterclockwise composition is not. Indeed, let M^5 be the Dold manifold $P(1,2) = (S^1 \times \mathbb{CP}^2)/(x,z) \sim (-x,\bar{z})$. Its cohomology ring with $\mathbb{Z}/_2$-coefficients is the same as the one of the untwisted product, that is, the truncated polynomial ring

$$\mathbb{Z}/_2[c,d]/(c^2 = 0, d^3 = 0),$$

where c has degree one and d has degree two. The total Stiefel–Whitney class of M^5 is

$$w(M) = (1+c)(1+c+d)^3,$$

so that the de Rham invariant $\mathrm{dR}(M)$ is given by $\mathrm{dR}(M) = w_2 w_3(M) = cd^2$, which is the generator. We also see that $w_1(M) = 0$, so that M is orientable. The counterclockwise composition maps the bordism class of a smooth 5-manifold to its de Rham invariant in $\mathbb{L}_5^\bullet(\mathrm{pt}) = L^5(\mathbb{Z}e) = \mathbb{Z}/2$. The Dold manifold M^5 represents the generator $[M^5] \in \Omega_5^{\mathrm{SO}}(\mathrm{pt}) = \mathbb{Z}/2$. Thus the counterclockwise composition is the identity map $\mathbb{Z}/2 \to \mathbb{Z}/2$ and the diagram does not commute. Mapping M^5 to a point and using the naturality of the assembly map induces a commutative diagram

$$
\begin{array}{ccccc}
\mathrm{Sig}_5(M) & \xrightarrow{\mu(M)} & \mathbb{L}_5^\bullet(M) & \xrightarrow{A} & L^5(\mathbb{Z}\pi_1 M) \\
\downarrow & & \downarrow & & \downarrow{\scriptstyle\varepsilon} \\
0 = \mathrm{Sig}_5(\mathrm{pt}) & \xrightarrow{\mu(\mathrm{pt})} & \mathbb{L}_5^\bullet(\mathrm{pt}) & \xrightarrow[A]{\cong} & L^5(\mathbb{Z}e) = \mathbb{Z}/2,
\end{array}
$$

which shows that the signature homology orientation class of M, $[M]_{\mathrm{Sig}} \in \mathrm{Sig}_5(M)$ does not hit the $\mathbb{L}^\bullet$-orientation of M, $[M]_{\mathbb{L}} \in \mathbb{L}_5^\bullet(M)$ under μ, for otherwise

$$0 = \varepsilon A \mu[M]_{\mathrm{Sig}} = \varepsilon A[M]_{\mathbb{L}} = \varepsilon \sigma^*(M) = \mathrm{dR}(M) \neq 0.$$

Thus one may take the viewpoint that it is perhaps not prudent to call $\mu[X]_{\mathrm{Sig}}$ an "$\mathbb{L}^\bullet$-orientation" of a pseudomanifold X with $SD(X)$ not empty. Nor might even its image under assembly deserve the title "symmetric signature" of X. On the other hand, one may wish to attach higher priority to the bordism invariance (in the singular world) of a concept such as the symmetric signature than to its compatibility with manifold invariants and nonsingular bordism invariance, and therefore deem such terminology justified.

We conclude with a brief remark on integral Novikov problems. Let π be a discrete group and let $K(\pi, 1)$ be the associated Eilenberg–Mac Lane space. The composition of the split inclusion $\mathrm{Sig}_n(K(\pi, 1)) \hookrightarrow \mathbb{L}_n^\bullet(K(\pi, 1))$ with the assembly map

$$A : \mathbb{L}_n^\bullet(K(\pi, 1)) \to L^n(\mathbb{Z}\pi)$$

yields what one may call a "signature homology assembly" map

$$A_{\mathrm{Sig}} : \mathrm{Sig}_n(K(\pi, 1)) \to L^n(\mathbb{Z}\pi),$$

which may be helpful in studying an integral refinement of the Novikov conjecture, as suggested by Matthias Kreck: When is the integral orientation class

$$\alpha_*[M]_{\text{Sig}} \in \text{Sig}_n(K(\pi, 1))$$

homotopy invariant? Here M^n is a closed smooth oriented manifold with fundamental group $\pi = \pi_1(M)$; the map $\alpha : M \to K(\pi, 1)$ classifies the universal cover of M. Note that when tensored with the rationals, one obtains the classical Novikov conjecture because rationally the signature homology orientation class $[M]_{\text{Sig}}$ is the L-class $L_*(M)$. One usually refers to integral refinements such as this one as "Novikov *problems*" because there are groups π for which they are known to be false.

References

[B⁺84] A. Borel et al., *Intersection cohomology*, Progr. Math. **50**, Birkhäuser, Boston, 1984.

[Ban02] M. Banagl, *Extending intersection homology type invariants to non-Witt spaces*, Memoirs Amer. Math. Soc. **160** (2002), 1–83.

[Ban06a] ———, *Computing twisted signatures and L-classes of non-Witt spaces*, Proc. London Math. Soc. (3) **92** (2006), 428–470.

[Ban06b] ———, *The L-class of non-Witt spaces*, Annals of Math. **163**:3 (2006), 743–766.

[Ban06c] ———, *On topological invariants of stratified maps with non-Witt target*, Trans. Amer. Math. Soc. **358**:5 (2006), 1921–1935.

[Ban07] ———, *Topological invariants of stratified spaces*, Springer Monographs in Mathematics, Springer, Berlin, 2007.

[Ban09] ———, *Intersection spaces, spatial homology truncation, and string theory*, Lecture Notes in Math. **1997** Springer, Berlin, 2010.

[BB08] B. Bleile and H. J. Baues, *Poincaré duality complexes in dimension four*, preprint arXiv:0802.3652v2, 2008.

[BCS03] M. Banagl, S. E. Cappell, and J. L. Shaneson, *Computing twisted signatures and L-classes of stratified spaces*, Math. Ann. **326**:3 (2003), 589–623.

[Bre97] G. E. Bredon, *Sheaf theory*, second ed., Grad. Texts in Math. **170**, Springer, New York, 1997.

[BSY] J.-P. Brasselet, J. Schürmann, and S. Yokura, *Hirzebruch classes and motivic Chern classes for singular spaces*, preprint math.AG/0503492.

[Che79] J. Cheeger, *On the spectral geometry of spaces with cone-like singularities*, Proc. Natl. Acad. Sci. USA **76** (1979), 2103–2106.

[Che80] ———, *On the Hodge theory of Riemannian pseudomanifolds*, Proc. Sympos. Pure Math. **36** (1980), 91–146.

[Che83] _____, *Spectral geometry of singular Riemannian spaces*, J. Differential Geom. **18** (1983), 575–657.

[CS91] S. E. Cappell and J. L. Shaneson, *Stratifiable maps and topological invariants*, J. Amer. Math. Soc. **4** (1991), 521–551.

[CSW91] S. E. Cappell, J. L. Shaneson, and S. Weinberger, *Classes topologiques caractéristiques pour les actions de groupes sur les espaces singuliers*, C. R. Acad. Sci. Paris Sér. I Math. **313** (1991), 293–295.

[Cur92] S. Curran, *Intersection homology and free group actions on Witt spaces*, Michigan Math. J. **39** (1992), 111–127.

[CW91] S. E. Cappell and S. Weinberger, *Classification de certaines espaces stratifiés*, C. R. Acad. Sci. Paris Sér. I Math. **313** (1991), 399–401.

[Epp07] T. Eppelmann, *Signature homology and symmetric L-theory*, Ph.D. thesis, Ruprecht-Karls Universität Heidelberg, 2007.

[GM80] M. Goresky and R. D. MacPherson, *Intersection homology theory*, Topology **19** (1980), 135–162.

[GM83] _____, *Intersection homology II*, Invent. Math. **71** (1983), 77–129.

[Gor78] M. Goresky, *Triangulation of stratified objects*, Proc. Amer. Math. Soc. **72** (1978), 193–200.

[Gor84] _____, *Intersection homology operations*, Comm. Math. Helv. **59** (1984), 485–505.

[GS83] R. M. Goresky and P. H. Siegel, *Linking pairings on singular spaces*, Comm. Math. Helv. **58** (1983), 96–110.

[Hen77] H. Hendriks, *Obstruction theory in 3-dimensional topology: An extension theorem*, J. London Math. Soc. (2) **16** (1977), 160–164.

[KW06] F. Kirwan and J. Woolf, *An introduction to intersection homology theory*, second ed., Chapman & Hall/CRC, 2006.

[Min04] A. Minatta, *Hirzebruch homology*, Ph.D. thesis, Ruprecht-Karls Universität Heidelberg, 2004.

[Min06] _____, *Signature homology*, J. Reine Angew. Math. **592** (2006), 79–122.

[Par90] W. L. Pardon, *Intersection homology, Poincaré spaces and the characteristic variety theorem*, Comment. Math. Helv. **65** (1990), 198–233.

[Ran79] A. A. Ranicki, *The total surgery obstruction*, Proceedings 1978 Århus Topology Conference, Lecture Notes in Math. **763**, Springer, 1979, pp. 275–316.

[Ran92] _____, *Algebraic L-theory and topological manifolds*, Cambridge Tracts in Math. **102**, Cambridge University Press, 1992.

[Sie83] P. H. Siegel, *Witt spaces: A geometric cycle theory for KO-homology at odd primes*, Amer. J. Math. **105** (1983), 1067–1105.

[Sul70] D. Sullivan, *A general problem*, p. 230, Manifolds, Amsterdam, Lecture Notes in Math. **197**, Springer, Berlin, 1970.

[Sul05] Dennis P. Sullivan, *Geometric topology: localization, periodicity and Galois symmetry*, *K*-Monographs in Mathematics **8**, Springer, Dordrecht, 2005 (reprint of 1970 MIT notes. with a preface by Andrew Ranicki).

[SY07] J. Schürmann and S. Yokura, *A survey of characteristic classes of singular spaces*, Singularity Theory, World Scientific, 2007, pp. 865–952.

[Tho58] R. Thom, *Les classes caractéristiques de Pontrjagin des variétés triangulées*, Symposium Internacional de Topologia Algebraica, La Universidad Nacional Autonoma de Mexico y la Unesco, 1958, pp. 54–67.

[Tur90] V. G. Turaev, *Three-dimensional Poincaré complexes: Homotopy classification and splitting*, Russ. Acad. Sci., Sb. **67** (1990), 261–282.

[TW79] L. Taylor and B. Williams, *Surgery spaces: Formulae and structure*, Proceedings 1978 Waterloo Algebraic Topology Conference, Lecture Notes in Math. **741**, Springer, 1979, pp. 170–195.

[Wei94] S. Weinberger, *The topological classification of stratified spaces*, Chicago Lectures in Math., Univ. of Chicago Press, Chicago, 1994.

[Whi40] J. H. C. Whitehead, *On C^1-complexes*, Ann. of Math. **41** (1940), 809–824.

[WJ66] R. E. Williamson Jr., *Cobordism of combinatorial manifolds*, Ann. of Math. **83** (1966), 1–33.

[Woo08] J. Woolf, *Witt groups of sheaves on topological spaces*, Comment. Math. Helv. **83**:2 (2008), 289–326.

MARKUS BANAGL
MATHEMATISCHES INSTITUT
UNIVERSITÄT HEIDELBERG
IM NEUENHEIMER FELD 288
69120 HEIDELBERG
GERMANY
banagl@mathi.uni-heidelberg.de

Topology of Stratified Spaces
MSRI Publications
Volume **58**, 2011

Intersection homology Wang sequence

FILIPP LEVIKOV

ABSTRACT. We prove the existence of a Wang-like sequence for intersection homology. A result is given on vanishing of the middle dimensional intersection homology group of "generalized Thom spaces", which naturally occur in the decomposition formula of S. Cappell and J. Shaneson. Based upon this result, consequences for the signature are drawn.

For non-Witt spaces X, signature and L-classes are defined via the hyper-cohomology groups $\mathcal{H}^i(X; \mathbf{IC}^{\bullet}_{\mathcal{L}})$, introduced in [Ban02]. A hypercohomology Wang sequence is deduced, connecting $\mathcal{H}^i(-; \mathbf{IC}^{\bullet}_{\mathcal{L}})$ of the total space with that of the fibre. Also here, a consequence for the signature under collapsing sphere-singularities is drawn.

1. Introduction

The goal of this article is to add to the intersection homology toolkit another useful long exact sequence. In [Wan49], H. C. Wang, calculating the homology of the total space of a fibre bundle over a sphere, actually proved an exact sequence, which is named after him today. It is a useful tool for dealing with fibre bundles over spheres and it is natural to ask: Is there a Wang sequence for intersection homology?

Given an appropriate notion of a stratified fibration, the natural framework for dealing with a question of the kind above would be an intersection homology analogue of a Leray–Serre spectral sequence. Greg Friedman has investigated this and established an appropriate framework in [Fri07]. For a simplified setting of a stratified bundle, however, i.e., a locally trivial bundle over a manifold with a stratified fibre, it seems more natural to explore the hypercohomology spectral sequence directly. In the following we are going to demonstrate this approach.

Mathematics Subject Classification: 55N33.
Keywords: intersection homology, Wang sequence, signature.

Section 3 is a kind of foretaste of what is to come. We prove the monodromy case by hand using only elementary intersection homology and apply it to calculate the intersection homology groups of neighbourhoods of circle singularities with toric links in a 4-dimensional pseudomanifold.

We recall the construction of induced maps in Section 4.1. Because of their central role in the application, the Cappell–Shaneson decomposition formula is explained in Section 4.2. Section 5 contains a proof of the Wang sequence for fibre bundles over simply connected spheres. It is shown that under a certain assumption the middle-dimensional middle perversity intersection homology of generalized Thom spaces of bundles over spheres vanish. The formula of Cappell and Shaneson then implies, that in this situation the signature does not change under the collapsing of the spherical singularities.

In Section 6, we demonstrate a second, concise proof — this is merely the sheaf-theoretic combination of the relative long exact sequence and the suspension isomorphism. However, this proof is mimicked in Section 7 to derive a Wang-like sequence for hypercohomology groups $\mathcal{H}^*(X; \mathbf{S}^\bullet)$ with values in a self-dual perverse sheaf complex $\mathbf{S}^\bullet \in SD(X)$. In Section 8, finally, together with Novikov additivity, this enables us to identify situations when collapsing spherical singularities in non-Witt spaces does not change the signature.

2. Basic notions

We will work in the framework of [GM83]. In the following $X = X_n \supset X_{n-2} \supset \cdots \supset X_0 \supset X_{-1} = \varnothing$ will denote an oriented *n-dimensional stratified topological pseudomanifold*. The intersection homology groups of X with respect to perversity $\bar{p}$ are denoted by $IH_i^{\bar{p}}(X)$, and the analogous compact-support homology groups by $IH_i^{c\,\bar{p}}(X)$. The indexing convention is also that of [GM83]. Most of the fibre bundles to be considered in the following are going to be *stratified* bundles in the following sense (see also [Fri07, Definition 5.6]):

DEFINITION 2.1. A projection $E \to B$ to a manifold is called a stratified bundle if for each point $b \in B$ there exist a neighbourhood $U \subset B$ and a stratum-preserving trivialization $p^{-1}(U) \cong U \times F$, where F is a topological stratified pseudomanifold.

We will also restrict the automorphism group of F to stratum preserving automorphisms and work with the corresponding fibre bundles in the usual sense. Since we will basically need the local triviality, Definition 2.1 is mostly sufficient. When we pass to applications for Whitney stratified pseudomanifolds, however, the considered bundles will actually be fibre bundles — this follows from the theory of Whitney stratifications. A stratification of the fibre induces an obvious stratification of the total space with the same *l*-codimensional links, namely by E_{k+n-l} — the total spaces of bundles with fibre F_{k-l} and n the dimension of B.

3. Mapping torus

PROPOSITION 3.1 (INTERSECTION HOMOLOGY WANG SEQUENCE FOR S^1). *Let* $F = F_n \supset F_{n-2} \supset \cdots \supset F_0$ *be a topological stratified pseudomanifold,* $\phi : F \to F$ *a stratum and codimension preserving automorphism, i.e., a stratum preserving homeomorphism with stratum preserving inverse such that both maps respect the codimension. Let* M_ϕ *be the mapping torus of* ϕ*, i.e., the quotient space* $F \times I / (y, 1) \sim (\phi(y), 0)$*. Denote by* $i : F = F \times 0 \hookrightarrow M_\phi$ *the inclusion. Then the sequence*

$$\cdots \longrightarrow IH_k^{c\bar{p}}(F) \xrightarrow{\mathrm{id} - \phi_*} IH_k^{c\bar{p}}(F) \xrightarrow{i_*} IH_k^{c\bar{p}}(M_\phi) \xrightarrow{\partial} IH_{k-1}^{c\bar{p}}(F) \longrightarrow \cdots$$

is exact.

PROOF. The proof is analogous to the one for ordinary homology. Start with the quotient map $q : (F \times I, F) \to (M_\phi, F)$ and look at the corresponding diagram of long exact sequences of pairs. The boundary of $F \times I$ is a codimension 1 stratum and hence not a pseudomanifold. We have either to introduce the notion of a pseudomanifold with boundary here or work with intersection homology for cs-sets [Kin85; HS91]. However, we can also manage with a work-around: Define

$$I_\varepsilon := (-\varepsilon, 1 + \varepsilon), \quad \partial I_\varepsilon := (-\varepsilon, \varepsilon) \cup (1 - \varepsilon, 1 + \varepsilon), \quad F_\varepsilon := F \times (-\varepsilon, \varepsilon).$$

We extend the identification $(y, 1) \sim (\phi(y), 0)$ to $F \times I_\varepsilon$ by introducing the quotient map $q : F \times I_\varepsilon \to M_\phi$,

$$q(y, t) = \begin{cases} (\phi^{-1}(y), 1 + t) & \text{if } t \in (-\varepsilon, 0] \\ (y, t) & \text{if } t \in (0, 1) \\ (\phi(y), t - 1) & \text{if } t \in [1, 1 + \varepsilon). \end{cases}$$

Evidently, $M_\phi = q(F \times I_\varepsilon)$. Now $F \times \partial I_\varepsilon = (F \times \partial I_\varepsilon)_{n+1} \supset (F \times \partial I_\varepsilon) \supset \cdots \supset (F \times \partial I_\varepsilon)_0$ is an open *sub*pseudomanifold of $F \times I_\varepsilon$ and $F = F_n \supset F_{n-2} \supset \cdots \supset F_0$ sits normally nonsingular in M_ϕ. Hence the inclusions induce morphisms on intersection homology and we get a morphism of the corresponding exact sequences of pairs:

$$\cdots \xrightarrow{0} IH_k^{c\bar{p}}(F \times I_\varepsilon, F \times \partial I_\varepsilon) \xrightarrow{\partial} IH_{k-1}^{c\bar{p}}(F \times \partial I_\varepsilon) \xrightarrow{j_*} IH_{k-1}^{c\bar{p}}(F \times I_\varepsilon) \xrightarrow{0} \cdots$$
$$\Big\downarrow q_* \qquad\qquad\qquad \Big\downarrow q_* \qquad\qquad\qquad \Big\downarrow q_*$$
$$\cdots \xrightarrow{} IH_k^{c\bar{p}}(M_\phi, F_\varepsilon) \xrightarrow{\partial} IH_{k-1}^{c\bar{p}}(F_\varepsilon) \xrightarrow{i_*} IH_{k-1}^{c\bar{p}}(M_\phi) \xrightarrow{} \cdots$$

The "boundary" of $F \times I_\varepsilon$ is the disjoint union of two components of the form $F \times \mathbb{R}$, so j_* is surjective and the outer arrows are zero maps. The connecting morphism ∂ is injective and therefore an isomorphism onto its image, i.e., onto

$$\ker j_*$$

$$= \{(\alpha, \beta) \mid \alpha \in IH_k^{c\bar{p}}(F \times (-\varepsilon, +\varepsilon)), \ \beta \in IH_k^{c\bar{p}}(F \times (1-\varepsilon, 1+\varepsilon)), \ [\alpha + \beta] = 0\}$$

$$= \{(\alpha, -\alpha)\} \cong IH_k^{c\bar{p}}(F \times \mathbb{R}) \cong IH_k^{c\bar{p}}(F).$$

The middle q_* maps $(\alpha, -\alpha)$ to $(\alpha - \phi_*(\alpha)) \in IH_k^{c\bar{p}}(F_\varepsilon) \cong IH_k^{c\bar{p}}(F)$. Since q commutes with ∂, one has

$$\partial \circ q_* \circ \partial|^{-1} = q_*|_{\ker j_* \cong IH_k^{c\bar{p}}(F)} = \mathrm{id} - \phi_*.$$

Hence, we have the diagram

$$
\begin{array}{ccccccc}
\cdots \to IH_k^{c\bar{p}}(F) & \xrightarrow{\ q_*|_{\ker j_*}\ } & IH_k^{c\bar{p}}(F) & \xrightarrow{i_*} & IH_k^{c\bar{p}}(M_\phi) & \to & IH_k^{c\bar{p}}(M_\phi, F_\varepsilon) \\
\| \wr & & \uparrow & & & & \\
\ker j_* & & & & & & \\
\wr \downarrow \partial|^{-1} & & \partial & & & & \\
IH_{k+1}^{c\bar{p}}(F \times I_\varepsilon, F \times \partial I_\varepsilon) & \xrightarrow{q_*} & IH_{k+1}^{c\bar{p}}(M_\phi, F_\varepsilon) & & & &
\end{array}
$$

where the top sequence is exact and the bottom square is commutative. As in ordinary homology one can show that $q_* : IH_k^{c\bar{p}}(F \times I_\varepsilon, F \times \partial I_\varepsilon) \to IH_k^{c\bar{p}}(M_\phi, F_\varepsilon)$ is an isomorphism. Finally, observe that on the right hand side $IH_k^{c\bar{p}}(M_\phi, F_\varepsilon) \cong IH_{k-1}^{c\bar{p}}(F)$ via $\partial \circ q_*^{-1}$. $\qquad \square$

Let $X = X_4 \supset X_1 \supset X_0$ be a compact stratified pseudomanifold with $X_0 = \varnothing$. Then, the stratum of codimension 3 is just a disjoint union of circles $X_1 = S^1 \sqcup \cdots \sqcup S^1$. If we assume X to be PL, the link L at a point $p \in X_1$ is independent of p within a connective component of X_1. Furthermore, in X_4, there is a neighbourhood U of the circle containing p, which is a fibre bundle over S^1 and hence homeomorphic to the mapping torus M_ϕ with

$$\phi : \mathring{c}(L) \xrightarrow{\cong} \mathring{c}(L).$$

Putting this data together and using the Wang sequence we can compute $IH_k^{c\bar{p}}(U)$, with U a neighbourhood of $X_1 \subset X_4$. The group $IH_k^{c\bar{p}}(X)$ can then be computed via the Mayer–Vietoris sequence.

In this section we restrict ourselves to the case of L being a torus T^2. While the orientation preserving mapping class group of the torus is known to be $\mathrm{SL}(2; \mathbb{Z})$, we have to make the following restriction on its cone: In the following, we look only at those automorphisms $\phi : \mathring{c}(T^2) \to \mathring{c}(T^2)$ which are induced by an automorphism of the underlying torus $\psi : T^2 \xrightarrow{\cong} T^2$.[1] It is given by a matrix $\alpha \in SL(2; \mathbb{Z})$ and by abuse of notation we will again write α for this torus automorphism.

Defining ϕ_k to be the map $IH_k^{c\bar{p}}(\mathring{c}(T^2)) \xrightarrow{\text{id} - (\mathring{c}(\alpha))_*} IH_k^{c\bar{p}}(\mathring{c}(T^2))$, we obtain the sequence

$$\cdots \longrightarrow IH_k^{c\bar{p}}(\mathring{c}(T^2)) \xrightarrow{\phi_k}$$
$$IH_k^{c\bar{p}}(\mathring{c}(T^2)) \xrightarrow{i_*} IH_k^{c\bar{p}}(M_\alpha) \xrightarrow{\partial} IH_{k-1}^{c\bar{p}}(\mathring{c}(T^2)) \longrightarrow \cdots$$

For the open cone we have

$$IH_k^{c\bar{p}}(\mathring{c}(T^2)) = \begin{cases} IH_0^{c\bar{p}}(T^2) & \text{for } k = 0, \\ IH_1^{c\bar{p}}(T^2) & \text{for } \bar{p} = \bar{0} \text{ and } k = 1, \\ 0 & \text{else.} \end{cases}$$

Clearly, $IH_k^{c\bar{p}}(M_\alpha) = 0$ for $k \geq 3$. Now examine the nontrivial part of the sequence

$$0 \longrightarrow IH_2^{c\bar{p}}(M_\alpha) \xrightarrow{\partial} IH_1^{c\bar{p}}(\mathring{c}(T^2)) \xrightarrow{\phi_1} IH_1^{c\bar{p}}(\mathring{c}(T^2)) \xrightarrow{i_*}$$
$$IH_1^{c\bar{p}}(M_\alpha) \xrightarrow{\partial} IH_0^{c\bar{p}}(\mathring{c}(T^2)) \xrightarrow{\phi_0} IH_0^{c\bar{p}}(\mathring{c}(T^2)) \xrightarrow{i_*} IH_0^{c\bar{p}}(M_\alpha) \longrightarrow 0.$$

Since $\mathring{c}(\alpha)_0$ maps a point to a point, clearly $\mathring{c}(\alpha)_0 = \text{id}$, hence $\phi_0 = \text{id} - \text{id} = 0$. It follows that $IH_0^{c\bar{p}}(M_\alpha) = \mathbb{Z}$.

Note that the only possible perversities in this example are $\bar{0}$ and $\bar{t}$. So far we have not distinguished between them. Due to $IH_1^{c\bar{t}}(\mathring{c}(T^2)) = 0$ there is $IH_2^{c\bar{t}}(M_\alpha) = 0$ and $IH_1^{c\bar{t}}(M_\alpha) = \mathbb{Z}$. For the zero perversity, we have

$$0 \longrightarrow IH_2^{c\bar{0}}(M_\alpha) \xrightarrow{\partial_*} IH_1^{c\bar{0}}(\mathring{c}(T^2)) \xrightarrow{\phi_1} IH_1^{c\bar{0}}(\mathring{c}(T^2)) \longrightarrow IH_1^{c\bar{0}}(M_\alpha) \longrightarrow$$
$$\| \wr$$
$$\mathbb{Z} \oplus \mathbb{Z} \qquad\qquad IH_0^{c\bar{0}}(\mathring{c}(T^2)) \longrightarrow 0$$

The group $IH_1^{c\bar{0}}(\mathring{c}(T^2))$ is isomorphic to $H_1(T^2)$ and is therefore generated by the corresponding homology classes of the torus. Hence, $(\mathring{c}(\alpha))_*$ is just the matrix α. If $\alpha = \text{id}$, $\phi_1 = 0$ and $IH_2^{c\bar{0}}(M_\alpha) \cong \mathbb{Z} \oplus \mathbb{Z}$. In the general case, $IH_2^{c\bar{0}}(M_\alpha)$ is isomorphic to $\text{im } \partial_* = \ker \phi_1$. We examine the determinant

---

[1] I believe that in the PL context this does not constitute a real restriction. With [Hud69, Theorem 3.6C] we can find an admissible triangulation of $\mathring{c}(T^2)$, such that ϕ becomes simplicial. Furthermore, it is not difficult to see, that the simplicial link of the cone point $L(c)$ is preserved under ϕ. Since the geometric realization of $L(c)$ is a torus, we get a candidate for ψ. By linearity, every slice between $L(c)$ and the cone point c is mapped by ψ. If we could extend the argument to the rest of $\mathring{c}(T^2)$ the goal would be achieved.

of the matrix: $\det \phi_1 = \det(\mathrm{id} - \alpha) = p_\alpha(1)$, where $p_\alpha(t)$ is the characteristic polynomial of α, which is

$$
\begin{aligned}
p_\alpha(t) = \det\left(t \,\mathrm{id} - \begin{pmatrix} a_{11} & a_{12} \\ a_{21} & a_{22} \end{pmatrix}\right) &= (t - a_{11})(t - a_{22}) - a_{21}a_{12} \\
&= t^2 - (a_{11} + a_{22})t + (a_{11}a_{22} - a_{21}a_{12}) \\
&= t^2 - \mathrm{tr}\,\alpha\, t + \det \alpha \\
&= t^2 - \mathrm{tr}\,\alpha\, t + 1.
\end{aligned}
$$

Here, $\mathrm{tr}\,\alpha$ is the trace of $\alpha \in \mathrm{SL}(2; \mathbb{Z})$. Thus we have $\det \phi_1 = 2 - \mathrm{tr}\,\alpha$ and get

$$
IH_2^{c\bar{0}}(M_\alpha) \cong \begin{cases} \mathbb{Z} \oplus \mathbb{Z} & \text{if } \alpha = \mathrm{id}, \\ \mathbb{Z} & \text{if } \mathrm{tr}\,\alpha = 2 \text{ and } \alpha \neq \mathrm{id}, \\ 0 & \text{if } \mathrm{tr}\,\alpha \neq 2. \end{cases}
$$

Since $IH_0^{c\bar{0}}(\mathring{c}(T^2))$ is free, the sequence above reduces to a split short exact sequence

$$
0 \longrightarrow \mathrm{coker}\,\phi_1 \longrightarrow IH_1^{c\bar{0}}(M_\phi) \longrightarrow IH_0^{c\bar{0}}(\mathring{c}(T^2)) \longrightarrow 0.
$$

Hence $IH_1^{c\bar{0}}(M_\phi) \cong \mathbb{Z} \oplus \mathrm{coker}\,\phi_1$. In this final case our interest reduces to a cokernel calculation of the 2×2-matrix $\phi_1 = \mathrm{id} - \alpha$. The image $\mathrm{im}\,\phi_1 \subset \mathbb{Z} \oplus \mathbb{Z}$ is of the form $n\mathbb{Z} \oplus m\mathbb{Z}$, $n, m \in \mathbb{Z}$ and so every group $\mathbb{Z} \oplus \mathbb{Z}/n\mathbb{Z} \oplus \mathbb{Z}/m\mathbb{Z}$ can be realized as $IH_1^{c\bar{0}}(M_\phi)$. In particular a torsion intersection homology group may appear. Using $\det \phi_1 = \det(\mathrm{id} - \alpha) = 2 - \mathrm{tr}\,\alpha$ as above, we immediately see that

$$
\mathrm{coker}\,\phi_1 \cong \begin{cases} \mathbb{Z} \oplus \mathbb{Z} & \text{if } \alpha = \mathrm{id}, \\ 0 & \text{if } \mathrm{tr}\,\alpha = 1, 3. \end{cases}
$$

Summarizing all these results we get:

PROPOSITION 3.2. *Let M_α be the mapping torus over the open cone $\mathring{c}(T^2)$ of a torus glued via $\alpha : T^2 \xrightarrow{\cong} T^2$. Then its intersection homology groups are*

$$
IH_k^{c\bar{1}}(M_\alpha) \cong \begin{cases} \mathbb{Z} & \text{if } k = 0, 1, \\ 0 & \text{if } k \geq 2, \end{cases}
$$

$$
IH_k^{c\bar{0}}(M_\alpha) \cong \begin{cases} \mathbb{Z} & \text{if } k = 0, \\ \mathbb{Z} \oplus \mathbb{Z} \oplus \mathbb{Z} & \text{if } k = 1 \text{ and } \alpha = \mathrm{id} \\ \mathbb{Z} & \text{if } k = 1 \text{ and } \mathrm{tr}\,\alpha = 1, 3, \\ \mathbb{Z} \oplus \mathrm{coker}(\mathrm{id} - \alpha) & \text{if } k = 1 \text{ (in general)}, \\ \mathbb{Z} & \text{if } k = 2, \mathrm{tr}\,\alpha = 2 \text{ and } \alpha \neq \mathrm{id}, \\ \mathbb{Z} \oplus \mathbb{Z} & \text{if } k = 2 \text{ and } \alpha = \mathrm{id}, \\ 0 & \text{if } k = 2 \text{ and } \mathrm{tr}\,\alpha \neq 2, \\ 0 & \text{if } k \geq 3. \end{cases}
$$

EXAMPLE 3.3. Let X be the fibre bundle over S^1 with fibre ΣT^2 and monodromy $\alpha \in \mathrm{SL}_2(\mathbb{Z})$. We assume $\Sigma\alpha$ to be orientation preserving, i.e., the suspension points are fixed under it. The space X has a filtration $X_4 \supset X_1 = S^1 \sqcup S^1$ and our situation applies. Let α be $\left(\begin{smallmatrix} 2 & -1 \\ 1 & 0 \end{smallmatrix}\right)$. First, using the ordinary Wang sequence, we compute the homology of the total space E of the fibre bundle $T^2 \to E \to S^1$ with the same monodromy:

$$0 \to H_3(E) \xrightarrow{\partial} H_2(T^2) \xrightarrow{\mathrm{id}-\alpha_*} H_2(T^2) \to H_2(E) \xrightarrow{\partial} H_1(T^2)$$
$$\xrightarrow{\mathrm{id}-\alpha_*} H_1(T^2) \to H_1(E) \xrightarrow{\partial} H_0(T^2) \xrightarrow{\mathrm{id}-\alpha_*} H_0(T^2) \to H_0(E) \to 0.$$

In degrees 2 and 0, the map α_* is the identity, so we substitute zeros for $\mathrm{id}-\alpha_*$ to see that $H_3(E) \cong \mathbb{Z} \cong H_0(E)$. In degree 1, the map α_* is just the matrix α. Using $\mathrm{im}\,\partial_2 = \ker(\mathrm{id}-\alpha_*) \cong \mathbb{Z}$, we get the sequence

$$0 \to H_2(T^2) \to H_2(E) \xrightarrow{\partial} \mathbb{Z} \to 0,$$

which yields $H_2(E) \cong \mathbb{Z} \oplus \mathbb{Z}$ and

$$0 \to \mathrm{coker}(\mathrm{id}-\alpha_*) \to H_1(E) \to \mathbb{Z} \to 0.$$

It follows by $\mathrm{coker}(\mathrm{id}-\alpha_*) \cong \mathbb{Z}$ that $H_1(E) \cong \mathbb{Z} \oplus \mathbb{Z}$. Let us now compute the intersection homology groups of X via the Mayer–Vietoris sequence. The neighbourhoods of the two S^1 are of the desired form, i.e., mapping tori over $\mathring{c}(T^2)$ and their intersection is a fibre bundle over S^1 with fibre $T^2 \times \mathbb{R}$, so that the intersection homology groups are just $H_*(E)$ from above. Looking at the exact sequence

$$0 \longrightarrow IH_4^{c\bar{p}}(X) \to IH_3^{c\bar{p}}(E) \to IH_3^{c\bar{p}}(M_\alpha) \oplus IH_3^{c\bar{p}}(M_\alpha) \to \cdots,$$
$$\qquad\qquad\qquad\quad \| \qquad\qquad\qquad \|$$
$$\qquad\qquad\qquad\quad 0 \qquad\qquad\qquad 0$$

we see that $IH_4^{c\bar{p}}(X) \cong H_3(E) \cong \mathbb{Z}$. In degree 0 the inclusion of E induces an injection on homology, i.e.,

$$0 \longrightarrow H_0(E) \to IH_0^{c\bar{p}}(M_\alpha) \oplus IH_0^{c\bar{p}}(M_\alpha) \to IH_0^{c\bar{p}}(X) \to 0,$$

and $IH_0^{c\bar{p}}(X) \cong \mathbb{Z}$ as it should be. Turning to the interesting degrees, we look at $\bar{p} = (0, 1, \ldots)$ first. Due to $IH_2^{c\bar{t}}(M_\alpha) = 0$, there is $IH_2^{c\bar{t}}(X) \cong \ker(i_* \oplus i_*)_1$. Finally $IH_1^{c\bar{t}}(X)$ is isomorphic to the cokernel of the inclusion $(i_* \oplus i_*)_1 : H_1(E) \to IH_1^{c\bar{t}}(M_\alpha) \oplus IH_1^{c\bar{t}}(M_\alpha) \cong \mathbb{Z} \oplus \mathbb{Z}$, which is the diagonal map; hence $IH_1^{c\bar{t}}(X) \cong \mathbb{Z}$ and $IH_2^{c\bar{t}}(X) \cong \ker(i_* \oplus i_*)_1 \cong \mathbb{Z}$. Similarly, for $\bar{p} = (0, 0, \ldots)$ we have $IH_3^{c\bar{0}}(X) \cong \ker(i_* \oplus i_*)_2 \cong \mathbb{Z}$; with $IH_1^{c\bar{0}}(M_\alpha) \cong \mathbb{Z} \oplus \mathbb{Z}$ it follows $IH_1^{c\bar{0}}(X) \cong \mathbb{Z} \oplus \mathbb{Z}$. And $IH_2^{c\bar{0}}(X) \cong \mathrm{coker}(i_* \oplus i_*)_2 \cong \mathbb{Z}$. Because all the groups are free, the duality is already seen working with integral coefficients, especially

$$IH_3^{c\bar{0}}(X) \cong IH_1^{c\bar{t}}(X) \cong \mathbb{Z} \oplus \mathbb{Z},$$
$$IH_1^{c\bar{0}}(X) \cong IH_3^{c\bar{t}}(X) \cong \mathbb{Z}.$$

4. Some more advanced tools

4A. Normally nonsingular maps. Intersection homology is not a functor on the full subcategory of **Top** consisting of pseudomanifolds, since induced maps do not exist in general. However, on the category of topological pseudomanifolds and normally nonsingular maps, intersection homology is a bivariant theory in the sense of [FM81]. This fact is often suppressed. Since most of the maps which we will encounter are normally nonsingular, we recall in this section how induced maps are constructed. See particularly [GM83, 5.4].

DEFINITION 4.1. A map $f : Y \to X$ between two pseudomanifolds is called normally nonsingular (nns) of relative dimension $c = c_1 - c_2$ if it is a composition of a nns inclusion of dimension c_1 — meanining that Y is sitting in a c_1-dimensional tubular neighbourhood in the target — and a nns projection, i.e., a bundle projection with c_2-dimensional manifold fibre.

EXAMPLE 4.2. An open inclusion $U \hookrightarrow X$ is normally nonsingular. The inclusion of the fibre $F = b \times F \hookrightarrow E$, where E is fibred over a manifold is normally nonsingular. The projection $\mathbb{R}^n \times X \to X$ is normally nonsingular.

PROPOSITION 4.3 [GM83, 5.4.1, 5.4.2]. *Let $f : Y \to X$ be normally nonsingular of codimension c. Then there are isomorphisms*

$$f^* \mathbf{IC}_{\bar{p}}^{\bullet}(X) \cong \mathbf{IC}_{\bar{p}}^{\bullet}(Y)[c] \qquad and \qquad f^! \mathbf{IC}_{\bar{p}}^{\bullet}(X) \cong \mathbf{IC}_{\bar{p}}^{\bullet}(Y).$$

DEFINITION 4.4. If $f : Y \to X$ is a proper normally nonsingular map of codimension c, we have induced homomorphisms

$$f_* : IH_k^{\bar{p}}(Y) \to IH_k^{\bar{p}}(X) \qquad \text{and} \qquad f^* : IH_k^{\bar{p}}(X) \to IH_{k-c}^{\bar{p}}(Y).$$

They are constructed by considering the adjunction morphisms of the adjoint pairs $(Rf_!, f^!)$ and (f^*, Rf_*),

$$Rf_! f^! \mathbf{IC}_{\bar{p}}^{\bullet}(X) \to \mathbf{IC}_{\bar{p}}^{\bullet}(X) \qquad \text{and} \qquad \mathbf{IC}_{\bar{p}}^{\bullet}(X) \to Rf_* f^* \mathbf{IC}_{\bar{p}}^{\bullet}(X),$$

by combining them with the proposition above and by finally applying hypercohomology.

We will also need induced maps on intersection homology with compact supports, which are not discussed in [GM83]. The above construction of Goresky and MacPherson works equally well for $IH_*^{c\bar{p}}$. If f is not proper, the map

f^* still exists. For the case of compact supports, $f_! : IH_k^{c\bar{p}}(Y) \to IH_k^{c\bar{p}}(X)$ can be constructed in the same manner. These different maps are listed in the following table — note that only f_* and f^* for proper f are explicitly mentioned in [GM83, 5.4].

f proper	f not proper
$f_* : IH_k^{\bar{p}}(Y) \to IH_k^{\bar{p}}(X)$	
$f_! : IH_k^{c\bar{p}}(Y) \to IH_k^{c\bar{p}}(X)$	$f_! : IH_k^{c\bar{p}}(Y) \to IH_k^{c\bar{p}}(X)$
$f^* : IH_k^{\bar{p}}(X) \to IH_{k-c}^{\bar{p}}(Y)$	$f^* : IH_k^{\bar{p}}(X) \to IH_{k-c}^{\bar{p}}(Y)$
$f^! : IH_k^{c\bar{p}}(X) \to IH_{k-c}^{c\bar{p}}(Y)$	

4B. Behaviour under stratified maps. Computing intersection homology invariants of one space out of the invariants of the other often relies on the decomposition formula of S. Cappell and J. Shaneson [CS91]. Since we will need it in the application below, we briefly recall it in this section.

Let $f : X^n \to Y^m$ be a stratified map between closed, oriented Whitney stratified sets of even relative dimension $2t = n - m$, Y having only even-codimensional strata. Let $\mathbf{S}^\bullet \in D_c^b(X)$ be a self-dual complex. Denote by $\mathcal{V}$ the set of components of pure strata of Y. For each $y \in V_y \in \mathcal{V}$, define[2]

$$E_y := f^{-1}(c\, L(y)) \cup_{f^{-1}L(y)} c\, f^{-1}(L(y)),$$

where $L(y)$ is the link of the stratum component V_y containing y. If y lies in the top stratum, we set $E_y = f^{-1}(y)$. We have the inclusions

$$E_y \overset{i_y}{\hookleftarrow} f^{-1}(\mathring{N}(y)) \overset{\rho_y}{\hookrightarrow} X,$$

where $N(y)$ is the normal slice of y. Note that $N(y) \cong c\, L$ and $\partial N(y) \cong L(y)$ (see [GM88] or [Ban07, 6.2]). Define now the complex

$$\mathbf{S}^\bullet(y) = \tau_{\leq -c-t-1}^{\mathrm{cone}} R i_{y*} \rho_y^! \mathbf{S}^\bullet,$$

where $\tau_{\leq}^{\mathrm{cone}}$ stands for truncation over the cone point[3] of $c\, f^{-1}(L(y))$ and $2c = 2c(V) = n - \dim V$ is the codimension of V.

---

[2] Here, $c\, L$ stands for the closed cone $L \times [0, 1] / L \times \{0\}$.

[3] There is a general notion of truncation over a closed subset in [GM83, 1.14]. Let c be the cone-point. For $\mathbf{A}^\bullet \in D_c^b(E_y)$, the derived stalks are

$$H^i(\tau_{\leq p}^{\mathrm{cone}} \mathbf{A}^\bullet)_x = \begin{cases} 0 & \text{if } x = c \text{ and } i > p, \\ H^i(\mathbf{A}^\bullet)_x & \text{otherwise.} \end{cases}$$

For $V \in \mathcal{V}$, let $\mathcal{S}_f^V$ be the local coefficient system over V with stalk $(\mathcal{S}_f^V)_z = \mathcal{H}^{-c-t}(E_z; \mathbf{S}^\bullet(z))$. There is an induced nondegenerate bilinear pairing

$$\phi_z : (\mathcal{S}_f^V)_z \times (\mathcal{S}_f^V)_z \to \mathbb{R}.$$

If $\mathbf{S}^\bullet$ is the intersection chain complex $\mathbf{IC}_{\bar{m}}^\bullet(X)$, the pull-back $\rho_y^! \mathbf{IC}_{\bar{m}}^\bullet(X)$ is clearly $\mathbf{IC}_{\bar{m}}^\bullet(E_y \setminus \{c\})$. Because of the stalk vanishing of $\mathbf{IC}_{\bar{m}}^\bullet(E_y \setminus \{c\})$, the truncation $\tau_{\le -c-t-1}^{\mathrm{cone}}$ is the usual truncation $\tau_{\le -c-t-1}$ and hence

$$\mathbf{S}^\bullet(y) = \tau_{\le -c-t-1} R i_{y*} \mathbf{IC}_{\bar{m}}^\bullet(E_y \setminus \{c\}),$$

which is simply the Deligne extension $\mathbf{IC}_{\bar{m}}^\bullet(E_y)$ of $\mathbf{IC}_{\bar{m}}^\bullet(E_y \setminus \{c\})$ to the point.

Denoting by $\mathbf{IC}_{\bar{m}}^\bullet(\bar{V}; \mathcal{S}_f^V)$ the lower-middle perversity intersection chain complex on the closure of V with coefficients in the local system $\mathcal{S}_f^V$, we can now formulate the important decomposition formula of Cappell and Shaneson.

THEOREM 4.5 [CS91, Theorem 4.2]. *There is an orthogonal decomposition up to algebraic bordism of self-dual complexes of sheaves*

$$Rf_* \mathbf{S}^\bullet[-t] \sim \bigoplus_{V \in \mathcal{V}} j_* \mathbf{IC}_{\bar{m}}^\bullet(\bar{V}, \mathcal{S}_f^V)[c(V)],$$

where $j : \bar{V} \hookrightarrow Y$ *is the inclusion.*

We abstain from giving the definition of algebraic bordism here and refer to the original paper or to Chapter 8 of [Ban07]. All we need for the application is the following, where for a self-dual sheaf $\mathbf{S}^\bullet$ over X, $\sigma(X, \mathbf{S}^\bullet)$ denotes the signature of the pairing on the middle-dimensional hypercohomology induced by self-duality.

PROPOSITION 4.6. *If two self-dual complexes over* X, $\mathbf{S}_1^\bullet$ *and* $\mathbf{S}_2^\bullet$ *are (algebraically) bordant, then* $\sigma(X, \mathbf{S}_1^\bullet) = \sigma(X, \mathbf{S}_2^\bullet)$.

PROOF. See [Ban07, Cor. 8.2.5], for example. □

PROPOSITION 4.7 [CS91, 5.5]. *If, in the setting above,* $L_i(X, \mathbf{A}^\bullet)$ *denotes the* i-*th L-class of the self-dual sheaf* $\mathbf{A}^\bullet$ *over a pseudomanifold* X, *we have*

$$L_i(Y, Rf_* \mathbf{S}^\bullet[-t]) = f_* L_i(X, \mathbf{S}^\bullet).$$

THEOREM 4.8. *With the notation* $L_i(\bar{V}, \mathcal{S}_f^V)$ *for* $L_i(\bar{V}, \mathbf{IC}_{\bar{m}}^\bullet(\bar{V}; \mathcal{S}_f^V))$ *we get*

$$f_* L_i(X, \mathbf{S}^\bullet) = \sum_{V \in \mathcal{V}} j_* L_i(\bar{V}, \mathcal{S}_f^V).$$

and bearing in mind that $\sigma(X) = \varepsilon_* L_0(X)$, where ε_* is the augmentation, we conclude:

COROLLARY 4.9. $\qquad \sigma(X, \mathbf{S}^\bullet) = \sigma(Y, Rf_* \mathbf{S}^\bullet[-t]) = \sum_{V \in \mathcal{V}} \sigma(\bar{V}, \mathcal{S}_f^V).$

In the case of simply connected components of Y, all the coefficient systems become constant and using multiplicativity formulae [Ban07, 8.2.19, 8.2.20], we get:

THEOREM 4.10. *Assume each $V \in \mathcal{V}$ to be simply connected and choose a basepoint y_V for every $V \in \mathcal{V}$. Then*

$$f_* L_i(X) = \sum_{V \in \mathcal{V}} \sigma(E_{y_V}) j_* L_i(\bar{V}).$$

And finally, for the signature:

COROLLARY 4.11. $\qquad \sigma(X) = \sum_{V \in \mathcal{V}} \sigma(E_{y_V}) \sigma(\bar{V}).$

5. The general simply connected case

PROPOSITION 5.1 (WANG SEQUENCE FOR $n \geq 2$). *Let $F \longrightarrow E \xrightarrow{\pi} S^n$ be a stratified bundle (2.1) with F a topological pseudomanifold with finitely generated cohomology, $n \geq 2$. Let $j : F \hookrightarrow E$ be the inclusion.*

(i) *For intersection homology the sequence*

$$\cdots \longrightarrow IH_k^{\bar{p}}(E) \xrightarrow{j^*} IH_{k-n}^{\bar{p}}(F) \longrightarrow IH_{k-1}^{\bar{p}}(F) \xrightarrow{j_*} IH_{k-1}^{\bar{p}}(E) \longrightarrow \cdots$$

is exact.

(ii) *For intersection homology with compact supports the sequence*

$$\cdots \longrightarrow IH_k^{c\bar{p}}(E) \xrightarrow{j^*} IH_{k-n}^{c\bar{p}}(F) \longrightarrow IH_{k-1}^{c\bar{p}}(F) \xrightarrow{j_*} IH_{k-1}^{c\bar{p}}(E) \longrightarrow \cdots$$

is exact.

(iii) *These sequences are natural with respect to fibre-preserving proper normally nonsingular maps between stratified bundles over S^n, i.e., let $F' \rightarrow E' \rightarrow S^n$ be another fibre bundle such that there is a commutative triangle*

$$
\begin{array}{ccc}
E' & \xrightarrow{\;f\;} & E \\
& \searrow{\scriptstyle \pi'} \quad \swarrow{\scriptstyle \pi} & \\
& S^n &
\end{array}
$$

with f proper normally nonsingular, then there is a commutative diagram of the corresponding Wang-sequences induced by f — both in a covariant and a contravariant way.

PROOF. (i) We begin with the hypercohomology spectral sequence ([Bry93]) for $\mathbf{A}^{\bullet} := R\pi_* \, \mathbf{IC}^{\bullet}_{\bar{p}}(E)$, which converges to $\mathcal{H}^{p+q}(S^n, \mathbf{A}^{\bullet}) \cong IH^{\bar{p}}_{-p-q}(E)$. Let $U \subset S^n$ be an open set such that $\pi^{-1}(U) \cong U \times F$, then by 4.3

$$\mathbf{IC}^{\bullet}_{\bar{p}}(E)|_{\pi^{-1}(U)} \cong \mathbf{IC}^{\bullet}_{\bar{p}}(U \times F) \cong pr^* \, \mathbf{IC}^{\bullet}_{\bar{p}}(F)[n].$$

By IV.7.3 of [Bre97], the sheaf $\mathbf{H}^q(\mathbf{A}^{\bullet})$, being the *Leray sheaf* of the fibration, is locally constant. Hence, by the assumption $n \geq 2$, it is constant with stalk

$$\mathcal{H}^q(F, \mathbf{IC}^{\bullet}_{\bar{p}}(F)[n]) = IH^{\bar{p}}_{-q-n}(F).$$

Finally

$$E_2^{p,q} \cong \begin{cases} IH^{\bar{p}}_{-q-n}(F) & \text{if } p = 0 \text{ or } p = n, \\ 0 & \text{else.} \end{cases}$$

Hence $E_2 \cong \ldots \cong E_n$ and the sequence collapses at $n + 1$. Now, the proof can be finished as in the ordinary case (see [Spa66, 8.5], for instance). In order to show that $IH^{\bar{p}}_k(F) \to IH^{\bar{p}}_k(E)$ is induced by the inclusion $j : F = b_0 \times F \hookrightarrow E$, look at the fibration

$$F \to b_0 \times F \xrightarrow{\pi'} b_0$$

for $b_0 \in S^n$ the north pole. We have a commutative diagram

$$
\begin{array}{ccc}
b_0 \times F & \xrightarrow{\pi'} & b_0 \\
\downarrow{\scriptstyle j} & & \downarrow{\scriptstyle j_0} \\
E & \xrightarrow{\pi} & S^n .
\end{array}
$$

For $R(j_0\pi')_* \, \mathbf{IC}^{\bullet}_{\bar{p}}(b_0 \times F)$ there is a corresponding spectral sequence converging to

$$\mathcal{H}^{p+q}(b_0 \times F, \mathbf{IC}^{\bullet}_{\bar{p}}(b_0 \times F)) \cong IH^{\bar{p}}_{-p-q}(F).$$

If we start with

$$R\pi_* R j_* j^! \, \mathbf{IC}^{\bullet}_{\bar{p}}(E) \to R\pi_* \, \mathbf{IC}^{\bullet}_{\bar{p}}(E)$$

and use the commutative square above, we get a morphism

$$R(j_0\pi')_* \, \mathbf{IC}^{\bullet}_{\bar{p}}(b_0 \times F) \to R\pi_* \, \mathbf{IC}^{\bullet}_{\bar{p}}(E).$$

This induces

$$IH^{\bar{p}}_i(F) \to IH^{\bar{p}}_i(E),$$

which is j_* by construction (cf. 4.3 and 4.4). The E_2-term of the spectral sequence associated to $R(j_0\pi')_* \, \mathbf{IC}^{\bullet}_{\bar{p}}(b_0 \times F)$ is

$$E_2^{p,q} = H^p(S^n, \mathbf{H}^q(R(j_0\pi')_* \, \mathbf{IC}^{\bullet}_{\bar{p}}(b_0 \times F))).$$

Since here j_{0*} is just extension by zero the group on the right is isomorphic to $H^p(b_0, IH^{\bar{p}}_{-q}(F))$. For both sequences, the differentials $E_r^{n,q} \to E_r^{n+r,q-r+1}$ are zero for all $r \geq 2$, and so we have epimorphisms $E_r^{n,q} \twoheadrightarrow E_\infty^{n,q}$. Finally by the commutative diagram (denoting by $'$ the terms of the spectral sequence associated to $R(j_0\pi')_* \mathbf{IC}^\bullet_{\bar{p}}(b_0 \times F)$)

$$
\begin{array}{ccccc}
E_n^{n,-i-n} & \longrightarrow & E_\infty^{n,-i-n} & \longrightarrow & \mathcal{H}^q(E, \mathbf{IC}^\bullet_{\bar{p}}(E)) \\
\big\Vert\wr & & \big\uparrow & & \big\uparrow{\scriptstyle j_*} \\
E_n^{'n,-i-n} & \stackrel{\cong}{\longrightarrow} & E_\infty^{'n,-i-n} & \stackrel{\cong}{\longrightarrow} & \mathcal{H}^q(F, \mathbf{IC}^\bullet_{\bar{p}}(F))
\end{array}
$$

we deduce that the upper composition is j_*, as stated. A very similar argument works for $IH^{\bar{p}}_k(E) \to IH^{\bar{p}}_{k-n}(F)$.

(ii) Consider the hypercohomology spectral sequence for the complex $\mathbf{B}^\bullet := R\pi_! \mathbf{IC}^\bullet_{\bar{p}}(E)$. The main argument is as before. The spectral sequence converges to $\mathcal{H}^{p+q}(S^n, \mathbf{B}^\bullet) = IH^{c\bar{p}}_{-p-q}(E)$. Being the qth derived functor of $\pi_!$, the stalk of the Leray sheaf $\mathbf{H}^q(\mathbf{B}^\bullet)$ is $\mathcal{H}^q_c(F, \mathbf{IC}^\bullet_{\bar{p}}(E)) \cong IH^{c\bar{p}}_{-q-n}(F)$ (see [Bor84, VI, 2.7], for instance). Hence the E_2-terms are:

$$E_2^{0,q} \cong IH^{c\bar{p}}_{-q-n}(F),$$
$$E_2^{n,q-n+1} \cong IH^{c\bar{p}}_{-q-1}(F).$$

These yield the second sequence. The proof that the maps involved in this sequence are j_* and j^* follows as in (i).

(iii) Again, we use the fact that the hypercohomology spectral sequence is natural with respect to morphisms of sheaves over the base space.

For the covariant case, we have to construct

$$f_* : R\pi'_* \mathbf{IC}^\bullet_{\bar{p}}(E') \to R\pi_* \mathbf{IC}^\bullet_{\bar{p}}(E) \quad \text{or} \quad f_* : R\pi'_! \mathbf{IC}^\bullet_{\bar{p}}(E') \to R\pi_! \mathbf{IC}^\bullet_{\bar{p}}(E),$$

as the case may be producing morphisms between the terms of the Wang sequences. Then the corresponding maps will commute. Take the adjunction morphism

$$Rf_! f^! \mathbf{IC}^\bullet_{\bar{p}}(E) \to \mathbf{IC}^\bullet_{\bar{p}}(E),$$

apply $R\pi_*$ and use functoriality. The case of $IH^{c\bar{p}}_*$ is analogous. Since f is proper, we have $Rf_! = Rf_*$. Observe that when working with intersection homology with compact supports f need not be proper[4]! In the contravariant case, we proceed as above, using the other adjunction morphism

$$\mathbf{IC}^\bullet_{\bar{p}}(E) \to Rf_* f^* \mathbf{IC}^\bullet_{\bar{p}}(E). \qquad \square$$

---

[4] See also the comment at the end of Section 4A.

Now we are going to use this sequence in a concrete computation.

PROPOSITION 5.2. *Let* $F^k \to E \xrightarrow{\pi} S^n$ *be a locally trivial fibre bundle with* F *a topological pseudomanifold,* $n \geq 2, n + k + 1$ *even. Define* $M := c\,E \cup_E \bar{E}$, *where* $\bar{E}$ *is the total space of the induced — meaning that the structure group acts levelwise — fibre bundle*

$$c\,F \to \bar{E} \xrightarrow{c(\pi)} S^n.$$

Suppose further that the following condition (S) is fulfilled for the Wang sequence of E:

(S)   *the map* $j^* : IH^{c\bar{m}}_{(n+k+1)/2}(E) \twoheadrightarrow IH^{c\bar{m}}_{(-n+k+1)/2}(F)$ *is surjective.*

Then

$$IH^{c\bar{m}}_{(n+k+1)/2}(M) = 0.$$

PROOF. Throughout the proof, the perversity shall be the lower middle perversity $\bar{m}$, unless stated otherwise. Assume for now that $n = 2b$, $k = 2a - 1$, with $a, b \geq 1$; and with the cone formula there holds

$$IH^c_i(\mathring{c}F) \cong \begin{cases} IH^c_i(F) & \text{if } i < a, \\ 0 & \text{if } i \geq a. \end{cases}$$

Using the Wang sequence of Proposition 5.1 for $\mathring{c}F \to \mathring{\bar{E}} \xrightarrow{\mathring{c}(\pi)} S^n$,

$$\cdots \to IH^c_{a+b}(\mathring{c}F) \to IH^c_{a+b}(\mathring{\bar{E}}) \to IH^c_{a-b}(\mathring{c}F) \to IH^c_{a+b-1}(\mathring{c}F) \to \cdots$$

we get

$$IH^c_{a+b}(\mathring{\bar{E}}) \cong IH^c_{a-b}(F).$$

For the cone on E, there is:

$$IH^c_i(\mathring{c}E) \cong \begin{cases} IH^c_i(E) & \text{if } i < a + b, \\ 0 & \text{if } i \geq a + b. \end{cases}$$

Now consider the Mayer–Vietoris sequence[5]

$$\cdots \to IH^c_{a+b}(E) \xrightarrow{i_{a+b}} IH^c_{a+b}(\mathring{c}E) \oplus IH^c_{a+b}(\mathring{\bar{E}}) \to IH^c_{a+b}(M) \to \cdots$$

which reduces to

$$\cdots \xrightarrow{i_{a+b}} IH^c_{a-b}(F) \longrightarrow IH^c_{a+b}(M) \longrightarrow$$
$$IH^c_{a+b-1}(E) \xrightarrow{i_{a+b-1}} IH^c_{a+b-1}(E) \oplus IH^c_{a+b-1}(\mathring{\bar{E}}) \longrightarrow \cdots$$

---

[5] To avoid pseudomanifolds with boundary, we take the open part $\mathring{\bar{E}}$ of the induced bundle $\bar{E}$ in the Mayer–Vietoris decomposition.

The map i_{a+b-1} is easily seen to be injective and due to (S), i_{a+b} is surjective. Finally, let $n = 2b + 1, k = 2a, a, b \geq 1$. By the cone formula we have:

$$IH_i^c(\overset{\circ}{c}E) \cong \begin{cases} IH_i^c(E) & i < a+b+1 \\ 0 & i \geq a+b+1 \end{cases}$$

and

$$IH_i^c(\overset{\circ}{c}F) \cong \begin{cases} IH_i^c(F) & i < a+1 \\ 0 & i \geq a+1. \end{cases}$$

Similarly, the Wang sequence yields

$$IH_{a+b+1}^c(\overset{\circ}{\tilde{E}}) \cong IH_{a-b}^c(F).$$

Now, as above the Mayer–Vietoris sequence gives

$$\xrightarrow{i_{a+b+1}} IH_{a-b}^c(F) \to IH_{a+b+1}^c(M) \to IH_{a+b}^c(E) \xrightarrow{i_{a+b-1}} IH_{a+b}^c(E) \oplus IH_{a+b}^c(\overset{\circ}{\tilde{E}})$$

where $\ker i_{a+b} = 0$ and $\operatorname{im} i_{a+b+1} = IH_{a-b}^c(F)$ due to (S). Hence,

$$IH_{a+b+1}^c(M) = 0. \qquad \square$$

COROLLARY 5.3. *If M is a Witt space and $n + k + 1$ is divisible by 4, the signature $\sigma(M)$ vanishes (of course it always vanishes if $n+k+1$ is not divisible by 4.)*

Let us now formulate an important consequence of the observations above:

PROPOSITION 5.4. *Let X be a Whitney stratified Witt space of dimension $4k$, with a disjoint union of spheres as the singular locus $\Sigma = S^{n_1} \sqcup \cdots \sqcup S^{n_l}$. Assume $n_j \geq 2$ for $1 \leq j \leq l$. Let Y be the space obtained from X by collapsing the spheres S^{n_j} to points y_j and let $f : X \to Y$ be the collapsing map. Given a fibre bundle neighbourhood of S^{n_j}, we denote by E_j the corresponding fibre bundle with fibre the link of S^{n_j}. If for all $1 \leq j \leq l$, E_j satisfies (S), the signature of X does not change under f, i.e.,*

$$\sigma(X) = \sigma(Y).$$

PROOF. Because of 4.11, we have

$$\sigma(X) = \sum_{V \in \mathcal{V}} \sigma(E_{y_V})\sigma(\bar{V})$$

where the sum is taken over all strata. When we isolate the contribution of the top stratum, this looks like

$$= \sum_{y_j} \sigma(E_{y_{j_V}})\sigma(\bar{V}) + \sigma(Y).$$

The $E_{y_j V}$, in turn, are of the form M of Proposition 5.2 and since the underlying fibrations satisfy (S), the resulting signatures vanish by 5.3. $\qquad\square$

Similarly for the L-classes we have:

PROPOSITION 5.5. *In the situation above*,

$$f_* L(X) = L(Y).$$

The following two examples show, that the introduced condition (S) is indeed fulfilled for certain fibre bundles.

EXAMPLE 5.6. In the setting above let the base sphere be of odd dimension $n = 2b + 1$. If we are interested in computing the signature of E, its dimension has to be divisible by 4 — otherwise it is trivial anyway. In this case the fibre F has even dimension $k = 2a$, so that $(k - n + 1)/2$ is odd. Thus, the vanishing of odd dimensional intersection homology of F would imply (S). See [Roy87] for examples of spaces, for which the intersection homology vanishes in odd degrees.

EXAMPLE 5.7. Let the dimension of the sphere be greater than the dimension of the fibre plus 1, i.e., $k + 1 < n$. Then $(k - n + 1)/2$ is negative and the corresponding homology group is zero, thereby (S) is fulfilled.

Since we have not studied the intersection pairing on M, the condition (S) is clearly only sufficient and not necessary. However, the following "counterexample" to the proposition is a case where (S) does not hold.

EXAMPLE 5.8. Let X be $\mathbb{C}P^2$ stratified as $\mathbb{C}P^2 \supset \mathbb{C}P^1 = S^2$ and f be the map, collapsing the 2-sphere to a point. So the target is $Y = S^4 \supset [S^2]$. Obviously, $\sigma(X) \neq \sigma(Y)$. The link of $\mathbb{C}P^1$ is a circle and the bundle we have to check (S) for is the Hopf bundle $S^3 \to S^2$. However

$$H_2(S^3) \to H_0(S^2)$$

is not onto and (S) fails.

6. A new proof

The application to the signature in the last section suggests a similar approach in the setting of spaces which no longer satisfy the Witt condition, however still posses a signature and L-classes. The suitable homology groups for defining these invariants are the hypercohomology groups $\mathcal{H}^i(-; \mathbf{IC}_{\mathcal{L}}^\bullet)$ of Banagl [Ban02]. In the next section we will establish a Wang-like exact sequence for these groups i.e., for hypercohomology with values in a self-dual sheaf complex arising from a Lagrangian structure along the odd-codimension strata. Compare also [Ban07] for a concise exposition. The proof will be modeled on another elegant proof of the Wang sequence without the usage of the spectral sequence. This will be demonstrated in the following.

Suspension isomorphisms in intersection homology are very familiar. For the functoriality, however, we would like to have explicit maps realizing these isomorphisms:

LEMMA 6.1 (SUSPENSION ISOMORPHISM). *Let F be a pseudomanifold. The inclusion* $l : F = 0 \times F \hookrightarrow \mathbb{R}^n \times F$ *induces isomorphisms*

(a) $l^* : IH_k^{\bar{p}}(\mathbb{R}^n \times F) \to IH_{k-n}^{\bar{p}}(F)$,
(b) $l_! : IH_k^{c\bar{p}}(F) \to IH_k^{c\bar{p}}(\mathbb{R}^n \times F)$.

PROOF. (a) Let $p : \mathbb{R}^n \times F \to F$ be the normally nonsingular projection. By [Bor84, V,3.13] $Rp_* \circ p^* \simeq$ id, so the adjunction morphism is an isomorphism

$$\mathbf{IC}_{\bar{p}}^{\bullet}(F) \xrightarrow{\cong} Rp_* p^* \mathbf{IC}_{\bar{p}}^{\bullet}(F) \cong Rp_* \mathbf{IC}_{\bar{p}}^{\bullet}(\mathbb{R}^n \times F)[-n].$$

Applying hypercohomology we get

$$p^* : IH_k^{\bar{p}}(F) \xrightarrow{\cong} IH_{k+n}^{\bar{p}}(\mathbb{R}^n \times F).$$

Now $p \circ l =$ id and hence $l^* \circ p^* \simeq$ id. Thereby, l^* is the inverse of p^* and the statement follows.

(b) is similar to (a), but uses the fact that $\mathcal{D}_X \mathcal{D}_X \mathbf{A}^{\bullet} \cong \mathbf{A}^{\bullet}$ for $\mathbf{A}^{\bullet} \in D_c^b(X)$, and the duality between p^* and $p^!$. □

PROPOSITION 6.2. *Let* $F \longrightarrow E \xrightarrow{\pi} S^n$ *be a stratified bundle with F a topological pseudomanifold. Denote by*

$$j : F = b_0 \times F \hookrightarrow E, \quad i : E \setminus b_0 \times F = U \times F \hookrightarrow E, \quad k : b_1 \times F \hookrightarrow E$$

the inclusions, where b_0 *is the north pole and* b_1 *the south pole. Then there are the following long exact sequences:*

$$\cdots \longrightarrow IH_k^{\bar{p}}(F) \xrightarrow{j_*} IH_k^{\bar{p}}(E) \xrightarrow{k^*} IH_{k-n}^{\bar{p}}(F) \longrightarrow IH_{k-1}^{\bar{p}}(F) \longrightarrow \cdots$$

$$\cdots \longrightarrow IH_k^{c\bar{p}}(F) \xrightarrow{k_!} IH_k^{c\bar{p}}(E) \xrightarrow{j^*} IH_{k-n}^{\bar{p}}(F) \longrightarrow IH_{k-1}^{c\bar{p}}(F) \longrightarrow \cdots.$$

PROOF. In the following, trivializations of the fibre bundle E are always involved. However, for every pseudomanifold $X, h : X \xrightarrow{\cong} X$ implies $\mathbf{IC}_{\bar{p}}^{\bullet}(X) \cong h_* \mathbf{IC}_{\bar{p}}^{\bullet}(X) \cong h^* \mathbf{IC}_{\bar{p}}^{\bullet}(X)$. Therefore, for the proof we can suppress them.

We begin with the distinguished triangle

$$Rj_* j^! \, \mathbf{IC}_{\bar{p}}^{\bullet}(E) \longrightarrow \mathbf{IC}_{\bar{p}}^{\bullet}(E)$$

$$[1] \diagdown \quad \diagup$$

$$Ri_* i^* \, \mathbf{IC}_{\bar{p}}^{\bullet}(E)$$

and keeping in mind that $j^! \, \mathbf{IC}_{\bar{p}}^{\bullet}(E) \cong \mathbf{IC}_{\bar{p}}^{\bullet}(F)$, $i^* \, \mathbf{IC}_{\bar{p}}^{\bullet}(E) \cong \mathbf{IC}_{\bar{p}}^{\bullet}(U \times F)$ we apply hypercohomology to get

$$\cdots \to IH_k^{\bar{p}}(F) \xrightarrow{j_*} IH_k^{\bar{p}}(E) \xrightarrow{i^*} IH_k^{\bar{p}}(U \times F) \to \cdots.$$

The third term is isomorphic to $IH_{k-n}^{\bar{p}}(F)$ under l^* by the preceding lemma. However by the commutative triangle

$$b_1 \times F \xrightarrow{\quad k \quad} E$$

$$l \diagdown \quad \diagup i$$

$$U \times F$$

we have $k^* \simeq l^* \circ i^*$ and the sequence for closed supports is proven.

Now turn to the case of compact supports.[6] Consider the triangle

$$Ri_! i^* \, \mathbf{IC}_{\bar{p}}^{\bullet}(E) \longrightarrow \mathbf{IC}_{\bar{p}}^{\bullet}(E)$$

$$[1] \diagdown \quad \diagup$$

$$Rj_* j^* \, \mathbf{IC}_{\bar{p}}^{\bullet}(E)$$

and apply hypercohomology with compact supports to get

$$\cdots \to \mathcal{H}_c^{-k}(E; Ri_! \, \mathbf{IC}_{\bar{p}}^{\bullet}(U \times F)) \xrightarrow{i_!} IH_k^{c\bar{p}}(E) \xrightarrow{j^*} \mathcal{H}_c^{-k}(E; Rj_* \, \mathbf{IC}_{\bar{p}}^{\bullet}(F)[n]) \to \cdots.$$

Now $Rj_* = Rj_!$ as j is a closed inclusion. Hence

$$\mathcal{H}_c^{-k}(E; Rj_* \, \mathbf{IC}_{\bar{p}}^{\bullet}(F)[n]) \cong \mathcal{H}_c^{-k}(F; \mathbf{IC}_{\bar{p}}^{\bullet}(F)[n]) \cong IH_{k-n}^{c\bar{p}}(F).$$

For the first term, we have

$$\mathcal{H}_c^{-k}(E; Ri_! \, \mathbf{IC}_{\bar{p}}^{\bullet}(U \times F)) \cong \mathcal{H}_c^{-k}(U \times F; \mathbf{IC}_{\bar{p}}^{\bullet}(U \times F)) \xleftarrow[\cong]{l_!} IH_k^{c\bar{p}}(F)$$

and with $i_! \circ l_! = k_!$ the assertion follows. $\square$

---

[6]Recall that for $f : X \to Y$, $\mathbf{A}^{\bullet} \in D_c^b(X)$ and $Z \subset X$, we have $\Gamma_c(Z, f_* \mathbf{A}^{\bullet}) \not\cong \Gamma_c(f^{-1}(Z), \mathbf{A}^{\bullet})$. However $\Gamma_c(Z, f_! \mathbf{A}^{\bullet}) \cong \Gamma_c(f^{-1}(Z), \mathbf{A}^{\bullet})$.

7. The non-Witt case

7A. The category $SD(X)$. Originally, Goresky and MacPherson defined the signature for spaces with only even-codimensional strata. In [Sie83], Siegel generalizes the definition to Witt spaces. If $IH_*^{\bar{m}}(X) \not\cong IH_*^{\bar{n}}(X)$, there still is a method to define a signature and L-classes for a pseudomanifold X compatible with the old definition. In his work [Ban02], Banagl establishes a corresponding framework and decomposition results similar to those of Cappell and Shaneson are presented in further papers. In this section we merely give the definition.

DEFINITION 7.1. Let $X = X_n \supset \cdots \supset X_0$ be an oriented pseudomanifold with orientation

$$\mathfrak{o} : \mathbb{D}_{U_2}^\bullet \xrightarrow{\cong} \mathbb{R}_{U_2}[n].$$

For $k \geq 2$, we write $U_k := X \setminus X_{n-k}$. Define $SD(X)$ as the full subcategory of $D_c^b(X)$ of those $\mathbf{S}^\bullet \in D_c^b(X)$ satisfying the following:

(SD1) Normalization: There is an isomorphism $\nu : \mathbb{R}_{U_2}[n] \xrightarrow{\cong} \mathbf{S}^\bullet|_{U_2}$.

(SD2) Lower bound: $\mathbf{H}^i(\mathbf{S}^\bullet) = 0$, for $i < -n$.

(SD3) Stalk condition for $\bar{n}$: $\mathbf{H}^i(\mathbf{S}^\bullet|_{U_{k+1}}) = 0$, for $i > \bar{n}(k) - n, k \geq 2$.

(SD4) Self-duality: There is an isomorphism $d : \mathcal{D}_X \mathbf{S}^\bullet[n] \to \mathbf{S}^\bullet$ compatible with the orientation, i.e., such that the square

$$
\begin{array}{ccc}
\mathbb{R}_{U_2}[n] & \xrightarrow[\cong]{\nu} & \mathbf{S}^\bullet|_{U_2} \\
\mathfrak{o} \Big\uparrow \cong & & \cong \Big\uparrow d|_{U_2} \\
\mathbb{D}_{U_2}^\bullet & \xrightarrow[\cong]{\mathcal{D}_X \nu^{-1}[n]} & \mathcal{D}_X \mathbf{S}^\bullet|_{U_2}[n]
\end{array}
\qquad \text{commutes.}
$$

We refer to [Ban02] for results on this category, especially for the structure theorem, establishing the relation between $\mathbf{S}^\bullet \in SD(X)$ and a choice of Lagrangian structures along odd-codimensional strata of X.

REMARK 7.2. If X is a Witt space, $SD(X)$ consists up to isomorphism only of $IC_{\bar{m}}^\bullet(X)$. On the other hand, $SD(X)$ might be empty — e.g., $SD(\Sigma \mathbb{C}P^2) = \varnothing$.

THEOREM 7.3 [Ban02, Theorem.2.2]. *For $\mathbf{S}^\bullet \in SD(X)$, there is a factorization*

$$IC_{\bar{m}}^\bullet(X) \xrightarrow{\alpha} \mathbf{S}^\bullet \xrightarrow{\beta} IC_{\bar{n}}^\bullet(X),$$

that is compatible with the normalization (and is unique with respect to this property) and such that

$$\mathbf{IC}^{\bullet}_{\bar{m}}(X) \xrightarrow{\quad\alpha\quad} \mathbf{S}^{\bullet}$$

$$\cong \Big\uparrow \qquad\qquad \cong \Big\uparrow d$$

$$\mathcal{D}_X \mathbf{IC}^{\bullet}_{\bar{n}}(X) \xrightarrow{\mathcal{D}_X \beta[n]} \mathcal{D}_X \mathbf{S}^{\bullet} \qquad\qquad commutes.$$

Thus, an object in $SD(X)$ is in fact a self-dual interpolation between $\mathbf{IC}^{\bullet}_{\bar{m}}(X)$ and $\mathbf{IC}^{\bullet}_{\bar{n}}(X)$. It is obvious that in the case of X being a Witt space, $SD(X)$ consists (up to quasi-isomorphism) only of $\mathbf{IC}^{\bullet}_{\bar{m}}(X)$.

DEFINITION 7.4. Let X^n be a closed stratified topological pseudomanifold, not necessarily Witt and $\mathbf{S}^{\bullet} \in SD(X)$. In case n is divisible by 4, define $\sigma(X^n, \mathbf{S}^{\bullet}, d)$ to be the signature on $\mathcal{H}^{-n/2}(X^n, \mathbf{S}^{\bullet})$ induced by the self-duality of $\mathbf{S}^{\bullet}$.

REMARK 7.5. If X happens to be a Witt space, $\sigma(X^n, \mathbf{S}^{\bullet}, d)$ is the usual intersection homology signature due to Theorem 7.3.

Finally, in order to speak of *the* signature of a pseudomanifold (as long as $SD(X) \neq \varnothing$), we need the following important result:

THEOREM 7.6 [Ban06, 4.1]. *Let X^n be an even-dimensional closed oriented pseudomanifold with $SD(X) \neq \varnothing$. For $(\mathbf{S}^{\bullet}_1, d_1), (\mathbf{S}^{\bullet}_2, d_2) \in SD(X)$ one has*

$$\sigma(X^n, \mathbf{S}^{\bullet}_1, d_1) = \sigma(X^n, \mathbf{S}^{\bullet}_2, d_2).$$

7B. Hypercohomology Wang sequence. Before we deduce the exact sequence for hypercohomology with values in SD-sheaves, we have to determine what the involved complexes of sheaves are going to be. Starting with a SD complex over the total space E we define a SD complex over the fibre F in a canonical way. We will need the following little lemma.

LEMMA 7.7. *For the inclusion $j : X^n \times 0 \hookrightarrow X^n \times \mathbb{R}^m$ with associated projection $p : X^n \times \mathbb{R}^m \to X^n$ we have*

$$j^! \simeq j^*[-m]$$

and thereby

$$p^! \circ j^! \simeq \mathrm{id}.$$

PROOF. By [Ban02, Lemma 5.2], $p^* \circ j^* \simeq \mathrm{id}$. Consequently, using $p^*[m] \simeq p^!$ ([Ban02, Lemma 4.2, Proof]), we get

$$j^! \simeq j^! \circ p^* \circ j^* \simeq j^! \circ p^! \circ j^*[-m] \simeq (p \circ j)^! \circ j^*[-m] \simeq j^*[-m].$$

The second identity is clear by using $p^*[m] \simeq p^!$ again. $\qquad\qquad\square$

LEMMA 7.8. *Let* $X^n \xrightarrow{j} X^n \times \mathbb{R}^m$ *be the standard inclusion. Given* $\mathbf{T}^\bullet \in SD(X^n \times \mathbb{R}^m)$, *the complex* $j^! \mathbf{T}^\bullet$ *is in* $SD(X)$.

PROOF. Looking at the commutative square

$$
\begin{array}{ccc}
\mathbb{R}_{U_2}[n+m] & \xrightarrow{\quad v \quad} & \mathbf{T}^\bullet|_{U_2} \\
{\scriptstyle j^!}\downarrow & & \downarrow{\scriptstyle j^!} \\
j^!\mathbb{R}_{U_2}[n+m] \cong \mathbb{R}_{U_2 \cap X}[n] & \xrightarrow{\;j^!(v)\;} & j^!\mathbf{T}^\bullet|_{U_2 \cap X},
\end{array}
$$

one checks that (SD1) is fulfilled because of the functoriality of $j^!$. Now using that the inverse image functor j^* is exact, look at

$$\mathbf{H}^i(j^!\mathbf{T}^\bullet) \cong \mathbf{H}^i(j^*\mathbf{T}^\bullet[-m]) \cong \mathbf{H}^{i-m}(j^*\mathbf{T}^\bullet) \cong j^*\mathbf{H}^{i-m}(\mathbf{T}^\bullet).$$

Observe that the last term is zero for $i - m < -(n+m)$ or $i < -n$, and so (SD2) holds. Now let $i > \bar{n}(k) - (n), k \geq 2$. We have

$$\mathbf{H}^i((j^!\mathbf{T}^\bullet)|_{U_{k+1} \cap X}) \cong \mathbf{H}^i((j^*\mathbf{T}^\bullet[-m])|_{U_{k+1} \cap X}) \cong j^*\mathbf{H}^{i-m}(\mathbf{T}^\bullet|_{U_{k+1}}),$$

where the last term is zero due to $i - m > \bar{n}(k) - (n+m)$ and hence (SD3) holds as well.

Finally by [Ban07, Proposition 3.4.5] we have an isomorphism

$$\mathcal{D}_X j^!\mathbf{T}^\bullet[n] \cong j^*\mathcal{D}_{X \times \mathbb{R}^m}\mathbf{T}^\bullet[n] \cong j^!(\mathcal{D}_{X \times \mathbb{R}^m}\mathbf{T}^\bullet[n+m]) \cong j^!\mathbf{T}^\bullet$$

which is compatible with the orientation, proving (SD4). $\qquad\qquad\square$

Let us now return to the original context. We start with the total space E of a fibre bundle over S^n — a topological pseudomanifold of dimension $k + n$ — and a complex $\mathbf{T}^\bullet \in SD(E)$. Given a trivializing neighbourhood $U \subset S^n$ of the north pole b_0 resp. the south pole b_1, the restriction of $\mathbf{T}^\bullet$ to $\pi^{-1}(U) \cong U \times F$ is clearly in $SD(U \times F)$ since $\pi^{-1}(U)$ is open. With the preceding lemma we can now define:

DEFINITION 7.9. Let $F \to E \to S^n$ be a fibre bundle as above and $\mathbf{T}^\bullet \in SD(E)$. For $i = 0, 1$, we have inclusions

$$
\begin{array}{ccc}
b_i \times F & \xrightarrow{\quad j_i \quad} & E \\
{\scriptstyle i_i^1}\downarrow & & \uparrow{\scriptstyle i_i^2} \\
U \times F & \xrightarrow[\;\cong\;]{\phi_i} & \pi^{-1}(U)
\end{array}
$$

where j_i is defined to be the composition $i_i^2 \circ \phi_i \circ i_i^1$.

Set $\mathbf{S}^\bullet := \mathbf{S}_N^\bullet := j_0^! \mathbf{T}^\bullet \cong i_0^{1!}((\phi_0^! \mathbf{T}^\bullet)|_{U \times F}) \in SD(F)$, with $j_0 : b_0 \times F \hookrightarrow$ the inclusion of the north pole fibre for some trivialization $\phi_0 : U \times F \xrightarrow{\cong} \pi^{-1}(U)$. Define $\mathbf{S}_S^\bullet$ in the same way using the inclusion of the south pole fibre.

In order for $\mathbf{S}^\bullet$ to be well defined over $F = b_i \times F$ with $b_i \in U_1 \cap U_2$, we have to make an extra assumption, a kind of homogeneity:

DEFINITION 7.10. We call the structure group G of a fibre bundle of the form above *adapted to* $\mathbf{A}^\bullet \in SD(F)$, if for all $h \in G$, $h^! \mathbf{A}^\bullet \cong h^* \mathbf{A}^\bullet \cong \mathbf{A}^\bullet$.

EXAMPLE 7.11. If F is a Witt space, $\mathbf{IC}_{\bar{m}}^\bullet(F) \in SD(F)$. For every stratum-preserving automorphism $h : F \xrightarrow{\cong} F$, we have $h^* \mathbf{IC}_{\bar{m}}^\bullet(F) \cong \mathbf{IC}_{\bar{m}}^\bullet(F)$.

REMARK 7.12. Assume G to be adapted to $\mathbf{S}^\bullet$. What if we are given two trivializing neighbourhoods $U_1, U_2 \subset S^n$ with $\phi_i : U_i \times F \to \pi^{-1}(U_i)$? We have a commutative diagram

$$
\begin{array}{ccccc}
F & \xrightarrow{i^2} & b_0 \times F & \longrightarrow & (U_1 \cap U_2) \times F \\
{\scriptstyle h_{21}(b_0)} \big\uparrow {\scriptstyle \cong} & & \big\uparrow {\scriptstyle \cong} & & {\scriptstyle \phi_2^{-1}\phi_1} \big\uparrow \\
F & \xrightarrow{i^1} & b_0 \times F & \longrightarrow & (U_1 \cap U_2) \times F
\end{array}
\quad
\begin{array}{c}
\xrightarrow{\phi_2} \\[2ex]
\pi^{-1}(U_1 \cap U_2) \longrightarrow E \\[2ex]
\xrightarrow{\phi_1}
\end{array}
$$

where the upper composition is j^1 and the lower composition is j^2. We have to show that $(j^1)^! \mathbf{T}^\bullet = (j^2)^! \mathbf{T}^\bullet$. Since G is adapted to $\mathbf{S}^\bullet$, the transition function $h_{21}(b_0)$ preserves $\mathbf{S}^\bullet$ and thereby $(j^1)^! \mathbf{T}^\bullet = h_{21}(b_0)^! (j^2)^! \mathbf{T}^\bullet \cong (j^2)^! \mathbf{T}^\bullet$.

We will need some form of suspension isomorphism for hypercohomology with values in a SD sheaf.

LEMMA 7.13. *Let F be a pseudomanifold and $\mathbf{S}^\bullet \in SD(F)$. Let $p : \mathbb{R}^n \times F \to F$ be the projection. The inclusion $l : 0 \times F \hookrightarrow \mathbb{R}^n \times F$ induces the following isomorphisms on hypercohomology:*

(a) $l^* : \mathcal{H}^k(\mathbb{R}^n \times F; \mathbf{S}^\bullet) \to \mathcal{H}^{k+n}(F; l^! \mathbf{S}^\bullet)$

(b) $l_! : \mathcal{H}_c^k(F; l^! \mathbf{S}^\bullet) \to \mathcal{H}_c^k(\mathbb{R}^n \times F; \mathbf{S}^\bullet)$

PROOF. Same as for Lemma 6.1, using $p^* \simeq p^![-n]$ for (a). Note that by [Bor84, V, 3.13] $Rp_* p^* \mathbf{A}^\bullet \cong \mathbf{A}^\bullet$ for all $\mathbf{A}^\bullet \in D^b(X)$.   □

Now we are able to formulate the next proposition and imitate the proof of the Wang sequence given in Section 6.

PROPOSITION 7.14. *Let $F \to E \to S^n$ be a fibre bundle with a suitable structure group G of automorphisms of the pseudomanifold F, which is adapted to*

$\mathbf{S}_N^\bullet$ and $\mathbf{S}_S^\bullet$. *Assume E to be canonically stratified. Given $\mathbf{T}^\bullet \in SD(E)$ there are long exact sequences*

(a) $\cdots \to \mathcal{H}^k(F; \mathbf{S}^\bullet) \to \mathcal{H}^k(E; \mathbf{T}^\bullet) \to \mathcal{H}^{k+n}(F; \mathbf{S}^\bullet) \to \cdots,$

(b) $\cdots \to \mathcal{H}_c^k(F; \mathbf{S}^\bullet) \to \mathcal{H}_c^k(E; \mathbf{T}^\bullet) \to \mathcal{H}_c^{k+n}(F; \mathbf{S}^\bullet) \to \cdots,$

where $\mathbf{S}^\bullet$ is the self-dual sheaf of Definition 7.9.

PROOF. Due to Remark 7.12 we need not pay attention to different trivializations. Therefore, in the following proof we will not mention them explicitly.

(a) Denote by j the inclusion of the north pole fibre and by i the inclusion of the complement $V \times F$. We begin with the distinguished triangle

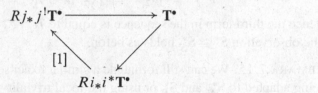

and apply hypercohomology to get

$$\cdots \to \mathcal{H}^k(F; j^!\mathbf{T}^\bullet) \to \mathcal{H}^k(E; \mathbf{T}^\bullet) \to \mathcal{H}^k(V \times F; i^*\mathbf{T}^\bullet) \to \cdots.$$

By construction, we have $\mathcal{H}^k(F; j^!\mathbf{T}^\bullet) \cong \mathcal{H}^k(F; \mathbf{S}^\bullet)$. If $i : V \times F \to E$ denotes the inclusion, then $i^*\mathbf{T}^\bullet$ is again isomorphic to a self-dual complex over the product bundle $V \times F$ containing the south pole fibre $b_1 \times F$. With the inclusion $j_1 : b_1 \times F \hookrightarrow E$ and using the preceding lemma, we finally get

$$\mathcal{H}^k(V \times F; i^*\mathbf{T}^\bullet) = \mathcal{H}^{k+n}(F; j_1^!\mathbf{T}^\bullet) \cong \mathcal{H}^{k+n}(F; \mathbf{S}_S^\bullet).$$

Now choose a trivializing neighbourhood O containing the north pole b_0 and the south pole b_1. Let l_0 and l_1 be the corresponding inclusions into $O \times F$. We have

$$\mathbf{S}^\bullet = j_0^!\mathbf{T}^\bullet = l_0^!(\mathbf{T}^\bullet|_{O \times F}) \cong l_1^!(\mathbf{T}^\bullet|_{O \times F}) \cong j_1^!\mathbf{T}^\bullet = \mathbf{S}_S^\bullet,$$

where the middle isomorphism holds because G is adapted to $\mathbf{S}_N^\bullet$ and $\mathbf{S}_S^\bullet$. This completes the proof.

(b) Begin with the triangle

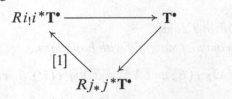

and apply hypercohomology with compact supports to get

$$\cdots \to \mathcal{H}_c^k(V \times F; i^*\mathbf{T}^\bullet) \to \mathcal{H}_c^k(E; \mathbf{T}^\bullet) \to \mathcal{H}_c^k(F; j^*\mathbf{T}^\bullet) \to \cdots.$$

Now, using part (b) of the preceding lemma, we identify

$$\mathcal{H}_c^k(V \times F; i^*\mathbf{T}^\bullet) \cong \mathcal{H}_c^k(F; j_1^! i^*\mathbf{T}^\bullet) \cong \mathcal{H}_c^k(F; \mathbf{S}^\bullet).$$

And again by using

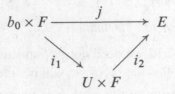

we obtain

$$j^*\mathbf{T}^\bullet \cong i_1^* i_2^* \mathbf{T}^\bullet \cong i_1^!(\mathbf{T}^\bullet|_{V \times F})[n] \cong \mathbf{S}^\bullet[n].$$

Hence the third term in the sequence is equal to $\mathcal{H}_c^k(F; j^*\mathbf{T}^\bullet) \cong \mathcal{H}_c^{k+n}(F; \mathbf{S}^\bullet)$. The observation $\mathbf{S}^\bullet \cong \mathbf{S}_S^\bullet$ holds as before.   □

REMARK 7.15. We can still formulate a similar exact sequence even without G being adapted to $\mathbf{S}_N^\bullet$ and $\mathbf{S}_S^\bullet$ or using the local triviality of the stratifold bundle only. Then, however, the involved SD-complexes over the fibre may be different and depend on the choice of trivializations:

$$\cdots \to \mathcal{H}^k(F; \mathbf{S}_N^\bullet) \to \mathcal{H}^k(E; \mathbf{T}^\bullet) \to \mathcal{H}^{k+n}(F; \mathbf{S}_S^\bullet) \to \cdots,$$

$$\cdots \to \mathcal{H}_c^k(F; \mathbf{S}_S^\bullet) \to \mathcal{H}_c^k(E; \mathbf{T}^\bullet) \to \mathcal{H}_c^{k+n}(F; \mathbf{S}_N^\bullet) \to \cdots.$$

8. Novikov additivity and collapsing of spheres

In [Sie83], Siegel generalizes the classical Novikov additivity of the signature for manifolds to pseudomanifolds satisfying the Witt condition. We want to make a further step forward by dropping the Witt condition in certain cases.[7]

PROPOSITION 8.1. *Let* $X = X_{2n} \supset X_{2n-k} \supset \varnothing$, $k \geq 2$, *be a Whitney stratified compact pseudomanifold with* $SD(X) \neq \varnothing$. *Given subspaces* $M, E, T \subset X$ *with* T *a closed neighbourhood of* X_{2n-k}, *such that*

(1) $X = M \cup T$,
(2) $M \cap T = E$,
(3) E *has a collar in* T, *and*
(4) (M, E) *is a compact manifold with boundary.*

Define $X^1 := M \cup_E c(E)$ *and* $X^2 := T \cup_E c(E)$. *Then we have the identity*

$$\sigma(X) = \sigma(X^1) + \sigma(X^2).$$

---

[7]A similar result is given in Theorem 3 of [Hun07]. Thanks to the referee for pointing this out to me.

PROOF. Once again, consider the formula of Cappell and Shaneson 4.5. Let $f : X \to Y$ be the map collapsing X_{2n-k} to a point. The push forward $Rf_*\mathbf{T}^\bullet$ of $\mathbf{T}^\bullet \in SD(E)$ is algebraically cobordant to

$$\mathbf{IC}_{\bar{m}}^\bullet(X^1, \mathcal{S}_f^{\dot{M}}) \oplus j_{c*} \mathbf{IC}_{\bar{m}}^\bullet(\{c\}, \mathcal{S}_f^{\{c\}})[n].$$

Here, c is the conepoint $f(X_{2n-k})$ and j_c its inclusion into Y. Since $\mathcal{S}_f^{\dot{M}}$ is constant with rank one, the first term is just equal to $\mathbf{IC}_{\bar{m}}^\bullet(X^1)$. Let us look at $(\mathcal{S}_f^{\{c\}})_c$. The link $L(c)$ of c is E, so we deduce from

$$E_c = f^{-1}(c\, L(c)) \cup_E c\, f^{-1}(L(c)) = X^2$$

and

$$E_c \overset{i_c}{\hookleftarrow} E_c \backslash \{c\} \cong f^{-1}(c\, L(c)) \overset{\rho_c}{\hookrightarrow} X$$

(see Section 4B) that $\mathbf{S}^\bullet(c)$ is a Deligne extension of $\rho_c^!\mathbf{S}^\bullet$ — which is just the restriction of the original self-dual complex. Hence[8] $\mathbf{S}^\bullet(c) \in SD(X_2)$. We have $(\mathcal{S}_f^{\{c\}})_c = \mathcal{H}^{-n}(X^2; \mathbf{S}^\bullet(c))$ and consequently

$$\mathcal{H}^{-n}(Y; j_{c*}\mathbf{IC}_{\bar{m}}^\bullet(\{c\}, \mathcal{S}_f^{\{c\}})[n]) \cong \mathcal{H}^{-n}(\{c\}; \mathbf{IC}_{\bar{m}}^\bullet(\{c\}, \mathcal{S}_f^{\{c\}})[n])$$
$$\cong \mathcal{H}^{-n}(X^2; \mathbf{S}^\bullet(c)).$$

Finally, combining these observations and passing to the signature we get

$$\sigma(X) = \sigma(X, Rf_*(\mathbf{T}^\bullet))$$
$$= \sigma(X^1, \mathbf{IC}_{\bar{m}}^\bullet(X^1)) + \sigma(X^2, \mathbf{S}^\bullet(c)) = \sigma(X^1) + \sigma(X^2). \qquad \square$$

Let E be the total space of a fibre bundle over S^m as before. We investigate, when the middle hypercohomology group $\mathcal{H}^{-n}(M, \mathbf{S}^\bullet)$ vanishes for a given complex $\mathbf{S}^\bullet$ over a non-Witt $M := c\, E \cup_E \bar{E}$ of dimension $2n$. Since we are only interested in computing the signature, only odd-dimensional spheres are considered here. The strategy is very similar to that of Section 5.

PROPOSITION 8.2. Let $F^{2a} \to E \overset{\pi}{\to} S^{2b+1}$ be a fibre bundle with F a compact topological pseudomanifold, $a \geq 1, b \geq 2$. Define $M := c\, E \cup_E \bar{E}$, where $\bar{E}$ is the total space of the induced fibre bundle

$$c\, F \to \bar{E} \overset{c(\pi)}{\longrightarrow} S^{2b+1}.$$

Given $\mathbf{S}^\bullet \in SD(M)$, denote by $\mathbf{T}^\bullet$ the induced element in $SD(E)$ (compare Section 7). Assume that the following condition (S) is fulfilled for the hypercohomology Wang sequence of E:

$$\mathcal{H}^{-(a+b+1)}(E; \mathbf{T}^\bullet) \twoheadrightarrow \mathcal{H}^{-(a-b)}(F; \mathbf{S}_{\mathbf{S}}^\bullet) \text{ is surjective,}$$

---

[8] All the axioms are clearly satisfied. The only "new" stalk to look at is the one at c, but $\{c\}$ has even codimension and the modified Deligne extension of $\rho_c^!\mathbf{S}^\bullet$ explained in 4B ensures that SD1-SD4 remain valid.

Then

$$\mathcal{H}^{-(a+b+1)}(M;\mathbf{S}^\bullet) = 0.$$

We will need the following vanishing lemma for the hypercohomology of the cone.

LEMMA 8.3. *Let X be a $2n$-dimensional or a $(2n+1)$-dimensional, compact Witt space such that the signature of the pairing over the middle-dimensional intersection homology vanishes. For $\mathbf{S}^\bullet \in SD(\mathring{c}X)$, we have*

$$\mathcal{H}_c^{-i}(\mathring{c}X;\mathbf{S}^\bullet) = 0 \quad for \ i \geq n+1.$$

PROOF. Since $\mathbf{S}^\bullet$ is constructible and $\mathring{c}X$ is a distinguished neighbourhood of the conepoint c, there is (by [Ban07, p. 97], for instance)

$$\mathcal{H}_c^{-i}(\mathring{c}X;\mathbf{S}^\bullet) = H^{-i}(j_c^!\mathbf{S}^\bullet)$$

where the latter is the costalk of $\mathbf{S}^\bullet$ at c. Because of self-duality, however, $\mathbf{S}^\bullet$ satisfies the costalk vanishing condition

$$H^{-i}(j_c^!\mathbf{S}^\bullet) = 0 \text{ for } -i \leq \begin{cases} \bar{m}(2n+1) - \dim \mathring{c}X + 1 = -(n+1), \\ \bar{m}(2n+2) - \dim \mathring{c}X + 1 = -(n+1), \end{cases}$$

which is equivalent to the statement. $\square$

PROOF OF PROPOSITION 8.2.. Look at the hypercohomology Wang sequence for $\mathring{c}F \to \mathring{\tilde{E}} \to S^{2b+1}$

$$\cdots \to \mathcal{H}_c^{-(a+b+1)}(\mathring{c}F;\mathbf{U}^\bullet) \to \mathcal{H}_c^{-(a+b+1)}(\mathring{\tilde{E}};\mathbf{T}^\bullet) \to \mathcal{H}_c^{-(a-b)}(\mathring{c}F;\mathbf{U}^\bullet) \to \cdots$$

where $\mathbf{T}^\bullet = \mathbf{S}^\bullet|_{\mathring{\tilde{E}}}$ is in $SD(\mathring{\tilde{E}})$ and $\mathbf{U}^\bullet \in SD(\mathring{c}F)$ is constructed as in 7.9. Since $b \geq 2$, we see from the preceding lemma that

$$\mathcal{H}_c^{-(a+b+1)}(\mathring{c}F;\mathbf{U}^\bullet) = \mathcal{H}_c^{-(a+b)}(\mathring{c}F;\mathbf{U}^\bullet) = 0$$

and hence

$$\mathcal{H}_c^{-(a+b+1)}(\mathring{\tilde{E}};\mathbf{T}^\bullet) \cong \mathcal{H}_c^{-(a-b)}(\mathring{c}F;\mathbf{U}^\bullet).$$

Decompose M into the open subsets $\mathring{\tilde{E}}$ and $\mathring{c}E$ with $\mathring{\tilde{E}} \cap \mathring{c}E = E \times (0,1)$. Consider the Mayer–Vietoris hypercohomology sequence (see [Ive86, III.7.5] or [Bre97, II, §13], for instance)

$$\cdots \longrightarrow \mathcal{H}_c^{-(a+b+1)}(E \times (0,1);\mathbf{S}^\bullet|) \xrightarrow{i_{a+b+1}}$$

$$\mathcal{H}_c^{-(a+b+1)}(\mathring{c}E;\mathbf{S}^\bullet|) \oplus \mathcal{H}_c^{-(a+b+1)}(\mathring{\tilde{E}};\mathbf{S}^\bullet|) \longrightarrow$$

$$\mathcal{H}_c^{-(a+b+1)}(M;\mathbf{S}^\bullet) \longrightarrow \mathcal{H}_c^{-(a+b)}(E \times (0,1);\mathbf{S}^\bullet|) \xrightarrow{i_{a+b}} \cdots$$

Let us first show that i_{a+b} is injective. We use the following exact sequence for hypercohomology with compact supports (see [Ive86, III.7.7])

$$\cdots \to \mathcal{H}_c^{-(a+b+1)}(\{c\};\mathbf{S}^\bullet) \to \mathcal{H}_c^{-(a+b)}(\mathring{c}E\backslash\{c\};\mathbf{S}^\bullet) \to \mathcal{H}_c^{-(a+b)}(\mathring{c}E;\mathbf{S}^\bullet) \to \cdots.$$

It suffices to show that

$$\mathcal{H}_c^{-(a+b+1)}(\{c\};\mathbf{S}^\bullet) \cong \mathcal{H}^{-(a+b+1)}(\{c\};\mathbf{S}^\bullet)$$

vanishes. The latter is isomorphic to $\mathbf{H}^{-(a+b+1)}(\mathbf{S}_c^\bullet)$ which is 0 because of (SD3). Due to the preceding lemma $\mathcal{H}_c^{-(a+b+1)}(\mathring{c}E;\mathbf{S}^\bullet|)$ is 0. The surjectivity of i_{a+b+1} follows now from the condition (S) using the naturality of the hypercohomology Wang sequence in completely the same manner as in the proof of Proposition 5.2. $\qquad\qquad\square$

REMARK 8.4. As you can see, we have only used the vanishing lemma 8.3, which is valid for every $\mathbf{S}^\bullet \in SD(F)$. Hence, in the proposition, the structure group G need not be adapted.

REMARK 8.5. In the case of a Witt fibre F the condition (S) is the one of Section 5. See the examples there, especially 5.6.

COROLLARY 8.6. *Let M be as in 8.2. The signature $\sigma(M)$ of M vanishes.*

COROLLARY 8.7. *Let X be as in 8.1 such that X_{n-k} is an odd-dimensional sphere. Assume E to satisfy* (S). *Then $\sigma(X) = \sigma(X^1) = \sigma(M,\partial M)$, where the latter is the Novikov signature of M.*

Acknowledgements

The article is based on the author's Diploma thesis at the University of Heidelberg [Lev07] and is motivated by the MSRI workshop on the Topology of Stratified Spaces. I want to express my gratitude to Greg Friedman, Eugénie Hunsicker, Anatoly Libgober and Laurentiu-George Maxim for the organisation of this event, the accompanying support and the opportunity for giving a talk. I am also indebted to my thesis advisor Markus Banagl for his help and advice during the development of the thesis. Last but not least I am grateful to the referee for the careful reading of the preliminary version.

References

[Ban02] M. Banagl, *Extending intersection homology type invariants to non-Witt spaces*, Mem. Amer. Math. Soc. **160** (2002), no. 760.

[Ban06] ———, *The L-class of non-Witt spaces*, Ann. of Math. **163** (2006), no. 3.

[Ban07] ———, *Topological invariants of stratified spaces*, Springer Monographs in Mathematics, Springer, 2007.

[Bor84] A. Borel et al., *Intersection cohomology*, Progr. Math., vol. 50, Birkhäuser, Boston, 1984.

[Bre97] G. E. Bredon, *Sheaf theory*, second ed., Grad. Texts in Math., vol. 170, Springer, New York, 1997.

[Bry93] J.-L. Brylinski, *Loop spaces, characteristic classes and geometric quantization*, Progr. Math., vol. 107, Birkhäuser, Boston, 1993.

[CS91] S. E. Cappell and J. Shaneson, *Stratifiable maps and topological invariants*, J. Amer. Math. Soc. **4** (1991), 521–551.

[FM81] W. Fulton and R. MacPherson, *Categorical framework for the study of singular spaces*, Mem. Amer. Math. Soc. **31** (1981), no. 243.

[Fri07] G. Friedman, *Intersection homology of stratified fibrations and neighborhoods*, Adv. Math. **215** (2007), 24–65.

[GM83] M. Goresky and R. Macpherson, *Intersection homology II*, Invent. Math. **72** (1983), no. 1, 77–129.

[GM88] M. Goresky and R. MacPherson, *Stratified morse theory*, Ergebnisse der Mathematischen Wissenschaften, vol. 14, Springer, Berlin Heidelberg, 1988.

[HS91] N. Habegger and L. Saper, *Intersection cohomology of cs-spaces and Zeeman's filtration*, Invent. Math. **105** (1991), no. 2, 247–272.

[Hud69] J. F. P. Hudson, *Piecewise linear topology*, Mathematics Lecture Notes Series, W. A. Benjamin, New York, 1969.

[Hun07] E. Hunsicker, *Hodge and signature theorems for a family of manifolds with fibration boundary*, Geom. Top. **11** (2007), 1581–1622.

[Ive86] B. Iverson, *Cohomology of sheaves*, Universitext, Springer, Berlin, 1986.

[Kin85] H. King, *Topological invariance of intersection homology without sheaves*, Topology Appl. **20** (1985), 149–160.

[Lev07] F. Levikov, *Wang sequences in intersection homology*, Diploma Thesis, Universität Heidelberg, 2007.

[Roy87] J. Roy, *Vanishing of odd-dimensional intersection cohomology*, Math. Z. **195** (1987), 239–253.

[Sie83] P.H. Siegel, *Witt spaces: A geometric cycle theory for KO-homology at odd primes*, American J. Math. **105** (1983), 1067–1105.

[Spa66] E. Spanier, *Algebraic topology*, McGraw-Hill, New York, 1966.

[Wan49] H. C. Wang, *The homology groups of fibre bundles over a sphere*, Duke Math. J. **16** (1949), 33–38.

FILIPP LEVIKOV
INSTITUTE OF MATHEMATICS
KING'S COLLEGE
ABERDEEN AB24 3UE
SCOTLAND
    f.levikov@abdn.ac.uk

Topology of Stratified Spaces
MSRI Publications
Volume **58**, 2011

An exponential history of functions with logarithmic growth

MATT KERR AND GREGORY PEARLSTEIN

ABSTRACT. We survey recent work on normal functions, including limits and
singularities of admissible normal functions, the Griffiths–Green approach to
the Hodge conjecture, algebraicity of the zero locus of a normal function,
Néron models, and Mumford–Tate groups. Some of the material and many
of the examples, especially in Sections 5 and 6, are original.

Introduction

CONTENTS

In a talk on the theory of motives, A. A. Beilinson remarked that according to
his time-line of results, advances in the (relatively young) field were apparently a
logarithmic function of t; hence, one could expect to wait 100 years for the next
significant milestone. Here we allow ourselves to be more optimistic: following
on a drawn-out history which begins with Poincaré, Lefschetz, and Hodge, the
theory of *normal functions* reached maturity in the programs of Bloch, Griffiths,

The second author was supported by NSF grant DMS-0703956.

Zucker, and others. But the recent blizzard of results and ideas, inspired by works of M. Saito on admissible normal functions, and Green and Griffiths on the Hodge Conjecture, has been impressive indeed. Besides further papers of theirs, significant progress has been made in work of P. Brosnan, F. Charles, H. Clemens, H. Fang, J. Lewis, R. Thomas, Z. Nie, C. Schnell, C. Voisin, A. Young, and the authors — much of this in the last 4 years. This seems like a good time to try to summarize the state of the art and speculate about the future, barring (say) 100 more results between the time of writing and the publication of this volume.

In the classical algebraic geometry of curves, Abel's theorem and Jacobi inversion articulate the relationship (involving rational integrals) between configurations of points with integer multiplicities, or zero-cycles, and an abelian variety known as the Jacobian of the curve: the latter algebraically parametrizes the cycles of degree 0 modulo the subgroup arising as divisors of meromorphic functions. Given a family $\mathcal{X}$ of algebraic curves over a complete base curve S, with smooth fibers over S^* (S minus a finite point set Σ over which fibers have double point singularities), Poincaré [P1; P2] defined *normal functions* as holomorphic sections of the corresponding family of Jacobians over S which behave normally (or logarithmically) in some sense near the boundary. His main result, which says essentially that they parametrize 1-dimensional cycles on $\mathcal{X}$, was then used by Lefschetz (in the context where $\mathcal{X}$ is a pencil of hyperplane sections of a projective algebraic surface) to prove his famous $(1, 1)$ theorem for algebraic surfaces [L]. This later became the basis for the Hodge conjecture, which says that certain *topological-analytic* invariants of an *algebraic* variety must come from *algebraic* subvarieties:

CONJECTURE 1. *For a smooth projective complex algebraic variety X, with* $\mathrm{Hg}^m(X)_{\mathbb{Q}}$ *the classes in* $H^{2m}_{\mathrm{sing}}(X^{\mathrm{an}}_{\mathbb{C}}, \mathbb{Q})$ *of type* (m, m), *and* $CH^m(X)$ *the Chow group of codimension-m algebraic cycles modulo rational equivalence, the fundamental class map* $CH^m(X) \otimes \mathbb{Q} \to \mathrm{Hg}^m(X)_{\mathbb{Q}}$ *is surjective.*

Together with a desire to learn more about the structure of Chow groups (the Bloch–Beilinson conjectures reviewed in §5), this can be seen as the primary motivation behind all the work described (as well as the new results) in this paper. In particular, in §1 (after mathematically fleshing out the Poincaré–Lefschetz story) we describe the attempts to directly generalize Lefschetz's success to higher-codimension cycles which led to Griffiths' Abel–Jacobi map (from the codimension m cycle group of a variety X to its m-th "intermediate" Jacobian), horizontality and variations of mixed Hodge structure, and S. Zucker's Theorem on Normal Functions. As is well-known, the breakdown (beyond codimension 1) of the relationship between cycles and (intermediate)

Jacobians, and the failure of the Jacobians to be algebraic, meant that the same game played in 1 parameter would not work outside very special cases.

It has taken nearly three decades to develop the technical underpinnings for a study of normal functions over a *higher*-dimensional base S: Kashiwara's work on admissible variations of mixed Hodge structure [K], M. Saito's introduction of mixed Hodge modules [S4], multivariable nilpotent and SL_2-orbit theorems ([KNU1],[Pe2]), and so on. And then in 2006, Griffiths and Green had a fundamental idea tying the Hodge conjecture to the presence of *nontorsion singularities* — nontrivial invariants in local intersection cohomology — for multiparameter normal functions arising from Hodge classes on algebraic varieties [GG]. We describe their main result and the follow-up work [BFNP] in §3. Prior to that the reader will need some familiarity with the boundary behavior of "admissible" normal functions arising from higher codimension algebraic cycles. The two principal invariants of this behavior are called *limits* and *singularities*, and we have tried in §2 to give the reader a geometric feel for these through several examples and an explanation of the precise sense in which the limit of Abel–Jacobi invariants (for a family of cycles) is again some kind of Abel–Jacobi invariant. In general throughout §§1–2 (and §4.5–6) normal functions are "of geometric origin" (arise from cycles), whereas in the remainder the formal Hodge-theoretic point of view dominates (though Conjecture 1 is always in the background). We should emphasize that the first two sections are intended for a broad audience, while the last four are of a more specialized nature; one might say that the difficulty level increases exponentially.

The transcendental (nonalgebraic) nature of intermediate Jacobians means that even for a normal function of geometric origin, algebraicity of its vanishing locus (as a subset of the base S), let alone its sensitivity to the field of definition of the cycle, is not a foreordained conclusion. Following a review of Schmid's nilpotent and SL_2-orbit theorems (which lie at the heart of the limit mixed Hodge structures introduced in §2), in §4 we explain how generalizations of those theorems to mixed Hodge structures (and multiple parameters) have allowed complex algebraicity to be proved for the zero loci of "abstract" admissible normal functions [BP1; BP2; BP3; S5]. We then address the field of definition in the geometric case, in particular the recent result of Charles [Ch] under a hypothesis on the VHS underlying the zero locus, the situation when the family of cycles is algebraically equivalent to zero, and what all this means for filtrations on Chow groups. Another reason one would want the zero locus to be algebraic is that the Griffiths–Green normal function attached to a nontrivial Hodge class can then be shown, by an observation of C. Schnell, to have a singularity in the intersection of the zero locus with the boundary $\Sigma \subset S$ (though this intersection could very well be empty).

Now, *a priori*, admissible normal functions (ANFs) are only horizontal and holomorphic sections of a Jacobian bundle over $S \backslash \Sigma$ which are highly constrained along the boundary. Another route (besides orbit theorems) that leads to algebraicity of their zero loci is the construction of a "Néron model" — a partial compactification of the Jacobian bundle satisfying a Hausdorff property (though not a complex analytic space in general) and graphing admissible normal functions over all of S. Néron models are taken up in §5; as they are better understood they may become useful in defining global invariants of (one or more) normal functions. However, unless the underlying variation of Hodge structure (VHS) is a nilpotent orbit the group of components of the Néron model (i.e., the possible singularities of ANFs at that point) over a codimension≥ 2 boundary point remains mysterious. Recent examples of M. Saito [S6] and the second author [Pe3] show that there are analytic obstructions which prevent ANFs from surjecting onto (or even mapping nontrivially to) the putative singularity group for ANFs (rational $(0, 0)$ classes in the local intersection cohomology). At first glance this appears to throw the existence of singularities for Griffiths–Green normal functions (and hence the Hodge conjecture) into serious doubt, but in §5.5 we show that this concern is probably ill-founded.

The last section is devoted to a discussion of Mumford–Tate groups of mixed Hodge structures (introduced by Y. André [An]) and variations thereof, in particular those attached to admissible normal functions. The motivation for writing this section was again to attempt to "force singularities to exist" via conditions on the normal function (e.g., involving the zero locus) which maximize the monodromy of the underlying local system inside the M-T group; we were able to markedly improve André's maximality result (but not to produce singularities). Since the general notion of (non)singularity of a VMHS at a boundary point is defined here (in §6.3), which generalizes the notion of singularity of a normal function, we should point out that there is another sense in which the word "singularity" is used in this paper. The "singularities" of a *period mapping associated to a VHS or VMHS* are points where the connection has poles or the local system has monodromy (Σ in the notation above), and at which one must compute a limit mixed Hodge structure (LMHS). These contain the "singularities of the VMHS", nearly always as a *proper* subset; indeed, pure VHS never have singularities (in the sense of §6.3), though their corresponding period mappings do.

This paper has its roots in the first author's talk at a conference in honor of Phillip Griffiths' 70th birthday at the IAS, and the second author's talk at MSRI during the conference on the topology of stratified spaces to which this volume is dedicated. The relationship between normal functions and stratifications occurs in the context of mixed Hodge modules and the Decomposition Theorem

[BBD], and is most explicitly on display in the construction of the multivariable Néron model in §5 as a topological group whose restrictions to the strata of a Whitney stratification are complex Lie groups. We want to thank the conference organizers and Robert Bryant for doing an excellent job at putting together and hosting a successful interdisciplinary meeting blending (amongst other topics) singularities and topology of complex varieties, L^2 and intersection cohomology, and mixed Hodge theory, all of which play a role below. We are indebted to Patrick Brosnan, Phillip Griffiths, and James Lewis for helpful conversations and sharing their ideas. We also want to thank heartily both referees as well as Chris Peters, whose comments and suggestions have made this a better paper.

One observation on notation is in order, mainly for experts: to clarify the distinction in some places between monodromy weight filtrations arising in LMHS and weight filtrations postulated as part of the data of an admissible variation of mixed Hodge structure (AVMHS), the former are always denoted $M_\bullet$ (and the latter $W_\bullet$) in this paper. In particular, for a degeneration of (pure) weight n HS with monodromy logarithm N, the weight filtration on the LMHS is written $M(N)_\bullet$ (and centered at n). While perhaps nontraditional, this is consistent with the notation $M(N, W)_\bullet$ for relative weight monodromy filtrations for (admissible) degenerations of MHS. That is, when W is "trivial" ($W_n = \mathcal{H}$, $W_{n-1} = \{0\}$) it is simply omitted.

Finally, we would like to draw attention to the interesting recent article [Gr4] of Griffiths which covers ground related to our §§2–5, but in a complementary fashion that may also be useful to the reader.

1. Prehistory and classical results

The present chapter is not meant to be heroic, but merely aims to introduce a few concepts which shall be used throughout the paper. We felt it would be convenient (whatever one's background) to have an up-to-date, "algebraic" summary of certain basic material on normal functions and their invariants in one place. For background or further (and better, but much lengthier) discussion of this material the reader may consult the excellent books [Le1] by Lewis and [Vo2] by Voisin, as well as the lectures of Green and Voisin from the "Torino volume" [GMV] and the papers [Gr1; Gr2; Gr3] of Griffiths.

Even experts may want to glance this section over since we have included some bits of recent provenance: the relationship between log-infinitesimal and topological invariants, which uses work of M. Saito; the result on inhomogeneous Picard–Fuchs equations, which incorporates a theorem of Müller-Stach and del Angel; the important example of Morrison and Walcher related to open mirror symmetry; and the material on K-motivation of normal functions (see §1.3 and §1.7), which will be used in Sections 2 and 4.

Before we begin, a word on the *currents* that play a rôle in the bullet-train proof of Abel's Theorem in § 1.1. These are differential forms with distribution coefficients, and may be integrated against C^∞ forms, with exterior derivative d defined by "integration by parts". They form a complex computing $\mathbb{C}$-cohomology (of the complex manifold on which they lie) and include C^∞ chains and log-smooth forms. For example, for a C^∞ chain Γ, the delta current δ_Γ has the defining property $\int \delta_\Gamma \wedge \omega = \int_\Gamma \omega$ for any C^∞ form ω. (For more details, see Chapter 3 of [GH].)

1.1. Abel's Theorem.

Our (historically incorrect) story begins with a divisor D of degree zero on a smooth projective algebraic curve $X/\mathbb{C}$; the associated analytic variety X^{an} is a Riemann surface. (Except when explicitly mentioned, we continue to work over $\mathbb{C}$.) Writing $D = \sum_{\mathrm{finite}} n_i \, p_i \in Z^1(X)_{\mathrm{hom}}$ ($n_i \in \mathbb{Z}$ such that $\sum n_i = 0$, $p_i \in X(\mathbb{C})$), by Riemann's existence theorem one has a meromorphic 1-form $\hat\omega$ with $\mathrm{Res}_{p_i}(\hat\omega) = n_i$ ($\forall i$). Denoting by $\{\omega_1, \ldots, \omega_g\}$ a basis for $\Omega^1(X)$, consider the map

$$
\overbrace{Z^1(X)_{\mathrm{hom}} \longrightarrow \frac{\Omega^1(X)^\vee}{\int_{H_1(X,\mathbb{Z})}(\cdot)} \xrightarrow[\cong]{\mathrm{ev}_{\{\omega_i\}}} \frac{\mathbb{C}^g}{\Lambda^{2g}} =: J^1(X)}^{\widetilde{AJ}}
\tag{1-1}
$$

$$
D \longmapsto \int_\Gamma \longmapsto \left(\int_\Gamma \omega_1, \ldots, \int_\Gamma \omega_g \right)
$$

where $\Gamma \in C_1(X^{\mathrm{an}})$ is any chain with $\partial\Gamma = D$ and $J^1(X)$ is the *Jacobian* of X. The 1-current $\kappa := \hat\omega - 2\pi i \delta_\Gamma$ is closed; moreover, if $\widetilde{AJ}(D) = 0$ then Γ may be chosen so that all $\int_\Gamma \omega_i = 0$ implies $\int_X \kappa \wedge \omega_i = 0$. We can therefore smooth κ in its cohomology class to $\omega = \kappa - d\eta$ ($\omega \in \Omega^1(X)$; $\eta \in D^0(X) = $ 0-currents), and

$$
f := \exp\left\{ \int (\hat\omega - \omega) \right\}
\tag{1-2}
$$

$$
= e^{2\pi i \int \delta_\Gamma} e^\eta
\tag{1-3}
$$

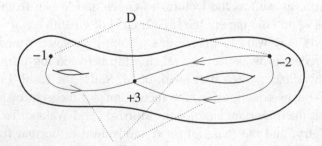

is single-valued — though possibly discontinuous — by (1-3), while being mero-morphic — though possibly multivalued — by (1-2). Locally at p_i, $e^{\int (n_i/z)\,dz} = Cz^{n_i}$ has the right degree; and so the divisor of f is precisely D. Conversely, if $D = (f) = f^{-1}(0) - f^{-1}(\infty)$ for $f \in \mathbb{C}(X)^*$, then

$$t \mapsto \int_{f^{-1}(\overrightarrow{0.t})} (\cdot)$$

induces a holomorphic map $\mathbb{P}^1 \to J^1(X)$. Such a map is necessarily constant (say, to avoid pulling back a nontrivial holomorphic 1-form), and by evaluating at $t = 0$ one finds that this constant is zero. So we have proved part (i) of

THEOREM 2. (i) [Abel] *Writing* $Z^1(X)_{\text{rat}}$ *for the divisors of functions* $f \in \mathbb{C}(X)^*$, $\widetilde{AJ}$ *descends to an injective homomorphism of abelian groups*

$$CH^1(X)_{\text{hom}} := \frac{Z^1(X)_{\text{hom}}}{Z^1(X)_{\text{rat}}} \xrightarrow{AJ} J^1(X).$$

(ii) [Jacobi inversion] AJ *is surjective; in particular, fixing* $q_1, \ldots, q_g \in X(\mathbb{C})$ *the morphism* $\mathrm{Sym}^g X \to J^1(X)$ *induced by* $p_1 + \cdots + p_g \mapsto \int_{\partial^{-1}(\sum p_i - q_i)}(\cdot)$ *is birational.*

Here $\partial^{-1} D$ means any 1-chain bounding on D. Implicit in (ii) is that $J^1(X)$ is an (abelian) algebraic variety; this is a consequence of ampleness of the theta line bundle (on $J^1(X)$) induced by the polarization

$$Q: H^1(X, \mathbb{Z}) \times H^1(X, \mathbb{Z}) \to \mathbb{Z}$$

(with obvious extensions to $\mathbb{Q}, \mathbb{R}, \mathbb{C}$) defined equivalently by cup product, inter-section of cycles, or integration $(\omega, \eta) \mapsto \int_X \omega \wedge \eta$. The ampleness boils down to the *second Riemann bilinear relation*, which says that $iQ(\cdot, \bar{\cdot})$ is positive definite on $\Omega^1(X)$.

1.2. Normal functions. We now wish to vary the Abel–Jacobi map in families. Until § 2, all our normal functions shall be over a curve S. Let $\mathcal{X}$ be a smooth projective surface, and $\bar{\pi}: \mathcal{X} \to S$ a (projective) morphism which is
   (a) smooth off a finite set $\Sigma = \{s_1, \ldots, s_e\} \subset S$, and
   (b) locally of the form $(x_1, x_2) \mapsto x_1 x_2$ at singularities (of $\bar{\pi}$).
   Write $X_s := \bar{\pi}^{-1}(s)$ $(s \in S)$ for the fibers. The singular fibers X_{s_i} $(i = 1, \ldots, e)$ then have only nodal (ordinary double point) singularities, and writing $\mathcal{X}^*$ for their complement we have $\pi: \mathcal{X}^* \to S^* := S \setminus \Sigma$. Fixing a general $s_0 \in S^*$, the local monodromies $T_{s_i} \in \mathrm{Aut}\left(H^1(X_{s_0}, \mathbb{Z}) =: \mathbb{H}_{\mathbb{Z}, s_0}\right)$ of the local system $\mathbb{H}_{\mathbb{Z}} := R^1 \pi_* \mathbb{Z}_{\mathcal{X}^*}$ are then computed by the Picard–Lefschetz formula

$$(T_{s_i} - I)\gamma = \sum_j (\gamma \cdot \delta_j)\delta_j. \tag{1-4}$$

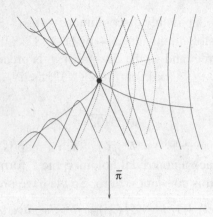

Here $\{\delta_j\}$ are the Poincaré duals of the (possibly nondistinct) vanishing cycle classes $\in \ker\{H_1(X_{s_0},\mathbb{Z}) \to H_1(X_{s_i},\mathbb{Z})\}$ associated to each node on X_{s_i}; we note $(T_{s_i} - I)^2 = 0$. For a family of elliptic curves, (1-4) is just the familiar *Dehn twist:*

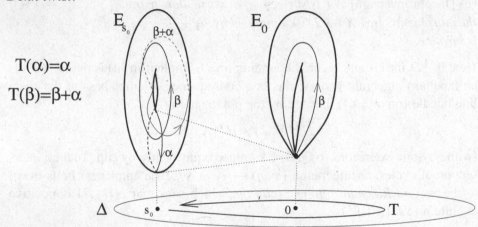

(For the reader new to such pictures, the two crossing segments in the "local real" picture at the top of the page become the two touching "thimbles", i.e., a small neighborhood of the singularity in E_0, in this diagram.)

Now, in our setting, the bundle of Jacobians $\mathcal{J} := \bigcup_{s\in S^*} J^1(X_s)$ is a complex (algebraic) manifold. It admits a partial compactification to a fiber space of complex abelian Lie groups, by defining

$$J^1(X_{s_i}) := \frac{H^0(\omega_{X_{s_i}})}{\text{im}\{H^1(X_{s_i},\mathbb{Z})\}}$$

(with ω_{x_s} the dualizing sheaf) and $\mathcal{J}_e := \bigcup_{s\in S} J^1(X_s)$. (How this is topologized will be discussed in a more general context in §5.) The same notation will denote their sheaves of sections,

$$0 \to \mathbb{H}_{\mathbb{Z}} \to \mathcal{F}^{\vee} \to \mathcal{J} \to 0 \qquad \text{(on } S^*\text{)} \qquad (1\text{-}5)$$

$$0 \to \mathbb{H}_{\mathbb{Z},e} \to (\mathcal{F}_e)^{\vee} \to \mathcal{J}_e \to 0 \qquad \text{(on } S\text{)}, \qquad (1\text{-}6)$$

with $\mathcal{F} := \pi_* \omega_{\mathcal{X}/S}$, $\mathcal{F}_e := \bar{\pi}_* \omega_{\mathcal{X}/S}$, $\mathbb{H}_{\mathbb{Z}} = R^1 \pi_* \mathbb{Z}$, $\mathbb{H}_{\mathbb{Z},e} = R^1 \bar{\pi}_* \mathbb{Z}$.

DEFINITION 3. A *normal function* (NF) is a holomorphic section (over S^*) of $\mathcal{J}$. An *extended* (*or Poincaré*) *normal function* (ENF) is a holomorphic section (over S) of $\mathcal{J}_e$. An NF is *extendable* if it lies in $\mathrm{im}\{H^0(S, \mathcal{J}_e) \to H^0(S^*, \mathcal{J})\}$.

Next consider the long exact cohomology sequence (sections over S^*)

$$0 \to H^0(\mathbb{H}_{\mathbb{Z}}) \to H^0(\mathcal{F}^{\vee}) \to H^0(\mathcal{J}) \to H^1(\mathbb{H}_{\mathbb{Z}}) \to H^1(\mathcal{F}^{\vee}); \qquad (1\text{-}7)$$

the *topological invariant* of a normal function $\nu \in H^0(\mathcal{J})$ is its image $[\nu] \in H^1(S^*, \mathbb{H}_{\mathbb{Z}})$. It is easy to see that the restriction of $[\nu]$ to $H^1(\Delta_i^*, \mathbb{H}_{\mathbb{Z}})$ (Δ_i a punctured disk about s_i) computes the local monodromy $(T_{s_i} - I)\tilde{\nu}$ (where $\tilde{\nu}$ is a multivalued local lift of ν to $\mathcal{F}^{\vee}$), modulo the monodromy of topological cycles. We say that ν is locally liftable if all these restrictions vanish, i.e., if

$$(T_{s_i} - I)\tilde{\nu} \in \mathrm{im}\{(T_{s_i} - I)\mathbb{H}_{\mathbb{Z},s_0}\}.$$

Together with the assumption that as a (multivalued, singular) "section" of $\mathcal{F}_e^{\vee}$, $\tilde{\nu}_e$ has at worst logarithmic divergence at s_i (the "logarithmic growth" in the title), this is equivalent to extendability.

1.3. Normal functions of geometric origin. Let $\mathfrak{Z} \in Z^1(\mathcal{X})_{\mathrm{prim}}$ be a divisor properly intersecting fibers of $\bar{\pi}$ and avoiding its singularities, and which is *primitive* in the sense that each $Z_s := \mathfrak{Z} \cdot X_s$ ($s \in S^*$) is of degree 0. (In fact, the intersection conditions can be done away with, by moving the divisor in a rational equivalence.) Then $s \mapsto AJ(Z_s)$ defines a section $\nu_{\mathfrak{Z}}$ of $\mathcal{J}$, and it can be shown that a multiple $N\nu_{\mathfrak{Z}} = \nu_{N\mathfrak{Z}}$ of $\nu_{\mathfrak{Z}}$ is always extendable. One says that $\nu_{\mathfrak{Z}}$ itself is *admissible*.

Now assume $\bar{\pi}$ has a section $\sigma : S \to \mathcal{X}$ (also avoiding singularities) and consider the analog of (1-7) for $\mathcal{J}_e$

$$0 \to \frac{H^0(\mathcal{F}_e^{\vee})}{H^0(\mathbb{H}_{\mathbb{Z},e})} \to H^0(\mathcal{J}_e) \to \ker\{H^1(\mathbb{H}_{\mathbb{Z},e}) \to H^1(\mathcal{F}_e^{\vee})\} \to 0.$$

With a bit of work, this becomes

$$0 \to J^1(\mathcal{X}/S)_{\mathrm{fix}} \xrightarrow{} ENF \xrightarrow{[\cdot]} \frac{\mathrm{Hg}^1(\mathcal{X})_{\mathrm{prim}}}{\mathbb{Z}\langle[X_{s_0}]\rangle} \to 0, \qquad (1\text{-}8)$$

where the Jacobian of the fixed part $J^1(\mathcal{X}/S)_{\mathrm{fix}} \hookrightarrow J^1(X_s)$ ($\forall s \in S$) gives a constant subbundle of $\mathcal{J}_e$ and the primitive Hodge classes $\mathrm{Hg}^1(\mathcal{X})_{\mathrm{prim}}$ are the

Q-orthogonal complement of a general fiber X_{s_0} of $\bar{\pi}$ in $\mathrm{Hg}^1(\mathcal{X}):=H^2(\mathcal{X},\mathbb{Z})\cap H^{1,1}(\mathcal{X},\mathbb{C})$.

PROPOSITION 4. *Let v be an ENF.*

(i) *If $[v]=0$ then v is a constant section of $\mathcal{J}_{\mathrm{fix}}:=\bigcup_{s\in S}J^1(\mathcal{X}/S)_{\mathrm{fix}}\subset\mathcal{J}_e$.*
(ii) *If $(v=)v_{\mathfrak{Z}}$ is of geometric origin, then $[v_{\mathfrak{Z}}]=\overline{[\mathfrak{Z}]}$ ($[\mathfrak{Z}]=$ fundamental class).*
(iii) *[Poincaré Existence Theorem] Every ENF is of geometric origin.*

We note that (i) follows from considering sections $\{\omega_1,\ldots,\omega_g\}(s)$ of $\mathcal{F}_e^\vee$ whose restrictions to general X_s are linearly independent (such do exist), evaluating a lift $\tilde{v}\in H^0(\mathcal{F}_e^\vee)$ against them, and applying Liouville's Theorem. The resulting constancy of the abelian integrals, by a result in Hodge Theory (cf. end of §1.6), implies the membership of $v(s)\in\mathcal{J}_{\mathrm{fix}}$. To see (iii), apply "Jacobi inversion with parameters" and $q_i(s)=\sigma(s)$ ($\forall i$) over S^* (really, over the generic point of S), and then take Zariski closure.[1] Finally, when v is geometric, the monodromies of a lift $\tilde{v}$ (to $\mathcal{F}_e^\vee$) around each loop in S (which determine $[v]$) are just the corresponding monodromies of a bounding 1-chain Γ_s ($\partial\Gamma_s=Z_s$), which identify with the Leray $(1,1)$ component of $[\mathfrak{Z}]$ in $H^2(\mathcal{X})$; this gives the gist of (ii).

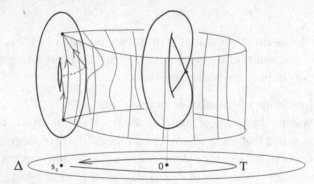

A normal function is said to be *motivated over K* ($K\subset\mathbb{C}$ a subfield) if it is of geometric origin as above, and if the coefficients of the defining equations of $\mathfrak{Z}$, $\mathcal{X}$, $\bar{\pi}$, and S belong to K.

1.4. Lefschetz (1,1) Theorem. Now take $X\subset\mathbb{P}^N$ to be a smooth projective surface of degree d, and $\{X_s:=X\cdot H_s\}_{s\in\mathbb{P}^1}$ a *Lefschetz pencil* of hyperplane sections: the singular fibers have exactly one (nodal) singularity. Let $\beta:\mathcal{X}\twoheadrightarrow X$ denote the blow-up at the base locus $B:=\bigcap_{s\in\mathbb{P}^1}X_s$ of the pencil, and $\bar{\pi}:\mathcal{X}\to\mathbb{P}^1=:S$ the resulting fibration. We are now in the situation considered above, with $\sigma(S)$ replaced by d sections $E_1\amalg\cdots\amalg E_d=\beta^{-1}(B)$, and fibers of genus $g=\binom{d-1}{2}$; and with the added bonus that there is no torsion in any

---

[1] Here the $q_i(s)$ are as in Theorem 2(ii) (but varying with respect to a parameter). If at a generic point $v(\eta)$ is a special divisor then additional argument is needed.

$H^1(\Delta_i^*, \mathbb{H}_\mathbb{Z})$, so that admissible $\Rightarrow$ extendable. Hence, given $Z \in Z^1(X)_{\text{prim}}$ $(\deg(Z \cdot X_{s_0}) = 0)$: $\beta^* Z$ is primitive, $v_Z := v_{\beta^* Z}$ is an ENF, and $[v_Z] = \beta^*[Z]$ under $\beta^* : \text{Hg}^1(X)_{\text{prim}} \hookrightarrow \text{Hg}^1(\mathcal{X})_{\text{prim}}/\mathbb{Z}\langle[X_{s_0}]\rangle$.

If, on the other hand, we start with a Hodge class $\xi \in \text{Hg}^1(X)_{\text{prim}}$, $\beta^* \xi$ is (by (1-8) + Poincaré existence) the class of a geometric ENF v_3; and $[3] \equiv [v_3] \equiv \beta^* \xi$ mod $\mathbb{Z}\langle[X_{s_0}]\rangle$ implies $\xi \equiv \beta_* \beta^* \xi \equiv [\beta_* 3 =: Z]$ in $\text{Hg}^1(X)/\mathbb{Z}\langle[X_{s_0}]\rangle$, which implies $\xi = [Z']$ for some $Z' \in Z^1(X)_{(\text{prim})}$. This is the gist of Lefschetz's original proof [L] of

THEOREM 5. *Let X be a (smooth projective algebraic) surface. The fundamental class map $CH^1(X) \xrightarrow{[\cdot]} \text{Hg}^1(X)$ is (integrally) surjective.*

This continues to hold in higher dimension, as can be seen from an inductive treatment with ENF's or (more easily) from the "modern" treatment of Theorem 5 using the exponential exact sheaf sequence

$$0 \to \mathbb{Z}_X \longrightarrow \mathcal{O}_X \xrightarrow{e^{2\pi i(\cdot)}} \mathcal{O}_X^* \to 0.$$

One simply puts the induced long exact sequence in the form

$$0 \to \frac{H^1(X, \mathcal{O})}{H^1(X, \mathbb{Z})} \to H^1(X, \mathcal{O}^*) \to \ker\{H^2(X, \mathbb{Z}) \to H^2(X, \mathcal{O})\} \to 0,$$

and interprets it as

$$0 \longrightarrow J^1(X) \longrightarrow \left\{\begin{array}{c}\text{holomorphic} \\ \text{line bundles}\end{array}\right\} \longrightarrow \text{Hg}^1(X) \longrightarrow 0 \qquad (1\text{-}9)$$

$$CH^1(X)$$

where the dotted arrow takes the divisor of a meromorphic section of a given bundle. Existence of the section is a standard but nontrivial result.

We note that for $\mathcal{X} \to \mathbb{P}^1$ a Lefschetz pencil of X, in (1-8) we have

$$J^1(\mathcal{X}/\mathbb{P}^1)_{\text{fix}} = J^1(X) := \frac{H^1(X, \mathbb{C})}{F^1 H^1(X, \mathbb{C}) + H^1(X, \mathbb{Z})},$$

which is zero if X is a complete intersection; in that case *ENF* is finitely generated and β^* embeds $\text{Hg}^1(\mathcal{X})_{\text{prim}}$ in *ENF*.

EXAMPLE 6. For X a cubic surface $\subset \mathbb{P}^3$, divisors with support on the 27 lines already surject onto $\text{Hg}^1(X) = H^2(X, \mathbb{Z}) \cong \mathbb{Z}^7$. Differences of these lines generate all primitive classes, hence all of $\text{im}(\beta^*)$ $(\cong \mathbb{Z}^6)$ in *ENF* $(\cong \mathbb{Z}^8)$.

Note that $\mathcal{J}_e$ is essentially an elliptic surface and ENF comprises the (holomorphic) sections passing through the $\mathbb{C}^*$'s over points of Σ. There are no torsion sections.

1.5. Griffiths' AJ map. A $\mathbb{Z}$-Hodge structure (HS) of weight m comprises a finitely generated abelian group $H_{\mathbb{Z}}$ together with a descending filtration $F^{\bullet}$ on $H_{\mathbb{C}} := H_{\mathbb{Z}} \otimes_{\mathbb{Z}} \mathbb{C}$ satisfying $F^p H_{\mathbb{C}} \oplus \overline{F^{m-p+1} H_{\mathbb{C}}} = H_{\mathbb{C}}$, the *Hodge filtration*; we denote the lot by H. Examples include the m-th (singular/Betti + de Rham) cohomology groups of smooth projective varieties over $\mathbb{C}$, with $F^p H_{dR}^m(X, \mathbb{C})$ being that part of the de Rham cohomology represented by C^{∞} forms on X^{an} with *at least* p holomorphic differentials wedged together in each monomial term. (These are forms of *Hodge type* $(p, m - p) + (p + 1, m - p - 1) + \cdots$; note that $H_{\mathbb{C}}^{p, m-p} := F^p H_{\mathbb{C}} \cap \overline{F^{m-p} H_{\mathbb{C}}}$.) To accommodate H^m of nonsmooth or incomplete varieties, the notion of a ($\mathbb{Z}$-)mixed Hodge structure (MHS) V is required: in addition to $F^{\bullet}$ on $V_{\mathbb{C}}$, introduce a decreasing *weight filtration* $W_{\bullet}$ on $V_{\mathbb{Q}}$ such that the $\left(\mathrm{Gr}_i^W V_{\mathbb{Q}}, (\mathrm{Gr}_i^W (V_{\mathbb{C}}, F^{\bullet}))\right)$ are $\mathbb{Q}$-HS of weight i. Mixed Hodge structures have Hodge group

$$\mathrm{Hg}^p(V) := \ker\{V_{\mathbb{Z}} \oplus F^p W_{2p} V_{\mathbb{C}} \to V_{\mathbb{C}}\}$$

(for for $V_{\mathbb{Z}}$ torsion-free becomes $V_{\mathbb{Z}} \cap F^p W_{2p} V_{\mathbb{C}}$) and Jacobian group

$$J^p(V) := \frac{W_{2p} V_{\mathbb{C}}}{F^p W_{2p} V_{\mathbb{C}} + W_{2p} V_{\mathbb{Q}} \cap V_{\mathbb{Z}}},$$

with special cases $\mathrm{Hg}^m(X) := \mathrm{Hg}^m(H^{2m} X))$ and $J^m(X) := J^m(H^{2m-1}(X))$. Jacobians of HS yield complex tori, and subtori correspond bijectively to sub-HS.

A *polarization* of a Hodge structure H is a morphism Q of HS (defined over $\mathbb{Z}$; complexification respects $F^{\bullet}$) from $H \times H$ to the trivial HS $\mathbb{Z}(-m)$ of weight $2m$ (and type (m, m)), such that viewed as a pairing Q is nondegenerate and satisfies a positivity constraint generalizing that in §1.1 (the *second Hodge–Riemann bilinear relation*). A consequence of this definition is that under Q, F^p is the annihilator of F^{m-p+1} (the *first Hodge–Riemann bilinear relation* in abstract form). If X is a smooth projective variety of dimension d, $[\Omega]$ the class of a hyperplane section, write (for $k \leq d$, say)

$$H^m(X, \mathbb{Q})_{\mathrm{prim}} := \ker\{H^m(X, \mathbb{Q}) \xrightarrow{\cup \Omega^{d-k+1}} H^{2d-m+2}(X, \mathbb{Q})\}.$$

This Hodge structure is then polarized by $Q(\xi, \eta) := (-1)^{\binom{m}{2}} \int_X \xi \wedge \eta \wedge \Omega^{d-k}$, $[\Omega]$ the class of a hyperplane section (obviously since this is a $\mathbb{Q}$-HS, the polarization is only defined over $\mathbb{Q}$).

Let X be a smooth projective $(2m-1)$-fold; we shall consider some equivalence relations on algebraic cycles of codimension m on X. Writing $Z^m(X)$ for the free abelian group on irreducible (complex) codimension p subvarieties of X, two cycles $Z_1, Z_2 \in Z^m(X)$ are homologically equivalent if their difference bounds a C^∞ chain $\Gamma \in C_{2m-1}^{\text{top}}(X^{\text{an}}; \mathbb{Z})$ (of real dimension $2m-1$). Algebraic equivalence is generated by (the projection to X of) differences of the form $W \cdot (X \times \{p_1\}) - W \cdot (X \times \{p_2\})$ where C is an algebraic curve, $W \in Z^m(X \times C)$, and $p_1, p_2 \in C(\mathbb{C})$ (or $C(K)$ if we are working over a subfield $K \subset \mathbb{C}$). Rational equivalence is obtained by taking C to be rational ($C \cong \mathbb{P}^1$), and for $m = 1$ is generated by divisors of meromorphic functions. We write $Z^m(X)_{\text{rat}}$ for cycles $\equiv_{\text{rat}} 0$, etc. Note that

$$CH^m(X) := \frac{Z^m(X)}{Z^m(X)_{\text{rat}}} \supset CH^m(X)_{\text{hom}} := \frac{Z^m(X)_{\text{hom}}}{Z^m(X)_{\text{rat}}}$$

and

$$CH^m(X)_{\text{hom}} \supset CH^m(X)_{\text{alg}} := \frac{Z^m(X)_{\text{alg}}}{Z^m(X)_{\text{rat}}}$$

are proper inclusions in general.

Now let $W \subset X \times C$ be an irreducible subvariety of codimension m, with π_X and π_C the projections from a desingularization of W to X and C. If we put $Z_i := \pi_{X_*} \pi_C^* \{p_i\}$, then $Z_1 \equiv_{\text{alg}} Z_2$ implies $Z_1 \equiv_{\text{hom}} Z_2$, which can be seen explicitly by setting $\Gamma := \pi_{X_*} \pi_C^* (\overrightarrow{q.p})$ (so that $Z_1 - Z_2 = \partial \Gamma$).

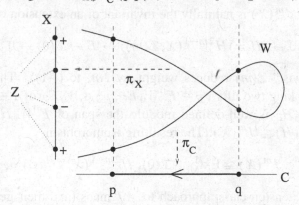

Let ω be a d-closed form of Hodge type $(j, 2m - j - 1)$ on X, for j at least m. Consider $\int_\Gamma \omega = \int_p^q \kappa$, where $\kappa := \pi_{C_*} \pi_X^* \omega$ is a d-closed 1-current of type $(j - m + 1, m - j)$ as integration along the $(m-1)$-dimensional fibers of π_C eats up $(m-1, m-1)$. So $\kappa = 0$ unless $j = m$, and by a standard regularity theorem in that case κ is holomorphic. In particular, if C is rational, we have $\int_\Gamma \omega = 0$. This is essentially the reasoning behind the following result:

PROPOSITION 7. *The Abel–Jacobi map*

$$CH^m(X)_{\text{hom}} \xrightarrow{\quad AJ \quad} \frac{\left(F^m H^{2m-1}(X, \mathbb{C})\right)^{\vee}}{\int_{H_{2m-1}(X, \mathbb{Z})}(\cdot)} \cong J^m(X) \qquad (1\text{-}10)$$

induced by $Z = \partial\Gamma \mapsto \int_{\Gamma}(\cdot)$, *is well-defined and restricts to*

$$CH^m(X)_{\text{alg}} \xrightarrow{\quad AJ_{\text{alg}} \quad} \frac{F^m H^{2m-1}_{\text{hdg}}(X, \mathbb{C})}{\int_{H_{2m-1}(X, \mathbb{Z})}(\cdot)} \cong J^m(H^{2m-1}_{\text{hdg}}(X)) =: J^m_h(X), \quad (1\text{-}11)$$

where $H^{2m-1}_{\text{hdg}}(X)$ *is the largest sub-HS of* $H^{2m-1}(X)$ *contained (after tensoring with* $\mathbb{C}$) *in* $H^{m-1,m}(X, \mathbb{C}) \oplus H^{m,m-1}(X, \mathbb{C})$. *While* $J^m(X)$ *is in general only a complex torus (with respect to the complex structure of Griffiths),* $J^m_h(X)$ *is an abelian variety. Further, assuming a special case of the generalized Hodge conjecture, if* X *is defined over* k *then* $J^m_h(X)$ *and* $AJ_{\text{alg}}(Z)$ *are defined over* $\bar{k}$.

REMARK 8. (i) To see that $J^m_h(X)$ is an abelian variety, one uses the Kodaira embedding theorem: by the Hodge–Riemann bilinear relations, the polarization of $H^{2m-1}(X)$ induces a Kähler metric $h(u, v) = -i Q(u, \bar{v})$ on $J^m_h(X)$ with rational Kähler class.

(ii) The mapping (1-10) is neither surjective nor injective in general, and (1-11) is not injective in general; however, (1-11) is conjectured to be surjective, and regardless of this $J^m_{\text{alg}}(X) := \text{im}(AJ_{\text{alg}}) \subseteq J^m_h(X)$ is in fact a subabelian variety.

(iii) A point in $J^m(X)$ is naturally the invariant of an extension of MHS

$$0 \to (H =) H^{2m-1}(X, \mathbb{Z}(m)) \to E \to \mathbb{Z}(0) \to 0$$

(where the "twist" $\mathbb{Z}(m)$ reduces weight by $2m$, to (-1)). The invariant is evaluated by taking two lifts $v_F \in F^0 W_0 E_{\mathbb{C}}$, $v_{\mathbb{Z}} \in W_0 E_{\mathbb{Z}}$ of $1 \in \mathbb{Z}(0)$, so that $v_F - v_{\mathbb{Z}} \in W_0 H_{\mathbb{C}}$ is well-defined modulo the span of $F^0 W_0 H_{\mathbb{C}}$ and $W_0 H_{\mathbb{Z}}$ hence is in $J^0(H) \cong J^m(X)$. The resulting isomorphism

$$J^m(X) \cong \text{Ext}^1_{\text{MHS}}(\mathbb{Z}(0), H^{2m-1}(X, \mathbb{Z}(m)))$$

is part of an extension-class approach to AJ maps (and their generalizations) due to Carlson [Ca].

(iv) The Abel–Jacobi map appears in [Gr3].

1.6. Horizontality. Generalizing the setting of § 1.2, let $\mathcal{X}$ be a smooth projective $2m$-fold fibered over a curve S with singular fibers $\{X_{s_i}\}$ each of either

(i) NCD (normal crossing divisor) type: locally $(x_1, \ldots, x_{2m}) \xmapsto{\pi} \prod_{j=1}^{k} x_j$; or

(ii) ODP (ordinary double point) type: locally $(\underline{x}) \mapsto \sum_{j=1}^{2m} x_j^2$.

An immediate consequence is that all $T_{s_i} \in \mathrm{Aut}\big(H^{2m-1}(X_{s_0}, \mathbb{Z})\big)$ are *unipotent*: $(T_{s_i} - I)^n = 0$ for $n \geq 2m$ in case (i) or $n \geq 2$ in case (ii). (If all fibers are of NCD type, then we say the family $\{X_s\}$ of $(2m-1)$-folds is *semistable*.)

The Jacobian bundle of interest is $\mathcal{J} := \bigcup_{s \in S^*} J^m(X_s) (\supset \mathcal{J}_{\mathrm{alg}})$. Writing

$$\{\mathcal{F}^{(m)} := \mathbb{R}^{2m-1}\pi_* \Omega^{\bullet \geq m}_{\mathcal{X}^*/S^*}\} \subset \{\mathcal{H} := \mathbb{R}^{2m-1}\pi_* \Omega^{\bullet}_{\mathcal{X}^*/S^*}\}$$
$$\supset \{\mathbb{H}_{\mathbb{Z}} := R^{2m-1}\pi_* \mathbb{Z}_{\mathcal{X}^*}\},$$

and noting $\mathcal{F}^{\vee} \cong \mathcal{H}/\mathcal{F}$ via $Q \colon \mathcal{H}^{2m-1} \times \mathcal{H}^{2m-1} \to \mathcal{O}_{S^*}$, the sequences (1-5) and (1-7), as well as the definitions of NF and topological invariant $[\cdot]$, all carry over. A normal function of geometric origin, likewise, comes from $\mathfrak{Z} \in Z^m(\mathcal{X})_{\mathrm{prim}}$ with $Z_{s_0} := \mathfrak{Z} \cdot X_{s_0} \equiv_{\mathrm{hom}} 0$ (on X_{s_0}), but now has an additional feature known as *horizontality*, which we now explain.

Working locally over an analytic ball $U \subset S^*$ containing s_0, let

$$\tilde{\omega} \in \Gamma(\mathcal{X}_U, F^{m+1}\Omega^{2m-1}_{\mathcal{X}^{\infty}})$$

be a "lift" of $\omega(s) \in \Gamma(U, \mathcal{F}^{m+1})$, and $\Gamma_s \in C^{\mathrm{top}}_{2m-1}(X_s; \mathbb{Z})$ be a continuous family of chains with $\partial \Gamma_s = Z_s$. Let P^{ε} be a path from s_0 to $s_0 + \varepsilon$; then $\hat{\Gamma}^{\varepsilon} := \bigcup_{s \in P^{\varepsilon}} \Gamma_s$ has boundary $\Gamma_{s_0+\varepsilon} - \Gamma_{s_0} + \bigcup_{s \in P^{\varepsilon}} Z_s$, and

$$
\begin{aligned}
\left(\frac{\partial}{\partial s} \int_{\Gamma_s} \omega(s)\right)_{s=s_0} &= \lim_{\varepsilon \to 0} \frac{1}{\varepsilon} \int_{\Gamma_{s_0+\varepsilon}-\Gamma_{s_0}} \tilde{\omega} \\
&= \lim_{\varepsilon \to 0} \frac{1}{\varepsilon}\left(\int_{\partial \hat{\Gamma}^{\varepsilon}} \tilde{\omega} - \int_{s_0}^{s_0+\varepsilon} \int_{Z_s} \omega(s)\right) \\
&= \int_{\Gamma_{s_0}} \langle \widetilde{d/dt}, d\tilde{\omega}\rangle - \int_{Z_{s_0}} \omega(s_0), \qquad (1\text{-}12)
\end{aligned}
$$

where $\pi_* \widetilde{d/dt} = d/dt$ (with $\widetilde{d/dt}$ tangent to $\hat{\Gamma}^{\varepsilon}, \hat{Z}^{\varepsilon}$).

The Gauss–Manin connection $\nabla \colon \mathcal{H} \to \mathcal{H} \otimes \Omega^1_{S^*}$ differentiates the periods of cohomology classes (against topological cycles) in families, satisfies Griffiths transversality $\nabla(\mathcal{F}^m) \subset \mathcal{F}^{m-1} \otimes \Omega^1_{S^*}$, and is computed by

$$\nabla \omega = [\langle \widetilde{d/dt}, d\tilde{\omega}\rangle] \otimes dt.$$

Moreover, the pullback of any form of type F^m to Z_{s_0} (which is of dimension $m-1$) is zero, so that $\int_{Z_{s_0}} \omega(s_0) = 0$ and $\int_{\Gamma_{s_0}} \nabla \omega$ is well-defined. If $\tilde{\Gamma} \in \Gamma(U, \mathcal{H})$ is any lift of $AJ(\Gamma_s) \in \Gamma(U, \mathcal{J})$, we therefore have

$$Q(\nabla_{d/dt}\tilde{\Gamma}, \omega) = \frac{d}{ds}Q(\tilde{\Gamma}, \omega) - Q(\tilde{\Gamma}, \nabla \omega) = \frac{d}{ds}\int_{\Gamma_s} \omega - \int_{\Gamma_s} \nabla \omega,$$

which is zero by (1-12) and the remarks just made. We have shown that $\nabla_{d/dt}\,\tilde{\Gamma}$ kills $\mathcal{F}^{m+1}$, and so $\nabla_{d/dt}\,\tilde{\Gamma}$ is a local section of $\mathcal{F}^{m-1}$.

DEFINITION 9. A normal function $\nu \in H^0(S^*, \mathcal{J})$ is *horizontal* if $\nabla\tilde{\nu} \in \Gamma(U, \mathcal{F}^{m-1} \otimes \Omega^1_U)$ for any local lift $\tilde{\nu} \in \Gamma(U, \mathcal{H})$. Equivalently, if we set $\mathcal{H}_{\text{hor}} := \ker\!\big(\mathcal{H} \to \frac{\mathcal{H}}{\mathcal{F}^{m-1}} \otimes \Omega^1_{S^*}\big) \supset \mathcal{F}^m =: \mathcal{F}$, $(\mathcal{F}^\vee)_{\text{hor}} := \frac{\mathcal{H}_{\text{hor}}}{\mathcal{F}}$, and $\mathcal{J}_{\text{hor}} := \frac{(\mathcal{F}^\vee)_{\text{hor}}}{\mathbb{H}_{\mathbb{Z}}}$, then $\mathrm{NF}_{\text{hor}} := H^0(S, \mathcal{J}_{\text{hor}})$.

Much as an AJ image was encoded in a MHS in Remark 8(ii), we may encode horizontal normal functions in terms of variations of MHS. A VMHS $\mathcal{V}/S^*$ consists of a $\mathbb{Z}$-local system $\mathbb{V}$ with an increasing filtration of $\mathbb{V}_{\mathbb{Q}} := \mathbb{V}_{\mathbb{Z}} \otimes_{\mathbb{Z}} \mathbb{Q}$ by sub- local systems $W_i\mathbb{V}_{\mathbb{Q}}$, a decreasing filtration of $\mathcal{V}_{(\mathcal{O})} := \mathbb{V}_{\mathbb{Q}} \otimes_{\mathbb{Q}} \mathcal{O}_{S^*}$ by holomorphic vector bundles $\mathcal{F}^j (= \mathcal{F}^j\mathcal{V})$, and a connection $\nabla : \mathcal{V} \to \mathcal{V} \otimes \Omega^1_{S^*}$ such that $\nabla(\mathbb{V}) = 0$, the fibers $(\mathbb{V}_s, W_\bullet, V_s, F^\bullet_s)$ yield $\mathbb{Z}$-MHS, and $\nabla(\mathcal{F}^j) \subset \mathcal{F}^{j-1} \otimes \Omega^1_{S^*}$ (transversality). (Of course, a VHS is just a VMHS with one nontrivial $\mathrm{Gr}^W_i\,\mathbb{V}_{\mathbb{Q}}$, and $((\mathbb{H}_{\mathbb{Z}}, \mathcal{H}, \mathcal{F}^\bullet), \nabla)$ in the geometric setting above gives one.) A horizontal normal function corresponds to an extension

$$0 \to \overbrace{\mathcal{H}(m)}^{\substack{\text{wt.}-1 \\ \text{VHS}}} \to \mathcal{E} \to \mathbb{Z}(0)_{S^*} \to 0 \tag{1-13}$$

"varying" the setup of Remark 8(iii), with the transversality of the lift of $\nu_F(s)$ (together with flatness of $\nu_{\mathbb{Z}}(s)$) reflecting horizontality.

REMARK 10. Allowing the left-hand term of (1-13) to have weight less than -1 yields "higher" normal functions related to families of *generalized* ("higher") algebraic cycles. These have been studied in [DM1; DM2; DK], and will be considered in later sections.

An important result on VHS over a smooth quasiprojective base is that the global sections $H^0(S^*, \mathbb{V})$ (resp. $H^0(S^*, \mathbb{V}_{\mathbb{R}})$, $H^0(S^*, \mathbb{V}_{\mathbb{C}})$) span the $\mathbb{Q}$-local system (resp. its tensor product with $\mathbb{R}$, $\mathbb{C}$) of a (necessarily constant) sub-VMHS $\subset \mathcal{V}$, called the *fixed part* $\mathcal{V}_{\text{fix}}$ (with constant Jacobian bundle $\mathcal{J}_{\text{fix}}$).

1.7. Infinitesimal invariant. Given $\nu \in \mathrm{NF}_{\text{hor}}$, the "$\nabla\tilde{\nu}$" for various local liftings patch together after going modulo $\nabla\mathcal{F}^m \subset \mathcal{F}^{m-1} \otimes \Omega^1_{S^*}$. If $\nabla\tilde{\nu} = \nabla f$ for $f \in \Gamma(U, \mathcal{F}^m)$, then the alternate lift $\tilde{\nu} - f$ is flat, i.e., equals $\sum_i c_i \gamma_i$ where $\{\gamma_i\} \subset \Gamma(U, \mathbb{V}_{\mathbb{Z}})$ is a basis and the c_i are complex constants. Since the composition $(s \in S^*)$ $H^{2m-1}(X_s, \mathbb{R}) \hookrightarrow H^{2m-1}(X_s, \mathbb{C}) \twoheadrightarrow \frac{H^{2m-1}(X_s, \mathbb{C})}{F^m}$ is an isomorphism, we may take the $c_i \in \mathbb{R}$, and then they are unique in $\mathbb{R}/\mathbb{Z}$. This implies that $[\nu]$ lies in the torsion group $\ker\big(H^1(\mathbb{H}_{\mathbb{Z}}) \to H^1(\mathbb{H}_{\mathbb{R}})\big)$, so that a multiple $N\nu$ lifts to $H^0(S^*, \mathbb{H}_{\mathbb{R}}) \subset \mathcal{H}_{\text{fix}}$. This motivates the definition of an infinitesimal invariant

$$\delta v \in \mathbb{H}^1\left(S^*, \mathcal{F}^m \xrightarrow{\nabla} \mathcal{F}^{m-1} \otimes \Omega^1_{S^*}\right) \xrightarrow[\text{affine}]{\text{if } S^*} H^0\left(S, \tfrac{\mathcal{F}^{m-1} \otimes \Omega^1}{\mathcal{F}^m}\right) \qquad (1\text{-}14)$$

as the image of $v \in H^0\left(S^*, \tfrac{\mathcal{H}_{\text{hor}}}{\mathcal{F}}\right)$ under the connecting homomorphism induced by

$$0 \to \mathrm{Cone}\left(\mathcal{F}^m \xrightarrow{\nabla} \mathcal{F}^{m-1} \otimes \Omega^1\right)[-1] \to \mathrm{Cone}\left(\mathcal{H} \xrightarrow{\nabla} \mathcal{H} \otimes \Omega^1\right)[-1]$$
$$\to \frac{\mathcal{H}_{\text{hor}}}{\mathcal{F}} \to 0. \quad (1\text{-}15)$$

PROPOSITION 11. *If $\delta v = 0$, then up to torsion, $[v] = 0$ and v is a (constant) section of $\mathcal{J}_{\text{fix}}$.*

An interesting application to the differential equations satisfied by normal functions is essentially due to Manin [Ma]. For simplicity let $S = \mathbb{P}^1$, and suppose $\mathcal{H}$ is generated by $\omega \in H^0(S^*, \mathcal{F}^{2m-1})$ as a D-module, with monic *Picard–Fuchs operator* $F(\nabla_{\delta_s := s\frac{d}{ds}}) \in \mathbb{C}(\mathbb{P}^1)^*[\nabla_{\delta_s}]$ killing ω. Then its periods satisfy the homogeneous P-F equation $F(\delta_s) \int_{\gamma_i} \omega = 0$, and one can look at the multi-valued holomorphic function $Q(\tilde{v}, \omega)$ (where Q is the polarization, and $\tilde{v}$ is a multivalued lift of v to $\mathcal{H}_{\text{hor}}/\mathcal{F}$), which in the geometric case is just $\int_{\Gamma_s} \omega(s)$. The resulting equation

$$(2\pi i)^m F(\delta_s) Q(\tilde{v}, \omega) =: G(s) \qquad (1\text{-}16)$$

is called the *inhomogeneous Picard–Fuchs equation* of v.

PROPOSITION 12. (i) [DM1] $G \in \mathbb{C}(\mathbb{P}^1)^*$ *is a rational function holomorphic on S^*; in the K-motivated setting (taking also $\omega \in H^0(\mathbb{P}^1, \bar{\pi}_* \omega_{\mathcal{X}/\mathbb{P}^1})$, and hence F, over K), $G \in K(\mathbb{P}^1)^*$.*

(ii) [Ma; Gr1] $G \equiv 0 \iff \delta v = 0$.

EXAMPLE 13. [MW] The solutions to

$$(2\pi i)^2 \left\{\delta_z^4 - 5z \prod_{\ell=1}^{4} (5\delta_z + \ell)\right\}(\cdot) = -\frac{15}{4}\sqrt{z}$$

are the membrane integrals $\int_{\Gamma_s} \omega(s)$ for a family of 1-cycles on the mirror quintic family of Calabi–Yau 3-folds. (The family of cycles is actually only well-defined on the double-cover of this family, as reflected by the $\sqrt{z}$.) What makes this example particularly interesting is the "mirror dual" interpretation of the solutions as generating functions of open Gromov–Witten invariants of a fixed Fermat quintic 3-fold.

The horizontality relation $\nabla \tilde{\nu} \in \mathcal{F}^{m-1} \otimes \Omega^1$ is itself a differential equation, and the constraints it puts on ν over higher-dimensional bases will be studied in §5.4–5.

Returning to the setting described in §1.6, there are *canonical extensions* $\mathcal{H}_e$ and $\mathcal{F}_e^\bullet$ of $\mathcal{H}, \mathcal{F}^\bullet$ across the s_i as holomorphic vector bundles or subbundles (reviewed in §2 below); for example, if all fibers are of NCD type then $\mathcal{F}_e^p \cong R^{2m-1}\bar{\pi}_* \Omega_{\mathcal{X}/S}^{\bullet \geq p}(\log(\mathcal{X} \setminus \mathcal{X}^*))$. Writing[2]

$$\mathbb{H}_{\mathbb{Z},e} := R^{2m-1}\bar{\pi}_* \mathbb{Z}_{\mathcal{X}} \quad \text{and} \quad \mathcal{H}_{e,\text{hor}} := \ker\left\{ \mathcal{H}_e \xrightarrow{\nabla} \frac{\mathcal{H}_e}{\mathcal{F}_e^{m-1}} \otimes \Omega_S^1(\log \Sigma) \right\},$$

we have short exact sequences

$$0 \to \mathbb{H}_{\mathbb{Z},e} \to \frac{\mathcal{H}_{e(,\text{hor})}}{\mathcal{F}_e^m} \to \mathcal{J}_{e(,\text{hor})} \to 0 \tag{1-17}$$

and set $ENF_{(hor)} := H^0(S, \mathcal{J}_{e(,\text{hor})})$.

THEOREM 14. (i) $\mathfrak{Z} \in Z^m(\mathcal{X})_{\text{prim}}$ *implies* $N\nu_{\mathfrak{Z}} \in ENF_{\text{hor}}$ *for some* $N \in \mathbb{N}$. (ii) $\nu \in ENF_{\text{hor}}$ *with* $[\nu]$ *torsion implies* $\delta\nu = 0$.

REMARK 15. (ii) is essentially a consequence of the proof of Corollary 2 in [S2]. For $\nu \in ENF_{\text{hor}}$, $\delta\nu$ lies in the subspace

$$\mathbb{H}^1\left(S, \mathcal{F}^m \xrightarrow{\nabla} \mathcal{F}_e^{m-1} \otimes \Omega_S^1(\log \Sigma)\right),$$

the restriction of

$$\mathbb{H}^1\left(S^*, \mathcal{F}^m \xrightarrow{\nabla} \mathcal{F}^{m-1} \otimes \Omega_{S^*}^1\right) \to H^1(S^*, \mathbb{H}_{\mathbb{C}})$$

to which is injective.

1.8. The Hodge Conjecture? Putting together Theorem 14(ii) and Proposition 12, we see that a horizontal ENF with trivial topological invariant lies in $H^0(S, \mathcal{J}_{\text{fix}}) =: J^m(\mathcal{X}/S)_{\text{fix}}$ (constant sections). In fact, the long exact sequence associated to (17) yields

$$0 \to J^m(\mathcal{X}/S)_{\text{fix}} \to ENF_{\text{hor}} \xrightarrow{[\cdot]} \frac{\text{Hg}^m(\mathcal{X})_{\text{prim}}}{\text{im}\{\text{Hg}^{m-1}(X_{s_0})\}} \to 0,$$

with $[\nu_{\mathfrak{Z}}] = \overline{[\mathfrak{Z}]}$ (if $\nu_{\mathfrak{Z}} \in ENF$) as before. If $\mathcal{X} \xrightarrow{\bar{\pi}} \mathbb{P}^1 = S$ is a Lefschetz pencil on a $2m$-fold X, this becomes

---

[2]Warning: while $\mathcal{H}_e$ has no jumps in rank, the stalk of $\mathbb{H}_{\mathbb{Z},e}$ at $s_i \in \Sigma$ is of strictly smaller rank than at $s \in S^*$.

$$
\begin{array}{ccc}
J^m(X) \hookrightarrow & ENF_{\mathrm{hor}} \xrightarrow[(*)]{[\cdot]} & \mathrm{Hg}^m(X)_{\mathrm{prim}} \oplus \ker\left\{\begin{matrix}\mathrm{Hg}^{m-1}(B)\\ \to \mathrm{Hg}^m(X)\end{matrix}\right\} \\
\uparrow \scriptstyle{AJ} & \uparrow \scriptstyle{v_{(\cdot)}} & \\
 & CH^m(\mathcal{X})_{\mathrm{prim}} \Big) v_{(\cdot)} & \uparrow \scriptstyle{(\mathrm{id},0)} \qquad\qquad (1\text{-}18) \\
 & \uparrow \scriptstyle{\beta^*} & \\
\ker([\cdot]) \hookrightarrow & CH^m(X)_{\mathrm{prim}} \xrightarrow[(**)]{[\cdot]} & \mathrm{Hg}^m(X)_{\mathrm{prim}}
\end{array}
$$

where the surjectivity of $(*)$ is due to Zucker (compare Theorems 31 and 32 in §3 below; his result followed on work of Griffiths and Bloch establishing the surjectivity for *sufficiently ample* Lefschetz pencils). What we are after (modulo tensoring with $\mathbb{Q}$) is surjectivity of the fundamental class map $(**)$. This would clearly follow from surjectivity of $v_{(\cdot)}$, i.e., a Poincaré existence theorem, as in §1.4. By Remark 8(ii) this cannot work in most cases; however we have this:

THEOREM 16. *The Hodge Conjecture $HC(m,m)$ is true for X if $J^m(X_{s_0}) = J^m(X_{s_0})_{\mathrm{alg}}$ for a general member of the pencil.*

EXAMPLE 17 [Zu1]. As $J^2 = J^2_{\mathrm{alg}}$ is true for cubic threefolds by the work of Griffiths and Clemens [GC], $HC(2,2)$ holds for cubic fourfolds in $\mathbb{P}^5$.

The Lefschetz paradigm, of taking a 1-parameter family of slices of a primitive Hodge class to get a normal function and constructing a cycle by Jacobi inversion, appears to have led us (for the most part) to a dead end in higher codimension. A beautiful new idea of Griffiths and Green, to be described in §3, replaces the Lefschetz pencil by a complete linear system (of higher degree sections of X) so that $\dim(S) \gg 1$, and proposes to recover algebraic cycles *dual* to the given Hodge class from features of the (admissible) normal function in codimension ≥ 2 on S.

1.9. Deligne cycle-class. This replaces the fundamental and AJ classes by one object. Writing $\mathbb{Z}(m) := (2\pi i)^m \mathbb{Z}$, define the Deligne cohomology of X (smooth projective of any dimension) by

$$
H^*_{\mathcal{D}}(X^{\mathrm{an}}, \mathbb{Z}(m)) :=
$$
$$
H^*\big(\mathrm{Cone}\{C^{\bullet}_{\mathrm{top}}(X^{\mathrm{an}}; \mathbb{Z}(m)) \oplus F^m \mathcal{D}^{\bullet}(X^{\mathrm{an}}) \to D^{\bullet}(X^{\mathrm{an}})\}[-1]\big),
$$

and $c_{\mathcal{D}} : CH^m(X) \to H^{2m}_{\mathcal{D}}(X, \mathbb{Z}(m))$ by $Z \mapsto (2\pi i)^m (Z_{\mathrm{top}}, \delta_Z, 0)$. One easily derives the exact sequence

$$
0 \to J^m(X) \to H^{2m}_{\mathcal{D}}(X, \mathbb{Z}(m)) \to \mathrm{Hg}^m(X) \to 0,
$$

which invites comparison to the top row of (1-18).

2. Limits and singularities of normal functions

Focusing on the geometric case, we now wish to give the reader a basic intuition for many of the objects — singularities, Néron models, limits of NF's and VHS — which will be treated from a more formal Hodge-theoretic perspective in later sections.[3] The first part of this section (§§ 2.2–8) considers a cohomologically trivial cycle on a 1-parameter semistably degenerating family of odd-dimensional smooth projective varieties. Such a family has two invariants "at" the central singular fiber:

- the limit of the Abel–Jacobi images of the intersections of the cycle with the smooth fibers, and
- the Abel–Jacobi image of the intersection of the cycle with the singular fiber.

We define what these mean and explain the precise sense in which they agree, which involves limit mixed Hodge structures and the Clemens–Schmid exact sequence, and links limits of AJ maps to the Bloch–Beilinson regulator on higher K-theory.

In the second part, we consider what happens if the cycle is only assumed to be homologically trivial *fiberwise*. In this case, just as the fundamental class of a cycle on a variety must be zero to define its AJ class, the family of cycles has a singularity class which must be zero in order to define the limit AJ invariant. Singularities are first introduced for normal functions arising from families of cycles, and then in the abstract setting of admissible normal functions (and higher normal functions). At the end we say a few words about the relation of singularities to the Hodge conjecture, their rôle in multivariable Néron models, and the analytic obstructions to singularities discovered by M. Saito, topics which § 3, § 5.1–2, and § 5.3–5, respectively, will elaborate extensively upon.

We shall begin by recasting $c_{\mathcal{D}}$ from § 1.9 in a more formal vein, which works $\otimes \mathbb{Q}$. The reader should note that henceforth in this paper, we have to introduce appropriate Hodge twists (largely suppressed in § 1) into VHS, Jacobians, and related objects.

2.1. AJ map. As we saw earlier (Section 1), the AJ map is the basic Hodge-theoretic invariant attached to a cohomologically trivial algebraic cycle on a smooth projective algebraic variety $X/\mathbb{C}$; say $\dim(X) = 2m-1$. In the diagram that follows, if $cl_{X,\mathbb{Q}}(Z) = 0$ then $Z = \partial \Gamma$ for Γ (say) a rational C^∞ $(2m-1)$-chain on X^{an}, and $\int_\Gamma \in (F^m H^{2m-1}(X, \mathbb{C}))^\vee$ induces $AJ_{X,\mathbb{Q}}(Z)$.

---

[3] Owing to our desire to limit preliminaries and/or notational complications here, there are a few unavoidable inconsistencies of notation between this and later sections.

$$\text{Hom}_{\text{MHS}}\left(\mathbb{Q}(0), H^{2m}(X, \mathbb{Q}(m))\right) = \!\!= \!\!= (H^{2m}(X))_{\mathbb{Q}}^{(m,m)}$$

$$CH^m(X) \xrightarrow{\quad\quad} \text{Ext}^1_{D^b\text{MHS}}\left(\mathbb{Q}(0), \mathcal{K}^\bullet[2m](m)\right)$$

$$\ker(cl_X) \xrightarrow{\;AJ_X\;} \text{Ext}^1_{\text{MHS}}\left(\mathbb{Q}(0), H^{2m-1}(X, \mathbb{Q}(m))\right) = \!\!= J^m(X)_{\mathbb{Q}} \cong \frac{\left(F^m H_{\mathbb{C}}^{2m-1}\right)^{\vee}}{H_{\mathbb{Q}(m)}^{2m-1}}$$

$$(2\text{-}1)$$

where the cl_X map is drawn as a diagonal arrow.

The middle term in the vertical short-exact sequence is isomorphic to Deligne cohomology and Beilinson's absolute Hodge cohomology $H_{\mathcal{H}}^{2m}(X^{\text{an}}, \mathbb{Q}(m))$, and can be regarded as the ultimate strange fruit of Carlson's work on extensions of mixed Hodge structures. Here $\mathcal{K}^\bullet$ is a canonical complex of MHS quasi-isomorphic (noncanonically) to $\bigoplus_i H^i(X)[-i]$, constructed from two general configurations of hyperplane sections $\{H_i\}_{i=0}^{2m-1}$, $\{\tilde{H}_j\}_{j=0}^{2m-1}$ of X. More precisely, looking (for $|I|, |J| > 0$) at the corresponding "cellular" cohomology groups

$$C_{H,\tilde{H}}^{I,J}(X) := H^{2m-1}\left(X \setminus \bigcup_{i \in I} H_i, \; \bigcup_{j \in J} H_j \setminus \cdots; \mathbb{Q}\right),$$

one sets

$$\mathcal{K}^\ell := \bigoplus_{\substack{I,J \\ |I|-|J|=\ell-2m+1}} C_{H,\tilde{H}}^{I,J}(X);$$

refer to [RS]. (Ignoring the description of $J^m(X)$ and AJ, and the comparisons to $c_{\mathcal{D}}$, $H_{\mathcal{D}}$, all of this works for smooth quasiprojective X as well; the vertical short-exact sequence is true even without smoothness.)

The reason for writing AJ in this way is to make plain the analogy to (2-9) below. We now pass back to $\mathbb{Z}$-coefficients.

2.2. AJ in degenerating families.

To let $AJ_X(Z)$ vary with respect to a parameter, consider a semistable degeneration (SSD) over an analytic disk

$$\begin{array}{ccccc}
\mathcal{X}^* & \hookrightarrow & \mathcal{X} & \xleftarrow{\iota_0} & X_0 = \bigcup_i Y_i \\
\downarrow \pi & & \downarrow \tilde{\pi} & & \downarrow \\
\Delta^* & \xhookrightarrow{\;J\;} & \Delta & \hookleftarrow & \{0\}
\end{array}$$

$$(2\text{-}2)$$

where X_0 is a reduced NCD with smooth irreducible components Y_i, $\mathcal{X}$ is smooth of dimension $2m$, $\tilde{\pi}$ is proper and holomorphic, and π is smooth. An algebraic cycle $\mathfrak{Z} \in Z^m(\mathcal{X})$ properly intersecting fibers gives rise to a family

$$Z_s := \mathfrak{Z} \cdot X_s \in Z^m(X_s), \quad s \in \Delta.$$

Assume $0 = [3] \in H^{2m}(\mathcal{X})$ [which implies $0 = [Z_s] \in H^{2m}(X_s)$]; then is there a sense in which

$$\lim_{s \to 0} AJ_{X_s}(Z_s) = AJ_{X_0}(Z_0)? \qquad (2\text{-}3)$$

(Of course, we have yet to say what either side means.)

2.3. Classical example. Consider a degeneration of elliptic curves E_s which pinches 3 loops in the same homology class to points, yielding for E_0 three $\mathbb{P}^1$'s joined at 0 and ∞ (called a "Néron 3-gon" or "Kodaira type I_3" singular fiber).

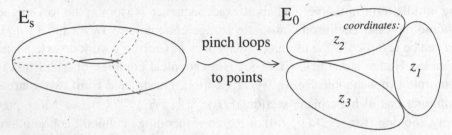

Denote the total space by $\mathcal{E} \xrightarrow{\bar{\pi}} \Delta$. One has a family of holomorphic 1-forms $\omega_s \in \Omega^1(E_s)$ limiting to $\{\mathrm{dlog}(z_j)\}_{j=1}^3$ on E_0; this can be thought of as a holomorphic section of $R^0\bar{\pi}_* \Omega^1_{\mathcal{E}/\Delta}(\log E_0)$.

There are two distinct possibilities for limiting behavior when $Z_s = p_s - q_s$ is a difference of points. (These do not include the case where one or both of p_0, q_0 lies in the intersection of two of the $\mathbb{P}^1$'s, since in that case 3 is not considered to properly intersect X_0.)

Case (I):

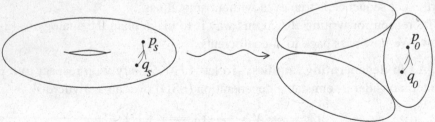

Here p_0 and q_0 lie in the same $\mathbb{P}^1$ (the $j = 1$ component, say): in which case

$$AJ_{E_s}(Z_s) = \int_{q_s}^{p_s} \omega_s \in \mathbb{C}/\mathbb{Z}\langle \int_{\alpha_s}\omega_s, \int_{\beta_s}\omega_s\rangle$$

limits to

$$\int_{p_0}^{q_0} \mathrm{dlog}(z_1) = \log\frac{z_1(p_0)}{z_1(q_0)} \in \mathbb{C}/2\pi i \mathbb{Z}.$$

Case (II):

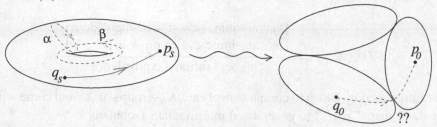

In this case, p_0 and q_0 lie in different $\mathbb{P}^1$ components, in which case $0 \neq [Z_0] \in H^2(X_0)$ [which implies [3] $\neq 0$] and we say that $AJ(Z_0)$ is "obstructed".

2.4. Meaning of the LHS of (2-3). If we assume only that $0 = [3^*] \in H^{2m}(\mathcal{X}^*)$, then

$$AJ_{X_s}(Z_s) \in J^m(X_s) \qquad (2\text{-}4)$$

is defined for each $s \in \Delta^*$. We can make this into a horizontal, holomorphic section of a bundle of intermediate Jacobians, which is what we shall mean henceforth by a *normal function* (on Δ^* in this case).

Recall the ingredients of a variation of Hodge structure (VHS) over Δ^*:

$$\mathcal{H} = ((\mathbb{H}, \mathcal{H}_{\mathcal{O}}, \mathcal{F}^\bullet), \nabla), \quad \nabla \mathcal{F}^p \subset \mathcal{F}^{p-1} \otimes \Omega_S^1, \quad 0 \to \mathbb{H} \to \frac{\mathcal{H}_{\mathcal{O}}}{\mathcal{F}^m} \to \mathcal{J} \to 0,$$

where $\mathbb{H} = R^{2m-1}\pi_*\mathbb{Z}(m)$ is a local system, $\mathcal{H}_{\mathcal{O}} = \mathbb{H} \otimes_{\mathbb{Z}} \mathcal{O}_{\Delta^*}$ is [the sheaf of sections of] a holomorphic vector bundle with holomorphic subbundles $\mathcal{F}^\bullet$, and these yield HS's H_s fiberwise (notation: $H_s = (\mathbb{H}_s, H_{s(,\mathbb{C})}, F_s^\bullet)$). Henceforth we shall abbreviate $\mathcal{H}_{\mathcal{O}}$ to $\mathcal{H}$.

Then (2-4) yields a section of the intermediate Jacobian bundle

$$v_3 \in \Gamma(\Delta^*, \mathcal{J}).$$

Any holomorphic vector bundle over Δ^* is trivial, each trivialization inducing an extension to Δ. The extensions we want are the "canonical" or "privileged" ones (denoted $(\cdot)_e$); as in §1.7, we define an extended Jacobian bundle $\mathcal{J}_e$ by

$$0 \to j_*\mathbb{H} \to \frac{\mathcal{H}_e}{\mathcal{F}_e^m} \to \mathcal{J}_e \to 0. \qquad (2\text{-}5)$$

THEOREM 18 [EZ]. *There exists a holomorphic $\bar{v}_3 \in \Gamma(\Delta, \mathcal{J}_e)$ extending v_3.*

Define $\lim_{s \to 0} AJ_{X_s}(Z_s) := \bar{v}_3(0)$ in $(\mathcal{J}_e)_0$, the fiber over 0 of the Jacobian bundle. To be precise: since $H^1(\Delta, j_*\mathbb{H}) = \{0\}$, we can lift the $\bar{v}_3$ to a section of the middle term of (2-5), i.e., of a vector bundle, evaluate at 0, then quotient by $(j_*\mathbb{H})_0$.

2.5. Meaning of the RHS of (2-3). Higher Chow groups

$$CH^p(X,n) := \frac{\left\{ \begin{array}{c} \text{``admissible, closed'' codimension } p \\ \text{algebraic cycles on } X \times \mathbb{A}^n \end{array} \right\}}{\text{``higher'' rational equivalence}}$$

were introduced by Bloch to compute algebraic K_n-groups of X, and come with "regulator maps" $\mathrm{reg}^{p,n}$ to generalized intermediate Jacobians

$$J^{p,n}(X) := \frac{H^{2p-n-1}(X,\mathbb{C})}{F^p H^{2p-n-1}(X,\mathbb{C}) + H^{2p-n-1}(X,\mathbb{Z}(p))}.$$

(Explicit formulas for $\mathrm{reg}^{p,n}$ have been worked out by the first author with J. Lewis and S. Müller-Stach in [KLM].) The singular fiber X_0 has motivic cohomology groups $H^*_{\mathcal{M}}(X_0, \mathbb{Z}(\cdot))$ built out of higher Chow groups on the substrata

$$Y^{[\ell]} := \amalg_{|I|=\ell+1} Y_I := \amalg_{|I|=\ell+1} \left(\bigcap_{i \in I} Y_i \right),$$

(which yield a semi-simplicial resolution of X_0). Inclusion induces

$$\iota_0^* : CH^m(\mathcal{X})_{\mathrm{hom}} \to H^{2m}_{\mathcal{M}}(X_0, \mathbb{Z}(m))_{\mathrm{hom}}$$

and we define $Z_0 := \iota_0^* 3$. The AJ map

$$AJ_{X_0} : H^{2m}_{\mathcal{M}}(X_0, \mathbb{Z}(m))_{\mathrm{hom}} \to J^m(X_0) := \frac{H^{2m-1}(X_0,\mathbb{C})}{\left\{ \begin{array}{c} F^m H^{2m-1}(X_0,\mathbb{C})+ \\ H^{2m-1}(X_0,\mathbb{Z}(m)) \end{array} \right\}}$$

is built out of regulator maps on substrata, in the sense that the semi-simplicial structure of X_0 induces "weight" filtrations $M_\bullet$ on both sides[4] and

$$\mathrm{Gr}^M_{-\ell} H^{2m}_{\mathcal{M}}(X_0, \mathbb{Z}(m))_{\mathrm{hom}} \xrightarrow{\mathrm{Gr}^M_{-\ell} AJ} \mathrm{Gr}^M_{-\ell} J^m(X_0)$$

boils down to

$$\{\text{subquotient of } CH^m(Y^{[\ell]}, \ell)\} \xrightarrow{\mathrm{reg}^{m,\ell}} \{\text{subquotient of } J^{m,\ell}(Y^{[\ell]})\}.$$

---

[4]For the advanced reader, we note that if $M_\bullet$ is Deligne's weight filtration on $H^{2m-1}(X_0, \mathbb{Z}(m))$, then $M_{-\ell} J^m(X_0) := \mathrm{Ext}^1_{\mathrm{MHS}}(\mathbb{Z}(0), M_{-\ell-1} H^{2m-1}(X_0, \mathbb{Z}(m)))$. The definition of the $M_\bullet$ filtration on motivic cohomology is much more involved, and we must refer the reader to [GGK, sec. III.A].

2.6. Meaning of equality in (2-3).

Specializing (2-5) to 0, we have

$$(\bar{\nu}_3(0) \in)\; J^m_{\lim}(X_s) := (\mathcal{J}_e)_0 = \frac{(\mathcal{H}_e)_0}{(\mathcal{F}^m_e)_0 + (J_*\mathbb{H})_0},$$

where $(J_*\mathbb{H})_0$ are the monodromy invariant cycles (and we are thinking of the fiber $(\mathcal{H}_e)_0$ over 0 as the limit MHS of $\mathcal{H}$, see next subsection). H. Clemens [Cl1] constructed a retraction map $\mathfrak{r} : \mathcal{X} \twoheadrightarrow X_0$ inducing

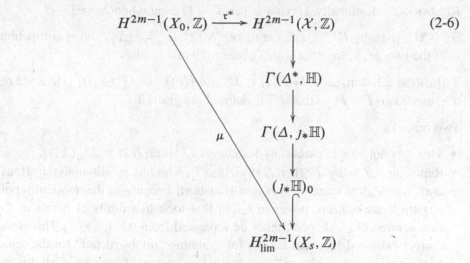

$$H^{2m-1}(X_0, \mathbb{Z}) \xrightarrow{\;\mathfrak{r}^*\;} H^{2m-1}(\mathcal{X}, \mathbb{Z}) \qquad (2\text{-}6)$$

$$\downarrow$$

$$\Gamma(\Delta^*, \mathbb{H})$$

$$\downarrow$$

$$\mu \searrow \qquad \Gamma(\Delta, J_*\mathbb{H})$$

$$\downarrow$$

$$(J_*\mathbb{H})_0$$

$$\hookleftarrow$$

$$H^{2m-1}_{\lim}(X_s, \mathbb{Z})$$

(where μ is a morphism of MHS), which in turn induces

$$J(\mu) : J^m(X_0) \to J^m_{\lim}(X_s).$$

THEOREM 19 [GGK]. $\lim_{s\to 0} AJ_{X_s}(Z_s) = J(\mu)\left(AJ_{X_0}(Z_0)\right).$

2.7. Graphing normal functions.

On Δ^*, let $T : \mathbb{H} \to \mathbb{H}$ be the counterclockwise monodromy transformation, which is unipotent since the degeneration is semistable. Hence the monodromy logarithm

$$N := \log(T) = \sum_{k=1}^{2m-1} \frac{(-1)^{k-1}}{k}(T - I)^k$$

is defined, and we can use it to "untwist" the local system $\otimes\mathbb{Q}$:

$$\mathbb{H}_\mathbb{Q} \mapsto \tilde{\mathbb{H}}_\mathbb{Q} := \exp\left(-\frac{\log s}{2\pi i} N\right) \mathbb{H}_\mathbb{Q} \hookrightarrow \mathcal{H}_e.$$

In fact, this yields a basis for, and defines, the privileged extension $\mathcal{H}_e$. Moreover, since N acts on $\tilde{\mathbb{H}}_\mathbb{Q}$, it acts on $\mathcal{H}_e$, and therefore on $(\mathcal{H}_e)_0 = H^{2m-1}_{\lim}(X_s)$, inducing a "weight monodromy filtration" $M_\bullet$. Writing $H = H^{2m-1}_{\lim}(X_s, \mathbb{Q}(m))$, this is the unique filtration $\{0\} \subset M_{-2m} \subset \cdots \subset M_{2m-2} = H$ satisfying

$N(M_k) \subset M_{k-2}$ and $N^k : \mathrm{Gr}^M_{-1+k} H \xrightarrow{\cong} \mathrm{Gr}^M_{-1-k} H$ for all k. In general it is centered about the weight of the original variation (cf. the convention in the Introduction).

EXAMPLE 20. In the "Dehn twist" example of §1.2, $N = T - I$ (with $N(\alpha) = 0$, $N(\beta) = \alpha$) so that $\tilde{\alpha} = \alpha$, $\tilde{\beta} = \beta - \frac{\log s}{2\pi i}\alpha$ are monodromy free and yield an $\mathcal{O}_\Delta$-basis of $\mathcal{H}_e$. We have $M_{-3} = \{0\}$, $M_{-2} = M_{-1} = \langle\alpha\rangle$, $M_0 = H$.

REMARK 21. Rationally, $\ker(N) = \ker(T - I)$ even when $N \neq T - I$.

By [Cl1], μ maps $H^{2m-1}(X_0)$ onto $\ker(N) \subset H^{2m-1}_{\lim}(X_s)$ and is compatible with the two $M_\bullet$'s; together with Theorem 19 this implies

THEOREM 22. $\lim_{s\to 0} AJ_{X_s}(Z_s) \in J^m\,(\ker(N))\ (\subset J^m_{\lim}(X_s))$. (Here we really mean $\ker(T - I)$ so that J^m is defined integrally.)

Two remarks:

- This was not visible classically for curves ($J^1(\ker(N)) = J^1_{\lim}(X_s)$).
- Replacing $(\mathcal{J}_e)_0$ by $J^m(\ker(N))$ yields $\mathcal{J}'_e$, which is a "slit-analytic[5] Hausdorff topological space" ($\mathcal{J}_e$ is non-Hausdorff because in the quotient topology there are nonzero points in $(\mathcal{J}_e)_0$ that look like limits of points in the zero-section of $\mathcal{J}_e$, hence cannot be separated from $0 \in (\mathcal{J}_e)_0$.[6]) This is the correct extended Jacobian bundle for graphing "unobstructed" (in the sense of the classical example) or "singularity-free" normal functions. Call this the "pre-Néron-model".

2.8. Nonclassical example.

Take a degeneration of Fermat quintic 3-folds

$$\mathcal{X} = \text{semistable reduction of} \left\{ s\sum_{j=1}^{4} z_j^5 = \prod_{k=0}^{4} z_k \right\} \subset \mathbb{P}^4 \times \Delta,$$

so that X_0 is the union of 5 $\mathbb{P}^3$'s blown up along curves isomorphic to $C = \{x^5 + y^5 + z^5 = 0\}$. Its motivic cohomology group $H^4_{\mathcal{M}}(X_0, \mathbb{Q}(2))_{\mathrm{hom}}$ has Gr^M_0 isomorphic to 10 copies of $\mathrm{Pic}^0(C)$, Gr^M_{-1} isomorphic to 40 copies of $\mathbb{C}^*$, $\mathrm{Gr}^W_{-2} = \{0\}$, and $\mathrm{Gr}^M_{-3} \cong K^{\mathrm{ind}}_3(\mathbb{C})$. One has a commuting diagram

$$
\begin{array}{ccc}
H^4_{\mathcal{M}}(X_0, \mathbb{Q}(2))_{\mathrm{hom}} \xrightarrow{\ AJ_{X_0}\ } & J^2(X_0)_{\mathbb{Q}} =\!\!=\!\!= & J^2(\ker(N))_{\mathbb{Q}} \\
\uparrow \qquad\qquad\qquad & \uparrow & \\
K^{\mathrm{ind}}_3(\mathbb{C}) \xrightarrow{\ \mathrm{reg}^{2,3}\ } & \mathbb{C}/(2\pi i)^2\mathbb{Q} \xrightarrow{\ \mathrm{Im}\ } & \mathbb{R}
\end{array}
\tag{2-7}
$$

---

[5]That is, each point has a neighborhood of the form: open ball about $\underline{0}$ in $\mathbb{C}^{a+b}$ intersected with $((\mathbb{C}^a\backslash\{\underline{0}\}) \times \mathbb{C}^b) \cup (\{\underline{0}\} \times \mathbb{C}^c)$, where $c \leq b$.

[6]See the example before Theorem II.B.9 in [GGK].

and explicit computations with higher Chow precycles in [GGK, §4] lead to the result:

THEOREM 23. *There exists a family of 1-cycles* $\mathfrak{Z} \in CH^2(\mathcal{X})_{\mathrm{hom},\mathbb{Q}}$ *such that* $Z_0 \in M_{-3}H^4_{\mathcal{M}}$ *and* $\mathrm{Im}(AJ_{X_0}(Z_0)) = D_2(\sqrt{-3})$, *where* D_2 *is the Bloch–Wigner function.*

Hence, $\lim_{s \to 0} AJ_{X_s}(Z_s) \neq 0$ and so the general Z_s in this family is not rationally equivalent to zero. The main idea is that the family of cycles limits to a (nontrivial) higher cycle in a substratum of the singular fiber.

2.9. Singularities in 1 parameter. If only $[Z_s] = 0$ $(s \in \Delta^*)$, and $[\mathfrak{Z}^*] = 0$ *fails*, then

$$\lim_{s \to 0} AJ \text{ is obstructed}$$

and we say $\bar{\nu}_3(s)$ has a singularity (at $s = 0$), measured by the finite group

$$G \cong \frac{\mathrm{Im}(T_{\mathbb{Q}} - I) \cap \mathbb{H}_{\mathbb{Z}}}{\mathrm{Im}(T_{\mathbb{Z}} - I)} = \begin{cases} \mathbb{Z}/3\mathbb{Z} \text{ in the classical example,} \\ (\mathbb{Z}/5\mathbb{Z})^3 \text{ in the nonclassical one.} \end{cases}$$

(The $(\mathbb{Z}/5\mathbb{Z})^3$ is generated by differences of lines limiting to distinct components of X_0.) The Néron model is then obtained by replacing $J(\ker(N))$ (in the pre-Néron-model) by its product with G (this will graph *all* admissible normal functions, as defined below).

The next example demonstrates the "finite-group" (or torsion) nature of singularities in the 1-parameter case. In §2.10 we will see how this feature disappears when there are many parameters.

EXAMPLE 24. Let $\xi \in \mathbb{C}$ be general and fixed. Then

$$C_s = \{x^2 + y^2 + s(x^2 y^2 + \xi) = 0\}$$

defines a family of elliptic curves (in $\mathbb{P}^1 \times \mathbb{P}^1$) over Δ^* degenerating to a Néron 2-gon at $s = 0$. The cycle

$$Z_s := \left(i \sqrt{\frac{1+\xi s}{1+s}}, 1 \right) - \left(-i \sqrt{\frac{1+\xi s}{1+s}}, 1 \right)$$

is nontorsion, with points limiting to distinct components. (See figure on next page.)

Hence, $AJ_{C_s}(Z_s) =: \nu(s)$ limits to the nonidentity component $(\cong \mathbb{C}^*)$ of the Néron model. The presence of the nonidentity component removes the obstruction (observed in case (II) of §2.3) to graphing ANFs with singularities.

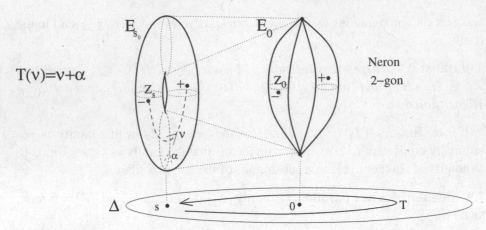

$T(\nu)=\nu+\alpha$

Two remarks:

- By tensoring with $\mathbb{Q}$, we can "correct" this: write α, β for a basis for $H^1(C_s)$ and N for the monodromy log about 0, which sends $\alpha \mapsto 0$ and $\beta \mapsto 2\alpha$. Since $N(\nu) = \alpha = N(\frac{1}{2}\beta)$, $\nu - \frac{1}{2}\beta$ will pass through the identity component (which becomes isomorphic to $\mathbb{C}/\mathbb{Q}(1)$ after tensoring with $\mathbb{Q}$, however).
- Alternately, to avoid tensoring with $\mathbb{Q}$, one can add a 2-torsion cycle like

$$T_s := (i\xi^{\frac{1}{4}}, \xi^{\frac{1}{4}}) - (-i\xi^{\frac{1}{4}}, -\xi^{\frac{1}{4}}).$$

2.10. Singularities in 2 parameters.

EXAMPLE 25. Now we will effectively allow ξ (from the last example) to vary: consider the smooth family

$$C_{s,t} := \{x^2 + y^2 + sx^2y^2 + t = 0\}$$

over $(\Delta^*)^2$. The degenerations $t \to 0$ and $s \to 0$ pinch physically distinct cycles in the same homology class to zero, so that $C_{0,0}$ is an I_2; we have obviously that $N_1 = N_2$ (both send $\beta \mapsto \alpha \mapsto 0$). Take

$$Z_{s,t} := \left(i\sqrt{\frac{1+t}{1+s}}, 1\right) - \left(-i\sqrt{\frac{1+t}{1+s}}, 1\right)$$

for our family of cycles, which splits between the two components of the I_2 at $(0, 0)$. See figure at top of next page.

Things go much more wrong here. Here are 3 ways to see this:

- try to correct monodromy (as we did in Example 24 with $-\frac{1}{2}\beta$): $N_1(\nu) = \alpha$, $N_1(\beta) = \alpha$, $N_2(\nu) = 0$, $N_2(\beta) = \alpha$ implies an impossibility;
- in T_s (from Example 1), $\xi^{1/4}$ becomes (here) $(t/s)^{1/4}$ — so its obvious extension isn't well-defined. In fact, there is *no* 2-torsion family of cycles with fiber over $(0, 0)$ a difference of two points in the two distinct components of

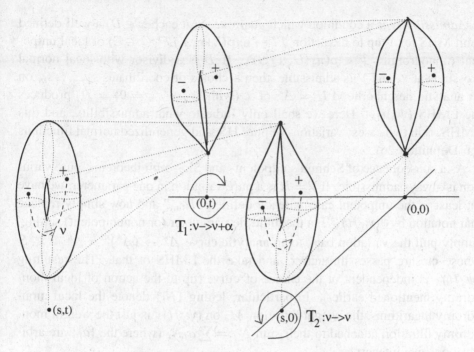

$C_{0,0}$ (that is, one that limits to have the same cohomology class in $H^2(C_{0,0})$ as $Z_{0,0}$).

- take the "motivic limit" of AJ at $t = 0$: under the uniformization of $C_{s,0}$ by

$$\mathbb{P}^1 \ni z \longmapsto \left(\frac{2z}{1 - sz^2}, \frac{2iz}{1 + sz^2} \right),$$

$$\left(\frac{i}{s}(1 + \sqrt{1+s}) \right) - \left(\frac{i}{s}(1 - \sqrt{1+s}) \right) \longmapsto Z_{s,0}.$$

Moreover, the isomorphism $\mathbb{C}^* \cong K_1(\mathbb{C}) \cong M_{-1} H^2_{\mathcal{M}}(C_{s,0}, \mathbb{Z}(1)) \, (\ni Z_{s,0})$ sends

$$\frac{1 + \sqrt{1+s}}{1 - \sqrt{1+s}} \in \mathbb{C}^*$$

to $Z_{s,0}$, and at $s = 0$ (considering it as a precycle in $Z^1(\Delta, 1)$) this obviously has a residue.

The upshot is that *nontorsion* singularities appear in codimension 2 and up.

2.11. Admissible normal functions. We now pass to the abstract setting of a complex analytic manifold $\bar{S}$ (for example a polydisk or smooth projective variety) with Zariski open subset S, writing $D = \bar{S} \setminus S$ for the complement. Throughout, we shall assume that $\pi_0(S)$ is finite and $\pi_1(S)$ is finitely generated. Let $\mathcal{V} = (\mathbb{V}, \mathcal{V}_{(\mathcal{O})}, \mathcal{F}^\bullet, W_\bullet)$ be a variation of MHS over S.

Admissibility is a condition which guarantees (at each $x \in D$) a well-defined limit MHS for $\mathcal{V}$ up to the action $\mathcal{F}^\bullet \mapsto \exp(\lambda \log T)\mathcal{F}^\bullet$ ($\lambda \in \mathbb{C}$) of local unipotent monodromies $T \in \rho(\pi_1(U_x \cap S))$. If D is a divisor with local normal crossings at x, and $\mathcal{V}$ is admissible, then a choice of coordinates $s_1, \ldots, s_m$ on an analytic neighborhood $U = \Delta^k$ of x (with $\{s_1 \cdots s_m = 0\} = D$) produces the LMHS $(\psi_{\underline{s}} \mathcal{V})_x$. Here we shall only indicate what admissibility, and this LMHS, is in two cases: variations of pure HS, and generalized normal functions (cf. Definition 26).

As a consequence of Schmid's nilpotent- and SL_2-orbit theorems, pure variation is always admissible. If $\mathcal{V} = \mathcal{H}$ is a pure variation in one parameter, we have (at least in the unipotent case) already defined "$H_{\lim}$" and now simply replace that notation by "$(\psi_{\underline{s}} \mathcal{H})_x$". In the multiple parameter (or nonunipotent) setting, simply pull the variation back to an analytic curve $\Delta^* \to (\Delta^*)^m \times \Delta^{k-m} \subset S$ whose closure passes through x, and take the LMHS of that. The resulting $(\psi_{\underline{s}} \mathcal{H})_x$ is independent of the choice of curve (up to the action of local monodromy mentioned earlier). In particular, letting $\{N_i\}$ denote the local monodromy logarithms, the weight filtration $M_\bullet$ on $(\psi_{\underline{s}} \mathcal{H})_x$ is just the weight monodromy filtration attached to their sum $N := \sum a_i N_i$ (where the $\{a_i\}$ are arbitrary positive integers).

Now let $r \in \mathbb{N}$.

DEFINITION 26. A *(higher) normal function* over S is a VMHS of the form $\mathcal{V}$ in (the short-exact sequence)

$$0 \to \mathcal{H} \longrightarrow \mathcal{V} \longrightarrow \mathbb{Z}_S(0) \to 0 \tag{2-8}$$

where $\mathcal{H}$ is a [pure] VHS of weight $(-r)$ and the [trivial, constant] variation $\mathbb{Z}_S(0)$ has trivial monodromy. (The terminology "higher" only applies when $r > 1$.) This is equivalent to a holomorphic, horizontal section of the generalized Jacobian bundle

$$J(\mathcal{H}) := \frac{\mathcal{H}}{\mathcal{F}^0 \mathcal{H} + \mathbb{H}_{\mathbb{Z}}}.$$

EXAMPLE 27. Given a smooth proper family $\mathcal{X} \xrightarrow{\pi} S$, with $x_0 \in S$. A higher algebraic cycle $\mathfrak{Z} \in CH^p(\mathcal{X}, r-1)_{\mathrm{prim}} := \ker\{CH^p(\mathcal{X}, r-1) \to CH^p(X_{x_0}, r-1) \to \mathrm{Hg}^{p,r-1}(X_{x_0})\}$ yields a section of $J(R^{2p-r}\pi_*\mathbb{C} \otimes \mathcal{O}_S) =: \mathcal{J}^{p,r-1}$; this is what we shall mean by a *(higher) normal function of geometric origin*.[7] (The notion of *motivation over K* likewise has an obvious extension from the classical 1-parameter case in §1.)

---

[7] Note that $\mathrm{Hg}^{p,r-1}(X_{x_0})_{\mathbb{Q}} := H^{2p-r+1}(X_{x_0}, \mathbb{Q}(p)) \cap F^p H^{2p-r+1}(X_{x_0}, \mathbb{C})$ is actually zero for $r > 1$, so that the "prim" comes for free for some multiple of $\mathfrak{Z}$.

We now give the definition of admissibility for VMHS of the form in Definition 26 (but simplifying to $D = \{s_1 \cdots s_k = 0\}$), starting with the local unipotent case. For this we need Deligne's definition [De1] of the $I^{p,q}(H)$ of a MHS H, for which the reader may refer to Theorem 68 (in §4) below. To simplify notation, we shall abbreviate $I^{p,q}(H)$ to $H^{(p,q)}$, so that, for instance, $H_{\mathbb{Q}}^{(p,p)} = I^{p,p}(H) \cap H_{\mathbb{Q}}$, and drop the subscript x for the LMHS notation.

DEFINITION 28. Let $S = (\Delta^*)^k$, $\mathcal{V} \in \mathrm{NF}^r(S, \mathcal{H})_{\mathbb{Q}}$ (i.e., as in Definition 26, $\otimes \mathbb{Q}$), and $x = (\underline{0})$.

(I) [unipotent case] Assume the monodromies T_i of $\mathbb{H}$ are unipotent, so that the logarithms N_i and associated monodromy weight filtrations $M_\bullet^{(i)}$ are defined. (Note that the $\{N_i\}$ resp. $\{T_i\}$ automatically commute, since any local system must be a representation of $\pi_1((\Delta^*)^k)$, an abelian group.) We may "untwist" the local system $\otimes \mathbb{Q}$ via

$$\tilde{\mathbb{V}} := \exp\left(\frac{-1}{2\pi\sqrt{-1}} \sum_i \log(s_i) N_i\right) \mathbb{V}_{(\mathbb{Q})},$$

and set $\mathcal{V}_e := \tilde{\mathbb{V}} \otimes \mathcal{O}_{\Delta^k}$ for the Deligne extension. Then $\mathcal{V}$ is $(\bar{S}\text{-})$admissible if and only if

(a) $\mathcal{H}$ is polarizable,

(b) there exists a lift $\nu_{\mathbb{Q}} \in (\tilde{\mathbb{V}})_0$ of $1 \in \mathbb{Q}(0)$ such that $N_i \nu_{\mathbb{Q}} \in M_{-2}^{(i)}(\psi_{\underline{s}}\mathcal{H})_{\mathbb{Q}}\ (\forall i)$, and

(c) there exists a lift $\nu_F(s) \in \Gamma(\bar{S}, \mathcal{V}_e)$ of $1 \in \mathbb{Q}_S(0)$ such that $\nu_F|_S \in \Gamma(S, \mathcal{F}^0)$.

(II) In general there exists a minimal finite cover $\zeta : (\Delta^*)^k \to (\Delta^*)^k$ (sending $\underline{s} \mapsto \underline{s}^{\mu}$) such that the $T_i^{\mu_i}$ are unipotent. $\mathcal{V}$ is admissible if and only if $\zeta^* \mathcal{V}$ satisfies (a), (b), and (c).

The main result [K; SZ] is then that $\mathcal{V} \in \mathrm{NF}^r(S, \mathcal{H})_{\bar{S}}^{\mathrm{ad}}$ has well-defined $\psi_{\underline{s}}\mathcal{V}$, given as follows. On the underlying rational structure $(\tilde{\mathbb{V}})_0$ we put the weight filtration $M_i = M_i \psi_{\underline{s}}\mathcal{H} + \mathbb{Q}\langle\nu_{\mathbb{Q}}\rangle$ for $i \geq 0$ and $M_i = M_i \psi_{\underline{s}}\mathcal{H}$ for $i < 0$; while on its complexification $(\cong (\mathcal{V}_e)_0)$ we put the Hodge filtration $F^j = F^j \psi_{\underline{s}}\mathcal{H}_{\mathbb{C}} + \mathbb{C}\langle\nu_F(0)\rangle$ for $j \leq 0$ and $F^j = F^j \psi_{\underline{s}}\mathcal{H}$ for $j > 0$. (Here we are using the inclusion $\tilde{\mathbb{H}} \subset \tilde{\mathbb{V}}$, and the content of the statement is that this actually does define a MHS.)

We can draw some further conclusions from (a)–(c) in case (I). With some work, it follows from (c) that

(c') $\nu_F(0)$ gives a lift of $1 \in \mathbb{Q}(0)$ satisfying $N_i \nu_F(0) \in (\psi_{\underline{s}}\mathcal{H})^{(-1,-1)}$;

and one can also show that $N_i \nu_{\mathbb{Q}} \in M_{-2}(\psi_{\underline{s}}\mathcal{H})_{\mathbb{Q}}\ (\forall i)$. Furthermore, if $r = 1$ then each $N_i \nu_{\mathbb{Q}}$ [resp. $N_i \nu_F(0)$] belongs to the image under $N_i : \psi_{\underline{s}}\mathcal{H} \to \psi_{\underline{s}}\mathcal{H}(-1)$

of a rational [resp. type-$(0,0)$] element. To see this, use the properties of N_i to deduce that $\mathrm{im}(N_i) \supseteq M^{(i)}_{-r-1}$; then for $r = 1$ we have, from (b) and (c), $N_i \nu_F(0)$, $N_i \nu_{\mathbb{Q}} \in M^{(i)}_{-2}$.

(III) The definition of admissibility over an arbitrary smooth base S together with good compactification $\bar{S}$ is then local, i.e., reduces to the $(\Delta^*)^k$ setting. Another piece of motivation for the definition of admissibility is this, for which we refer the reader to [BZ, Theorem 7.3]:

THEOREM 29. *Any (higher) normal function of geometric origin is admissible.*

2.12. Limits and singularities of ANFs. Now the idea of the "limit of a normal function" should be to interpret $\psi_s \mathcal{V}$ as an extension of $\mathbb{Q}(0)$ by $\psi_s \mathcal{H}$. The obstruction to being able to do this is the singularity, as we now explain. All MHS in this section are $\mathbb{Q}$-MHS.

According to [BFNP, Corollary 2.9], we have

$$\mathrm{NF}^r(S, \mathcal{H})^{\mathrm{ad}}_{\bar{S}} \otimes \mathbb{Q} \cong \mathrm{Ext}^1_{\mathrm{VMHS}(S)^{\mathrm{ad}}_{\bar{S}}}(\mathbb{Q}(0), \mathcal{H}),$$

as well as an equivalence of categories $\mathrm{VMHS}(S)^{\mathrm{ad}}_{\bar{S}} \simeq \mathrm{MHM}(S)^{\mathrm{ps}}_{\bar{S}}$. We want to push (in a sense canonically extend) our ANF $\mathcal{V}$ into $\bar{S}$ and restrict the result to x. Of course, writing $J : S \hookrightarrow \bar{S}$, J_* is not right exact; so to preserve our extension, we take the derived functor RJ_* and land in the derived category $D^b\mathrm{MHM}(\bar{S})$. Pulling back to $D^b\mathrm{MHM}(\{x\}) \cong D^b\mathrm{MHS}$ by ι_x^*, we have defined an invariant $(\iota_x^* RJ_*)^{\mathrm{Hdg}}$:

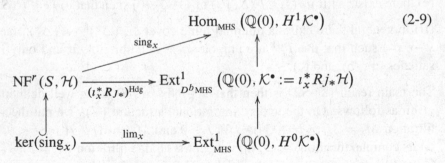

$$\mathrm{Hom}_{\mathrm{MHS}}\left(\mathbb{Q}(0), H^1\mathcal{K}^\bullet\right) \qquad (2\text{-}9)$$

where the diagram makes a clear analogy to (2-1).

For $S = (\Delta^*)^k$ and $\mathbb{H}_{\mathbb{Z}}$ unipotent we have

$$\mathcal{K}^\bullet \simeq \left\{ \psi_{\underline{s}}\mathcal{H} \xrightarrow{\oplus N_i} \bigoplus_i \psi_{\underline{s}}\mathcal{H}(-1) \longrightarrow \bigoplus_{i<j} \psi_{\underline{s}}\mathcal{H}(-2) \longrightarrow \cdots \right\},$$

and the map

$$\mathrm{sing}_x : \mathrm{NF}^r((\Delta^*)^k, \mathcal{H})^{\mathrm{ad}}_{\Delta^k} \to (H^1\mathcal{K}^\bullet)^{(0,0)}_{\mathbb{Q}} \quad (\cong \mathrm{coker}(N)(-1) \text{ for } k = 1)$$

is induced by $\mathcal{V} \mapsto \{N_i v_{\mathbb{Q}}\} \equiv \{N_i v_F(0)\}$. The limits, which are computed by

$$\lim_x : \ker(\text{sing}_x) \to J\big(\textstyle\bigcap_i \ker(N_i)\big),$$

more directly generalize the 1-parameter picture. The target $J(\bigcap \ker(N_i))$ is exactly what to put in over $\underline{0}$ to get the multivariable pre-Néron-model.

We have introduced the general case $r \geq 1$ because of interesting applications of higher normal functions to irrationality proofs, local mirror symmetry [DK]. In case $r = 1$ — we are dealing with classical normal functions — we can replace R_{J*} in the above by perverse intermediate extension $_{J!*}$ (which by a lemma in [BFNP] preserves the extension in this case: see Theorem 46 below). Correspondingly, $\mathcal{K}^\bullet$ is replaced by the local intersection cohomology complex

$$\mathcal{K}^\bullet_{\text{red}} \simeq \Big\{ \psi_{\underline{s}}\mathcal{H} \xrightarrow{\oplus N_i} \bigoplus_i \text{Im}(N_i)(-1) \longrightarrow \bigoplus_{i<j} \text{Im}(N_i N_j)(-2) \to \cdots \Big\};$$

while the target for $\lim_x$ is unchanged, the one for sing_x is reduced to 0 if $k = 1$ and to

$$\left(\frac{\ker(N_1) \cap \text{im}(N_2)}{N_2(\ker N_1)} \right)^{(-1,-1)}_{\mathbb{Q}} \tag{2-10}$$

if $k = 2$.

2.13. Applications of singularities. We hint at some good things to come:

(i) Replacing the sing_x-target (e.g., (2-10)) by actual *images* of ANFs, and using their differences to glue pre-Néron components together yields a generalized Néron model (over Δ^r, or $\bar{S}$ more generally) graphing ANFs. Again over x one gets an extension of a discrete (but not necessarily finite) singularity group by the torus $J(\cap \ker(N_i))$. A. Young [Yo] did this for abelian varieties, then [BPS] for general VHS. This will be described more precisely in §5.2.

(ii) (Griffiths and Green [GG]) The Hodge conjecture (HC) on a $2p$-dimensional smooth projective variety X is equivalent to the following statement for *each* primitive Hodge (p, p) class ζ and very ample line bundle $\mathcal{L} \to X$: there exists $k \gg 0$ such that the natural normal function[8] ν_ζ over $|\mathcal{L}^k| \setminus \hat{X}$ (the complement of the dual variety in the linear system) has a nontorsion singularity at some point of $\hat{X}$. So, in a *sense*, the analog of HC for $(\Delta^*)^k$ is surjectivity of sing_x onto $(H^1 \mathcal{K}^\bullet_{\text{red}})^{(0,0)}_{\mathbb{Q}}$, and this *fails*:

(iii) (M. Saito [S6], Pearlstein [Pe3]) Let $\mathcal{H}_0/\Delta^*$ be a VHS of weight 3 rank 4 with nontrivial Yukawa coupling. Twisting it into weight -1, assume the LMHS is of type II_1: $N^2 = 0$, with Gr^M_{-2} of rank 1. Take for $\mathcal{H}/(\Delta^*)^2$ the pullback of $\mathcal{H}_0$ by $(s, t) \mapsto st$. Then $(2\text{-}10) \neq \{0\} = \text{sing}_{\underline{0}}\{\text{NF}^1((\Delta^*)^2, \mathcal{H})^{\text{ad}}_{\Delta^2}\}$. The

---

[8]cf. § 3.2–3, especially (3-5).

obstruction to the existence of normal functions with nontrivial singularity is analytic; and comes from a differential equation produced by the horizontality condition (see § 5.4–5).

(iv) One can explain the meaning of the residue of the limit K_1 class in Example 25 above: writing $\jmath^1 : (\Delta^*)^2 \hookrightarrow \Delta^* \times \Delta$, $\jmath^2 : \Delta^* \times \Delta \hookrightarrow \Delta^2$, factor $(\iota_x^* R\jmath_{!*})^{\mathrm{Hdg}}$ by $(\iota_x^* R\jmath_*^2)^{\mathrm{Hdg}} \circ (\iota_{\Delta^*}^* \jmath_{!*}^1)^{\mathrm{Hdg}}$ (where the $\iota^* R\jmath_*^2$ corresponds to the residue). That is, limit a normal function (or family of cycles) to a higher normal function (or family of higher Chow cycles) over a codimension-1 boundary component; the latter can then have (unlike normal functions) a singularity in codimension 1 — i.e., in codimension 2 with respect to the original normal function.

This technique gives a quick proof of the existence of singularities for the Ceresa cycle by limiting it to an Eisenstein symbol (see [Co] and the Introduction to [DK]). Additionally, one gets a geometric explanation of why one does not expect the singularities in (ii) to be supported in high-codimension substrata of $\hat{X}$ (supporting very degenerate hypersurfaces of X): along these substrata one may reach (in the sense of (iv)) higher Chow cycles with rigid AJ invariants, hence no residues. For this reason codimension 2 tends to be a better place to look for singularities than in much higher codimension. These "shallow" substrata correspond to hypersurfaces with ordinary double points, and it was the original sense of [GG] that such points should trace out an algebraic cycle "dual" to the original Hodge class, giving an *effective* proof of the HC.

3. Normal functions and the Hodge conjecture

In this section, we discuss the connection between normal functions and the Hodge conjecture, picking up where § 1 left off. We begin with a review of some properties of the Abel–Jacobi map. Unless otherwise noted, all varieties are defined over $\mathbb{C}$.

3.1. Zucker's Theorem on Normal Functions. Let X be a smooth projective variety of dimension d_X. Recall that $J_h^p(X)$ is the intermediate Jacobian associated to the maximal rationally defined Hodge substructure H of $H^{2p-1}(X)$ such that $H_{\mathbb{C}} \subset H^{p,p-1}(X) \oplus H^{p-1,p}(X)$, and that (by a result of Lieberman [Li])

$$J^p(X)_{\mathrm{alg}} = \mathrm{im}\left\{ AJ_X : CH^p(X)_{\mathrm{alg}} \to J^p(X) \right\} \tag{3-1}$$
$$\text{is a subabelian variety of } J^p(X)_h.$$

NOTATION 30. If $f : X \to Y$ is a projective morphism then f^{sm} denotes the restriction of f to the largest Zariski open subset of Y over which f is smooth. Also, unless otherwise noted, in this section, the underlying lattice $\mathbb{H}_{\mathbb{Z}}$ of every variation of Hodge structure is assumed to be torsion free, and hence for a geometric family $f : X \to Y$, we are really considering $\mathbb{H}_{\mathbb{Z}} = (R^k f_*^{\mathrm{sm}} \mathbb{Z}) / \{\text{torsion}\}$.

As reviewed in § 1, Lefschetz proved that every integral $(1, 1)$ class on a smooth projective surface is algebraic by studying Poincaré normal functions associated to such cycles. We shall begin here by revisiting Griffiths' program (also recalled in § 1) to prove the Hodge conjecture for higher codimension classes by extending Lefschetz's methods: By induction on dimension, the Hodge conjecture can be reduced to the case of middle-dimensional Hodge classes on even-dimensional varieties [Le1, Lecture 14]. Suppose therefore that $X \subseteq \mathbb{P}^k$ is a smooth projective variety of dimension $2m$. Following [Zu2, §4], let us pick a Lefschetz pencil of hyperplane sections of X, i.e., a family of hyperplanes $H_t \subseteq \mathbb{P}^k$ of the form $t_0 w_0 + t_1 w_1 = 0$ parametrized by $t = [t_0, t_1] \in \mathbb{P}^1$ relative to a suitable choice of homogeneous coordinates $w = [w_0, \ldots, w_k]$ on $\mathbb{P}^k$ such that:

- for all but finitely many points $t \in \mathbb{P}^1$, the corresponding hyperplane section of $X_t = X \cap H_t$ is smooth;
- the base locus $B = X \cap \{w \in \mathbb{P}^k \mid w_0 = w_1 = 0\}$ is smooth; and
- each singular hyperplane section of X has exactly one singular point, which is an ordinary double point.

Given such a Lefschetz pencil, let

$$Y = \{ (x, t) \in X \times \mathbb{P}^1 \mid x \in H_t \}$$

and let $\pi : Y \to \mathbb{P}^1$ denote projection onto the second factor. Let U denote the set of points $t \in \mathbb{P}^1$ such that X_t is smooth and $\mathcal{H}$ be the variation of Hodge structure over U with integral structure $\mathbb{H}_{\mathbb{Z}} = R^{2m-1} \pi_*^{\mathrm{sm}} \mathbb{Z}(m)$. Furthermore, by Schmid's nilpotent orbit theorem [Sc], the Hodge bundles $\mathcal{F}^\bullet$ have a canonical extension to a system of holomorphic bundles $\mathcal{F}_e^\bullet$ over $\mathbb{P}^1$. Accordingly, we have a short exact sequence of sheaves

$$0 \to j_* \mathbb{H}_{\mathbb{Z}} \to \mathcal{H}_e / \mathcal{F}_e^m \to \mathcal{J}_e^m \to 0, \tag{3-2}$$

where $j : U \to \mathbb{P}^1$ is the inclusion map. As before, let us call an element $\nu \in H^0(\mathbb{P}^1, \mathcal{J}_e^m)$ a Poincaré normal function. Then, we have the following two results [Zu2, Thms. 4.57, 4.17], the second of which is known as the Theorem on Normal Functions:

THEOREM 31. *Every Poincaré normal function satisfies Griffiths horizontality.*

THEOREM 32. *Every primitive integral Hodge class on X is the cohomology class of a Poincaré normal function.*

The next step in the proof of the Hodge conjecture via this approach is to show that for $t \in U$, the Abel–Jacobi map

$$AJ : CH^m(X_t)_{\text{hom}} \to J^m(X_t)$$

is surjective. However, for $m > 1$ this is rarely true (even granting the conjectural equality of $J^m(X)_{\text{alg}}$ and $J_h^m(X)$) since $J^m(X_t) \neq J_h^m(X_t)$ unless $H^{2m-1}(X_t, \mathbb{C}) = H^{m,m-1}(X_t) \oplus H^{m-1,m}(X_t)$. In plenty of cases of interest $J_h^m(X)$ is in fact trivial; Theorem 33 and Example 35 below give two different instances of this.

THEOREM 33 [Le1, Example 14.18]. *If $X \subseteq \mathbb{P}^k$ is a smooth projective variety of dimension $2m$ such that $H^{2m-1}(X) = 0$ and $\{X_t\}$ is a Lefschetz pencil of hyperplane sections of X such that $F^{m+1} H^{2m-1}(X_t) \neq 0$ for every smooth hyperplane section, then for generic $t \in U$, $J_h^m(X_t) = 0$.*

THEOREM 34. *If $J_h^p(X) = 0$, then the image of $CH^m(W)_{\text{hom}}$ in $J^p(X)$ under the Abel–Jacobi map is countable.*

SKETCH OF PROOF. As a consequence of (3-1), if $J_h^p(X) = 0$ the Abel–Jacobi map vanishes on $CH^p(X)_{\text{alg}}$. Therefore, the cardinality of the image of the Abel–Jacobi map on $CH^p(X)_{\text{hom}}$ is bounded by the cardinality of the Griffiths group $CH^p(X)_{\text{hom}}/CH^p(X)_{\text{alg}}$, which is known to be countable.   □

EXAMPLE 35. Specific hypersurfaces with $J_h^p(X) = 0$ were constructed by Shioda [Sh]: Let Z_m^n denote the hypersurface in $\mathbb{P}^{n+1}$ defined by the equation

$$\sum_{i=0}^{n+1} x_i x_{i+1}^{m-1} = 0 \qquad (x_{n+2} = x_0).$$

Suppose that $n = 2p - 1 > 1$, $m \geq 2 + 3/(p-1)$ and

$$d_0 = \{(m-1)^{n+1} + (-1)^{n+1}\}/m$$

is prime. Then $J_h^p(Z_m^n) = 0$.

3.2. Singularities of admissible normal functions. In [GG], Griffiths and Green proposed an alternative program for proving the Hodge conjecture by studying the singularities of normal functions over higher-dimensional parameter spaces. Following [BFNP], let S a complex manifold and $\mathcal{H} = (\mathbb{H}_\mathbb{Z}, \mathcal{F}^\bullet \mathcal{H}_\mathcal{O})$ be a variation of polarizable Hodge structure of weight -1 over S. Then, we have the short exact sequence

$$0 \to \mathbb{H}_\mathbb{Z} \to \mathcal{H}/\mathcal{F}^0 \to J(\mathcal{H}) \to 0$$

of sheaves and hence an associated long exact sequence in cohomology. In particular, the cohomology class $cl(\nu)$ of a normal function $\nu \in H^0(S, J(\mathcal{H}))$ is just the image of ν under the connecting homomorphism

$$\partial : H^0(S, J(\mathcal{H})) \to H^1(S, \mathbb{H}_{\mathbb{Z}}).$$

Suppose now that S is a Zariski open subset of a smooth projective variety $\bar{S}$. Then the singularity of ν at $p \in \bar{S}$ is the quantity

$$\sigma_{\mathbb{Z},p}(\nu) = \varinjlim_{p \in U} cl(\nu|_{U \cap S}) \in \varinjlim_{p \in U} H^1(U \cap S, \mathbb{H}_{\mathbb{Z}}) = (R^1 j_* \mathbb{H}_{\mathbb{Z}})_p$$

where the limit is taken over all analytic open neighborhoods U of p, and $j : S \to \bar{S}$ is the inclusion map. The image of $\sigma_{\mathbb{Z},p}(\nu)$ in cohomology with rational coefficients will be denoted by $sing_p(\nu_\zeta)$.

REMARK 36. If $p \in S$ then $\sigma_{\mathbb{Z},p}(\nu) = 0$.

THEOREM 37 [S1]. *Let ν be an admissible normal function on a Zariski open subset of a curve $\bar{S}$. Then, $\sigma_{\mathbb{Z},p}(\nu)$ is of finite order for each point $p \in \bar{S}$.*

PROOF. By [S1], an admissible normal function $\nu : S \to J(\mathcal{H})$ is equivalent to an extension

$$0 \to \mathcal{H} \to \mathcal{V} \to \mathbb{Z}(0) \to 0 \tag{3-3}$$

in the category of admissible variations of mixed Hodge structure. By the monodromy theorem for variations of pure Hodge structure, the local monodromy of $\mathcal{V}$ about any point $p \in \bar{S} - S$ is always quasi-unipotent. Without loss of generality, let us assume that it is unipotent and that $T = e^N$ is the local monodromy of $\mathcal{V}$ at p acting on some fixed reference fiber with integral structure $V_{\mathbb{Z}}$. Then, due to the length of the weight filtration W, the existence of the relative weight filtration of W and N is equivalent to the existence of an N-invariant splitting of W [SZ, Proposition 2.16]. In particular, let $e_{\mathbb{Z}} \in V_{\mathbb{Z}}$ project to $1 \in Gr_0^W \cong \mathbb{Z}(0)$. Then, by admissibility, there exists an element $h_{\mathbb{Q}} \in H_{\mathbb{Q}} = W_{-1} \cap V_{\mathbb{Q}}$ such that

$$N(e_{\mathbb{Z}} + h_{\mathbb{Q}}) = 0$$

and hence $(T - I)(e_{\mathbb{Z}} + h_{\mathbb{Q}}) = 0$.[9] Any two such choices of $e_{\mathbb{Z}}$ differ by an element $h_{\mathbb{Z}} \in W_{-1} \cap V_{\mathbb{Z}}$. Therefore, an admissible normal function ν determines a class

$$[\nu] = [(T - I)e_{\mathbb{Z}}] \in \frac{(T - I)(H_{\mathbb{Q}})}{(T - I)(H_{\mathbb{Z}})}$$

Tracing through the definitions, one finds that the left-hand side of this equation can be identified with $\sigma_{\mathbb{Z},p}(\nu)$, whereas the right-hand side is exactly the torsion subgroup of $(R^1 j_* \mathbb{H}_{\mathbb{Z}})_p$. $\qquad \square$

---

[9] Alternatively, one can just derive this from Definition 28(I).

DEFINITION 38 [BFNP]. An admissible normal function ν defined on a Zariski open subset of $\bar{S}$ is singular on $\bar{S}$ if there exists a point $p \in \bar{S}$ such that $\text{sing}_p(\nu) \neq 0$.

Let S be a complex manifold and $f : X \to S$ be a family of smooth projective varieties over S. Let $\mathcal{H}$ be the variation of pure Hodge structure of weight -1 over S with integral structure $\mathbb{H}_\mathbb{Z} = R^{2p-1} f_* \mathbb{Z}(p)$. Then, an element $w \in J^p(X) (= J^0(H^{2p-1}(X, \mathbb{Z}(p))))$ defines a normal function $\nu_w : S \to J(\mathcal{H})$ by the rule

$$\nu_w(s) = i_s^*(w), \tag{3-4}$$

where i_s denotes inclusion of the fiber $X_s = f^{-1}(s)$ into X. More generally, let $H_\mathcal{D}^{2p}(X, \mathbb{Z}(p))$ denote the Deligne cohomology of X, and recall that we have a short exact sequence

$$0 \to J^p(X) \to H_\mathcal{D}^{2p}(X, \mathbb{Z}(p)) \to H^{p,p}(X, \mathbb{Z}(p)) \to 0.$$

Call a Hodge class

$$\zeta \in H^{p,p}(X, \mathbb{Z}(p)) := H^{p,p}(X, \mathbb{C}) \cap H^{2p}(X, \mathbb{Z}(p))$$

primitive with respect to f if $i_s^*(\zeta) = 0$ for all $s \in S$, and let $H_{\text{prim}}^{p,p}(X, \mathbb{Z}(p))$ denote the group of all such primitive Hodge classes. Then, by the functoriality of Deligne cohomology, a choice of lifting $\tilde{\zeta} \in H_\mathcal{D}^{2p}(X, \mathbb{Z}(p))$ of a primitive Hodge class ζ determines a map $\nu_{\tilde{\zeta}} : S \to J(\mathcal{H})$. A short calculation (cf. [CMP, Ch. 10]) shows that $\nu_{\tilde{\zeta}}$ is a (horizontal) normal function over S. Furthermore, in the algebraic setting (meaning that X, S, f are algebraic), $\nu_{\tilde{\zeta}}$ is an admissible normal function [S1]. Let $\text{ANF}(S, \mathcal{H})$ denote the group of admissible normal functions with underlying variation of Hodge structure $\mathcal{H}$. By abuse of notation, let $J^p(X) \subset \text{ANF}(S, \mathcal{H})$ denote the image of the intermediate Jacobian $J^p(X)$ in $\text{ANF}(S, \mathcal{H})$ under the map $w \mapsto \nu_w$. Then, since any two lifts $\tilde{\zeta}$ of ζ to Deligne cohomology differ by an element of the intermediate Jacobian $J^p(X)$, it follows that we have a well-defined map

$$AJ : H_{\text{prim}}^{p,p}(X, \mathbb{Z}(p)) \to \text{ANF}(S, \mathcal{H})/J^p(X). \tag{3-5}$$

REMARK 39. We are able to drop the notation $\text{NF}(S, \mathcal{H})_{\bar{S}}^{\text{ad}}$ used in §2, because in the global algebraic case it can be shown that admissibility is independent of the choice of compactification $\bar{S}$.

3.3. The Main Theorem. Returning to the program of Griffiths and Green, let X be a smooth projective variety of dimension $2m$ and $L \to X$ be a very ample line bundle. Let $\bar{P} = |L|$ and

$$\mathcal{X} = \{(x, s) \in X \times \bar{P} \mid s(x) = 0\} \tag{3-6}$$

be the incidence variety associated to the pair (X, L). Let $\pi : \mathcal{X} \to \bar{P}$ denote projection on the second factor, and let $\hat{X} \subset \bar{P}$ denote the dual variety of X (the points $s \in \bar{P}$ such that $X_s = \pi^{-1}(s)$ is singular). Let $\mathcal{H}$ be the variation of Hodge structure of weight -1 over $P = \bar{P} - \hat{X}$ attached to the local system $R^{2m-1}\pi_*^{\mathrm{sm}}\mathbb{Z}(m)$.

For a pair (X, L) as above, an integral Hodge class ζ of type (m, m) on X is primitive with respect to π^{sm} if and only if it is primitive in the usual sense of being annihilated by cup product with $c_1(L)$. Let $H_{\mathrm{prim}}^{m,m}(X, \mathbb{Z}(m))$ denote the group of all such primitive Hodge classes, and note that $H_{\mathrm{prim}}^{m,m}(X, \mathbb{Z}(m))$ is unchanged upon replacing L by $L^{\otimes d}$ for $d > 0$. Given $\zeta \in H_{\mathrm{prim}}^{m,m}(X, \mathbb{Z}(m))$, let

$$\nu_\zeta = AJ(\zeta) \in \mathrm{ANF}(P, \mathcal{H})/J^m(X)$$

be the associated normal function (3-5).

LEMMA 40. *If $\nu_w : P \to J(\mathcal{H})$ is the normal function (3-4) associated to an element $w \in J^m(X)$ then $\mathrm{sing}_p(\nu_w) = 0$ at every point $p \in \hat{X}$.*

Accordingly, for any point $p \in \hat{X}$ we have a well defined map

$$\overline{\mathrm{sing}_p} : \mathrm{ANF}(P, \mathcal{H})/J^m(X) \to (R^1 j_*\mathbb{H}_\mathbb{Q})_p$$

which sends the element $[\nu] \in \mathrm{ANF}(P, \mathcal{H})/J^m(X)$ to $\mathrm{sing}_p(\nu)$. In keeping with our prior definition, we say that ν_ζ is singular on $\bar{P}$ if there exists a point $p \in \hat{X}$ such that $\mathrm{sing}_p(\nu) \neq 0$.

CONJECTURE 41 [GG; BFNP]. *Let L be a very ample line bundle on a smooth projective variety X of dimension $2m$. Then, for every nontorsion class ζ in $H_{\mathrm{prim}}^{m,m}(X, \mathbb{Z}(m))$ there exists an integer $d > 0$ such that $AJ(\zeta)$ is singular on $\bar{P} = |L^{\otimes d}|$.*

THEOREM 42 [GG; BFNP; dCM]. *Conjecture 41 holds (for every even-dimensional smooth projective variety) if and only if the Hodge conjecture is true.*

To outline the proof of Theorem 42, observe that for any point $p \in \hat{X}$, we have the diagram

$$
\begin{array}{ccc}
H_{\mathrm{prim}}^{m,m}(X, \mathbb{Z}(m)) & \xrightarrow{\ AJ\ } & \mathrm{ANF}(P, \mathcal{H})/J^m(X) \\
\alpha_p \downarrow & & \downarrow \overline{\mathrm{sing}_p} \\
H^{2m}(X_p, \mathbb{Q}(m)) & \overset{\beta_p}{\underset{??}{\dashrightarrow}} & (R^1 j_*\mathbb{H}_\mathbb{Q})_p
\end{array}
\qquad (3\text{-}7)
$$

where $\alpha_p : H_{\mathrm{prim}}^{m,m}(X, \mathbb{Z}(m)) \to H^{2m}(X_p, \mathbb{Q}(m))$ is the restriction map.

Suppose that there exists a map

$$\beta_p : H^{2m}(X_p, \mathbb{Q}(m)) \to (R^1 j_* \mathbb{H}_\mathbb{Q})_p, \tag{3-8}$$

which makes the diagram (3-7) commute, and that after replacing L by $L^{\otimes d}$ for some $d > 0$ the restriction of β_p to the image of α_p is injective. Then, existence of a point $p \in \hat{X}$ such that $\mathrm{sing}_p(\nu_\zeta) \neq 0$ implies that the Hodge class ζ restricts nontrivially to X_p. Now recall that by Poincaré duality and the Hodge–Riemann bilinear relations, the Hodge conjecture for a smooth projective variety Y is equivalent to the statement that for every rational (q, q) class on Y there exists an algebraic cycle W of dimension $2q$ on Y such that $\gamma \cup [W] \neq 0$.

Let $f : \tilde{X}_p \to X_p$ be a resolution of singularities of X_p and $g = i \circ f$, where $i : X_p \to X$ is the inclusion map. By a weight argument $g^*(\zeta) \neq 0$, and so there exists a class $\xi \in \mathrm{Hg}^{m-1}(\tilde{X}_p)$ with $\xi \cup \zeta \neq 0$. Embedding $\tilde{X}_p$ in some projective space, and inducing on *even* dimension, we can assume that the Hodge conjecture holds for a general hyperplane section $\mathcal{I} : \mathcal{Y} \hookrightarrow \tilde{X}_p$. This yields an algebraic cycle $\mathcal{W}$ on $\mathcal{Y}$ with $[\mathcal{W}] = \mathcal{I}^*(\xi)$. Varying $\mathcal{Y}$ in a pencil, and using weak Lefschetz, $\mathcal{W}$ traces out[10] a cycle $W = \sum_j a_j W_j$ on $\tilde{X}_p$ with $[W] = \xi$, so that $g^*(\zeta) \cup [W] \neq 0$; in particular, $\zeta \cup g_*[W_j] \neq 0$ for some j.

Conversely, by the work of Thomas [Th], if the Hodge conjecture is true then the Hodge class ζ must restrict nontrivially to some singular hyperplane section of X (again for some $L^{\otimes d}$ for d sufficiently large). Now one uses the injectivity of β_p on $\mathrm{im}(\alpha_p)$ to conclude that ν_ζ has a singularity.

EXAMPLE 43. Let $X \subset \mathbb{P}^3$ be a smooth projective surface. For every $\zeta \in H^{1,1}_{\mathrm{prim}}(X, \mathbb{Z}(1))$, there is a reducible hypersurface section $X_p \subset X$ and component curve W of X_p such that $\deg(\zeta|_W) \neq 0$. (Note that $\deg(\zeta|_{X_p})$ is necessarily 0.) As the reader should check, this follows easily from Lefschetz $(1,1)$. Moreover (writing d for the degree of X_p), p is a point in a codimension ≥ 2 substratum S' of $\hat{X} \subset \mathbb{P}H^0(\mathcal{O}(d))$ (since fibers over codimension-one substrata are irreducible), and $\mathrm{sing}_q(\nu_\zeta) \neq 0 \ \forall q \in S'$.

REMARK 44. There is a central geometric issue lurking in Conjecture 41:

If the HC holds, and $L = \mathcal{O}_X(1)$ (for some projective embedding of X), is there some minimum d_0 — uniform in some sense — for which $d \geq d_0$ implies that ν_ξ is singular?

In [GG] it is established that, at best, such a d_0 could only be uniform in moduli of the pair (X, ζ). (For example, in the case $\dim(X) = 2$, d_0 is of the form $C \times |\zeta \cdot \zeta|$, for C a constant. Since the self-intersection numbers of integral

---

[10] More precisely, one uses here a spread or Hilbert scheme argument. See for example the beginning of Chapter 14 of [Le1].

classes becoming Hodge in various Noether–Lefschetz loci increase without bound, there is certainly not any d_0 uniform in moduli of X.) Whether there is some such "lower bound" of this form remains an open question in higher dimension.

3.4. Normal functions and intersection cohomology. The construction of the map β_p depends on the decomposition theorem of Beilinson, Bernstein, and Deligne [BBD] and Morihiko Saito's theory of mixed Hodge modules [S4]. As first step in this direction, recall [CKS2] that if $\mathcal{H}$ is a variation of pure Hodge structure of weight k defined on the complement $S = \bar{S} - D$ of a normal crossing divisor on a smooth projective variety $\bar{S}$ then

$$H^\ell_{(2)}(S, \mathbb{H}_\mathbb{R}) \cong IH^\ell(\bar{S}, \mathbb{H}_\mathbb{R}),$$

where the left-hand side is L^2-cohomology and the right-hand side is intersection cohomology. Furthermore, via this isomorphism $IH^\ell(\bar{S}, \mathbb{H}_\mathbb{C})$ inherits a canonical Hodge structure of weight $k + \ell$.

REMARK 45. If Y is a complex algebraic variety, $\mathrm{MHM}(Y)$ is the category of mixed Hodge modules on Y. The category $\mathrm{MHM}(Y)$ comes equipped with a functor

$$\mathrm{rat} : \mathrm{MHM}(Y) \to \mathrm{Perv}(Y)$$

to the category of perverse sheaves on Y. If Y is smooth and $\mathcal{V}$ is a variation of mixed Hodge structure on Y then $\mathcal{V}[d_Y]$ is a mixed Hodge module on Y, and $\mathrm{rat}(\mathcal{V}[d_Y]) \cong \mathbb{V}[d_Y]$ is just the underlying local system $\mathbb{V}$ shifted into degree $-d_Y$.

If Y° is a Zariski open subset of Y and $\mathcal{P}$ is a perverse sheaf on Y° then

$$IH^\ell(Y, \mathcal{P}) = \mathbb{H}^{\ell - d_Y}(Y, j_{!*}\mathcal{P}[d_Y])$$

where $j_{!*}$ is the middle extension functor [BBD] associated to the inclusion map $j : Y^\circ \to Y$. Likewise, for any point $y \in Y$, the local intersection cohomology of $\mathcal{P}$ at y is defined to be

$$IH^\ell(Y, \mathcal{P})_y = \mathbb{H}^{k - d_Y}(\{y\}, i^* j_{!*}\mathcal{P}[d_Y])$$

where $i : \{y\} \to Y$ is the inclusion map. If $\mathcal{P}$ underlies a mixed Hodge module, the theory of MHM puts natural MHS on these groups, which in particular is how the pure HS on $IH^\ell(\bar{S}, \mathbb{H}_\mathbb{C})$ comes about.

THEOREM 46 [BFNP, Theorem 2.11]. *Let $\bar{S}$ be a smooth projective variety and $\mathcal{H}$ be a variation of pure Hodge structure of weight -1 on a Zariski open subset $S \subset \bar{S}$. Then, the group homomorphism*

$$\mathrm{cl} : \mathrm{ANF}(S, \mathcal{H}) \to H^1(S, \mathbb{H}_\mathbb{Q})$$

factors through $IH^1(\bar{S}, \mathbb{H}_{\mathbb{Q}})$.

SKETCH OF PROOF. Let $v \in \mathrm{ANF}(S, \mathcal{H})$ be represented by an extension

$$0 \to \mathcal{H} \to \mathcal{V} \to \mathbb{Z}(0) \to 0$$

in the category of admissible variations of mixed Hodge structure on S. Let $j : S \to \bar{S}$ be the inclusion map. Then, because $\mathcal{V}$ has only two nontrivial weight graded quotients which are adjacent, it follows by [BFNP, Lemma 2.18] that

$$0 \to j_{!*}\mathcal{H}[d_S] \to j_{!*}\mathcal{V}[d_S] \to \mathbb{Q}(0)[d_S] \to 0$$

is exact in $\mathrm{MHM}(\bar{S})$.   □

REMARK 47. In this particular context, $j_{!*}\mathcal{V}[d_S]$ can be described as the unique prolongation of $\mathcal{V}[d_S]$ to $\bar{S}$ with no nontrivial sub or quotient object supported on the essential image of the functor $i : \mathrm{MHM}(Z) \to \mathrm{MHM}(\bar{S})$ where $Z = \bar{S} - S$ and $i : Z \to \bar{S}$ is the inclusion map.

In the local case of an admissible normal function on a product of punctured polydisks $(\Delta^*)^r$ with unipotent monodromy, the fact that $\mathrm{sing}_0(v)$ (where 0 is the origin of $\Delta^r \supseteq (\Delta^*)^r$) factors through the local intersection cohomology groups can be seen as follows: Such a normal function v gives a short exact sequence of local systems

$$0 \to \mathbb{H}_{\mathbb{Q}} \to \mathbb{V}_{\mathbb{Q}} \to \mathbb{Q}(0) \to 0$$

over $(\Delta^*)^r$. Fix a reference fiber $V_{\mathbb{Q}}$ of $\mathbb{V}_{\mathbb{Q}}$ and let $N_j \in \mathrm{Hom}(V_{\mathbb{Q}}, V_{\mathbb{Q}})$ denote the monodromy logarithm of $\mathbb{V}_{\mathbb{Q}}$ about the j-th punctured disk. Then [CKS2], we get a complex of finite-dimensional vector spaces

$$B^p(V_{\mathbb{Q}}) = \bigoplus_{i_1 < i_2 < \cdots < i_p} N_{i_1} N_{i_2} \cdots N_{i_p}(V_{\mathbb{Q}})$$

with differential d, which acts on the summands of $B^p(V_{\mathbb{Q}})$ by the rule

$$N_{i_1} \cdots \hat{N}_{i_\ell} \cdots N_{i_{p+1}}(V_{\mathbb{Q}}) \xrightarrow{(-1)^{\ell-1} N_{i_\ell}} N_{i_1} \cdots N_{i_\ell} \cdots N_{i_{p+1}}(V_{\mathbb{Q}})$$

(and taking the sum over all insertions). Let $B^*(H_{\mathbb{Q}})$ and $B^*(\mathbb{Q}(0))$ denote the analogous complexes attached to the local systems $\mathbb{H}_{\mathbb{Q}}$ and $\mathbb{Q}(0)$. By [GGM], the cohomology of the complex $B^*(H_{\mathbb{Q}})$ computes the local intersection cohomology of $\mathbb{H}_{\mathbb{Q}}$. In particular, since the complexes $B^*(\mathbb{Q}(0))$ and $B^*(H_{\mathbb{Q}})$ sit inside the standard Koszul complexes which compute the ordinary cohomology of $\mathbb{Q}(0)$ and $H_{\mathbb{Q}}$, in order show that sing_0 factors through $IH^1(\mathbb{H}_{\mathbb{Q}})$ it is sufficient to show that $\partial \mathrm{cl}(v) \in H^1((\Delta^*)^r, \mathbb{H}_{\mathbb{Q}})$ is representable by an element of $B^1(H_{\mathbb{Q}})$. Indeed, let v be an element of $V_{\mathbb{Q}}$ which maps to $1 \in \mathbb{Q}(0)$. Then,

$$\partial \mathrm{cl}(v) = \partial 1 = [(N_1(v), \ldots, N_r(v))]$$

By admissibility and the short length of the weight filtration, for each j there exists an element $h_j \in H_{\mathbb{Q}}$ such that $N_j(h_j) = N_j(v)$, which is exactly the condition that

$$(N_1(v), \ldots, N_r(v)) \in B^1(V_{\mathbb{Q}}).$$

THEOREM 48 [BFNP, Theorem 2.11]. *Under the hypothesis of Theorem 46, for any point $p \in \bar{S}$ the group homomorphism* $\mathrm{sing}_p : \mathrm{ANF}(S, \mathcal{H}) \to (R^1 j_* \mathbb{H}_{\mathbb{Q}})_p$ *factors through the local intersection cohomology group* $IH^1(\mathbb{H}_{\mathbb{Q}})_p$.

To continue, we need to pass from Deligne cohomology to absolute Hodge cohomology. Recall that $\mathrm{MHM}(\mathrm{Spec}(\mathbb{C}))$ is the category MHS of graded-polarizable $\mathbb{Q}$ mixed Hodge structures. Let $\mathbb{Q}(p)$ denote the Tate object of type $(-p, -p)$ in MHS and $\mathbb{Q}_Y(p) = a_Y^* \mathbb{Q}(p)$ where $a_Y : Y \to \mathrm{Spec}(\mathbb{C})$ is the structure morphism. Let $\mathbb{Q}_Y = \mathbb{Q}_Y(0)$.

DEFINITION 49. Let M be an object of $\mathrm{MHM}(Y)$. Then,

$$H^n_{A\mathcal{H}}(Y, M) = \mathrm{Hom}_{D^b \mathrm{MHM}}(\mathbb{Q}_Y, M[n])$$

is the absolute Hodge cohomology of M.

The functor $\mathrm{rat} : \mathrm{MHM}(Y) \to \mathrm{Perv}(Y)$ induces a "cycle class map"

$$\mathrm{rat} : H^n_{A\mathcal{H}}(Y, M) \to \mathbb{H}^n(Y, \mathrm{rat}(M))$$

from the absolute Hodge cohomology of M to the hypercohomology of $\mathrm{rat}(M)$. In the case where Y is smooth and projective, $H^{2p}_{A\mathcal{H}}(Y, \mathbb{Q}_Y(p))$ is the Deligne cohomology group $H^{2p}_D(Y, \mathbb{Q}(p))$ and rat is the cycle class map on Deligne cohomology.

DEFINITION 50. Let $\bar{S}$ be a smooth projective variety and $\mathcal{V}$ be an admissible variation of mixed Hodge structure on a Zariski open subset S of $\bar{S}$. Then,

$$IH^n_{A\mathcal{H}}(\bar{S}, \mathcal{V}) = \mathrm{Hom}_{D^b \mathrm{MHM}(\bar{S})}(\mathbb{Q}_{\bar{S}}[d_S - n], j_{!*}\mathcal{V}[d_S]),$$

$$IH^n_{A\mathcal{H}}(\bar{S}, \mathcal{V})_s = \mathrm{Hom}_{D^b \mathrm{MHS}}(\mathbb{Q}[d_S - n], i^* j_{!*}\mathcal{V}[d_S]),$$

where $j : S \to \bar{S}$ and $i : \{s\} \to \bar{S}$ are inclusion maps.

The following lemma links absolute Hodge cohomology and admissible normal functions:

LEMMA 51. [BFNP, Proposition 3.3] *Let $\mathcal{H}$ be a variation of pure Hodge structure of weight -1 defined on a Zariski open subset S of a smooth projective variety $\bar{S}$. Then,* $IH^1_{A\mathcal{H}}(\bar{S}, \mathcal{H}) \cong \mathrm{ANF}(S, \mathcal{H}) \otimes \mathbb{Q}$.

3.5. Completion of the diagram (3-7). Let $f : X \to Y$ be a projective morphism between smooth algebraic varieties. Then, by the work of Morihiko Saito [S4], there is a direct sum decomposition

$$f_* \mathbb{Q}_X[d_X] = \bigoplus_i H^i (f_* \mathbb{Q}_X[d_X]) [-i] \qquad (3\text{-}9)$$

in MHM(Y). Furthermore, each summand $H^i(f_* \mathbb{Q}_X[d_X])$ is pure of weight $d_X + i$ and admits a decomposition according to codimension of support:

$$H^i (f_* \mathbb{Q}_X[d_X]) [-i] = \bigoplus_j E_{ij}[-i]; \qquad (3\text{-}10)$$

i.e., $E_{ij}[-i]$ is a sum of Hodge modules supported on codimension j subvarieties of Y. Accordingly, we have a system of projection operators (inserting arbitrary twists)

$$\bigoplus \Pi_{ij} : H^n_{\mathcal{AH}}(X, \mathbb{Q}(\ell)[d_X]) \;\xrightarrow{\cong}\; \bigoplus_{ij} H^{n-i}_{\mathcal{AH}}(Y, E_{ij}(\ell)),$$

$$\bigoplus \Pi_{ij} : H^n_{\mathcal{AH}}(X_p, \mathbb{Q}(\ell)[d_X]) \;\xrightarrow{\cong}\; \bigoplus_{ij} H^{n-i}_{\mathcal{AH}}(Y, \iota^* E_{ij}(\ell)),$$

$$\bigoplus \Pi_{ij} : \mathbb{H}^n(X, \mathrm{rat}(\mathbb{Q}(\ell)[d_X])) \;\xrightarrow{\cong}\; \bigoplus_{ij} \mathbb{H}^{n-i}(Y, \mathrm{rat}(E_{ij}(\ell))),$$

$$\bigoplus \Pi_{ij} : \mathbb{H}^n(X_p, \mathrm{rat}(\mathbb{Q}(\ell)[d_X])) \;\xrightarrow{\cong}\; \bigoplus_{ij} \mathbb{H}^{n-i}(Y, \iota^* \mathrm{rat}(E_{ij}(\ell))),$$

where $p \in Y$ and $\iota : \{p\} \to Y$ is the inclusion map.

LEMMA 52 [BFNP, Equation 4.12]. *Let $\mathcal{H}^q = R^q f^{\mathrm{sm}}_* \mathbb{Q}_X$ and recall that we have a decomposition*

$$\mathcal{H}^{2k-1} = \mathcal{H}^{2k-1}_{\mathrm{van}} \oplus \mathcal{H}^{2k-1}_{\mathrm{fix}}$$

where $\mathcal{H}^{2k-1}_{\mathrm{fix}}$ is constant and $\mathcal{H}^{2k-1}_{\mathrm{van}}$ has no global sections. For any point $p \in Y$, we have a commutative diagram

$$
\begin{array}{ccc}
H^{2k}_{\mathcal{AH}}(X, \mathbb{Q}(k)) & \xrightarrow{\;\Pi\;} & \mathrm{ANF}(Y^{\mathrm{sm}}, \mathcal{H}^{2k-1}_{\mathrm{van}}(k)) \\
\Big\downarrow{i^*} & & \Big\downarrow{i^*} \\
H^{2k}(X_p, \mathbb{Q}(k)) & \xrightarrow{\;\Pi\;} & IH^1(\mathcal{H}^{2k-1}(k))_p
\end{array}
\qquad (3\text{-}11)
$$

where Y^{sm} is the largest Zariski open set over which f is smooth and Π is induced by Π_{r0} for $r = 2k - 1 - d_X + d_Y$.

We now return to the setting of Conjecture 41: X is a smooth projective variety of dimension $2m$, L is a very ample line bundle on X and $\mathcal{X}$ is the associated incidence variety (3-6), with projections $\pi : \mathcal{X} \to \bar{P}$ and $\mathrm{pr} : \mathcal{X} \to X$. Then, we have the following "Perverse weak Lefschetz theorem":

THEOREM 53 [BFNP, Theorem 5.1]. *Let $\mathcal{X}$ be the incidence variety associated to the pair (X, L) and $\pi_* \mathbb{Q}_{\mathcal{X}} = \bigoplus_{ij} E_{ij}$ in accord with (3-9) and (3-10). Then:*

(i) $E_{ij} = 0$ *unless* $i \cdot j = 0$.

(ii) $E_{i0} = H^i(X, \mathbb{Q}_X[2m-1]) \otimes \mathbb{Q}_{\bar{P}}[d_{\bar{P}}]$. *for* $i < 0$.

Note that by hard Lefschetz, $E_{ij} \cong E_{-i,j}(-i)$ [S4].

To continue, recall that given a Lefschetz pencil $\Lambda \subset \bar{P}$ of hyperplane sections of X, we have an associated system of vanishing cycles $\{\delta_p\}_{p \in \Lambda \cap \hat{X}} \subset H^{2m-1}(X_t, \mathbb{Q})$ on the cohomology of the smooth hyperplane sections X_t of X with respect to Λ. As one would expect, the vanishing cycles of Λ are *nonvanishing* if for some (hence all) $p \in \Lambda \cap \hat{X}$, $\delta_p \neq 0$ (in $H^{2m-1}(X_t, \mathbb{Q})$). Furthermore, this property depends only on L and not the particular choice of Lefschetz pencil Λ. This property can always be arranged by replacing L by $L^{\otimes d}$ for some $d > 0$.

THEOREM 54. *If all vanishing cycles are nonvanishing then $E_{01} = 0$. Otherwise, E_{01} is supported on a dense open subset of $\hat{X}$.*

Using the Theorems 53 and 54, we now prove that the diagram

$$
\begin{array}{ccc}
H_{\mathcal{D}}^{2m}(X, \mathbb{Z}(m))_{\text{prim}} & \xrightarrow{\;AJ\;} & \text{ANF}(P, \mathcal{H})/J^m(X) \\
{\scriptstyle \text{pr}^*}\downarrow & & \downarrow{\scriptstyle \otimes \mathbb{Q}} \\
H_{\mathcal{AH}}^{2m}(\mathcal{X}, \mathbb{Q}(m)) & \xrightarrow{\;\Pi\;} & \text{ANF}(P, \mathcal{H}_{\text{van}}) \otimes \mathbb{Q}
\end{array}
\tag{3-12}
$$

commutes, where $H_{\mathcal{D}}^{2m}(X, \mathbb{Z}(m))_{\text{prim}}$ is the subgroup of $H_{\mathcal{D}}^{2m}(X, \mathbb{Z}(m))$ whose elements project to primitive Hodge classes in $H^{2m}(X, \mathbb{Z}(m))$, and Π is induced by Π_{00} together with projection onto $\mathcal{H}_{\text{van}}$. Indeed, by the decomposition theorem,

$$
H_{\mathcal{AH}}^{2m}(\mathcal{X}, \mathbb{Q}(m)) = H_{\mathcal{AH}}^{1-d_{\bar{P}}}(\mathcal{X}, \mathbb{Q}(m)[2m + d_{\bar{P}} - 1])
$$
$$
= \bigoplus H_{\mathcal{AH}}^{1-d_{\bar{P}}}(\bar{P}, E_{ij}(m)[-i]).
$$

Let $\tilde{\xi} \in H_{\mathcal{D}}^{2m}(X, \mathbb{Z}(m))$ be a primitive Deligne class and $\omega = \bigoplus_{ij} \omega_{ij}$ denote the component of $\omega = \text{pr}^*(\tilde{\xi})$ with respect to $E_{ij}(m)[-i]$ in accord with the previous equation. Then, in order to prove the commutativity of (3-12) it is sufficient to show that $(\omega)_q = (\omega_{00})_q$ for all $q \in P$. By Theorem 53, we know that $\omega_{ij} = 0$ unless $ij = 0$. Furthermore, by [BFNP, Lemma 5.5], $(\omega_{0j})_q = 0$ for $j > 1$. Likewise, by Theorem 54, $(\omega_{01})_q = 0$ for $q \in P$ since E_{01} is supported on $\hat{X}$.

Thus, in order to prove the commutativity of (3-12), it is sufficient to show that $(\omega_{i0})_q = 0$ for $i > 0$. However, as a consequence of Theorem 53(ii), $E_{i0}(m) =$

$K[d_{\bar{P}}]$, where K is a constant variation of Hodge structure on $\bar{P}$; and hence

$$H^{1-d_{\bar{P}}}(\mathcal{X}, E_{i0}(m)[-i]) = \mathrm{Ext}^{1-d_{\bar{P}}}_{D^b\mathrm{MHM}(\bar{P})}(\mathbb{Q}_{\bar{P}}, K[d_{\bar{P}} - i])$$
$$= \mathrm{Ext}^{1-i}_{D^b\mathrm{MHM}(\bar{P})}(\mathbb{Q}_{\bar{P}}, K).$$

Therefore, $(\omega_{i0})_q = 0$ for $i > 1$ while $(\omega_{10})_q$ corresponds to an element of $Hom(\mathbb{Q}(0), K_q)$ where K is the constant variation of Hodge structure with fiber $H^{2m}(X_q, \mathbb{Q}(m))$ over $q \in P$. It therefore follows from the fact that $\tilde{\zeta}$ is primitive that $(\omega_{10})_q = 0$. Splicing diagram (3-12) together with (3-11) (and replacing $f : X \to Y$ by $\pi : \mathcal{X} \to \bar{P}$, etc.) now gives the diagram (3-7).

REMARK 55. The effect of passing from $\mathcal{H}$ to $\mathcal{H}^{\mathrm{van}}$ in the constructions above is to annihilate $J^m(X) \subseteq H^{2m}_{\mathcal{D}}(X, \mathbb{Z}(m))_{\mathrm{prim}}$. Therefore, in (3-12) we can replace $H^{2m}_{\mathcal{D}}(X, \mathbb{Z}(m))_{\mathrm{prim}}$ by $H^{m,m}_{\mathrm{prim}}(X, \mathbb{Z}(m))$.

Finally, if all the vanishing cycles are nonvanishing, $E_{01} = 0$. Using this fact, we then get the injectivity of β_p on the image of α_p.

Returning to the beginning of this section, we now see that although extending normal functions along Lefschetz pencils is insufficient to prove the Hodge conjecture for higher codimension cycles, the Hodge conjecture is equivalent to a statement about the behavior of normal functions on the complement of the dual variety of X inside $|L|$ for $L \gg 0$. We remark that an interpretation of the GHC along similar lines has been done recently by the authors in [KP].

4. Zeroes of normal functions

4.1. Algebraicity of the zero locus. Some of the deepest evidence to date in support of the Hodge conjecture is the following result of Cattani, Deligne and Kaplan on the algebraicity of the Hodge locus:

THEOREM 56 [CDK]. *Let $\mathcal{H}$ be a variation of pure Hodge structure of weight 0 over a smooth complex algebraic variety S. Let α_{s_o} be an integral Hodge class of type $(0, 0)$ on the fiber of $\mathcal{H}$ at s_0. Let U be a simply connected open subset of S containing s_o and α be the section of $\mathbb{H}_{\mathbb{Z}}$ over U defined by parallel translation of α_{s_o}. Let T be the locus of points in U such that $\alpha(s)$ is of type $(0, 0)$ on the fiber of $\mathcal{H}$ over s. Then, the analytic germ of T at p is the restriction of a complex algebraic subvariety of S.*

More precisely, as explained in the introduction of [CDK], in the case where $\mathcal{H}$ arises from the cohomology of a family of smooth projective varieties $f : X \to S$, the algebraicity of the germ of T follows from the Hodge conjecture. A natural analog of this result for normal functions is this:

THEOREM 57. *Let S be a smooth complex algebraic variety, and $v : S \rightarrow J(\mathcal{H})$ be an admissible normal function, where $\mathcal{H}$ is a variation of pure Hodge structure of weight -1. Then, the zero locus*

$$\mathcal{Z}(v) = \{ s \in S \mid v(s) = 0 \}$$

is a complex algebraic subvariety of S.

This theorem was still a conjecture when the present article was submitted, and has just been proved by the second author in work with P. Brosnan [BP3]. It is of particular relevance to the Hodge conjecture, due to the following relationship between the algebraicity of $\mathcal{Z}(v)$ and the existence of singularities of normal functions. Say $\dim(X) = 2m$, and let (X, L, ζ) be a triple consisting of a smooth complex projective variety X, a very ample line bundle L on X and a primitive integral Hodge class ζ of type (m, m). Let v_ζ (assumed nonzero) be the associated normal function on the complement of the dual variety $\hat{X}$ constructed in §3, and $\mathcal{Z}$ be its zero locus. Then, assuming that $\mathcal{Z}$ is algebraic and positive-dimensional, the second author conjectured that v should have singularities along the intersection of the closure of $\mathcal{Z}$ with $\hat{X}$.

THEOREM 58 [Sl1]. *Let (X, L, ζ) be a triple as above, and assume that L is sufficiently ample that, given any point $p \in \hat{X}$, the restriction of β_p to the image of α_p in diagram (3-7) is injective. Suppose that $\mathcal{Z}$ contains an algebraic curve. Then, v_ζ has a nontorsion singularity at some point of the intersection of the closure of this curve with $\hat{X}$.*

SKETCH OF PROOF. Let C be the normalization of the closure of the curve in $\mathcal{Z}$. Let $\mathcal{X} \rightarrow \bar{P}$ be the universal family of hyperplane sections of X over $\bar{P} = |L|$ and W be the pullback of X to C. Let $\pi : W \rightarrow C$ be the projection map, and U the set of points $c \in C$ such that $\pi^{-1}(c)$ is smooth and $W_U = \pi^{-1}(U)$. Via the Leray spectral sequence for π, it follows that restriction of ζ to W_U is zero because $U \subseteq \mathcal{Z}$ and ζ is primitive. On the other hand, since $W \twoheadrightarrow X$ is finite, ζ must restrict (pull back) nontrivially to W, and hence ζ must restrict nontrivially to the fiber $\pi^{-1}(c)$ for some point $c \in C$ in the complement of U. $\quad\square$

Unfortunately, crude estimates for the expected dimension of the zero locus $\mathcal{Z}$ arising in this context appear to be negative. For instance, take X to be an abelian surface in the following:

THEOREM 59. *Let X be a surface and $L = \mathcal{O}_X(D)$ be an ample line bundle on X. Then, for n sufficiently large, the expected dimension of the zero locus of the normal function v_ζ attached to the triple $(X, L^{\otimes n}, \zeta)$ as above is*

$$h^{2,0} - h^{1,0} - n(D.K) - 1,$$

where K is the canonical divisor of X.

SKETCH OF PROOF. Since Griffiths' horizontality is trivial in this setting, computing the expected dimension boils down to computing the dimension of $|L|$ and genus of a smooth hyperplane section of X with respect to L.   □

REMARK 60. In Theorem 59, we construct ν_ξ from a choice of lift to Deligne cohomology (or an algebraic cycle) to get an element of $\mathrm{ANF}(P, \mathcal{H})$. But this is disingenuous, since we are starting with a Hodge class. It is more consistent to work with $\nu_\xi \in \mathrm{ANF}(P, \mathcal{H})/J^1(X)$ as in equation (3-5), and then the dimension estimate improves by $\dim(J^1(X)) = h^{1,0}$ to $h^{2,0} - n(D.K) - 1$. Notice that this salvages at least the abelian surface case (though it is still a crude estimate). For surfaces of general type, one is still in trouble without more information, like the constant C in Remark 44.

We will not attempt to describe the proof of Theorem 57 in general, but we will explain the following special case:

THEOREM 61 [BP2]. *Let S be a smooth complex algebraic variety which admits a projective completion $\bar{S}$ such that $D = \bar{S} - S$ is a smooth divisor. Let $\mathcal{H}$ be a variation of pure Hodge structure of weight -1 on S and $\nu : S \to J(\mathcal{H})$ be an admissible normal function. Then, the zero locus $\mathcal{Z}$ of ν is an complex algebraic subvariety of S.*

REMARK 62. This result was obtained contemporaneously by Morihiko Saito in [S5].

In analogy with the proof of Theorem 56 on the algebraicity of the Hodge locus, which depends heavily on the several variable SL_2-orbit theorem for nilpotent orbits of pure Hodge structure [CKS1], the proof of Theorem 57 depends upon the corresponding result for nilpotent orbits of mixed Hodge structure. For simplicity of exposition, we will now review the 1-variable SL_2-orbit theorem in the pure case (which is due to Schmid [Sc]) and a version of the SL_2-orbit theorem in the mixed case [Pe2] sufficient to prove Theorem 61. For the proof of Theorem 57, we need the full strength of the several variable SL_2-orbit theorem of Kato, Nakayama and Usui [KNU1].

4.2. The classical nilpotent and SL_2-orbit theorems. To outline the proof of Theorem 61, we now recall the theory of degenerations of Hodge structure: Let $\mathcal{H}$ be a variation of pure Hodge structure of weight k over a simply connected complex manifold S. Then, via parallel translation back to a fixed reference fiber $H = H_{s_o}$ we obtain a period map

$$\varphi : S \to \mathcal{D}, \qquad\qquad (4\text{-}1)$$

where $\mathcal{D}$ is Griffiths' classifying space of pure Hodge structures on H with fixed Hodge numbers $\{h^{p,k-p}\}$ which are polarized by the bilinear form Q of H. The set $\mathcal{D}$ is a complex manifold upon which the Lie group

$$G_{\mathbb{R}} = \mathrm{Aut}_{\mathbb{R}}(Q)$$

acts transitively by biholomorphisms, and hence $\mathcal{D} \cong G_{\mathbb{R}}/G_{\mathbb{R}}^{F_o}$, where $G_{\mathbb{R}}^{F_o}$ is the isotropy group of $F_o \in \mathcal{D}$. The compact dual of $\mathcal{D}$ is the complex manifold

$$\check{\mathcal{D}} \cong G_{\mathbb{C}}/G_{\mathbb{C}}^{F_o}$$

where F_o is any point in $\mathcal{D}$. (In general, $F = F^{\bullet}$ denotes a Hodge filtration.) If S is not simply connected, then the period map (4-1) is replaced by

$$\varphi : S \to \Gamma \backslash \mathcal{D} \qquad (4\text{-}2)$$

where Γ is the monodromy group of $\mathbb{H} \to S$ acting on the reference fiber H.

For variations of Hodge structure of geometric origin, S will typically be a Zariski open subset of a smooth projective variety $\bar{S}$. By Hironaka's resolution of singularities theorem, we can assume $D = \bar{S} - S$ to be a divisor with normal crossings. The period map (4-2) will then have singularities at the points of D about which $\mathbb{H}$ has nontrivial local monodromy. A precise local description of the singularities of the period map of a variation of Hodge structure was obtained by Schmid [Sc]: Let $\varphi : (\Delta^*)^r \to \Gamma \backslash \mathcal{D}$ be the period map of variation of pure polarized Hodge structure over the product of punctured disks. First, one knows that φ is locally liftable with quasi-unipotent monodromy. After passage to a finite cover, we therefore obtain a commutative diagram

$$\begin{array}{ccc} U^r & \xrightarrow{\ \ F\ \ } & \mathcal{D} \\ \downarrow & & \downarrow \\ (\Delta^*)^r & \xrightarrow{\ \ \varphi\ \ } & \Gamma \backslash \mathcal{D} \end{array} \qquad (4\text{-}3)$$

where U^r is the r-fold product of upper half-planes and $U^r \to (\Delta^*)^r$ is the covering map

$$s_j = e^{2\pi i z_j}, \qquad j = 1, \dots, r$$

with respect to the standard Euclidean coordinates $(z_1, \dots, z_r)$ on $U^r \subset \mathbb{C}^r$ and $(s_1, \dots, s_r)$ on $(\Delta^*)^r \subset \mathbb{C}^r$.

Let $T_j = e^{N_j}$ denote the monodromy of $\mathcal{H}$ about $s_j = 0$. Then,

$$\psi(z_1, \dots, z_r) = e^{-\sum_j z_j N_j} . F(z_1, \dots, z_r)$$

is a holomorphic map from U^r into $\check{\mathcal{D}}$ which is invariant under the transformation $z_j \mapsto z_j + 1$ for each j, and hence drops to a map $(\Delta^*)^r \to \check{\mathcal{D}}$ which we continue to denote by ψ.

DEFINITION 63. Let $\mathcal{D}$ be a classifying space of pure Hodge structure with associated Lie group $G_{\mathbb{R}}$. Let $\mathfrak{g}_{\mathbb{R}}$ be the Lie algebra of $G_{\mathbb{R}}$. Then, a holomorphic, horizontal map $\theta : \mathbb{C}^r \to \check{\mathcal{D}}$ is a nilpotent orbit if

(a) there exists $\alpha > 0$ such that $\theta(z_1, \ldots, z_r) \in \mathcal{D}$ if $\mathrm{Im}(z_j) > \alpha \ \forall j$; and

(b) there exist commuting nilpotent endomorphisms $N_1, \ldots, N_r \in \mathfrak{g}_{\mathbb{R}}$ and a point $F \in \check{\mathcal{D}}$ such that $\theta(z_1, \ldots, z_r) = e^{\sum_j z_j N_j}.F$.

THEOREM 64 (NILPOTENT ORBIT THEOREM [Sc]). *Let $\varphi : (\Delta^*)^r \to \Gamma \backslash \mathcal{D}$ be the period map of a variation of pure Hodge structure of weight k with unipotent monodromy. Let $d_{\mathcal{D}}$ be a $G_{\mathbb{R}}$-invariant distance on $\mathcal{D}$. Then:*

(a) $F_\infty = \lim_{s \to 0} \psi(s)$ *exists, i.e., $\psi(s)$ extends to a map $\Delta^r \to \check{\mathcal{D}}$;*

(b) $\theta(z_1, \ldots, z_r) = e^{\sum_j z_j N_j}.F_\infty$ *is a nilpotent orbit; and*

(c) *there exist constants C, α and $\beta_1, \ldots, \beta_r$ such that if $\mathrm{Im}(z_j) > \alpha \ \forall j$ then $\theta(z_1, \ldots, z_r) \in \mathcal{D}$ and*

$$d_{\mathcal{D}}(\theta(z_1, \ldots, z_r), F(z_1, \ldots, z_r)) < C \sum_j \mathrm{Im}(z_j)^{\beta_j} e^{-2\pi \mathrm{Im}(z_j)}.$$

REMARK 65. Another way of stating part (a) of this theorem is that the Hodge bundles $\mathcal{F}^p$ of $\mathcal{H}_{\mathcal{O}}$ extend to a system of holomorphic subbundles of the canonical extension of $\mathcal{H}_{\mathcal{O}}$. Indeed, recall from §2.7 that one way of constructing a model of the canonical extension in the unipotent monodromy case is to take a flat, multivalued frame $\{\sigma_1, \ldots, \sigma_m\}$ of $\mathbb{H}_{\mathbb{Z}}$ and twist it to form a single valued holomorphic frame $\{\tilde{\sigma}_1, \ldots, \tilde{\sigma}_m\}$ over $(\Delta^*)^r$ where $\tilde{\sigma}_j = e^{-\frac{1}{2\pi i} \sum_j \log(s_j) N_j} \sigma_j$, and then declaring this twisted frame to define the canonical extension.

Let N be a nilpotent endomorphism of a finite-dimensional vector space over a field k. Then, N can be put into Jordan canonical form, and hence (by considering a Jordan block) it follows that there is a unique, increasing filtration $W(N)$ of V, such that, for each index j,

(a) $N(W(N)_j) \subseteq W(N)_{j-2}$ and

(b) $N^j : \mathrm{Gr}_j^{W(N)} \to \mathrm{Gr}_{-j}^{W(N)}$ is an isomorphism.

If ℓ is an integer then $(W(N)[\ell])_j = W(N)_{j+\ell}$.

THEOREM 66. *Let $\varphi : \Delta^* \to \Gamma \backslash \mathcal{D}$ be the period map of a variation of pure Hodge structure of weight k with unipotent monodromy $T = e^N$. Then, the limit Hodge filtration F_∞ of φ pairs with the weight monodromy filtration $M(N) := W(N)[-k]$ to define a mixed Hodge structure relative to which N is a $(-1, -1)$-morphism.*

REMARK 67. The limit Hodge filtration F_∞ depends upon the choice of local coordinate s, or more precisely on the value of $(ds)_0$. Therefore, unless one has a preferred coordinate system (say, if the field of definition matters), in order to extract geometric information from the limit mixed Hodge structure $H_\infty = (F_\infty, M(N))$ one usually has to pass to the mixed Hodge structure induced by H_∞ on the kernel or cokernel of N. In particular, if $X \to \Delta$ is a semistable degeneration, the local invariant cycle theorem asserts that we have an exact sequence

$$H^k(X_0) \to H_\infty \xrightarrow{N} H_\infty,$$

where the map $H^k(X_0) \to H_\infty$ is obtained by first including the reference fiber X_{s_0} into X and then retracting X onto X_0.

The proof of Theorem 66 depends upon Schmid's SL_2-orbit theorem. Informally, this result asserts that any 1-parameter nilpotent orbit is asymptotic to a nilpotent orbit arising from a representation of $SL_2(\mathbb{R})$. In order to properly state Schmid's results we need to discuss splittings of mixed Hodge structures.

THEOREM 68 (DELIGNE [De1]). *Let (F, W) be a mixed Hodge structure on V. There exists a unique, functorial bigrading*

$$V_\mathbb{C} = \bigoplus_{p,q} I^{p,q}$$

such that

(a) $F^p = \bigoplus_{a \geq p} I^{a,b}$;

(b) $W_k = \bigoplus_{a+b \leq k} I^{a,b}$;

(c) $\overline{I^{p,q}} = I^{q,p} \mod \bigoplus_{r<q,s<p} I^{r,s}$.

In particular, if (F, W) is a mixed Hodge structure on V then (F, W) induces a mixed Hodge structure on $gl(V) \cong V \otimes V^*$ with bigrading

$$gl(V_\mathbb{C}) = \bigoplus_{r,s} gl(V)^{r,s}$$

where $gl(V)^{r,s}$ is the subspace of $gl(V)$ which maps $I^{p,q}$ to $I^{p+r,q+s}$ for all (p, q). In the case where (F, W) is graded-polarized, we have an analogous decomposition $\mathfrak{g}_\mathbb{C} = \bigoplus_{r,s} \mathfrak{g}^{r,s}$ of the Lie algebra of $G_\mathbb{C}(= \mathrm{Aut}(V_\mathbb{C}, Q))$. For future use, we define

$$\Lambda_{(F,W)}^{-1,-1} = \bigoplus_{r,s<0} gl(V)^{r,s} \tag{4-4}$$

and note that by properties (a)–(c) of Theorem 68

$$\lambda \in \Lambda_{(F,W)}^{-1,-1} \quad \Rightarrow \quad I_{(e^\lambda.F,W)}^{p,q} = e^\lambda.I_{(F,W)}^{p,q}. \tag{4-5}$$

A mixed Hodge structure (F, W) is split over $\mathbb{R}$ if $\bar{I}^{p,q} = I^{q,p}$ for (p, q). In general, a mixed Hodge structure (F, W) is not split over $\mathbb{R}$. However, by a theorem of Deligne [CKS1], there is a functorial splitting operation

$$(F, W) \mapsto (\hat{F}_\delta, W) = (e^{-i\delta}.F, W)$$

which assigns to any mixed Hodge structure (F, W) a split mixed Hodge structure $(\hat{F}_\delta, W)$, such that

(a) $\delta = \bar{\delta}$,

(b) $\delta \in \Lambda_{(F,W)}^{-1,-1}$, and

(c) δ commutes with all (r, r)-morphisms of (F, W).

REMARK 69. $\Lambda_{(F,W)}^{-1,-1} = \Lambda_{(\hat{F}_\delta,W)}^{-1,-1}$.

A nilpotent orbit $\hat{\theta}(z) = e^{zN}.F$ is an SL_2-orbit if there exists a group homomorphism $\rho : SL_2(\mathbb{R}) \to G_\mathbb{R}$ such that

$$\hat{\theta}(g.\sqrt{-1}) = \rho(g).\hat{\theta}(\sqrt{-1})$$

for all $g \in SL_2(\mathbb{R})$. The representation ρ is equivalent to the data of an sl_2-triple (N, H, N^+) of elements in $G_\mathbb{R}$ such that

$$[H, N] = -2N, \qquad [N^+, N] = H, \qquad [H, N^+] = 2N^+$$

We also note that, for nilpotent orbits of pure Hodge structure, the statement that $e^{zN}.F$ is an SL_2-orbit is equivalent to the statement that the limit mixed Hodge structure $(F, M(N))$ is split over $\mathbb{R}$ [CKS1].

THEOREM 70. (SL_2-ORBIT THEOREM, [Sc]). *Let $\theta(z) = e^{zN}.F$ be a nilpotent orbit of pure Hodge structure. Then, there exists a unique SL_2-orbit $\hat{\theta}(z) = e^{zN}.\hat{F}$ and a distinguished real-analytic function*

$$g(y) : (a, \infty) \to G_\mathbb{R}$$

(for some $a \in \mathbb{R}$) such that

(a) $\theta(iy) = g(y).\hat{\theta}(iy)$ *for $y > a$, and*

(b) *both $g(y)$ and $g^{-1}(y)$ have convergent series expansions about ∞ of the form*

$$g(y) = 1 + \sum_{k>0} g_j y^{-k}, \qquad g^{-1}(y) = 1 + \sum_{k>0} f_k y^{-k}$$

with $g_k, f_k \in \ker(\operatorname{ad} N)^{k+1}$.

Furthermore, the coefficients g_k and f_k can be expressed in terms of universal Lie polynomials in the Hodge components of δ with respect to $(\hat{F}, M(N))$ and $\operatorname{ad} N^+$.

REMARK 71. The precise meaning of the statement that $g(y)$ is a distinguished real-analytic function, is that $g(y)$ arises in a specific way from the solution of a system of differential equations attached to θ.

REMARK 72. If θ is a nilpotent orbit of pure Hodge structures of weight k and $\hat{\theta} = e^{zN}.\hat{F}$ is the associated SL_2-orbit then $(\hat{F}, M(N))$ is split over $\mathbb{R}$. The map $(F, M(N)) \mapsto (\hat{F}, M(N))$ is called the sl_2-splitting of $(F, M(N))$. Furthermore, $\hat{F} = e^{-\xi}.F$ where ξ is given by universal Lie polynomials in the Hodge components of δ. In this way, one obtains an sl_2-splitting $(F, W) \mapsto (\hat{F}, W)$ for any mixed Hodge structure (F, W).

4.3. Nilpotent and SL_2-orbit theorems in the mixed case.

In analogy to the theory of period domains for pure HS, one can form a classifying space of graded-polarized mixed Hodge structure $\mathcal{M}$ with fixed Hodge numbers. Its points are the decreasing filtrations F of the reference fiber V which pair with the weight filtration W to define a graded-polarized mixed Hodge structure (with the given Hodge numbers). Given a variation of mixed Hodge structure $\mathcal{V}$ of this type over a complex manifold S, one obtains a period map

$$\phi : S \to \Gamma \backslash \mathcal{M}.$$

$\mathcal{M}$ is a complex manifold upon which the Lie group G, consisting of elements of $\mathrm{GL}(V_{\mathbb{C}})$ which preserve W and act by real isometries on Gr^W, acts transitively. Next, let $G_{\mathbb{C}}$ denote the Lie group consisting of elements of $\mathrm{GL}(V_{\mathbb{C}})$ which preserve W and act by *complex* isometries on Gr^W. Then, in analogy with the pure case, the "compact dual" $\check{\mathcal{M}}$ of $\mathcal{M}$ is the complex manifold

$$\check{\mathcal{M}} \cong G_{\mathbb{C}}/G_{\mathbb{C}}^{F_o}$$

for any base point $F_o \in \mathcal{M}$. The subgroup $G_{\mathbb{R}} = G \cap \mathrm{GL}(V_{\mathbb{R}})$ acts transitively on the real-analytic submanifold $\mathcal{M}_{\mathbb{R}}$ consisting of points $F \in \mathcal{M}$ such that (F, W) is split over $\mathbb{R}$.

EXAMPLE 73. Let $\mathcal{M}$ be the classifying space of mixed Hodge structures with Hodge numbers $h^{1,1} = h^{0,0} = 1$. Then, $\mathcal{M} \cong \mathbb{C}$.

The proof of Schmid's nilpotent orbit theorem depends critically upon the fact that the classifying space $\mathcal{D}$ has negative holomorphic sectional curvature along horizontal directions [GS]. Thus, although one can formally carry out all of the constructions leading up to the statement of the nilpotent orbit theorem in the mixed case, in light of the previous example it follows that one can not have negative holomorphic sectional curvature in the mixed case, and hence there is no reason to expect an analog of Schmid's Nilpotent Orbit Theorem in the mixed case. Indeed, for this classifying space $\mathcal{M}$, the period map $\varphi(s) = \exp(s)$ gives an example of a period map with trivial monodromy which has an essential singularity at ∞. Some additional condition is clearly required, and this is where admissibility comes in.

In the geometric case of a degeneration of pure Hodge structure, Steenbrink [St] gave an alternative construction of the limit Hodge filtration that can be extended to variations of mixed Hodge structure of geometric origin [SZ]. More generally, given an *admissible* variation of mixed Hodge structure $\mathcal{V}$ over a smooth complex algebraic variety $S \subseteq \bar{S}$ such that $D = \bar{S} - S$ is a normal crossing divisor, and any point $p \in D$ about which $\mathbb{V}$ has unipotent local monodromy, one has an associated nilpotent orbit $(e^{\sum_j z_j N_j}.F_\infty, W)$ with limit mixed Hodge structure (F_∞, M) where M is the *relative weight filtration* of $N = \sum_j N_j$ and W.[11] Furthermore, one has the following "group theoretic" version of the nilpotent orbit theorem: As in the pure case, a variation of mixed Hodge structure $\mathcal{V} \to (\Delta^*)^r$ with unipotent monodromy gives a holomorphic map

$$\psi : (\Delta^*)^r \to \check{M},$$
$$\underline{z} \mapsto e^{-\sum z_j N_j} F(\underline{z}),$$

and this extends to Δ^r if $\mathcal{V}$ is admissible. Let

$$\mathfrak{q}_\infty = \bigoplus_{r<0} \mathfrak{g}^{r,s}$$

where $\mathfrak{g}_{\mathbb{C}} = \mathrm{Lie}(G_{\mathbb{C}}) = \bigoplus_{r,s} \mathfrak{g}^{r,s}$ relative to the limit mixed Hodge structure (F_∞, M). Then $\mathfrak{q}_\infty$ is a nilpotent Lie subalgebra of $\mathfrak{g}_{\mathbb{C}}$ which is a vector space complement to the isotropy algebra $\mathfrak{g}_{\mathbb{C}}^{F_\infty}$ of F_∞. Consequently, there exists an open neighborhood $\mathcal{U}$ of zero in $\mathfrak{g}_{\mathbb{C}}$ such that

$$\mathcal{U} \to \check{M},$$
$$u \mapsto e^u.F_\infty$$

is a biholomorphism, and hence after shrinking Δ^r as necessary we can write

$$\psi(s) = e^{\Gamma(s)}.F_\infty$$

relative to a unique $\mathfrak{q}_\infty$-valued holomorphic function Γ on Δ^r which vanishes at 0. Recalling the construction of ψ from the lifted period map F, it follows that

$$F(z_1, \ldots, z_r) = e^{\sum_j z_j N_j} e^{\Gamma(s)}.F_\infty.$$

This is called the *local normal form* of $\mathcal{V}$ at p and will be used in the calculations of §5.4–5.

There is also a version of Schmid's SL_2-orbit theorem for admissible nilpotent orbits. In the case of 1-variable and weight filtrations of short length, the is due to the second author in [Pe2]. More generally, Kato, Nakayama and Usui

---

[11] Recall [SZ] that in general the relative weight filtration $M = M(N, W)$ is the unique filtration (if it exists) such that $N(M_k) \subset M_{k-2}$ and M induces the monodromy weight filtration of N on each Gr_l^W (centered about i).

proved a several variable SL_2-orbit theorem with arbitrary weight filtration in [KNU1]. Despite the greater generality of [KNU1], in this paper we are going to stick with the version of the SL_2-orbit theorem from [Pe2] as it is sufficient for our needs and has the advantage that for normal functions, mutatis mutandis, it is identical to Schmid's result.

4.4. Outline of proof of Theorem 61. Let us now specialize to the case of an admissible normal function $\nu : S \rightarrow J(\mathcal{H})$ over a curve and outline the proof [BP1] of Theorem 61. Before proceeding, we do need to address one aspect of the SL_2-orbit theorem in the mixed case. Let $\hat{\theta} = (e^{zN}.F, W)$ be an admissible nilpotent orbit with limit mixed Hodge structure (F, M) which is split over $\mathbb{R}$. Then, $\hat{\theta}$ induces an SL_2-orbit on each Gr_k^W, and hence a corresponding sl_2-representation ρ_k.

DEFINITION 74. Let W be an increasing filtration, indexed by $\mathbb{Z}$, of a finite dimensional vector space V. A *grading* of W is a direct sum decomposition $W_k = V_k \oplus W_{k-1}$ for each index k.

In particular, a mixed Hodge structure (F, W) on V gives a grading of W by the rule $V_k = \bigoplus_{p+q=k} I^{p,q}$. Furthermore, if the ground field has characteristic zero, a grading of W is the same thing as a semisimple endomorphism Y of V which acts as multiplication by k on V_k. If (F, W) is a mixed Hodge structure we let $Y_{(F,W)}$ denote the grading of W which acts on $I^{p,q}$ as multiplication by $p + q$, the *Deligne grading* of (F, W).

Returning to the admissible nilpotent orbit $\hat{\theta}$ considered above, we now have a system of representations ρ_k on Gr_k^W. To construct an sl_2-representation on the reference fiber V, we need to pick a grading Y of W. Clearly for each Hodge flag $F(z)$ in the orbit we have the Deligne grading $Y_{(F(z),W)}$; but we are after something more canonical. Now we also have the Deligne grading $Y_{(\hat{F},M)}$ of M associated to the sl_2-splitting of the LMHS. In the unpublished letter [De3], Deligne observed that:

THEOREM 75. *There exists a unique grading Y of W which commutes with $Y_{(\hat{F},M)}$ and has the property that if (N_0, H, N_0^+) denote the liftings of the sl_2-triples attached to the graded representations ρ_k via Y then $[N - N_0, N_0^+] = 0$.*

With this choice of sl_2-triple, and $\hat{\theta}$ an admissible nilpotent orbit in 1-variable of the type arising from an admissible normal function, the main theorem of [Pe2] asserts that one has a direct analog of Schmid's SL_2-orbit theorem as stated above for $\hat{\theta}$.

REMARK 76. More generally, given an admissible nilpotent orbit $(e^{zN}F, W)$ with relative weight filtration $M = M(N, W)$, Deligne shows that there exists a grading $Y = Y(N, Y_{(F,M)})$ with similar properties. See [BP1] for details and further references.

REMARK 77. In the case of a normal function, if we decompose N according to ad Y we have $N = N_0 + N_{-1}$ where N_{-1} must be either zero or a highest weight vector of weight -1 for the representation of $sl_2(\mathbb{R})$ defined by (N_0, H, N_0^+). Accordingly, since there are no vectors of highest weight -1, we have $N = N_0$ and hence $[Y, N] = 0$.

The next thing that we need to recall is that if $v : S \to J(\mathcal{H})$ is an admissible normal function which is represented by an extension

$$0 \to H \to \mathcal{V} \to \mathbb{Z}(0) \to 0$$

in the category of admissible variations of mixed Hodge structure on S then the zero locus $\mathcal{Z}$ of v is exactly the set of points where the corresponding Deligne grading $Y_{(F,W)}$ is integral. In the case where $S \subset \bar{S}$ is a curve, in order to prove the algebraicity of $\mathcal{Z}$, all we need to do is show that $\mathcal{Z}$ cannot contain a sequence of points $s(m)$ which accumulate to a puncture $p \in \bar{S} - S$ unless v is identically zero. The first step towards the proof of Theorem 61 is the following result [BP1]:

THEOREM 78. *Let* $\varphi : \Delta^* \to \Gamma \backslash \mathcal{M}$ *denote the period map of an admissible normal function* $v : \Delta^* \to J(\mathcal{H})$ *with unipotent monodromy, and* Y *be the grading of* W *attached to the nilpotent orbit* θ *of* φ *by Deligne's construction* (Theorem 75). *Let* $F : U \to \mathcal{M}$ *denote the lifting of* φ *to the upper half-plane. Then, for* $\mathrm{Re}(z)$ *restricted to an interval of finite length, we have*

$$\lim_{\mathrm{Im}(z) \to \infty} Y_{(F(z),W)} = Y.$$

SKETCH OF PROOF. Using [De3], one can prove this result in the case where φ is a nilpotent orbit with limit mixed Hodge structure which is split over $\mathbb{R}$. Let $z = x + iy$. In general, one writes

$$F(z) = e^{zN} e^{\Gamma(s)}.F_\infty = e^{xN} e^{iyN} e^{\Gamma(s)} e^{-iyN} e^{iyN}.F_\infty$$

where e^{xN} is real, $e^{iyN}.F_\infty$ can be approximated by an SL_2-orbit and the function $e^{iyN} e^{\Gamma(s)} e^{-iyN}$ decays to 1 very rapidly. $\square$

In particular, if there exists a sequence $s(m)$ which converges to p along which $Y_{(F,W)}$ is integral it then follows from the previous theorem that Y is integral. An explicit computation then shows that the equation of the zero locus near p is given by the equation

$$\mathrm{Ad}(e^{\Gamma(s)})Y = Y,$$

which is clearly holomorphic on a neighborhood of p in $\bar{S}$.

That concludes the proof for S a curve. In the case where S has a compactification $\bar{S}$ such that $\bar{S} - S$ is a smooth divisor, one can prove Theorem 61 by the same techniques by studying the dependence of the preceding constructions on holomorphic parameters, i.e., at a point in D we get a nilpotent orbit

$$\theta(z; s_2, \ldots, s_r) = e^{zN}.F_\infty(s_2, \ldots, s_r),$$

where $F_\infty(s_2, \ldots, s_r)$ depend holomorphically on the parameters $(s_2, \ldots, s_r)$.

4.5. Zero loci and filtrations on Chow groups. Returning now to the algebraicity of the Hodge locus discussed at the beginning of this section, the Hodge Conjecture would further imply that if $f : X \to S$ can be defined over an algebraically closed subfield of $\mathbb{C}$ then so can the germ of T. Claire Voisin [Vo1] gave sufficient conditions for T to be defined over $\bar{\mathbb{Q}}$ if $f : X \to S$ is defined over $\mathbb{Q}$. Very recently F. Charles [Ch] carried out an analogous investigation of the field of definition of the zero locus $\mathcal{Z}$ of a normal function motivated over $\mathbb{F}$. We reprise this last notion (from Sections 1 and 2):

DEFINITION 79. Let S be a smooth quasiprojective variety defined over a subfield $\mathbb{F}_0 \subset \mathbb{C}$, and let $\mathbb{F} \subset \mathbb{C}$ be a finitely generated extension of $\mathbb{F}_0$. An admissible normal function $v \in \text{ANF}(S, \mathcal{H})$ is *motivated over* $\mathbb{F}$ if there exists a smooth quasiprojective variety $\mathcal{X}$, a smooth projective morphism $f : \mathcal{X} \to S$, and an algebraic cycle $\mathfrak{Z} \in Z^m(\mathcal{X})_{\text{prim}}$, all defined over $\mathbb{F}$, such that $\mathcal{H}$ is a subVHS of $(R^{2m-1} f_* \mathbb{Z}) \otimes \mathcal{O}_S$ and $v = v_\mathfrak{Z}$.

REMARK 80. Here $Z^m(\mathcal{X})_{\text{prim}}$ denotes algebraic cycles with homologically trivial restriction to fibers. One traditionally also assumes $\mathfrak{Z}$ is flat over S, but this can always be achieved by restricting to $U \subset S$ sufficiently small (Zariski open); and then by [S1] *(i)* $v_{\mathfrak{Z}_U}$ is $\bar{S}$ *admissible*. Next, for any $s_0 \in S$ one can move $\mathfrak{Z}$ by a rational equivalence to intersect X_{s_0} (hence the $\{X_s\}$ for s in an analytic neighborhood of s_0) properly, and then use the remarks at the beginning of [Ki] or [GGK, §III.B] to see that *(ii)* $v_\mathfrak{Z}$ *is defined and holomorphic over all of S*. Putting *(i)* and *(ii)* together with [BFNP, Lemma 7.1], we see that $v_\mathfrak{Z}$ is itself admissible.

Recall that the level of a VHS $\mathcal{H}$ is (for a generic fiber H_s) the maximum difference $|p_1 - p_2|$ for H^{p_1, q_1} and H^{p_2, q_2} both nonzero. A fundamental open question about motivic normal functions is then:

CONJECTURE 81. (i) $[3\mathfrak{L}(D, E)]$ *For every $\mathbb{F} \subset \mathbb{C}$ finitely generated over $\bar{\mathbb{Q}}$, $S/\mathbb{F}$ smooth quasiprojective of dimension D, and $\mathcal{H} \to S$ VHS of weight (-1) and level $\leq 2E - 1$, the following holds: v motivated over $\mathbb{F}$ implies that $\mathcal{Z}(v)$ is an at most countable union of subvarieties of S defined over (possibly different) finite extensions of $\mathbb{F}$.*

(ii) $[\widetilde{3\mathfrak{L}}(D, E)]$ *Under the same hypotheses,* $\mathcal{Z}(\nu)$ *is an algebraic subvariety of* S *defined over an algebraic extension of* $\mathbb{F}$.

Clearly Theorem 57 and Conjecture $3\mathfrak{L}(D, E)$ together imply $\widetilde{3\mathfrak{L}}(D, E)$, but it is much more natural to phrase some statements (especially Proposition 86 below) in terms of $3\mathfrak{L}(D, E)$. If true even for $D = 1$ (but general E), Conjecture 81(i) would resolve a longstanding question on the structure of Chow groups of complex projective varieties. To wit, the issue is whether the second Bloch–Beilinson filtrand and the kernel of the AJ map must agree; we now wish to describe this connection. We shall write $3\mathfrak{L}(D, 1)_{\mathrm{alg}}$ for the case when ν is motivated by a family of cycles algebraically equivalent to zero.

Let X be smooth projective and $m \in \mathbb{N}$. Denoting "$\otimes \mathbb{Q}$" by a subscript $\mathbb{Q}$, we have the two "classical" invariants $cl_{X,\mathbb{Q}} : CH^m(X)_{\mathbb{Q}} \to \mathrm{Hg}^m(X)_{\mathbb{Q}}$ and $AJ_{X,\mathbb{Q}} :$ $\ker(cl_{X,\mathbb{Q}}) \to J^m(X)_{\mathbb{Q}}$. It is perfectly natural both to ask for further Hodge-theoretic invariants for cycle-classes in $\ker(AJ_{X,\mathbb{Q}})$, and inquire as to what sort of filtration might arise from their successive kernels. The idea of a (conjectural) *system* of decreasing filtrations on the rational Chow groups of *all* smooth projective varieties over $\mathbb{C}$, compatible with the intersection product, morphisms induced by correspondences, and the algebraic Künneth components of the diagonal Δ_X, was introduced by A. Beilinson [Be], and independently by S. Bloch. (One has to assume something like the Hard Lefschetz Conjecture so that these Künneth components exist; the compatibility roughly says that $\mathrm{Gr}^i CH^m(X)_{\mathbb{Q}}$ is controlled by $H^{2m-i}(X)$.) Such a filtration $F_{\mathrm{BB}}^{\bullet}$ is unique if it exists and is universally known as a *Bloch–Beilinson filtration* (BBF); there is a wide variety of constructions which yield a BBF under the assumption of various more-or-less standard conjectures. The one which is key for the filtration (due to Lewis [Le2]) we shall consider is the *arithmetic Bloch–Beilinson Conjecture* (BBC):

CONJECTURE 82. *If* $\mathcal{X}/\bar{\mathbb{Q}}$ *is a quasiprojective variety, the absolute-Hodge cycle-class map*

$$c_{\mathcal{H}} : CH^m(\mathcal{X})_{\mathbb{Q}} \to H_{\mathcal{H}}^{2m}(\mathcal{X}_{\mathbb{C}}^{\mathrm{an}}, \mathbb{Q}(m)) \tag{4-6}$$

is injective. (Here $CH^m(\mathcal{X})$ denotes $\equiv_{\mathrm{rat}}$-classes of cycles over $\bar{\mathbb{Q}}$, and differs from $CH^m(\mathcal{X}_{\mathbb{C}})$.)

Now for $X/\mathbb{C}$, $c_{\mathcal{H}}$ on $CH^m(X)_{\mathbb{Q}}$ is far from injective (the kernel usually not even being parametrizable by an algebraic variety); but any given cycle $Z \in Z^m(X)$ (a priori defined over $\mathbb{C}$) is in fact defined over a subfield $K \subset \mathbb{C}$ finitely generated over $\bar{\mathbb{Q}}$, say of transcendence degree t. Considering X, Z over K, the $\bar{\mathbb{Q}}$-spread then provides

- a smooth projective variety $\bar{S}/\bar{\mathbb{Q}}$ of dimension t, with $\bar{\mathbb{Q}}(\bar{S}) \overset{\cong}{\to} K$ and $s_0 :$ $\mathrm{Spec}(K) \to \bar{S}$ the corresponding generic point;

- a smooth projective variety $\bar{\mathcal{X}}$ and projective morphism $\bar{\pi} : \bar{\mathcal{X}} \to \bar{S}$, both defined over $\bar{\mathbb{Q}}$, such that $X = X_{s_0} := \mathcal{X} \times_{s_0} \mathrm{Spec}(K)$; and
- an algebraic cycle $\bar{\mathfrak{Z}} \in Z^m(\mathcal{X}_{(\bar{\mathbb{Q}})})$ with $Z = \bar{\mathfrak{Z}} \times_{s_0} \mathrm{Spec}(K)$.

Writing $\bar{\pi}^{\mathrm{sm}} =: \pi : \mathcal{X} \to S$ (and $\mathfrak{Z} := \bar{\mathfrak{Z}} \cap \mathcal{X}$), we denote by $U \subset S$ any affine Zariski open subvariety defined over $\bar{\mathbb{Q}}$, and put $\mathcal{X}_U := \pi^{-1}(U)$, $\mathfrak{Z}_U := \bar{\mathfrak{Z}} \cap \mathcal{X}_U$; note that s_0 factors through all such U.

The point is that exchanging the field of definition for additional geometry allows $c_{\mathcal{H}}$ to "see" more; in fact, since we are over $\bar{\mathbb{Q}}$, it should now (by BBC) see everything. Now $c_{\mathcal{H}}(\mathfrak{Z}_U)$ packages cycle-class and Abel–Jacobi invariants together, and the idea behind Lewis's filtration (and filtrations of M. Saito and Green/Griffiths) is to split the whole package up into Leray graded pieces with respect to π. Miraculously, the 0-th such piece turns out to agree with the fundamental class of Z, and the next piece is the normal function generated by $\mathfrak{Z}_U$. The pieces after that define the so-called *higher* cycle-class and AJ maps. More precisely, we have

$$
\Psi := \quad
\begin{array}{c}
CH^m(X_{(K)})_{\mathbb{Q}} \\
\cong \downarrow \mathrm{spread} \\
\mathrm{im}\{CH^m(\bar{\mathcal{X}})_{\mathbb{Q}} \to \varinjlim_{U} CH^m(\mathcal{X}_U)_{\mathbb{Q}}\} \\
\downarrow c_{\mathcal{H}} \\
\underline{H}_{\mathcal{H}}^{2m} : = = \mathrm{im}\left(H_{\mathcal{D}}^{2m}(\bar{\mathcal{X}}_{\mathbb{C}}^{\mathrm{an}}, \mathbb{Q}(m)) \to \varinjlim_{U} H_{\mathcal{H}}^{2m}((\mathcal{X}_U)_{\mathbb{C}}^{\mathrm{an}}, \mathbb{Q}(m)) \right)
\end{array}
\tag{4-7}
$$

with $c_{\mathcal{H}}$ (hence Ψ) conjecturally injective. Lewis [Le2] defines a Leray filtration $\mathcal{L}^{\bullet} \underline{H}_{\mathcal{H}}^{2m}$ with graded pieces

$$
\begin{array}{c}
0 \\
\downarrow \\
\dfrac{J^0\left(\varinjlim_{U} W_{-1} H^{i-1}(U, R^{2m-i}\pi_*\mathbb{Q}(m)) \right)}{\mathrm{im} \varinjlim_{U} \mathrm{Hg}^0\left(\mathrm{Gr}_0^W H^i(U, R^{2m-i}\pi_*\mathbb{Q}(m)) \right)} \\
\downarrow \beta \\
\mathrm{Gr}_{\mathcal{L}}^i \underline{H}_{\mathcal{H}}^{2m} \\
\downarrow \alpha \\
\mathrm{Hg}^0\left(\varinjlim_{U} W_0 H^i(U, R^{2m-i}\pi_*\mathbb{Q}(m)) \right) \\
\downarrow \\
0
\end{array}
\tag{4-8}
$$

and sets $\mathcal{L}^i CH^m(X_K)_\mathbb{Q} := \Psi^{-1}(\mathcal{L}^i \underline{H}^{2m}_\mathcal{H})$. For $Z \in \mathcal{L}^i CH^m(X_K)_\mathbb{Q}$, we put $cl^i_X(Z) := \alpha(\mathrm{Gr}^i_\mathcal{L} \Psi(Z))$; if this vanishes then $\mathrm{Gr}^i_\mathcal{L} \Psi(Z) =: \beta(aj^{i-1}_X(Z))$, and vanishing of $cl^i(Z)$ and $aj^{i-1}(Z)$ implies membership in $\mathcal{L}^{i+1}$. One easily finds that $cl^0_X(Z)$ identifies with $cl_{X,\mathbb{Q}}(Z) \in \mathrm{Hg}^0(X)_\mathbb{Q}$.

REMARK 83. The arguments of Hg^0 and J^0 in (4-8) have canonical and functorial MHS by [Ar]. One should think of the top term as $\mathrm{Gr}^{i-1}_\mathcal{L}$ of the lowest-weight part of $J^m(\mathcal{X}_U)$ and the bottom as $\mathrm{Gr}^i_\mathcal{L}$ of the lowest-weight part of $\mathrm{Hg}^m(\mathcal{X}_U)$ (both in the limit over U).

Now to get a candidate BBF, Lewis takes

$$\mathcal{L}^i CH^m(X_\mathbb{C})_\mathbb{Q} := \varinjlim_{\substack{K \subset \mathbb{C} \\ \text{f.g.}/\bar{\mathbb{Q}}}} \mathcal{L}^i CH^m(X_K)_\mathbb{Q}.$$

Some consequences of the definition of a BBF mentioned above, specifically the compatibility with the Künneth components of Δ_X, include these:

(a) $\begin{cases} F^0_{\mathrm{BB}} CH^m(X)_\mathbb{Q} = CH^m(X)_\mathbb{Q}, \\ F^1_{\mathrm{BB}} CH^m(X)_\mathbb{Q} = CH^m_{\mathrm{hom}}(X)_\mathbb{Q}, \\ F^2_{\mathrm{BB}} CH^m(X)_\mathbb{Q} \subseteq \ker(AJ_{X,\mathbb{Q}}), \end{cases}$

(b) $F^{m+1}_{\mathrm{BB}} CH^m(X) = \{0\}$.

These are sometimes stated as additional requirements for a BBF.

THEOREM 84 [Le2]. $\mathcal{L}^\bullet$ is intersection- and correspondence-compatible, and satisfies (a). Assuming BBC, $\mathcal{L}^\bullet$ satisfies (b); and additionally assuming HLC, $\mathcal{L}^\bullet$ is a BBF.

The limits in (4-8) inside J^0 and Hg^0 stabilize for sufficiently small U; replacing S by such a U, we may consider the normal function $\nu_3 \in \mathrm{ANF}(S, \mathcal{H}^{2m-1}_{\mathcal{X}/S})$ attached to the $\bar{\mathbb{Q}}$-spread of Z.

PROPOSITION 85. (i) For $i = 1$, (4-8) becomes

$$0 \to J^m_{\mathrm{fix}}(\mathcal{X}/S)_\mathbb{Q} \to \mathrm{Gr}^1_\mathcal{L} \underline{H}^{2m}_\mathcal{H} \to \left(H^1(S, R^{2m-1}\pi_*\mathbb{Q})\right)^{(0,0)} \to 0.$$

(ii) For $Z \in CH^m_{\mathrm{hom}}(X_K)_\mathbb{Q}$, we have $cl^1_X(Z) = [\nu_3]_\mathbb{Q}$. If this vanishes, then $aj^0_X(Z) = AJ_X(Z)_\mathbb{Q} \in J^m_{\mathrm{fix}}(\mathcal{X}/S)_\mathbb{Q} \subset J^m(X)_\mathbb{Q}$ (implying $\mathcal{L}^2 \subset \ker AJ_\mathbb{Q}$).

So for $Z \in CH^m_{\mathrm{hom}}(X_K)$ with $\bar{\mathbb{Q}}$-spread $\mathfrak{Z}$ over S, the information contained in $\mathrm{Gr}^1_\mathcal{L} \Psi(Z)$ is (up to torsion) precisely ν_3. Working over $\mathbb{C}$, $\mathfrak{Z} \cdot X_{s_0} = Z$ is the fiber of the spread at a *very general point* $s_0 \in S(\mathbb{C})$: $\mathrm{trdeg}(\bar{\mathbb{Q}}(s_0)/\bar{\mathbb{Q}})$ is maximal, i.e., equal to the dimension of S. Since AJ is a *transcendental* (rather than algebraic) invariant, there is no outright guarantee that vanishing of

$AJ_X(Z) \in J^m(X)$ — or equivalently, of the normal function at a very general point — implies the identical vanishing of ν_3 or even $[\nu_3]$. To display explicitly the depth of the question:

PROPOSITION 86. (i) $3\mathfrak{L}(1, E) \ (\forall E \in \mathbb{N}) \iff \mathcal{L}^2 CH^m(X)_\mathbb{Q} = \ker(AJ_{X,\mathbb{Q}})$ ($\forall$ sm. proj. $X/\mathbb{C}$).

(ii) $3\mathfrak{L}(1, 1)_{\mathrm{alg}} \iff \mathcal{L}^2 CH^m(X)_\mathbb{Q} \cap CH^m_{\mathrm{alg}}(X)_\mathbb{Q} = \ker(AJ_{X,\mathbb{Q}}) \cap CH^m_{\mathrm{alg}}(X)_\mathbb{Q}$ ($\forall$ sm. proj. $X/\mathbb{C}$).

Roughly speaking, these statements say that "sensitivity of the zero locus (of a cycle-generated normal function) to field of definition" is equivalent to "spreads of homologically and AJ-trivial cycles give trivial normal functions". In (ii), the cycles in both statements are assumed algebraically equivalent to zero.

PROOF. We first remark that for any variety S with field of definition $\mathbb{F}$ of minimal transcendence degree, no proper $\bar{\mathbb{F}}$-subvariety of S contains (in its complex points) a very general point of S.

(i) ($\Rightarrow$) : Let 3 be the $\bar{\mathbb{Q}}$-spread of Z with $AJ(Z)_\mathbb{Q} = 0$, and suppose $\mathrm{Gr}^1_\mathcal{L} \Psi(Z) = \mathrm{Gr}^1_\mathcal{L} c_\mathcal{H}(3)$ does not vanish. Taking a 1-dimensional very general multiple hyperplane section $S_0 \subset S$ through s_0 (S_0 is "minimally" defined over $k \overset{\mathrm{trdeg.1}}{\subseteq} K$), the restriction $\mathrm{Gr}^1_\mathcal{L} c_\mathcal{H}(3_0) \neq 0$ by weak Lefschetz. Since each $\mathcal{Z}(\nu_{N3_0}) \subseteq S_0$ is a union of subvarieties defined over $\bar{k}$ and contains s_0 for some $N \in \mathbb{N}$, one of these is all of S_0 (which implies $\mathrm{Gr}^1_\mathcal{L} \Psi(Z) = 0$), a contradiction. So $Z \in \mathcal{L}^2$.

($\Leftarrow$) : Let $\mathcal{X}_0 \to S_0, 3_0 \in Z^m(\mathcal{X}_0)_{\mathrm{prim}}$, $\dim(S_0) = 1$, all be defined over k and suppose $\mathcal{Z}(\nu_{3_0})$ contains a point s_0 not defined over $\bar{k}$. Spreading this out over $\bar{\mathbb{Q}}$ to $3, \mathcal{X}, S \supset S_0 \ni s_0$, we have: $s_0 \in S$ is very general, 3 is the $\bar{\mathbb{Q}}$-spread of $Z = 3_0 \cdot X_{s_0}$, and $AJ(Z)_\mathbb{Q} = 0$. So $Z \in \mathcal{L}^2$ implies ν_3 is torsion, which implies ν_{3_0} is torsion. But then ν_{3_0} is zero since it is zero somewhere (at s_0). So $\mathcal{Z}(\nu_{3_0})$ is either S_0 or a (necessarily countable) union of $\bar{k}$-points of S_0.

(ii) The spread 3 of $Z_{(s_0)} \equiv_{\mathrm{alg}} 0$ has every fiber $Z_s \equiv_{\mathrm{alg}} 0$, hence ν_3 is a section of $J(\mathcal{H})$, $\mathcal{H} \subset (R^{2m-1} \pi_* \mathbb{Q}(m)) \otimes \mathcal{O}_S$ a subVHS of level one (which can be taken to satisfy $H_s = (H^{2m-1}(X_s))_h$ for a.e. $s \in S$). The rest is as in (i). $\square$

REMARK 87. A related candidate BBF which occurs in work of the first author with J. Lewis [KL, §4], is defined via successive kernels of *generalized* normal functions (associated to the $\bar{\mathbb{Q}}$-spread 3 of a cycle). These take values on very general $(i - 1)$-dimensional subvarieties of S (rather than at points), and have the above $cl^i(Z)$ as their topological invariants.

4.6. Field of definition of the zero locus. We shall begin by showing that the equivalent conditions of Proposition 86(ii) are satisfied; the idea of the argument is due in part to S. Saito [Sa]. The first paragraph in the following in fact gives more:

THEOREM 88. $\widetilde{\mathfrak{Z}\mathfrak{L}}(D, 1)_{\mathrm{alg}}$ *holds for all* $D \in \mathbb{N}$. *That is, the zero locus of any normal function motivated by a family of cycles over* $\mathbb{F}$ *algebraically equivalent to zero, is defined over an algebraic extension of* $\mathbb{F}$.

Consequently, cycles algebraically and Abel–Jacobi-equivalent to zero on a smooth projective variety over $\mathbb{C}$, *lie in the second Lewis filtrand.*

PROOF. Consider $\mathfrak{Z} \in Z^m(\mathcal{X})_{\mathrm{prim}}$ and $f : \mathcal{X} \to S$ defined over K (K being finitely generated over $\bar{\mathbb{Q}}$), with $Z_s \equiv_{\mathrm{alg}} 0 \ \forall s \in S$; and let $s_0 \in \mathcal{Z}(\nu_3)$. (Note: s_0 is just a complex point of S.) We need to show:

$$\exists N \in \mathbb{N} \text{ such that } \sigma(s_0) \in \mathcal{Z}(\nu_{N3}) \text{ for any } \sigma \in \mathrm{Gal}(\mathbb{C}/K). \qquad (4\text{-}9)$$

Here is why (4-9) will suffice: the analytic closure of the set of all conjugate points is simply the point's K-spread $S_0 \subset S$, a (possibly reducible) algebraic subvariety defined over K. Clearly, on the s_0-connected component of S_0, ν_3 itself then vanishes; and this component is defined over an algebraic extension of K. Trivially, $\mathcal{Z}(\nu_3)$ is the union of such connected spreads of its points s_0; and since K is finitely generated over $\bar{\mathbb{Q}}$, there are only countably many subvarieties of S_0 defined over K or algebraic extensions thereof. This proves $\mathfrak{Z}\mathfrak{L}(D, 1)_{\mathrm{alg}}$, hence (by Theorem 57) $\widetilde{\mathfrak{Z}\mathfrak{L}}(D, 1)_{\mathrm{alg}}$.

To show (4-9), write $X = X_{s_0}$, $Z = Z_{s_0}$, and $L(/K)$ for their field of definition. There exist over L

- a smooth projective curve C and points $0, q \in C(L)$;
- an algebraic cycle $W \in Z^m(C \times X)$ such that $Z = W_*(q - 0)$; and
- another cycle $\Gamma \in Z^1(J(C) \times C)$ defining Jacobi inversion.

Writing $\Theta := W \circ \Gamma \in Z^m(J(C) \times X)$, the induced map

$$[\Theta]_* : J(C) \to J^m(X)_{\mathrm{alg}} \ (\subseteq J^m(X)_h)$$

is necessarily a morphism of abelian varieties over L; hence the identity connected component of $\ker([\Theta]_*)$ is a subabelian variety of $J(C)$ defined over an algebraic extension $L' \supset L$. Define $\theta := \Theta|_B \in Z^m(B \times X)$, and observe that $[\theta]_* : B \to J^m(X)_{\mathrm{alg}}$ is zero by construction, so that $cl(\theta) \in \mathcal{L}^2 H^{2m}(B \times X)$.

Now, since $AJ_X(Z) = 0$, a multiple $b := N.AJ_C(q - 0)$ belongs to B, and then $N.Z = \theta_* b$. This "algebraizes" the AJ-triviality of $N.Z$: conjugating the 6-tuple $(s_0, Z, X, B, \theta, b)$ to $(\sigma(s_0), Z^\sigma[= Z_{\sigma(s_0)}], X^\sigma[= X_{\sigma(s_0)}], B^\sigma, \theta^\sigma, b^\sigma)$, we still have $N.Z^\sigma = \theta_*^\sigma b^\sigma$ and $cl(\theta^\sigma) \in \mathcal{L}^2 H^{2m}(B^\sigma \times X^\sigma)$ by motivicity of the Leray filtration [Ar], and this implies $N.AJ(Z^\sigma) = [\theta^\sigma]_* b^\sigma = 0$ as desired. $\qquad \square$

We now turn to the result of [Ch] indicated at the outset of §4.5. While interesting, it sheds no light on $3\mathfrak{L}(1, E)$ or filtrations, since the hypothesis that the VHS $\mathcal{H}$ have no global sections is untenable over a point.

THEOREM 89 [Ch, Theorem 3]. *Let $\mathcal{Z}$ be the zero locus of a k-motivated normal function $\nu : S \to J(\mathcal{H})$. Assume that $\mathcal{Z}$ is algebraic and $\mathbb{H}_\mathbb{C}$ has no nonzero global sections over $\mathcal{Z}$. Then $\mathcal{Z}$ is defined over a finite extension of k.*

PROOF. Charles's proof of this result uses the ℓ-adic Abel–Jacobi map. Alternatively, we can proceed as follows (using, with $\mathbb{F} = k$, the notation of Definition 79): take $\mathcal{Z}_0 \subset \mathcal{Z}(\nu)$ to be an irreducible component (without loss of generality assumed smooth), and $3_{\mathcal{Z}_0}$ the restriction of 3 to $\mathcal{Z}_0$. Let $[3_{\mathcal{Z}_0}]$ and $[3_{\mathcal{Z}_0}]_{dR}$ denote the Betti and de Rham fundamental classes of $3_{\mathcal{Z}_0}$, and $\mathcal{L}$ the Leray filtration. Then, $\mathrm{Gr}_\mathcal{L}^1[3_{\mathcal{Z}_0}]$ is the topological invariant of $[3_{\mathcal{Z}_0}]$ in $H^1(\mathcal{Z}_0, R^{2m-1} f_* \mathbb{Z})$, whereas $\mathrm{Gr}_\mathcal{L}^1[3_{\mathcal{Z}_0}]_{dR}$ is the infinitesimal invariant of ν_3 over $\mathcal{Z}_0$. In particular, since $\mathcal{Z}_0$ is contained in the zero locus of ν_3,

$$\mathrm{Gr}_\mathcal{L}^j[3_{\mathcal{Z}_0}]_{dR} = 0, \quad j = 0, 1. \tag{4-10}$$

Furthermore, by the algebraicity of the Gauss–Manin connection, (4-10) is invariant under conjugation:

$$\mathrm{Gr}_\mathcal{L}^j[3_{\mathcal{Z}_0^\sigma}]_{dR} = (\mathrm{Gr}_\mathcal{L}^j[3_{\mathcal{Z}_0}]_{dR})^\sigma$$

and hence $\mathrm{Gr}_\mathcal{L}^j[3_{\mathcal{Z}_0^\sigma}]_{dR} = 0$ for $j = 0, 1$. Therefore, $\mathrm{Gr}_\mathcal{L}^j[3_{\mathcal{Z}_0^\sigma}] = 0$ for $j = 0, 1$, and hence $AJ(Z_s)$ takes values in the fixed part of $J(\mathcal{H})$ for $s \in \mathcal{Z}_0^\sigma$. By assumption, $\mathbb{H}_\mathbb{C}$ has no fixed part over $\mathcal{Z}_0$, and hence no fixed part over $\mathcal{Z}_0^\sigma$ (since conjugation maps ∇-flat sections to ∇-flat sections by virtue of the algebraicity of the Gauss–Manin connection). As such, conjugation must take us to another component of $\mathcal{Z}$, and hence (since $\mathcal{Z}$ is algebraic over $\mathbb{C}$ implies $\mathcal{Z}$ has only finitely many components), $\mathcal{Z}_0$ must be defined over a finite extension of k. $\qquad\qquad\square$

We conclude with a more direct analog of Voisin's result [Vo1, Theorem 0.5(2)] on the algebraicity of the Hodge locus. If $\mathcal{V}$ is a variation of mixed Hodge structure over a complex manifold and

$$\alpha \in (\mathcal{F}^p \cap \mathcal{W}_{2p} \cap V_\mathbb{Q})_{s_o}$$

for some $s_o \in S$, then the Hodge locus T of α is the set of points in S where some parallel translate of α belongs to $\mathcal{F}^p$.

REMARK 90. If (F, W) is a mixed Hodge structure on V and $v \in F^p \cap W_{2p} \cap V_\mathbb{Q}$ then v is of type (p, p) with respect to Deligne's bigrading of (F, W).

THEOREM 91. *Let S be a smooth complex algebraic variety defined over a subfield k of $\mathbb{C}$, and $\mathcal{V}$ be an admissible variation of mixed Hodge structure of geometric origin over S. Suppose that T is an irreducible subvariety of S over $\mathbb{C}$ such that*:

(a) *T is an irreducible component of the Hodge locus of some*

$$\alpha \in (\mathcal{F}^p \cap \mathcal{W}_{2p} \cap \mathbb{V}_\mathbb{Q})_{t_0};$$

(b) *$\pi_1(T, t_0)$ fixes only the line generated by α.*

Then, T is defined over $\bar{k}$.

PROOF. If $\mathcal{V} \cong \mathbb{Q}(p)$ for some p then $T = S$. Otherwise, T cannot be an isolated point without violating (b). Assume therefore that $\dim T > 0$. Over T, we can extend α to a flat family of de Rham classes. By the algebraicity of the Gauss–Manin connection, the conjugate α^σ is flat over T^σ. Furthermore, if T^σ supports any additional flat families of de Rham classes, conjugation by σ^{-1} gives a contradiction to (b). Therefore, $\alpha^\sigma = \lambda\beta$, where β is a $\pi_1(T^\sigma)$-invariant Betti class on T^σ which is unique up to scaling. Moreover,

$$Q(\alpha, \alpha) = Q(\alpha^\sigma, \alpha^\sigma) = \lambda^2 Q(\beta, \beta)$$

and hence there are countably many Hodge classes that one can conjugate α to via $\text{Gal}(\mathbb{C}/k)$. Accordingly, T must be defined over $\bar{k}$. $\qquad\square$

5. The Néron model and obstructions to singularities

The unifying theme of the previous sections is the study of algebraic cycles via degenerations using the Abel–Jacobi map. In particular, in the case of a semistable degeneration $\pi : X \to \Delta$ and a *cohomologically trivial* cycle Z which properly intersects the fibers, we have

$$\lim_{s \to 0} AJ_{X_s}(Z_s) = AJ_{X_0}(Z_0)$$

as explained in detail in § 2. In general however, the existence of the limit Abel–Jacobi map is obstructed by the existence of the singularities of the associated normal function. Nonetheless, using the description of the asymptotic behavior provided by the nilpotent and SL_2-orbit theorems, we can define the limits of admissible normal functions along curves and prove the algebraicity of the zero locus.

5.1. Néron models in one parameter. In this section we consider the problem of geometrizing these constructions (ANFs and their singularities, limits and zeroes) by constructing a Néron model which graphs admissible normal functions. The quest to construct such objects has a long history which traces back to the work of Néron on minimal models for abelian varieties A_K defined over the field of fractions K of a discrete valuation ring R. In [Na], Nakamura proved the existence of an analytic Néron model for a family of abelian varieties $\mathcal{A} \to \Delta^*$ arising from a variation of Hodge structure $\mathcal{H} \to \Delta^*$ of level 1 with unipotent monodromy. With various restrictions, this work was then extended to normal functions arising from higher codimension cycles in the work of Clemens [Cl2], El Zein and Zucker [EZ], and Saito [S1].

REMARK 92. Unless otherwise noted, throughout this section we assume that the local monodromy of the variation of Hodge structure $\mathcal{H}$ under consideration is unipotent, and the local system $\mathbb{H}_{\mathbb{Z}}$ is torsion free.

A common feature in all of these analytic constructions of Néron models for variations of Hodge structure over Δ^* is that the fiber over $0 \in \Delta$ is a complex Lie group which has only finitely many components. Furthermore, the component into which a given normal function ν extends is determined by the value of $\sigma_{\mathbb{Z},0}(\nu)$. Using the methods of the previous section, one way to see this is as follows: Let

$$0 \to \mathcal{H} \to \mathcal{V} \to \mathbb{Z}(0) \to 0$$

represent an admissible normal function $\nu : \Delta^* \to J(\mathcal{H})$ and $F : U \to \mathcal{M}$ denote the lifting of the period map of $\mathcal{V}$ to the upper half-plane, with monodromy $T = e^N$. Then, using the SL_2-orbit theorem of the previous section, it follows (cf. Theorem 4.15 of [Pe2]) that

$$Y_{\text{Hodge}} = \lim_{\text{Im}(z) \to \infty} e^{-zN}.Y_{(F(z),W)}$$

exists, and is equal to the grading $Y(N, Y_{(F_\infty, M)})$ constructed in the previous section; recall also that $Y(N, Y_{(F_\infty, M)}) \in \ker(\text{ad } N)$ due to the short length of the weight filtration. Suppose further that there exists an integral grading $Y_{\text{Betti}} \in \ker(\text{ad } N)$ of the weight filtration W. Let $j : \Delta^* \to \Delta$ and $i : \{0\} \to \Delta$ denote the inclusion maps. Then, $Y_{\text{Hodge}} - Y_{\text{Betti}}$ defines an element in

$$J(H_0) = \text{Ext}^1_{\text{MHS}}(\mathbb{Z}(0), H^0(i^* Rj_* \mathcal{H})) \qquad (5\text{-}1)$$

by simply applying $Y_{\text{Hodge}} - Y_{\text{Betti}}$ to any lift of $1 \in \mathbb{Z}(0) = \text{Gr}_0^W$. Reviewing §2 and §3, we see that the obstruction to the existence of such a grading Y_{Betti} is exactly the class $\sigma_{\mathbb{Z},0}(\nu)$.

REMARK 93. More generally, if $\mathcal{H}$ is a variation of Hodge structure of weight -1 over a smooth complex algebraic variety S and $\bar{S}$ is a good compactification of S, given a point $s \in \bar{S}$ we define

$$J(H_s) = \text{Ext}^1_{\text{MHS}}(\mathbb{Z}, H_s), \qquad (5\text{-}2)$$

where $H_s = H^0(i_s^* R_{j*} \mathcal{H})$ and $j : S \to \bar{S}$, $i_s : \{s\} \to \bar{S}$ are the inclusion maps. In case $\bar{S} \backslash S$ is a NCD in a neighborhood of S, with $\{N_i\}$ the logarithms of the unipotent parts of the local monodromies, then $H_s \cong \bigcap_j \ker(N_j)$.

In general, except in the classical case of degenerations of Hodge structure of level 1, the dimension of $J(H_0)$ is usually strictly less than the dimension of the fibers of $J(\mathcal{H})$ over Δ^*. Therefore, any generalized Néron model $J_\Delta(\mathcal{H})$ of $J(\mathcal{H})$ which graphs admissible normal functions cannot be a complex analytic space. Rather, in the terminology of Kato and Usui [KU; GGK], we obtain a "slit analytic fiber space". In the case where the base is a curve, the observations above can be combined into the following result:

THEOREM 94. *Let $\mathcal{H}$ be a variation of pure Hodge structure of weight -1 over a smooth algebraic curve S with smooth projective completion $\bar{S}$. Let $j : S \to \bar{S}$ denote the inclusion map. Then, there exists a Néron model for $J(\mathcal{H})$, i.e., a topological group $J_{\bar{S}}(\mathcal{H})$ over $\bar{S}$ satisfying the following two conditions*:

(i) *$J_{\bar{S}}(\mathcal{H})$ restricts to $J(\mathcal{H})$ over S.*

(ii) *There is a one-to-one correspondence between the set of admissible normal functions $v : S \to J(\mathcal{H})$ and the set of continuous sections $\bar{v} : \bar{S} \to J_{\bar{S}}(\mathcal{H})$ which restrict to holomorphic, horizontal sections of $J(\mathcal{H})$ over S.*

Furthermore:

(iii) *There is a short exact sequence of topological groups*

$$0 \to J_{\bar{S}}(\mathcal{H})^0 \to J_{\bar{S}}(\mathcal{H}) \to G \to 0,$$

where G_s is the torsion subgroup of $(R_{j*}^1 \mathcal{H}_{\mathbb{Z}})_s$ for any $s \in \bar{S}$.

(iv) *$J_{\bar{S}}(\mathcal{H})^0$ is a slit analytic fiber space, with fiber $J(H_s)$ over $s \in \bar{S}$.*

(v) *If $v : S \to J(\mathcal{H})$ is an admissible normal function with extension $\bar{v}$ then the image of $\bar{v}$ in G_s at the point $s \in \bar{S} - S$ is equal to $\sigma_{\mathbb{Z},s}(v)$. Furthermore, if $\sigma_{\mathbb{Z},s}(v) = 0$ then the value of $\bar{v}$ at s is given by the class of $Y_{\mathrm{Hodge}} - Y_{\mathrm{Betti}}$ as in (5-1). Equivalently, in the geometric setting, if $\sigma_{\mathbb{Z},s}(v) = 0$ then the value of $\bar{v}$ at s is given by the limit Abel–Jacobi map.*

Regarding the topology of the Néron model, let us consider more generally the case of smooth complex variety S with good compactification $\bar{S}$, and recall from §2 that we have also have the Zucker extension $J_{\bar{S}}^Z(\mathcal{H})$ obtained by starting from the short exact sequence of sheaves

$$0 \to \mathbb{H}_{\mathbb{Z}} \to \mathcal{H}_\mathcal{O}/F^0 \to J(\mathcal{H}) \to 0$$

and replacing $\mathbb{H}_{\mathbb{Z}}$ by $j_* \mathbb{H}_{\mathbb{Z}}$ and $\mathcal{H}_\mathcal{O}/F^0$ by its canonical extension. Following [S5], let us suppose that $D = \bar{S} - S$ is a smooth divisor, and let $J_{\bar{S}}^Z(\mathcal{H})_D^{\mathrm{Inv}}$ be the subset of $J_{\bar{S}}^Z(\mathcal{H})$ defined by the local monodromy invariants.

THEOREM 95 [S5]. *The Zucker extension $J_{\bar{S}}^{Z}(\mathcal{H})$ has the structure of a complex Lie group over $\bar{S}$, and it is a Hausdorff topological space on a neighborhood of $J_{\bar{S}}^{Z}(\mathcal{H})_{D}^{\text{Inv}}$.*

Specializing this result to the case where S is a curve, we then recover the result of the first author together with Griffiths and Green that $J_{\bar{S}}(\mathcal{H})^{0}$ is Hausdorff, since in this case we can identify $J_{\bar{S}}(\mathcal{H})^{0}$ with $J_{\bar{S}}^{Z}(\mathcal{H})_{D}^{\text{Inv}}$.

REMARK 96. Using this Hausdorff property, Saito was able to prove in [S5] the algebraicity of the zero locus of an admissible normal function in this setting (i.e., D smooth).

5.2. Néron models in many parameters. To extend this construction further, we have to come to terms with the fact that unless S has a compactification $\bar{S}$ such that $D = \bar{S} - S$ is a smooth divisor, the normal functions that we consider may have nontorsion singularities along the boundary divisor. This will be reflected in the fact that the fibers G_s of G need no longer be finite groups. The first test case is when $\mathcal{H}$ is a Hodge structure of level 1. In this case, a Néron model for $J(\mathcal{H})$ was constructed in the thesis of Andrew Young [Yo]. More generally, in joint work with Patrick Brosnan and Morihiko Saito, the second author proved the following result:

THEOREM 97 [BPS]. *Let S be a smooth complex algebraic variety and $\mathcal{H}$ be a variation of Hodge structure of weight -1 over S. Let $j : S \to \bar{S}$ be a good compactification of $\bar{S}$ and $\{S_{\alpha}\}$ be a Whitney stratification of $\bar{S}$ such that*

(a) *S is one of the strata of $\bar{S}$, and*

(b) *the $R^{k} j_{*} \mathbb{H}_{\mathbb{Z}}$ are locally constant on each stratum.*

Then, there exists a generalized Néron model for $J(\mathcal{H})$, i.e., a topological group $J_{\bar{S}}(\mathcal{H})$ over $\bar{S}$ which extends $J(\mathcal{H})$ and satisfies these two conditions:

(i) *The restriction of $J_{\bar{S}}(\mathcal{H})$ to S is $J(\mathcal{H})$.*

(ii) *Any admissible normal function $\nu : S \to J(\mathcal{H})$ has a unique extension to a continuous section $\bar{\nu}$ of $J_{\bar{S}}(\mathcal{H})$.*

Furthermore:

(iii) *There is a short exact sequence of topological groups*

$$0 \to J_{\bar{S}}(\mathcal{H})^{0} \to J_{\bar{S}}(\mathcal{H}) \to G \to 0$$

over $\bar{S}$ such that G_s is a discrete subgroup of $(R^{1} j_{*} \mathbb{H}_{\mathbb{Z}})_{s}$ for any point $s \in \bar{S}$.

(iv) *The restriction of $J_{\bar{S}}(\mathcal{H})^{0}$ to any stratum S_{α} is a complex Lie group over S_{α} with fiber $J(H_s)$ over $s \in \bar{S}$.*

(v) *If $\nu : S \to J(\mathcal{H})$ is an admissible normal function with extension $\bar{\nu}$ then the image of $\bar{\nu}(s)$ in G_s is equal to $\sigma_{\mathbb{Z},s}(\nu)$ for all $s \in \bar{S}$. If $\sigma_{\mathbb{Z},s}(\nu) = 0$ for all $s \in \bar{S}$ then $\bar{\nu}$ restricts to a holomorphic section of $J_{\bar{S}}(\mathcal{H})^{0}$ over each strata.*

REMARK 98. More generally, this is true under the following hypothesis:

(1) S is a complex manifold and $j : S \to \bar{S}$ is a partial compactification of S as an analytic space;

(2) $\mathcal{H}$ is a variation of Hodge structure on S of negative weight, which need not have unipotent monodromy.

To construct the identity component $J_{\bar{S}}(\mathcal{H})^0$, let $\nu : S \to J(\mathcal{H})$ be an admissible normal function which is represented by an extension

$$0 \to \mathcal{H} \to \mathcal{V} \to \mathbb{Z}(0) \to 0 \tag{5-3}$$

and $j : S \to \bar{S}$ denote the inclusion map. Also, given $s \in \bar{S}$ let $i_s : \{s\} \to \bar{S}$ denote the inclusion map. Then, the short exact sequence (5-3) induces an exact sequence of mixed Hodge structures

$$0 \to H_s \to H^0(i_s^* Rj_* \mathcal{V}) \to \mathbb{Z}(0) \to H^1(i_s^* Rj_* \mathcal{H}), \tag{5-4}$$

where the arrow $\mathbb{Z}(0) \to H^1(i_s^* Rj_* \mathcal{H})$ is given by $1 \mapsto \sigma_{\mathbb{Z},s}(\nu)$. Accordingly, if $\sigma_{\mathbb{Z},s}(\nu) = 0$ then (5-4) determines a point $\bar{\nu}(s) \in J(H_s)$. Therefore, as a set, we define

$$J_{\bar{S}}(\mathcal{H})^0 = \coprod_{s \in \bar{S}} J(H_s)$$

and topologize by identifying it with a subspace of the Zucker extension $J_{\bar{S}}^Z(\mathcal{H})$.

Now, by condition (b) of Theorem 97 and the theory of mixed Hodge modules[S4], it follows that if $i_\alpha : S_\alpha \to \bar{S}$ are the inclusion maps then $H^k(i_\alpha^* Rj_* \mathcal{H})$ are admissible variations of mixed Hodge structure over each stratum S_α. In particular, the restriction of $J_{\bar{S}}(\mathcal{H})^0$ to S_α is a complex Lie group.

Suppose now that $\nu : S \to J(\mathcal{H})$ is an admissible normal function with extension $\bar{\nu} : \bar{S} \to J_{\bar{S}}(\mathcal{H})$ such that $\sigma_{\mathbb{Z},s}(\nu) = 0$ for each $s \in \bar{S}$. Then, in order to prove that $\bar{\nu}$ is a continuous section of $J_{\bar{S}}(\mathcal{H})^0$ which restricts to a holomorphic section over each stratum, it is sufficient to prove that $\bar{\nu}$ coincides with the section of the Zucker extension (cf. [S1, Proposition 2.3]). For this, it is in turn sufficient to consider the curve case by restriction to the diagonal curve $\Delta \to \Delta^r$ by $t \mapsto (t, \ldots, t)$; see [BPS, §1.4].

It remains now to construct $J_{\bar{S}}(\mathcal{H})$ via the following gluing procedure: Let U be an open subset of $\bar{S}$ and $\nu : U \to J(\mathcal{H})$ be an admissible normal function with cohomological invariant

$$\sigma_{\mathbb{Z},U}(\nu) = \partial(1) \in H^1(U, \mathbb{H}_{\mathbb{Z}})$$

defined by the map

$$\partial : H^0(U, \mathbb{Z}(0)) \to H^1(U, \mathbb{H}_{\mathbb{Z}})$$

induced by the short exact sequence (5-3) over U. Then, we declare $J_U(\mathcal{H}_{U \cap S})^\nu$ to be the component of $J_{\bar{S}}(\mathcal{H})$ over U, and equip $J_U(\mathcal{H}_{U \cap S})^\nu$ with a canonical morphism

$$J_U(\mathcal{H}_{U \cap S})^\nu \to J_U(\mathcal{H}_{U \cap S})^0$$

which sends ν to the zero section. If μ is another admissible normal function over U with $\sigma_{\mathbb{Z},U}(\nu) = \sigma_{\mathbb{Z},U}(\mu)$ then there is a canonical isomorphism

$$J_U(\mathcal{H}_{U \cap S})^\nu \cong J_U(\mathcal{H}_{U \cap S})^\mu$$

which corresponds to the section $\nu - \mu$ of $J_U(\mathcal{H}_{U \cap S})^0$ over U.

Addendum to §5.2. Since the submission of this article, there have been several important developments in the theory of Néron models for admissible normal functions on which we would like to report here. To this end, let us suppose that $\mathcal{H}$ is a variation of Hodge structure of level 1 over a smooth curve $S \subset \bar{S}$. Let $\mathcal{A}_S$ denote the corresponding abelian scheme with Néron model $\mathcal{A}_{\bar{S}}$ over $\bar{S}$. Then, we have a canonical morphism

$$\mathcal{A}_{\bar{S}} \to J_{\bar{S}}(\mathcal{H})$$

which is an isomorphism over S. However, unless $\mathcal{H}$ has unipotent local monodromy about each point $s \in \bar{S} - S$, this morphism is not an isomorphism [BPS]. Recently however, building upon his work on local duality and mixed Hodge modules [Sl2], Christian Schnell has found an alternative construction of the identity component of a Néron model which contains the construction of [BPS] in the case of unipotent local monodromy and agrees [SS] with the classical Néron model for VHS of level 1 in the case of nonunipotent monodromy. In the paragraphs below, we reproduce a summary of this construction which has been generously provided by Schnell for inclusion in this article.

The genesis of the new construction is in unpublished work of Clemens on normal functions associated to primitive Hodge classes. When Y is a smooth hyperplane section of a smooth projective variety X of dimension $2n$, and $H_\mathbb{Z} = H^{2n-1}(Y, \mathbb{Z})_{\text{van}}$ its vanishing cohomology modulo torsion, the intermediate Jacobian $J(Y)$ can be embedded into a bigger object, $K(Y)$ in Clemens's notation, defined as

$$K(Y) = \frac{\left(H^0\left(X, \Omega_X^{2n}(nY)\right)\right)^\vee}{H^{2n-1}(Y, \mathbb{Z})_{\text{van}}}.$$

The point is that the vanishing cohomology of Y is generated by residues of meromorphic $2n$-forms on X, with the Hodge filtration determined by the order of the pole (provided that $\mathcal{O}_X(Y)$ is sufficiently ample). Clemens introduced $K(Y)$ with the hope of obtaining a weak, topological form of Jacobi inversion for its points, and because of the observation that the numerator in its definition makes sense *even when Y becomes singular*. In his Ph.D. thesis [Sl3], Schnell proved that residues and the pole order filtration actually give a filtered holonomic $\mathcal{D}$-module on the projective space parametrizing hyperplane sections of X; and that this $\mathcal{D}$-module underlies the polarized Hodge module corresponding to the vanishing cohomology by Saito's theory. At least in the geometric case, therefore, there is a close connection between the question of extending intermediate Jacobians, and filtered $\mathcal{D}$-modules (with the residue calculus providing the link).

The basic idea behind Schnell's construction is to generalize from the geometric setting above to arbitrary bundles of intermediate Jacobians. As before, let $\mathcal{H}$ be a variation of polarized Hodge structure of weight -1 on a complex manifold S, and M its extension to a polarized Hodge module on $\bar{S}$. Let $(\mathcal{M}, F)$ be its underlying filtered left $\mathcal{D}$-module: $\mathcal{M}$ is a regular holonomic $\mathcal{D}$-module, and $F = F_{\bullet}\mathcal{M}$ a good filtration by coherent subsheaves. In particular, $F_0\mathcal{M}$ is a coherent sheaf on $\bar{S}$ that naturally extends the Hodge bundle $F^0\mathcal{H}_{\mathcal{O}}$. Now consider the analytic space over $\bar{S}$, given by

$$T = T(F_0\mathcal{M}) = \mathrm{Spec}_{\bar{S}}\big(\mathrm{Sym}_{\mathcal{O}_{\bar{S}}}(F_0\mathcal{M})\big),$$

whose sheaf of sections is $(F_0\mathcal{M})^{\vee}$. (Over S, it is nothing but the vector bundle corresponding to $(F^0\mathcal{H}_{\mathcal{O}})^{\vee}$.) It naturally contains a copy $T_{\mathbb{Z}}$ of the étale space of the sheaf $j_*\mathbb{H}_{\mathbb{Z}}$; indeed, every point of that space corresponds to a local section of $\mathbb{H}_{\mathbb{Z}}$, and it can be shown that every such section defines a map of $\mathcal{D}$-modules $\mathcal{M} \to \mathcal{O}_{\bar{S}}$ via the polarization.

Schnell proves that $T_{\mathbb{Z}} \subseteq T$ is a closed analytic subset, discrete on fibers of $T \to \bar{S}$. This makes the fiberwise quotient space $\bar{J} = T/T_{\mathbb{Z}}$ into an analytic space, naturally extending the bundle of intermediate Jacobians for H. He also shows that admissible normal functions with no singularities extend uniquely to holomorphic sections of $\bar{J} \to \bar{S}$. To motivate the extension process, note that the intermediate Jacobian of a polarized Hodge structure of weight -1 has two models,

$$\frac{H_{\mathbb{C}}}{F^0 H_{\mathbb{C}} + H_{\mathbb{Z}}} \simeq \frac{(F^0 H_{\mathbb{C}})^{\vee}}{H_{\mathbb{Z}}},$$

with the isomorphism coming from the polarization. An extension of mixed Hodge structure of the form

$$0 \to H \to V \to \mathbb{Z}(0) \to 0 \tag{5-5}$$

gives a point in the second model in the following manner.

Let $H^* = \mathrm{Hom}(H, \mathbb{Z}(0))$ be the dual Hodge structure, isomorphic to $H(-1)$ via the polarization. After dualizing, we have

$$0 \to \mathbb{Z}(0) \to V^* \to H^* \to 0,$$

and thus an isomorphism $F^1 V_{\mathbb{C}}^* \simeq F^1 H_{\mathbb{C}}^* \simeq F^0 H_{\mathbb{C}}$. Therefore, any $v \in V_{\mathbb{Z}}$ lifting $1 \in \mathbb{Z}$ gives a linear map $F^0 H_{\mathbb{C}} \to \mathbb{C}$, well-defined up to elements of $H_{\mathbb{Z}}$; this is the point in the second model of $J(H)$ that corresponds to the extension in (5-5).

It so happens that this second construction is the one that extends to all of $\bar{S}$. Given a normal function v on S, let

$$0 \to \mathbb{H}_{\mathbb{Z}} \to \mathbb{V}_{\mathbb{Z}} \to \mathbb{Z}_S \to 0$$

be the corresponding extension of local systems. By applying j_*, it gives an exact sequence

$$0 \to j_* \mathbb{H}_{\mathbb{Z}} \to j_* \mathbb{V}_{\mathbb{Z}} \to \mathbb{Z}_{\bar{S}} \to R^1 j_* \mathbb{H}_{\mathbb{Z}},$$

and when v has no singularities, an extension of sheaves

$$0 \to j_* \mathbb{H}_{\mathbb{Z}} \to j_* \mathbb{V}_{\mathbb{Z}} \to \mathbb{Z}_{\bar{S}} \to 0.$$

Using duality for filtered $\mathcal{D}$-modules, one obtains local sections of $(F_0 \mathcal{M})^\vee$ from local sections of $j_* \mathbb{V}_{\mathbb{Z}}$, just as above, and thus a well-defined holomorphic section of $\bar{J} \to \bar{S}$ that extends v.

As in the one-variable case, where the observation is due to Green, Griffiths, and Kerr, horizontality constrains such extended normal functions to a certain subset of $\bar{J}$; Schnell proves that this subset is precisely the identity component of the Néron model constructed by Brosnan, Pearlstein, and Saito. With the induced topology, the latter is therefore a Hausdorff space, as expected. This provides an additional proof for the algebraicity of the zero locus of an admissible normal function, similar in spirit to the one-variable result in Saito's paper, in the case when the normal function has no singularities.

The other advance, is the recent construction [KNU2] of log intermediate Jacobians by Kato, Nakayama and Usui. Although a proper exposition of this topic would take us deep into logarithmic Hodge theory [KU], the basic idea is as follows: Let $\mathcal{H} \to \Delta^*$ be a variation of Hodge structure of weight -1 with unipotent monodromy. Then, we have a commutative diagram

$$
\begin{array}{ccc}
J(\mathcal{H}) & \xrightarrow{\tilde{\varphi}} & \tilde{\Gamma} \backslash \mathcal{M} \\
\downarrow & & \downarrow{\scriptstyle \mathrm{Gr}_{-1}^W} \\
\Delta^* & \xrightarrow{\varphi} & \Gamma \backslash \mathcal{D}
\end{array}
\qquad (5\text{-}6)
$$

where $\tilde{\varphi}$ and φ are the respective period maps. In [KU], Kato and Usui explained how to translate the bottom row of this diagram into logarithmic Hodge theory. More generally, building on the ideas of [KU] and the several variable SL_2-orbit theorem [KNU1], Kato, Nakayama and Usui are able to construct a theory of logarithmic mixed Hodge structures which they can then apply to the top row of the previous diagram. In this way, they obtain a log intermediate Jacobian which serves the role of a Néron model and allows them to give an alternate proof of Theorem 57 [KNU3].

5.3. Singularities of normal functions overlying nilpotent orbits.

We now consider the group of components G_s of $J_{\bar{s}}(\mathcal{H})$ at $s \in \bar{S}$. For simplicity, we first consider the case where $\mathcal{H}$ is a nilpotent orbit $\mathcal{H}_{\text{nilp}}$ over $(\Delta^*)^r$. To this end, we recall that in the case of a variation of Hodge structure $\mathcal{H}$ over $(\Delta^*)^r$ with unipotent monodromy, the intersection cohomology of $\mathbb{H}_{\mathbb{Q}}$ is computed by the cohomology of a complex $(B^\bullet(N_1, \ldots, N_r), d)$ (cf. §3.4). Furthermore, the short exact sequence of sheaves

$$0 \to \mathbb{H}_{\mathbb{Q}} \to \mathbb{V}_{\mathbb{Q}} \to \mathbb{Q}(0) \to 0$$

associated to an admissible normal function $\nu : (\Delta^*)^r \to J(\mathcal{H})$ with unipotent monodromy gives a connecting homomorphism

$$\partial : IH^0(\mathbb{Q}(0)) \to IH^1(\mathbb{H}_{\mathbb{Q}})$$

such that

$$\partial(1) = [(N_1(e_o^{\mathbb{Q}}), \ldots, N_r(e_o^{\mathbb{Q}}))] = \text{sing}_0(\nu),$$

where $e_o^{\mathbb{Q}}$ is an element in the reference fiber $V_{\mathbb{Q}}$ of $\mathbb{V}_{\mathbb{Q}}$ over $s_o \in (\Delta^*)^r$ which maps to $1 \in \mathbb{Q}(0)$. After passage to complex coefficients, the admissibility of $\mathcal{V}$ allows us to pick an alternate lift $e_o \in V_{\mathbb{C}}$ to be of type $(0,0)$ with respect to the limit MHS of $\mathcal{V}$. It also forces $h_j = N_j(e_o)$ to equal $N_j(f_j)$ for some element $f_j \in H_{\mathbb{C}}$ of type $(0,0)$ with respect to the limit MHS of $\mathcal{H}$. Moreover, $e_0^{\mathbb{Q}} - e_0 =: h$ maps to $0 \in \text{Gr}_0^W$, hence lies in $H_{\mathbb{C}}$, so

$$(N_1(e_0^{\mathbb{Q}}), \ldots, N_r(e_0^{\mathbb{Q}})) \equiv (N_1(e_0), \ldots, N_r(e_0)) \quad \text{modulo } d(B^0) = \text{im} \bigoplus_{j=1}^r N_j$$

(i.e., up to $(N_1(h), \ldots, N_r(h))$).

COROLLARY 99. $\text{sing}_0(\nu)$ is a rational class of type $(0,0)$ in $IH^1(\mathbb{H}_{\mathbb{Q}})$.

SKETCH OF PROOF. This follows from the previous paragraph together with the explicit description of the mixed Hodge structure on the cohomology of $B^\bullet(N_1, \ldots, N_r)$ given in [CKS2]. $\square$

Conversely, we have:

LEMMA 100. Let $\mathcal{H}_{\text{nilp}} = e^{\sum_j z_j N_j}.F_\infty$ be a nilpotent orbit of weight -1 over Δ^{*r} with rational structure $\mathbb{H}_{\mathbb{Q}}$. Then, any class β of type $(0,0)$ in $IH^1(\mathbb{H}_{\mathbb{Q}})$ is representable by a $\mathbb{Q}$-normal function ν which is an extension of $\mathbb{Q}(0)$ by $\mathcal{H}_{\text{nilp}}$ such that $\text{sing}_0(\nu) = \beta$.

PROOF. By the remarks above, β corresponds to a collection of elements $h_j \in N_j(H_{\mathbb{C}})$ such that

(a) $h_1, \ldots, h_r$ are of type $(-1, -1)$ with respect to the limit mixed Hodge structure of $\mathcal{H}_{\mathrm{nilp}}$,

(b) $d(h_1, \ldots, h_r) = 0$, i.e., $N_j(h_k) - N_k(h_j) = 0$, and

(c) There exists $h \in H_{\mathbb{C}}$ such that $N_j(h) + h_j \in H_{\mathbb{Q}}$ for each j, i.e., the class of $(h_1, \ldots, h_r)$ in $IH^1(\mathbb{H}_{\mathbb{C}})$ belongs to the image $IH^1(\mathbb{H}_{\mathbb{Q}}) \to IH^1(\mathbb{H}_{\mathbb{C}})$.

We now define the desired nilpotent orbit by formally setting $V_{\mathbb{C}} = \mathbb{C}e_o \oplus H_{\mathbb{C}}$, where e_o is of type $(0, 0)$ with respect to the limit mixed Hodge structure and letting $V_{\mathbb{Q}} = \mathbb{Q}(e_o + h) \oplus H_{\mathbb{Q}}$. We define $N_j(e_o) = h_j$. Then, following Kashiwara [Ka]:

(a) The resulting nilpotent orbit $\mathcal{V}_{\mathrm{nilp}}$ is pre-admissible.

(b) The relative weight filtration of

$$W_{-2} = 0, \qquad W_{-1} = H_{\mathbb{Q}}, \qquad W_0 = V_{\mathbb{Q}}$$

with respect to each N_j exists.

Consequently $\mathcal{V}_{\mathrm{nilp}}$ is admissible, and the associated normal function ν has singularity β at 0. $\qquad\square$

5.4. Obstructions to the existence of normal functions with prescribed singularity class.

Thus, in the case of a nilpotent orbit, we have a complete description of the group of components of the Néron model $\otimes \mathbb{Q}$. In analogy with nilpotent orbits, one might expect that given a variation of Hodge structure $\mathcal{H}$ of weight -1 over $(\Delta^*)^r$ with unipotent monodromy, the group of components of the Néron model $\otimes \mathbb{Q}$ to equal the classes of type $(0, 0)$ in $IH^1(\mathbb{H}_{\mathbb{Q}})$. However, Saito [S6] has managed to construct examples of variations of Hodge structure over $(\Delta^*)^r$ which do not admit any admissible normal functions with nontorsion singularities. We now want to describe Saito's class of examples. We begin with a discussion of the deformations of an admissible nilpotent orbit into an admissible variation of mixed Hodge structure over $(\Delta^*)^r$.

Let $\varphi : (\Delta^*)^r \to \Gamma \backslash \mathcal{D}$ be the period map of a variation of pure Hodge structure with unipotent monodromy. Then, after lifting the period map of $\mathcal{H}$ to the product of upper half-planes U^r, the work of Cattani, Kaplan and Schmid on degenerations of Hodge structure gives us a local normal form of the period map

$$F(z_1, \ldots, z_r) = e^{\sum_j z_j N_j} e^{\Gamma(s)}. F_{\infty}.$$

Here, $(s_1, \ldots, s_r)$ are the coordinates on Δ^r, $(z_1, \ldots, z_r)$ are the coordinates on U^r relative to which the covering map $U^r \to (\Delta^*)^r$ is given by $s_j = e^{2\pi i z_j}$;

$$\Gamma : \Delta^r \to \mathfrak{g}_{\mathbb{C}}$$

is a holomorphic function which vanishes at the origin and takes values in the subalgebra

$$\mathfrak{q} = \bigoplus_{p<0} \mathfrak{g}^{p,q};$$

and $\bigoplus_{p,q} \mathfrak{g}^{p,q}$ denotes the bigrading of the MHS induced on $\mathfrak{g}_{\mathbb{C}}$ (cf. §4.2) by the limit MHS $(F_\infty, W(N_1 + \cdots N_r)[1])$ of $\mathcal{H}$. The subalgebra $\mathfrak{q}$ is graded nilpotent

$$\mathfrak{q} = \bigoplus_{a<0} \mathfrak{q}_a, \qquad \mathfrak{q}_a = \bigoplus_b \mathfrak{g}^{a,b},$$

with $N_1, \ldots, N_r \in \mathfrak{q}_{-1}$. Therefore,

$$e^{\sum_j z_j N_j} e^{\Gamma(s)} = e^{X(z_1,\ldots,z_r)},$$

where X takes values in $\mathfrak{q}$, and hence the horizontality of the period map becomes

$$e^{-X} \partial e^X = \partial X_{-1},$$

where $X = X_{-1} + X_{-2} + \cdots$ relative to the grading of $\mathfrak{q}$. Equality of mixed partial derivatives then forces

$$\partial X_{-1} \wedge \partial X_{-1} = 0$$

Equivalently,

$$\left[N_j + 2\pi i s_j \frac{\partial \Gamma_{-1}}{\partial s_j}, \ N_k + 2\pi i s_k \frac{\partial \Gamma_{-1}}{\partial s_k} \right] = 0. \qquad (5\text{-}7)$$

REMARK 101. The function Γ and the local normal form of the period map appear in [CK].

In his letter to Morrison [De4], Deligne showed that for VHS over $(\Delta^*)^r$ with maximal unipotent boundary points, one could reconstruct the VHS from data equivalent to the nilpotent orbit and the function Γ_{-1}. More generally, one can reconstruct the function Γ starting from ∂X_{-1} using the equation

$$\partial e^X = e^X \partial X_{-1}$$

subject to the integrability condition $\partial X_{-1} \wedge \partial X_{-1} = 0$. This is shown by Cattani and Javier Fernandez in [CF].

The above analysis applies to VMHS over $(\Delta^*)^r$ as well: As discussed in the previous section, a VMHS is given by a period map from the parameter space into the quotient of an appropriate classifying space of graded-polarized mixed Hodge structure $\mathcal{M}$. As in the pure case, we have a Lie group G which acts on $\mathcal{M}$ by biholomorphisms and a complex Lie group $G_{\mathbb{C}}$ which acts on the "compact dual" $\check{\mathcal{M}}$.

As in the pure case (and also discussed in §4), an admissible VMHS with nilpotent orbit $(e^{\sum_j z_j N_j}.F_\infty, W)$ will have a local normal form

$$F(z_1, \ldots, z_r) = e^{\sum_j z_j N_j} e^{\Gamma(s)}.F_\infty,$$

where $\Gamma : \Delta^r \to \mathfrak{g}_\mathbb{C}$ takes values in the subalgebra

$$\mathfrak{q} = \bigoplus_{p<0} \mathfrak{g}^{p,q}.$$

Conversely (given an admissible nilpotent orbit), subject to the integrability condition (5-7) above, any function Γ_{-1} determines a corresponding admissible VMHS; see [Pe1, Theorem 6.16].

Returning to Saito's examples (which for simplicity we only consider in the two-dimensional case), let $\mathcal{H}$ be a variation of Hodge structure of weight -1 over Δ^* with local normal form $F(z) = e^{zN} e^{\Gamma(s)}.F_\infty$. Let $\pi : \Delta^2 \to \Delta$ by $\pi(s_1, s_2) = s_1 s_2$. Then, for $\pi^*(\mathcal{H})$, we have

$$\Gamma_{-1}(s_1, s_2) = \Gamma_{-1}(s_1 s_2).$$

In order to construct a normal function, we need to extend $\Gamma_{-1}(s_1, s_2)$ and $N_1 = N_2 = N$ on the reference fiber $H_\mathbb{C}$ of $\mathcal{H}$ to include a new class u_0 of type $(0,0)$ which projects to 1 in $\mathbb{Z}(0)$. Set

$$N_1(u_0) = h_1, \qquad N_2(u_0) = h_2, \qquad \Gamma_{-1}(s_1, s_2)u_0 = \alpha(s_1, s_2).$$

Note that (h_1, h_2) determines the cohomology class of the normal function so constructed, and that $h_2 - h_1$ depends only on the cohomology class, and not the particular choice of representative (h_1, h_2).

In order to construct a normal function in this way, we need to check horizontality. This amounts to checking the equation

$$N\left(s_2 \frac{\partial \alpha}{\partial s_2} - s_1 \frac{\partial \alpha}{\partial s_1}\right) + s_1 s_2 \Gamma'_{-1}(s_1 s_2)(h_2 - h_1)$$
$$+ 2\pi i s_1 s_2 \Gamma'_{-1}(s_1 s_2)\left(s_2 \frac{\partial \alpha}{\partial s_2} - s_1 \frac{\partial \alpha}{\partial s_1}\right) = 0.$$

Computation shows that the coefficient of $(s_1 s_2)^m$ on the left side is

$$\frac{1}{(m-1)!} \Gamma_{-1}^{(m)}(0)(h_2 - h_1). \tag{5-8}$$

Hence, a necessary condition for the cohomology class represented by (h_1, h_2) to arise from an admissible normal function is for $h_2 - h_1$ to belong to the kernel of $\Gamma_{-1}(t)$. This condition is also sufficient since, under this hypothesis, one can simply set $\alpha = 0$.

EXAMPLE 102. Let $\mathcal{X} \xrightarrow{\rho} \Delta$ be a family of Calabi–Yau 3-folds (smooth over Δ^*, smooth total space) with Hodge numbers $h^{3,0} = h^{2,1} = h^{1,2} = h^{0,3} = 1$ and central singular fiber having an ODP. Setting $\mathcal{H} := \mathcal{H}^3_{\mathcal{X}^*/\Delta^*}(2)$, the LMHS has as its nonzero $I^{p,q}$'s $I^{-2,1}$, $I^{-1,-1}$, $I^{0,0}$, and $I^{1,-2}$. Assume that the Yukawa coupling $(\nabla_{\delta_s})^3 \in \mathrm{Hom}_{\mathcal{O}_\Delta}(\mathcal{H}^{3,0}_e, \mathcal{H}^{0,3}_e)$ is nonzero ($\delta_s = s\,d/ds$), and thus the restriction of $\Gamma_{-1}(s)$ to $\mathrm{Hom}_{\mathcal{O}_\Delta}(I^{-1,-1}, I^{-2,1})$, does not vanish identically. Then, for any putative singularity class, $0 \neq h_2 - h_1 \in (I^{-1,-1})_\mathbb{Q} \cong \ker(N)^{(-1,-1)}_\mathbb{Q}$ (this being isomorphic to (2-10) in this case, which is just one-dimensional) for admissible normal functions overlying $\pi^*\mathcal{H}$, nonvanishing of $\Gamma_{-1}(s)(h_2 - h_1)$ on Δ implies that (5-8) cannot be zero for every m.

5.5. Implications for the Griffiths–Green conjecture. Returning now to the work of Griffiths and Green on the Hodge conjecture via singularities of normal functions, it follows using the work of Richard Thomas that for a sufficiently high power of L, the Hodge conjecture implies that one can force ν_ζ to have a singularity at a point $p \in \hat{X}$ such that $\pi^{-1}(p)$ has only ODP singularities. In general, on a neighborhood of such a point $\hat{X}$ need not be a normal crossing divisor. However, the image of the monodromy representation is nevertheless abelian. Using a result of Steenbrink and Némethi [NS], it then follows from the properties of the monodromy cone of a nilpotent orbit of pure Hodge structure that $\mathrm{sing}_p(\nu_\zeta)$ persists under blowup. Therefore, it is sufficient to study ODP degenerations in the normal crossing case (cf. [BFNP, sec. 7]). What we will find below is that the "infinitely many" conditions above (vanishing of (5-8) for all m) are replaced by surjectivity of a single logarithmic Kodaira–Spencer map at each boundary component. Consequently, as suggested in the introduction, it appears that M. Saito's examples are not a complete show-stopper for existence of singularities for Griffiths–Green normal functions.

The resulting limit mixed Hodge structure is of the form

$$I^{0,0}$$
$$\cdots \quad I^{-2,1} \quad\quad I^{-1,0} \quad\quad I^{0,-1} \quad\quad I^{1,-2} \quad \cdots$$
$$I^{-1,-1}$$

and $N^2 = 0$ for every element of the monodromy cone $\mathcal{C}$. The weight filtration is given by

$$M_{-2}(N) = \sum_j N_j(H_\mathbb{C}), \quad M_{-1}(N) = \bigcap_j \ker(N_j), \quad M_0(N) = H_\mathbb{C}.$$

For simplicity of notation, let us restrict to a two parameter version of such a degeneration, and consider the obstruction to constructing an admissible normal function with cohomology class represented by (h_1, h_2) as above. As in Saito's example, we need to add a class u_o of type $(0,0)$ such that $N_j(u_o) = h_j$ and construct $\alpha = \Gamma_{-1}(u_o)$. Then, the integrability condition $\partial X_{-1} \wedge \partial X_{-1} = 0$ becomes

$$-(2\pi i s_2)\frac{\partial \Gamma_{-1}}{\partial s_2}(h_1) + (2\pi i s_1)\frac{\partial \Gamma_{-1}}{\partial s_1}(h_2)$$
$$+ (2\pi i s_1)(2\pi i s_2)\left(\frac{\partial \Gamma_{-1}}{\partial s_1}\frac{\partial \alpha}{\partial s_2} - \frac{\partial \Gamma_{-1}}{\partial s_2}\frac{\partial \alpha}{\partial s_1}\right) = 0, \quad (5\text{-}9)$$

since $\alpha = \Gamma_{-1}(u_o)$ takes values in $M_{-1}(N)$.

Write $\alpha = \sum_{j,k} s_1^j s_2^k \alpha_{jk}$ and $\Gamma_{-1} = \sum_{p,q} s_1^p s_2^q \gamma_{pq}$ on $H_{\mathbb{C}}$. Then, for ab nonzero, the coefficient of $s_1^a s_2^b$ on the left side of equation (5-9) is

$$-2\pi i b \gamma_{ab}(h_2) + 2\pi i a \gamma_{ab}(h_1) + (2\pi i)^2 \sum_{\substack{p+j=a \\ q+k=b}} (pk - qj)\gamma_{pq}(\alpha_{jk}).$$

Define

$$\zeta_{ab} = 2\pi i b \gamma_{ab}(h_2) - 2\pi i a \gamma_{ab}(h_1) - (2\pi i)^2 \sum_{\substack{p+j=a \\ q+k=b \\ pq \neq 0}} (pk - qj)\gamma_{pq}(\alpha_{jk}).$$

Then, equation (5-9) is equivalent to

$$(2\pi i)^2 b \gamma_{10}(\alpha_{(a-1)b}) - (2\pi i)^2 a \gamma_{01}(\alpha_{a(b-1)}) = \zeta_{ab},$$

where α_{jk} occurs in ζ_{ab} only in total degree $j + k < a + b - 1$. Therefore, *provided* that

$$\gamma_{10}, \gamma_{01} : F_\infty^{-1}/F_\infty^0 \to F_\infty^{-2}/F_\infty^{-1}$$

are surjective, we can always solve (nonuniquely!) for the coefficients α_{jk}, and hence formally (i.e., modulo checking convergénce of the resulting series) construct the required admissible normal function with given cohomology class.

REMARK 103. (i) Of course, it is not necessary to have only ODP singularities for the analysis above to apply. It is sufficient merely that the limit mixed Hodge structure have the stated form. In particular, this is always true for degenerations of level 1. Furthermore, in this case $\mathrm{Gr}_{F_\infty}^{-2} = 0$, and hence, after tensoring with $\mathbb{Q}$, the group of components of the Néron model surjects onto the Tate classes of type $(0,0)$ in $IH^1(\mathbb{H}_{\mathbb{Q}})$.

(ii) In Saito's examples from §5.4, even if $\Gamma'_{-1}(0) \neq 0$, we will have $\gamma_{01} = 0 = \gamma_{10}$, since the condition of being a pullback via $(s_1, s_2) \mapsto s_1 s_2$ means $\Gamma_{-1}(s_1, s_2) = \sum_{p,q} s_1^p s_2^q \gamma_{pq} = \sum_r s_1^r s_2^r \gamma_{rr}$.

EXAMPLE 104. In the case of a degeneration of Calabi–Yau threefolds with limit mixed Hodge structure on the middle cohomology (shifted to weight -1)

$$I^{0,0}$$
$$I^{-2,1} \qquad I^{-1,0} \qquad I^{0,-1} \qquad I^{1,-2}$$
$$I^{-1,-1}$$

the surjectivity of the partial derivatives of Γ_{-1} are related to the Yukawa coupling as follows: Let

$$F(z) = e^{\sum_j z_j N_j} e^{\Gamma(s)} . F_\infty$$

be the local normal form of the period map as above. Then, a global nonvanishing holomorphic section of the canonical extension of $\mathcal{F}^1$ (i.e., of $\mathcal{F}^3$ before we shift to weight -1) is of the form

$$\Omega = e^{\sum_j z_j N_j} e^{\Gamma(s)} \sigma_\infty(s),$$

where $\sigma_\infty : \Delta^r \to I^{1,-2}$ is holomorphic and nonvanishing. Then, the Yukawa coupling of Ω is given by

$$Q(\Omega, D_j D_k D_\ell \, \Omega), \qquad D_a = \frac{\partial}{\partial z_a}.$$

In keeping with the notation above, let $e^X = e^{\sum_j z_j N_j} e^{\Gamma(s)}$ and $A_j = D_j X_{-1}$. Using the first Hodge–Riemann bilinear relation and the fact that e^X is an automorphism of Q, it follows that

$$Q(\Omega, D_j D_k D_\ell \, \Omega) = Q(\sigma_\infty(s), A_j A_k A_\ell \, \sigma_\infty(s)).$$

Moreover (cf. [CK; Pe1]), the horizontality of the period map implies that

$$\left[\Gamma_{-1}|_{s_k=0}, N_k \right] = 0$$

Using this relation, it then follows that

$$\lim_{s \to 0} \frac{Q(\Omega, D_j D_k D_\ell \, \Omega)}{(2\pi i s_j)(2\pi i s_k)(2\pi i s_\ell)} = Q(\sigma_\infty(0), G_j G_k G_\ell \sigma_\infty(0))$$

for $j \neq k$, where

$$G_a = \frac{\partial \Gamma_{-1}}{\partial s_a}(0).$$

In particular, if for each index j there exist indices k and ℓ with $k \neq \ell$ such that the left-hand side of the previous equation is nonzero then $G_j : (F_\infty^{-1}/F_\infty^0) \to (F_\infty^{-2}/F_\infty^{-1})$ is surjective.

6. Global considerations: monodromy of normal functions

Returning to a normal function $V \in \mathrm{NF}^1(S, \mathcal{H})^{\mathrm{ad}}_{\underline{S}}$ over a *complete* base, we want to speculate a bit about how one might "force" singularities to exist. The (inconclusive) line of reasoning we shall pursue rests on two basic principles:

(i) maximality of the geometric (global) monodromy group of $\mathbb{V}$ may be deduced from hypotheses on the torsion locus of $\mathcal{V}$; and

(ii) singularities of $\mathcal{V}$ can be interpreted in terms of the local monodromy of $\mathbb{V}$ being sufficiently large.

While it is unclear what hypotheses (if any) would allow one to pass from global to local monodromy-largeness, the proof of the first principle is itself of interest as a first application of algebraic groups (the algebraic variety analog of Lie groups, originally introduced by Picard) to normal functions.

6.1. Background. Mumford–Tate groups of Hodge structures were introduced by Mumford [Mu] for pure HS and by André [An] in the mixed setting. Their power and breadth of applicability is not well-known, so we will first attempt a brief summary. They were first brought to bear on $H^1(A)$ for A an abelian variety, which has led to spectacular results:

- Deligne's theorem [De2] that $\mathbb{Q}$-Bettiness of a class in $F^p H^{2p}_{dR}(A_k)$ for k algebraically closed is independent of the embedding of k into $\mathbb{C}$ ("Hodge implies absolute Hodge");
- the proofs by Hazama [Ha] and Murty [Mr] of the HC for A "nondegenerate" (MT of $H^1(A)$ is maximal in a sense to be defined below); and
- the density of special (Shimura) subvarieties in Shimura varieties and the partial resolution of the André–Oort Conjecture by Klingler and Yafaev [KY].

More recently, MT groups have been studied for higher weight HS's; one can still use them to define special $\bar{\mathbb{Q}}$-subvarieties of (non-Hermitian-symmetric) period domains D, which classify polarized HS's with fixed Hodge numbers (and polarization). In particular, the 0-dimensional subdomains — still dense in D — correspond to HS with CM (complex multiplication); that is, with abelian MT group. One understands these HS well: their irreducible subHS may be constructed directly from primitive CM types (and have endomorphism algebra equal to the underlying CM field), which leads to a complete classification; and their Weil and Griffiths intermediate Jacobians are CM abelian varieties [Bo]. Some further applications of MT groups include:

- Polarizable CM-HS are motivic [Ab]; when they come from a CY variety, the latter often has good modularity properties;

- Given H^* of a smooth projective variety, the level of the MT Lie algebra furnishes an obstruction to the variety being dominated by a product of curves [Sc];
- Transcendence degree of the space of periods of a VHS (over a base S), viewed as a field extension of $\mathbb{C}(S)$ [An];

and specifically in the mixed case:

- the recent proof [AK] of a key case of the Beilinson–Hodge Conjecture for semiabelian varieties and products of smooth curves.

The latter paper, together with [An] and [De2], are the best references for the definitions and properties we now summarize.

To this end, recall that an algebraic group G over a field k is an algebraic variety over k together with k-morphisms of varieties $1_G : \mathrm{Spec}(k) \to G$, "multiplication" $\mu_G : G \times G \to G$, and "inversion" $\iota_G : G \to G$ satisfying obvious compatibility conditions. The latter ensure that for any extension K/k, the K-points $G(K)$ form a group.

DEFINITION 105. (i) A $(k\text{-})$closed algebraic subgroup $M \leq G$ is one whose underlying variety is $(k\text{-})$Zariski closed.

(ii) Given a subgroup $\mathcal{M} \leq G(K)$, the k-closure of $\mathcal{M}$ is the smallest k-closed algebraic subgroup M of G with K-points $M(K) \geq \mathcal{M}$.

If $\mathcal{M} := M(K)$ for an algebraic k-subgroup $M \leq G$, then the k-closure of $\mathcal{M}$ is just the k-Zariski closure of $\mathcal{M}$ (i.e., the algebraic variety closure).

But in general, this is not true: instead, M may be obtained as the k-Zariski (algebraic variety) closure of the group generated by the k-spread of $\mathcal{M}$.

We refer the reader to [Sp] (especially Chapter 6) for the definitions of reductive, semisimple, unipotent, etc. in this context (which are less crucial for the sequel). We will write $DG := [G, G] \, (\unlhd G)$ for the derived group.

6.2. Mumford–Tate and Hodge groups.

Let V be a (graded-polarizable) mixed Hodge structure with dual $V^\vee$ and tensor spaces

$$T^{m,n}V := V^{\otimes m} \otimes (V^\vee)^{\otimes n}$$

$(n, m \in \mathbb{Z}_{\geq 0})$. These carry natural MHS, and any $g \in \mathrm{GL}(V)$ acts naturally on $T^{m,n}V$.

DEFINITION 106. (i) A *Hodge (p, p)-tensor* is any $\tau \in (T^{m,n}V)_{\mathbb{Q}}^{(p,p)}$.

(ii) The *MT group* M_V of V is the (largest) $\mathbb{Q}$-algebraic subgroup of $\mathrm{GL}(V)$ fixing[12] the Hodge $(0, 0)$-tensors for all m, n. The weight filtration $W_\bullet$ on V is preserved by M_V.

---

[12] "Fixing" means fixing pointwise; the term for "fixing as a set" is "stabilizing".

Similarly, the *Hodge group* M_V° of V is the $\mathbb{Q}$-algebraic subgroup of $\mathrm{GL}(V)$ fixing the Hodge (p, p)-tensors for all m, n, p. (In an unfortunate coincidence of terminology, these are completely different objects from — though not unrelated to — the finitely generated abelian groups $\mathrm{Hg}^m(H)$ discussed in §1.)

(iii) The weight filtration on V induces one on MT/Hodge:

$$W_{-i} M_V^{(\circ)} := \{g \in M_V^{(\circ)} \mid (g - \mathrm{id}) W_\bullet V \subset W_{\bullet - i} V\} \trianglelefteq M_V^{(\circ)}.$$

One has: $W_0 M_V^{(\circ)} = M_V^{(\circ)}$; $W_{-1} M_V^{(\circ)}$ is unipotent; and $\mathrm{Gr}_0^W M_V^{(\circ)} \cong M_{V^{\mathrm{split}}}^{(\circ)}$ ($V^{\mathrm{split}} := \bigoplus_{\ell \in \mathbb{Z}} \mathrm{Gr}_\ell^W V$), cf. [An].

Clearly $M_V^\circ \trianglelefteq M_V$; and unless V is pure of weight 0, we have $M_V / M_V^\circ \cong \mathbb{G}_m$. If V has polarization $Q \in \mathrm{Hom}_{\mathrm{MHS}}(V \otimes V, \mathbb{Q}(-k))$ for $k \in \mathbb{Z} \setminus \{0\}$, then M_V° is of finite index in $M_V \cap \mathrm{GL}(V, Q)$ (where $g \in \mathrm{GL}(V, Q)$ means $Q(gv, gw) = Q(v, w)$), and if in addition $V (= H)$ is pure (or at least split) then both are reductive. One has in general that $W_{-1} M_V \subseteq DM_V \subseteq M_V^\circ \subseteq M_V$.

DEFINITION 107. (i) If M_V is abelian ($\Longleftrightarrow M_V(\mathbb{C}) \cong (\mathbb{C}^*)^{\times a}$), V is called a *CM-MHS*. (A subMHS of a CM-MHS is obviously CM.)

(ii) The endomorphisms $\mathrm{End}_{\mathrm{MHS}}(V)$ can be interpreted as the $\mathbb{Q}$-points of the algebra $(\mathrm{End}(V))^{M_V} =: E_V$. One always has $M_V \subset \mathrm{GL}(V, E_V)$ (=centralizer of E_V); if this is an equality, then V is said to be *nondegenerate*.

Neither notion implies the other; however: any CM or nondegenerate MHS is ($\mathbb{Q}$-)*split*, i.e., $V (= V^{\mathrm{split}})$ is a direct sum of pure HS in different weights.

REMARK 108. (a) We point out why CM-MHS are split. If M_V is abelian, then $M_V \subset E_V$ and so $M_V(\mathbb{Q})$ consists of morphisms of MHS. But then any $g \in W_{-1} M_V(\mathbb{Q})$, hence $g - \mathrm{id}$, is a morphism of MHS with $(g - \mathrm{id}) W_\bullet \subset W_{\bullet - 1}$; so $g = \mathrm{id}$, and $M_V = M_{V^{\mathrm{split}}}$, which implies $V = V^{\mathrm{split}}$.

(b) For an arbitrary MHS V, the subquotient tensor representations of M_V killing DM_V (i.e., factoring through the abelianization) are CM-MHS. By (a), they are split, so that $W_{-1} M_V$ acts trivially; this gives $W_{-1} M_V \subseteq DM_V$.

Now we turn to the representation-theoretic point of view on MHS. Define the algebraic $\mathbb{Q}$-subgroups $U \subset S \subset \mathrm{GL}_2$ via their complex points:

$$
S(\mathbb{C}) : \!=\!=\! \left\{ \begin{pmatrix} \alpha & \beta \\ -\beta & \alpha \end{pmatrix} \,\middle|\, \begin{matrix} \alpha, \beta \in \mathbb{C} \\ (\alpha, \beta) \neq (0,0) \end{matrix} \right\} \xrightarrow[\text{eigenvalues}]{\cong} \mathbb{C}^* \times \mathbb{C}^* \qquad \left(z, \frac{1}{z} \right)
$$

$$
\begin{array}{ccc}
\uparrow & & \uparrow \\
\end{array} \qquad (6\text{-}1)
$$

$$
U(\mathbb{C}) : \!=\!=\! \left\{ \begin{pmatrix} \alpha & \beta \\ -\beta & \alpha \end{pmatrix} \,\middle|\, \begin{matrix} \alpha, \beta \in \mathbb{C} \\ \alpha^2 + \beta^2 = 1 \end{matrix} \right\} \xrightarrow{\cong} \mathbb{C}^* \qquad z
$$

where the top map sends $\left(\begin{smallmatrix} \alpha & \beta \\ -\beta & \alpha \end{smallmatrix}\right) \mapsto (\alpha + i\beta, \alpha - i\beta) =: (z, w)$. (Points in $S(\mathbb{C})$ will be represented by the "eigenvalues" (z, w).) Let

$$\varphi : S(\mathbb{C}) \to GL(V_\mathbb{C})$$

be given by

$$\varphi(z, w)|_{I^{p,q}(H)} := \text{multiplication by } z^p w^q \quad (\forall p, q).$$

Note that this map is in general only defined over $\mathbb{C}$, though in the pure case it is defined over $\mathbb{R}$ (and as $S(\mathbb{R}) \subset S(\mathbb{C})$ consists of tuples $(z, \bar{z})$, one tends not to see precisely the approach above in the literature). The following useful result[13] allows one to compute MT groups in some cases.

PROPOSITION 109. M_V is the $\mathbb{Q}$-closure of $\varphi(S(\mathbb{C}))$ in $GL(V)$.

REMARK 110. In the pure $(V = H)$ case, this condition can be replaced by $M_H(\mathbb{R}) \supset \varphi(S(\mathbb{R}))$, and M_H° defined similarly as the $\mathbb{Q}$-closure of $\varphi(U(\mathbb{R}))$; unfortunately, for V a non-$\mathbb{Q}$-split MHS the $\mathbb{Q}$-closure of $\varphi(U(\mathbb{C}))$ is smaller than M_H°.

Now let H be a pure polarizable HS with Hodge numbers $h^{p,q}$, and take D (with compact dual $\check{D}$) to be the classifying space for such. We may view $\check{D}$ as a quasiprojective variety over $\mathbb{Q}$ in a suitable flag variety. Consider the subgroup $M_{H,\varphi}^\circ \subset M_H^\circ$ with real points $M_{H,\varphi}^\circ(\mathbb{R}) := (M_H^\circ(\mathbb{R}))^{\varphi(S(\mathbb{R}))}$. If we view M_H° as acting on a Hodge flag of $H_\mathbb{C}$ with respect to a (fixed) basis of $H_\mathbb{Q}$, then $M_{H,\varphi}^\circ$ is the stabilizer of the Hodge flag. This leads to a Noether–Lefschetz-type substratum in D:

PROPOSITION 111. *The MT domain*

$$D_H := \frac{M_H^\circ(\mathbb{R})}{M_{H,\varphi}^\circ(\mathbb{R})} \left(\subset \frac{M_H^\circ(\mathbb{C})}{M_{H,\varphi}^\circ(\mathbb{C})} =: \check{D}_H \right)$$

classifies HS with Hodge group contained in M_H, or equivalently with Hodge-tensor set containing that of H. The action of M_H° upon H embeds $\check{D}_H \hookrightarrow \check{D}$ as a quasiprojective subvariety, defined over an algebraic extension of $\mathbb{Q}$. The $GL(H_\mathbb{Q}, Q)$-translates of $\check{D}_H$ give isomorphic subdomains (with conjugate MT groups) dense in $\check{D}$.

A similar definition works for certain kinds of MHS. The trouble with applying this in the variational setting (which is our main concern here), is that the "tautological VHS" (or VMHS) over such domains (outside of a few classical cases in low weight or level) violate Griffiths transversality hence are not actually VHS.

---

[13]Proof of this, and of Proposition 111 below, will appear in a work of the first author with P. Griffiths and M. Green.

Still, it can happen that MT domains in non-Hermitian symmetric period domains are themselves Hermitian symmetric. For instance, taking Sym^3 of HS's embeds the classifying space ($\cong \mathfrak{H}$) of (polarized) weight-1 Hodge structures with Hodge numbers $(1, 1)$ into that of weight-3 Hodge structures with Hodge numbers $(1, 1, 1, 1)$.

6.3. MT groups in the variational setting. Let S be a smooth quasiprojective variety with good compactification $\bar{S}$, and $\mathcal{V} \in \mathrm{VMHS}(S)^{\mathrm{ad}}_{\bar{S}}$; assume $\mathcal{V}$ is graded-polarized, which means we have

$$Q \in \bigoplus_i \mathrm{Hom}_{\mathrm{VMHS}(S)}\left((\mathrm{Gr}_i^W \mathcal{V})^{\otimes 2}, \mathbb{Q}(-i)\right)$$

satisfying the usual positivity conditions. The Hodge flag embeds the universal cover $\hat{S}(\twoheadrightarrow S)$ in a flag variety; let the *image-point* of $\hat{s}_0 (\mapsto s_0)$ be of maximal transcendence degree. (One might say $s_0 \in S(\mathbb{C})$ is a "very general point in the sense of Hodge"; we are *not* saying s_0 is of maximal transcendence degree.) Parallel translation along the local system $\mathbb{V}$ gives rise to the monodromy representation $\rho : \pi_1(S, s_0) \to \mathrm{GL}(V_{s_0, \mathbb{Q}}, W_\bullet, Q)$. Moreover, taking as basis for $V_{s, \mathbb{Q}}$ the parallel translate of one for $V_{s_0, \mathbb{Q}}$, M_{V_s} is constant on paths (from s_0) avoiding a countable union T of proper analytic subvarieties of S, where in fact $S^\circ := S \setminus T$ is pathwise connected. (At points $t \in T$, $M_{V_t} \subset M_{V_s}$; and even the MT group of the LMHS $\psi_{\underline{s}} \mathcal{V}$ at $x \in \bar{S} \setminus S$ naturally includes in M_{V_s}.)

DEFINITION 112. (i) We call $M_{V_{s_0}} =: M_{\mathcal{V}}$ the *MT group*, and $M^\circ_{V_{s_0}} =: M^\circ_{\mathcal{V}}$ the *Hodge group*, of $\mathcal{V}$. One has $\mathrm{End}_{\mathrm{MHS}}(V_{s_0}) \cong \mathrm{End}_{\mathrm{VMHS}(S)}(\mathcal{V})$; see [PS2].

(ii) The identity connected component $\Pi_\mathcal{V}$ of the $\mathbb{Q}$-closure of $\rho(\pi_1(S, s_0))$ is the geometric monodromy group of $\mathcal{V}$; it is invariant under finite covers $\tilde{S} \twoheadrightarrow S$ (and semisimple in the split case).

PROPOSITION 113. *(André)* $\Pi_\mathcal{V} \trianglelefteq DM_\mathcal{V}$.

SKETCH OF PROOF. By a theorem of Chevalley, any closed $\mathbb{Q}$-algebraic subgroup of $\mathrm{GL}(V_{s_0})$ is the stabilizer, for some multitensor $\mathbf{t} \in \bigoplus_i T^{m_i, n_i}(V_{s_0, \mathbb{Q}})$ of $\mathbb{Q}\langle \mathbf{t}\rangle$. For $M_\mathcal{V}$, we can arrange for this $\mathbf{t}_\mathcal{V}$ to be *itself* fixed and to lie in $\bigoplus_i \left(T^{m_i, n_i}(V_{s_0})\right)^{(0,0)}_{\mathbb{Q}}$. By genericity of s_0, $\mathbb{Q}\langle \mathbf{t}_\mathcal{V}\rangle$ extends to a subVMHS with (again by $\exists$ of Q) finite monodromy group, and so $\mathbf{t}_\mathcal{V}$ is fixed by $\Pi_\mathcal{V}$. This proves $\Pi_\mathcal{V} \subset M_\mathcal{V}$ (in fact, $\subset M^\circ_\mathcal{V}$ since monodromy preserves Q). Normality of this inclusion then follows from the Theorem of the Fixed Part: the largest constant sublocal system of any $T^{m,n}(\mathbb{V})$ (stuff fixed by $\Pi_\mathcal{V}$) is a subVMHS, hence subMHS at s_0 and stable under $M_\mathcal{V}$.

Now let

$$M^{\mathrm{ab}}_\mathcal{V} := \frac{M_\mathcal{V}}{DM_\mathcal{V}}, \qquad \Pi^{\mathrm{ab}}_\mathcal{V} := \frac{\Pi_\mathcal{V}}{\Pi_\mathcal{V} \cap DM_\mathcal{V}} \subset M^{\mathrm{ab}}_\mathcal{V},$$

(which is a connected component of the $\mathbb{Q}$-closure of some $\pi^{ab} \subset M_{\mathcal{V}}^{ab,\circ}(\mathbb{Z})$), and (taking a more exotic route than André) V^{ab} be the (CM)MHS corresponding to a faithful representation of $M_{\mathcal{V}}^{ab}$. For each irreducible $H \subset V^{ab}$, the image $\overline{M_{\mathcal{V}}^{ab}}$ has integer points $\cong \mathcal{O}_L^*$ for some CM field L, and $\overline{M_{\mathcal{V}}^{ab,\circ}}(\mathbb{Q}) \subset L$ consists of elements of norm 1 under any embedding. The latter generate L (a well-known fact for CM fields) but, by a theorem of Kronecker, have $\underline{\text{finite}}$ intersection with $\mathcal{O}_L^*$: the roots of unity. It easily follows from this that $\Pi_{\mathcal{V}}^{ab}$, hence $\Pi_{\mathcal{V}}^{ab}$, is trivial. $\qquad\square$

DEFINITION 114. Let $x \in \bar{S}$ with neighborhood $(\Delta^*)^k \times \Delta^{n-k}$ in S and local (commuting) monodromy logarithms $\{N_i\}$;[14] define the weight monodromy filtration $M_\bullet^x := M(N, W)_\bullet$ where $N := \sum_{i=1}^k N_i$. In the following we assume a choice of path from s_0 to x:

(a) Write $\pi_{\mathcal{V}}^x$ for the *local monodromy group* in $\text{GL}(V_{s_0,\mathbb{Z}}, W_\bullet, Q)$ generated by the $T_i = (T_i)_{ss}e^{N_i}$, and ρ^x for the corresponding representation.

(b) We say that $\mathcal{V}$ is *nonsingular at* x if $V_{s_0} \cong \bigoplus_j \text{Gr}_j^W V_{s_0}$ as ρ^x-modules. In this case, the condition that $\psi_{\underline{s}}\mathcal{V} \cong \bigoplus_j \psi_{\underline{s}} \text{Gr}_j^W \mathcal{V}$ is independent of the choice of local coordinates $(s_1, \ldots, s_n)$ at x, and $\mathcal{V}$ is called *semisplit (nonsingular) at* x when this is satisfied.

(c) The $\text{Gr}_i^{M^x} \psi_{\underline{s}}\mathcal{V}$ are always independent of $\underline{s}$. We say that $\mathcal{V}$ is *totally degenerate (TD)* at x if these Gr_i^M are (pure) Tate and *strongly degenerate* (SD) at x if they are CM-HS. Note that the SD condition is interesting already for the nonboundary points ($x \in S$, $k = 0$).

We can now generalize results of André [An] and Mustafin [Ms].

THEOREM 115. *If $\mathcal{V}$ is semisplit TD (resp. SD) at a point $x \in \bar{S}$, then $\Pi_{\mathcal{V}} = M_{\mathcal{V}}^\circ$ (resp. $DM_{\mathcal{V}}^\circ$).*

REMARK 116. Note that semisplit SD at $x \in S$ simply means that V_x is a CM-MHS (this case is done in [An]). Also, if $\Pi_{\mathcal{V}} = M_{\mathcal{V}}^\circ$ then in fact $\Pi_{\mathcal{V}} = DM_{\mathcal{V}}^\circ = M_{\mathcal{V}}^\circ$.

PROOF. Passing to a finite cover to identify $\Pi_{\mathcal{V}}$ and $\overline{\rho(\pi_1)}$, if we can show that any invariant tensor $\mathsf{t} \in (T^{m,n}V_{s_0,\mathbb{Q}})^{\Pi_{\mathcal{V}}}$ is also fixed by $M_{\mathcal{V}}^\circ$ (resp. $DM_{\mathcal{V}}^\circ$), we are done by Chevalley. Now the span of $M_{\mathcal{V}}^\circ\mathsf{t}$ is (since $\Pi_{\mathcal{V}} \trianglelefteq M_{\mathcal{V}}^\circ$) fixed by $\rho(\pi_1)$, and (using the Theorem of the Fixed Part) extends to a constant subVMHS $\mathcal{U} \subset T^{m,n}\mathcal{V} =: \mathcal{T}$. Now the hypotheses on $\mathcal{V}$ carry over to $\mathcal{T}$ and taking LMHS at x, $\mathcal{U} = \psi_{\underline{s}}\mathcal{U} = \bigoplus_i \psi_{\underline{s}} \text{Gr}_i^W \mathcal{U} = \bigoplus_i \text{Gr}_i^W \mathcal{U}$, we see that $\mathcal{U}$ splits (as VMHS). As $\mathcal{T}$ is TD (resp. SD) at x, $\mathcal{U}$ is split Hodge–Tate (resp. CM-MHS).

---

[14]Though this has been suppressed so far throughout this paper, one has $\{N_i\}$ and LMHS even in the general case where the local monodromies T_i are only quasi-unipotent, by writing $T_i =: (T_i)_{ss}(T_i)_u$ uniquely as a product of semisimple and unipotent parts (Jordan decomposition) and setting $N_i := \log((T_i)_u)$.

If $\mathcal{U}$ is H-T then it consists of Hodge tensors; so $M_{\mathcal{V}}^{\circ}$ acts trivially on $\mathcal{U}$ hence on t.

If $\mathcal{U}$ is CM then $M_{\mathcal{V}}^{\circ}|_{\mathcal{U}} = M_{\mathcal{U}}^{\circ}$ is abelian; and so the action of $M_{\mathcal{V}}^{\circ}$ on $\mathcal{U}$ factors through $M_{\mathcal{V}}^{\circ}/DM_{\mathcal{V}}^{\circ}$, so that $DM_{\mathcal{V}}^{\circ}$ fixes t.   □

A reason why one would want this "maximality" result $\Pi_{\mathcal{V}} = M_{\mathcal{V}}^{\circ}$ is to satisfy the hypothesis of the following interpretation of Theorem 91 (which was a partial generalization of results of [Vo1] and [Ch]). Recall that a VMHS $\mathcal{V}/S$ is k-motivated if there is a family $\mathcal{X} \to S$ of quasiprojective varieties defined over k with $V_s = $ the canonical (Deligne) MHS on $H^r(X_s)$ for each $s \in S$.

PROPOSITION 117. *Suppose $\mathcal{V}$ is motivated over k with trivial fixed part, and let $T_0 \subset S$ be a connected component of the locus where $M_{V_s}^{\circ}$ fixes some vector (in V_s). If T_0 is algebraic (over $\mathbb{C}$), $M_{\mathcal{V}_{T_0}}^{\circ}$ has only one fixed line, and $\Pi_{\mathcal{V}_{T_0}} = M_{\mathcal{V}_{T_0}}^{\circ}$, then T_0 is defined over $\bar{k}$.*

Of course, to be able to use this one also needs a result on algebraicity of T_0, i.e., a generalization of the theorems of [CDK] and [BP3] to arbitrary VMHS. One *now has this* by work of Brosnan, Schnell, and the second author:

THEOREM 118. *Given any integral, graded-polarized $\mathcal{V} \in \mathrm{VMHS}(S)_{\bar{S}}^{\mathrm{ad}}$, the components of the Hodge locus of any $\alpha \in V_s$ yield complex algebraic subvarieties of S.*

6.4. MT groups of (higher) normal functions.

We now specialize to the case where $\mathcal{V} \in \mathrm{NF}^r(S, \mathcal{H})_{\bar{S}}^{\mathrm{ad}}$, with $\mathcal{H} \to S$ the underlying VHS of weight $-r$. $M_{\mathcal{V}}^{\circ}$ is then an extension of $M_{\mathcal{H}}^{\circ} \cong M_{\mathcal{V}^{\mathrm{split}}(=\mathcal{H} \oplus \mathbb{Q}_S(0))}^{\circ}$ by (and a semidirect product with) an additive (unipotent) group

$$U := W_{-r} M_{\mathcal{V}}^{\circ} \cong \mathbb{G}_a^{\times \mu},$$

with $\mu \leq \mathrm{rank}\,\mathbb{H}$. Since $M_{\mathcal{V}}^{\circ}$ respects weights, there is a natural map $\eta : M_{\mathcal{V}}^{\circ} \twoheadrightarrow M_{\mathcal{H}}^{\circ}$ and one might ask when this is an isomorphism.

PROPOSITION 119. *$\mu = 0 \iff \mathcal{V}$ is torsion.*

PROOF. First we note that $\mathcal{V}$ is torsion if and only if, for some finite cover $\tilde{S} \twoheadrightarrow S$, we have

$$\{0\} \neq \mathrm{Hom}_{\mathrm{VMHS}(\tilde{S})}(\mathbb{Q}_S(0), \mathcal{V}) = \mathrm{End}_{\mathrm{VMHS}(\tilde{S})}(\mathcal{V}) \cap \mathrm{ann}(\mathcal{H})$$
$$= \mathrm{End}_{\mathrm{MHS}}(V_{s_0}) \cap \mathrm{ann}(H_{s_0}) = \left(\mathrm{Hom}_{\mathbb{Q}}((V_{s_0}/H_{s_0}), V_{s_0})\right)^{M_{\mathcal{V}}^{\circ}}.$$

The last expression can be interpreted as consisting of vectors $\underline{w} \in H_{s_0,\mathbb{Q}}$ that satisfy $(\mathrm{id} - M)\underline{w} = \underline{u}$ whenever $\left(\begin{smallmatrix} 1 & 0 \\ \underline{u} & M \end{smallmatrix}\right) \in M_{\mathcal{V}}^{\circ}$. This is possible only if there is one $\underline{u}$ for each M, i.e., if $\eta : M_{\mathcal{V}}^{\circ} \to M_{\mathcal{H}}^{\circ}$ is an isomorphism. Conversely, assuming this, write $\underline{u} = \eta^{-1}(M)$ [noting

$$\tilde{\eta}^{-1}(M_1 M_2) = \tilde{\eta}^{-1}(M_1) + M_1 \tilde{\eta}^{-1}(M_2)] \tag{$*$}$$

and set $\underline{w} := \tilde{\eta}^{-1}(0)$. Taking $M_2 = 0$ and $M_1 = M$ in ($*$), we get $(\mathrm{id} - M)\underline{w} = (\mathrm{id} - M)\eta^{-1}(0) = \tilde{\eta}^{-1}(M) = \underline{u}$ for all $M \in M_{\mathcal{H}}^{\circ}$. $\square$

We can now address the problem which lies at the heart of this section: what can one say about the monodromy of the normal function above and beyond that of the underlying VHS — for example, about the kernel of the natural map $\Theta : \Pi_{\mathcal{V}} \twoheadrightarrow \Pi_{\mathcal{H}}$? One can make some headway simply by translating Definition 114 and Theorem 115 into the language of normal functions; all vanishing conditions are $\otimes \mathbb{Q}$.

PROPOSITION 120. *Let $\mathcal{V}$ be an admissible higher normal function over S, and let $x \in \bar{S}$ with local coordinate system $\underline{s}$.*

(i) $\mathcal{V}$ is nonsingular (as AVMHS) at x if and only if $\mathrm{sing}_x(\mathcal{V}) = 0$. Assuming this, $\mathcal{V}$ is semi-simple at x if and only if $\lim_x(\mathcal{V}) = 0$. (In case $x \in S$, $\mathrm{sing}_x(\mathcal{V}) = 0$ is automatic and $\lim_x(\mathcal{V}) = 0$ if and only if x is in the torsion locus of $\mathcal{V}$.)

(ii) $\mathcal{V}$ is TD (resp. SD) at x if and only if the underlying VHS $\mathcal{H}$ is. (For $x \in S$, this just means that H_x is CM.)

(iii) If $\mathrm{sing}_x(\mathcal{V})$, $\lim_x(\mathcal{V})$ vanish and $\psi_{\underline{s}}\mathcal{H}$ is graded CM, then $\Pi_{\mathcal{V}} = DM_{\mathcal{V}}$. (For $x \in S$, we are just hypothesizing that the torsion locus of $\mathcal{V}$ contains a CM point of $\mathcal{H}$.)

(iv) Let $x \in \bar{S} \setminus S$. If $\mathrm{sing}_x(\mathcal{V})$, $\lim_x(\mathcal{V})$ vanish and $\psi_{\underline{s}}\mathcal{H}$ is Hodge–Tate, then $\Pi_{\mathcal{V}} = {}^{\cdot}M_{\mathcal{V}}^{\circ}$.

(v) Under the hypotheses of (iii) and (iv), $\dim(\ker(\Theta)) = \mu$. (In general one has $\leq$.)

PROOF. All parts are self-evident except for (v), which follows from observing (in both cases (iii) and (iv)) via the diagram

$$\mathbb{G}_a^{\times \mu} \cong W_{-1} M_{\mathcal{V}}^{(\circ)} = \ker(\eta) \subseteq DM_{\mathcal{V}} =\!=\!= \Pi_{\mathcal{V}} \hookrightarrow M_{\mathcal{V}}^{(\circ)}$$
$$\left\downarrow{\scriptstyle\Theta}\qquad\qquad\quad \left\downarrow{\scriptstyle\eta} \qquad\qquad \tag{6-2}$$
$$\Pi_{\mathcal{H}} \hookrightarrow M_{\mathcal{H}}^{(\circ)}$$

that $\ker(\eta) = \ker(\Theta)$. $\square$

EXAMPLE 121. The Morrison–Walcher normal function from §1.7 (Example 13) lives "over" the VHS $\mathcal{H}$ arising from $R^3 \pi_* \mathbb{Z}(2)$ for a family of "mirror quintic" CY 3-folds, and vanishes at $z = \infty$. (One should take a suitable, e.g., order 2 or 10 pullback so that $\mathcal{V}$ is well-defined.) The underlying HS H at this point is of CM type (the fiber is the usual $(\mathbb{Z}/5\mathbb{Z})^3$ quotient of $\{\sum_{i=0}^4 Z_i^5 = 0\} \subset \mathbb{P}^4$), with $M_H(\mathbb{Q}) \cong \mathbb{Q}(\zeta_5)$. So $\mathcal{V}$ would satisfy the conditions of Proposition 120(iii). It should be interesting to work out the consequences of the resulting equality $\Pi_{\mathcal{V}} = DM_{\mathcal{V}}$.

There is a different aspect to the relationship between local and global behavior of $\mathcal{V}$. Assuming for simplicity that the local monodromies at x are unipotent, let $\kappa_x := \ker(\pi_{\mathcal{V}}^x \twoheadrightarrow \pi_{\mathcal{H}}^x)$ denote the local monodromy kernel, and μ_x the dimensions of its $\mathbb{Q}$-closure $\overline{\kappa_x}$. This is an additive (torsion-free) subgroup of $\ker(\Theta)$, and so $\dim(\ker(\Theta)) \geq \mu_x$ ($\forall x \in \bar{S} \backslash S$). Writing $\{N_i\}$ for the local monodromy logarithms at x, we have the

PROPOSITION 122. (i) $\mu_x > 0$ *implies* $\operatorname{sing}_x(\mathcal{V}) \neq 0$ (*nontorsion singularity*)
(ii) *The converse holds assuming* $r = 1$ *and* $\operatorname{rank}(N_i) = 1$ ($\forall i$).

PROOF. Let $g \in \pi_{\mathcal{V}}^x$, and define $\underline{m} \in \mathbb{Q}^{\oplus k}$ by $\log(g) =: \sum_{i=1}^k m_i N_i$. Writing $\bar{g}, \bar{N}_i$ for $g|_{\mathbb{H}}, N_i|_H$, consider the (commuting) diagram of morphisms of MHS

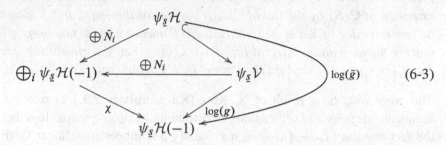

$$\tag{6-3}$$

where $\chi(\underline{w}_1, \ldots, \underline{w}_k) = \sum_{i=1}^k m_i \underline{w}_i$ and $\log(g) = \sum_{i=1}^k m_i \bar{N}_i$. Then $\operatorname{sing}_x(\mathcal{V})$ is nonzero if and only if $(\bigoplus N_i)v_\mathbb{Q}$ does not lie in $\operatorname{im}(\oplus \bar{N}_i)$, where $v_\mathbb{Q}$ (see Definition 2(b)) generates $\psi_{\underline{s}} \mathcal{V}/\psi_{\underline{s}} \mathcal{H}$.

(i) Suppose $g \in \kappa_x \backslash \{1\}$. Then $0 = \log(\bar{g})$ implies $0 = \chi(\operatorname{im}(\oplus \bar{N}_i))$ while $0 \neq \log g$ implies $0 \neq (\log(g))v_\mathbb{Q} = \chi((\bigoplus N_i)v_\mathbb{Q})$. So χ "detects" a singularity.

(ii) If $r = 1$ we may replace $\bigoplus_{i=1}^k \psi_{\underline{s}} \mathcal{H}(-1)$ in the diagram by the subspace $\bigoplus_{i=1}^k (N_i(\psi_{\underline{s}} \mathcal{H}))$. Since each summand is of dimension 1, and

$$(\bigoplus N_i)v_\mathbb{Q} \notin \operatorname{im}(\bigoplus \bar{N}_i)$$

(by assumption), we can choose $\underline{m} = \{m_i\}$ in order that χ kill $\operatorname{im}(\bigoplus \bar{N}_i)$ but not $(\bigoplus N_i)v_\mathbb{Q}$. Using the diagram, $\log(\bar{g}) = 0 \neq \log(g)$ implies $g \in \kappa_x \backslash \{1\}$. $\square$

REMARK 123. (a) The existence of a singularity *always* implies that $\mathcal{V}$ is nontorsion, hence $\mu > 0$.

(b) In the situation of [GG], we have $r = 1$ and rank 1 local monodromy logarithms; hence, by Proposition 122(ii), the existence of a singularity implies $\dim(\ker(\Theta)) > 0$, consistent with (a).

(c) By Proposition 122(i), in the normal function case ($r = 1$), $\mu_x = 0$ along codimension-1 boundary components.

(d) In the "maximal geometric monodromy" situation of Proposition 120(v), $\mu \geq \mu_x \ \forall x \in \bar{S} \backslash S$.

Obviously, for the purpose of forcing singularities to exist, the inequality in (d) points in the wrong direction. One wonders if some sort of cone or spread on a VMHS might be used to translate global into local monodromy kernel, but this seems unlikely to be helpful.

We conclude with an amusing application of differential Galois theory related to a result of André [An]:

PROPOSITION 124. *Consider a normal function $\mathcal{V}$ of geometric origin together with an $\mathcal{O}_S$-basis $\{\omega_i\}$ of holomorphic sections of $\mathcal{F}^0 \mathcal{H}$. (That is, V_s is the extension of MHS corresponding to $AJ(Z_s) \in J^p(X_s)$ for some flat family of cycles on a family of smooth projective varieties over S.) Let K denote the extension of $\mathbb{C}(S)$ by the (multivalued) periods of the $\{\omega_i\}$; and L denote the further extension of K via the (multivalued) Poincaré normal functions given by pairing the ω_i with an integral lift of $1 \in \mathbb{Q}_S(0)$ (i.e., the membrane integrals $\int_{\Gamma_s} \omega_i(s)$ where $\partial \Gamma_s = Z_s$). Then $\mathrm{trdeg}(L/K) = \dim(\ker(\Theta))$.*

The proof rests on a result of N. Katz [Ka, Corollary 2.3.1.1] relating transcendence degrees and dimensions of differential Galois groups, together with the fact that the $\{\int_{\Gamma_s} \omega_i\}$ (for each i) satisfy a homogeneous linear ODE with regular singular points [Gr1]. (This fact implies equality of differential Galois and geometric monodromy groups, since monodromy invariant solutions of such an ODE belong to $\mathbb{C}(S)$ which is the fixed field of the Galois group.) In the event that $\mathcal{H}$ has no fixed part (so that L can introduce no new constants and one has a "Picard–Vessiot field extension") and the normal function is motivated over $k = \bar{k}$, one can probably replace $\mathbb{C}$ by k in the statement.

References

[Ab] S. Abdulali, *Hodge structures of CM type*, J. Ramanujan Math. Soc. 20 (2005), no. 2, 155–162.

[An] Y. Andre, *Mumford–Tate groups of mixed Hodge structures and the theorem of the fixed part*, Compositio Math. 82 (1992), no. 1, 1–24.

[Ar] D. Arapura, *The Leray spectral sequence is motivic*, Invent. Math. 160 (2005), no. 3, 567–589.

[AK] D. Arapura and M. Kumar, *Beilinson–Hodge cycles on semiabelian varieties*, Math. Res. Lett. 16 (2009), no. 4, 557–562.

[BBD] A. Beilinson, J. Bernstein, and P. Deligne, *Faisceaux pervers*, in "Analysis and topology of singular spaces, I" (Luminy, 1981), pp. 5–171, Astérisque 100, Soc. Math. France, Paris, 1982.

[Be] A. A. Beilinson, *Height pairing between algebraic cycles*, in "*K*-theory, arithmetic, and geometry (Moscow, 1984–6)", pp. 1–25, LNM 1289, Springer, Berlin, 1987.

[Bl] S. Bloch, *Algebraic cycles and higher K-theory*, Adv. Math. 61 (1986), no. 3, 267–304.

[Bo] C. Borcea, *Calabi–Yau threefolds and complex multiplication*, in "Essays on mirror manifolds", 489–502, Intl. Press, Hong Kong, 1992.

[BP1] P. Brosnan and G. Pearlstein, *The zero-locus of an admissible normal function*, Ann. Math. 170 (2009), 883–897.

[BP2] ———, *Zero loci of normal functions with torsion singularities*, Duke Math. J. 150 (2009), no. 1, 77–100.

[BP3] ———, *On the algebraicity of the zero-locus of an admissible normal function*, arXiv:0910.0628v1.

[BFNP] P. Brosnan, H. Fang, Z. Nie, and G. Pearlstein, *Singularities of admissible normal functions*, Inventiones Math. 117 (2009), 599–629.

[BPS] P. Brosnan, G. Pearlstein and M. Saito, *A generalization of the Néron models of Griffiths, Green and Kerr*, arXiv:0809.5185, preprint.

[BPSc] P. Brosnan, G. Pearlstein and C. Schnell, *The locus of Hodge classes in an admissible variation of mixed Hodge structure*, C. R. Acad. Sci. Paris, Ser. I 348 (2010), 657–660.

[BZ] J.-L. Brylinski and S. Zucker, *An overview of recent advances in Hodge theory*, in "Several complex variables, VI", 39–142, Encyclopedia Math. Sci. 69, Springer, Berlin, 1990.

[Ca] J. Carlson, *Extensions of mixed Hodge structures*, in "Journées de Géometrie Algébrique d'Angers" (1979), pp. 107–127, Sijthoff & Noordhoff, Alphen aan def Rijn—Germantown, Md., 1980.

[CDK] E. Cattani, P. Deligne, and A Kaplan, *On the locus of Hodge classes*, JAMS 8 (1995), 483–506.

[CF] E. Cattani and J. Fernandez, *Asymptotic Hodge theory and quantum products*, math.AG/001113, in "Advances in algebraic geometry motivated my physics" (Lowell, MA), pp. 115–136, Contemp. Math. 276, AMS, Providence, RI, 2001.

[CK] E. Cattani and A. Kaplan, *Degenerating variations of Hodge structure*, in "Actes du Colloque de Théorie de Hodge" (Luminy), Astérisque 179–180 (1989), 9, 67–96.

[CKS1] E. Cattani, A. Kaplan, and W. Schmid, *Degeneration of Hodge structures*, Ann. of Math. 123 (1986), 457–535.

[CKS2] ———, *L^2 and intersection cohomologies for a polarizable variation of Hodge structure*, Invent. Math. 87 (1987), 217–252.

[CMP] J. Carlson, S. Müller-Stach, and C. Peters, "Period mappings and period domains", Cambridge Stud. Adv. Math. 85, Cambridge University Press, Cambridge,
2003.

[Ch] F. Charles, *On the zero locus of normal functions and the étale Abel–Jacobi map*,
IMRN vol. 2010, no. 12, 2283–2304.

[Cl1] H. Clemens, *Degeneration of Kähler manifolds*, Duke Math. J. 44 (1977), 212–
290.

[Cl2] ———, The Néron model for families of intermediate Jacobians acquiring
"algebraic" singularities, Publ. IHES 58 (1983), 5–18.

[Co] A. Collino, *Griffiths's infinitesimal invariant and higher K-theory on hyperelliptic
Jacobians*, J. Alg. Geom. 6 (1997), no.3, 393–415.

[dCM] M. A. de Cataldo and L. Migliorini, *On singularities of primitive cohomology
classes*, Proc. Amer. Math. Soc. 137 (2009), 3593–3600.

[DM1] P. del Angel and S. Müller-Stach, *Differential equations associated to families
of algebraic cycles*, Ann. Inst. Fourier (Grenoble) 58 (2008), no. 6, 2075–2085.

[DM2] ———, *Picard–Fuchs equations, integrable systems, and higher algebraic K-
theory*, in "Calabi–Yau Varieties and Mirror Symmetry" (Yui and Lewis, editors),
pp. 43–56, AMS, 2003.

[De1] P. Deligne, *Théorie de Hodge II, III* , Inst. Hautes Etudes Sci. Publ. Math. 40
(1971), 5–57; 44 (1974), 5–77.

[De2] P. Deligne, *Hodge cycles on abelian varieties* (notes by J. Milne), in "Hodge
cycles, motives, and Shimura varieties", LNM 900, Springer, New York, 1982.

[De3] ———, letter to Cattani and Kaplan (unpublished).

[De4] ———, *Local behavior of Hodge structures at infinity*, AMS/IP Stud. Adv.
Math. 1 (1997), AMS.

[DK] C. Doran and M. Kerr, *Algebraic K-theory of toric hypersurfaces*, preprint,
arXiv:0809.5279.

[EZ] F. El Zein and S. Zucker, *Extendability of normal functions associated to algebraic cycles*, in "Topics in transcendental algebraic geometry", Ann. Math. Stud.
106, Princeton Univ. Press, Princeton, NJ, 1984.

[GGM] A. Galligo, M. Granger, P. Maisonobe, *D-modules at faisceaux pervers dont
le support singulier est un croisement normal*, Ann. Inst. Fourier 35 (1985), no. 1,
1–48.

[GMV] M. Green, J. Murre, and C. Voisin, "Algebraic cycles and Hodge theory"
(Albano and Bardelli, editors), LNM 1594, Springer, 1994.

[Gr1] P. Griffiths, *A theorem concerning the differential equations satisfied by normal
functions attached to algebraic cycles*, Amer. J. Math. 101 (1979), no. 1, 94–131.

[Gr2] ———, *Poincaré and algebraic geometry*, Bull. AMS 6 (1982), no. 2, 147–159.

[Gr3] ———, *On the periods of certain rational integrals II*, Ann. of Math. 90 (1969),
no. 2, 496–541.

[Gr4] ———, *Singularities of admissible normal functions*, to appear in "Cycles, motives, and Shimura varieties" (Bombay, 2008).

[GC] P. Griffiths and H. Clemens, *The intermediate Jacobian of the cubic threefold*, Ann. of Math. (2) 95 (1972), 281–356.

[GG] P. Griffiths and M. Green, *Algebraic cycles and singularities of normal functions*, in "Algebraic cycles and motives", pp. 206–263, LMS Lect. Not. Ser. 343, Cambridge Univ. Press, Cambridge, 2007.

[GGK] P. Griffiths, M. Green and M. Kerr, *Néron models and limits of Abel–Jacobi mappings*, Compositio Math. 146 (2010), 288–366.

[GH] P. Griffiths and J. Harris, "Principles of algebraic geometry", Wiley, New York, 1978.

[GS] P. Griffiths and W. Schmid, *Locally homogeneous complex manifolds*, Acta Math. 123 (1969), 253–302.

[Ha] F. Hazama, *Hodge cycles on certain abelian varieties and powers of special surfaces*, J. Fac. Sci. Univ. Tokyo, Sect. 1a, 31 (1984), 487–520

[K] M. Kashiwara, *A study of variation of mixed Hodge structure*, Publ. Res. Inst. Math. Sci. 22 (1986), no. 5, 991–1024.

[KNU1] K. Kato, C. Nakayama and S. Usui, SL(2)-*orbit theorem for degeneration of mixed Hodge structure*, J. Algebraic Geom. 17 (2008), 401–479.

[KNU2] ———, *Log intermediate Jacobians*, Proc. Japan Acad. Ser. A Math. Sci. Volume 86, Number 4 (2010), 73–78.

[KNU3] ———, private communication.

[KU] K. Kato and S. Usui, "Classifying spaces of degenerating polarized Hodge structures", Ann. Math. Stud. 169, 2009.

[Ka] N. Katz, "Exponential sums and differential equations", Ann. of Math. Study 124, Princeton University Press, Princeton, NJ, 1990.

[Ki] J. King, *Log complexes of currents and functorial properties of the Abel–Jacobi map*, Duke Math. J. 50 (1983), no. 1, 1–53.

[KL] M. Kerr and J. Lewis, The Abel–Jacobi map for higher Chow groups, II, Invent. Math 170 (2007), no. 2, 355–420.

[KLM] M. Kerr, J. Lewis and S. Muller-Stach, *The Abel–Jacobi map for higher Chow groups*, Compositio Math. 142 (2006), no. 2, 374–396.

[KP] M. Kerr and G. Pearlstein, Normal functions and the GHC, preprint, available at http://www.math.wustl.edu/~matkerr/KPGHC.pdf.

[KY] B. Klingler and A. Yafaev, *The André–Oort Conjecture*, preprint, available at http://www.math.jussieu.fr/~klingler/papiers/KY12.pdf.

[L] S. Lefschetz, "L'Analysis situs et la géometrie algébrique", Gauthier-Villars, Paris, 1924.

[Le1] J. Lewis, "A survey of the Hodge Conjecture" (2nd ed.), CRM Monograph Ser. 10, AMS, Providence, RI, 1999.

[Le2] ———, *A filtration on the Chow groups of a complex projective variety*, Compositio Math. 128 (2001), no. 3, 299–322.

[Li] D. Lieberman, *On the module of intermediate Jacobians*, Amer. J. Math. 91 (1969), 671–682.

[Ma] Y. I. Manin, *Rational points on algebraic curves over function fields*, Izv. Akad. Nauk SSSR Ser. Mat. 27 (1963), 1395–1440.

[Mo] D. Morrison, *The Clemens–Schmid exact sequence and applications*, in "Topics in transcendental algebraic geometry", pp. 101–119, Ann. Math. Studies, Princeton Univ. Press, Princeton, NJ, 1984.

[MW] D. Morrison and J. Walcher, *D-branes and normal functions*, Adv. Theor. Math. Phys. 13 (2009), no. 2, 553–598.

[Mu] D. Mumford, *Families of abelian varieties*, in "Algebraic groups and discontinuous subgroups" (Boulder, CO, 1965), pp. 347–351, AMS, Providence, RI, 1966.

[Mr] V. K. Murty, *Exceptional Hodge classes on certain abelian varieties*, Math. Ann. 268 (1984), 197–206.

[Ms] G. A. Mustafin, *Families of algebraic varieties and invariant cycles*, Izv. Akad. Mauk. SSSR Ser. Mat. 49 (1985), no. 5, pp. 948–978.

[Na] I. Nakamura, *Relative compactification of the Néron model and its application*, in "Complex analysis and algebraic geometry", pp. 207–225, Iwanami Shoten, Tokyo, 1977.

[NS] A. Némethi and J. Steenbrink, *Extending Hodge bundles for abelian variations*, Annals of Math. 142 (1996), 1–18.

[Pe1] G. Pearlstein, *Variations of mixed Hodge structure, Higgs fields and quantum cohomology*, Manuscripta Math. 102 (2000), 269–310.

[Pe2] ———, *SL₂-orbits and degenerations of mixed Hodge structure*, J. Differential Geom. 74 (2006), no. 1, 1–67.

[Pe3] ———, *Hodge conjecture and the period map at infinity*, preprint.

[PS1] C. Peters and J. Steenbrink, "Mixed Hodge structures", Ergebnisse der Mathematik ser. 3 vol. 52, Springer, Berlin, 2008.

[PS2] ———, *Monodromy of variations of Hodge structure*, Acta Appl. Math. 75 (2003), no. 1–3, 183–194.

[P1] H. Poincaré, *Sur les courbes tracées sur les surfaces algébriques*, Ann. Sci. de l'École Norm. Sup. 27 (1910), 55–108.

[P2] ———, *Sur les courbes tracées sur une surface algébrique*, Sitz. der Berliner Math. Gesellschaft 10 (1911), 28–55.

[RS] A. Rosenschon and M. Saito, *Cycle map for strictly decomposable cycles*, Amer. J. Math. 125 (2003), 773–790.

[S1] M. Saito, *Admissible normal functions*, J. Alg. Geom. 5 (1996), no. 2, 235–276.

[S2] ———, *Direct image of logarithmic complexes and infinitesimal invariants of cycles*, in "Algebraic cycles and motives", vol. 2, pp. 304–318, LMS Lect. Not. Ser. 344, Cambridge Univ. Press, Cambridge, 2007.

[S3] ———, *Cohomology classes of admissible normal functions*, arXiv:0904.1593, preprint.

[S4] ———, *Mixed Hodge modules*, Publ. Res. Inst. Math. Sci. 26 (1990), no. 2, 221–333.

[S5] ———, *Hausdorff property of the Zucker extension at the monodromy invariant subspace*, arXiv:0803.2771, preprint.

[S6] ———, *Variations of Hodge structures having no cohomologically nontrivial admissible normal functions*, preprint.

[S7] ———, *On the theory of mixed Hodge modules*, AMS Translations (Series 2) 160 (1994), 47–61. Translated from Sūgaku.

[S8] ———, *Modules de Hodge polarisables*, Publ. Res. Inst. Math. Sci. 24 (1988), no. 6, 849–995.

[SS] M. Saito and C. Schnell, *A variant of Néron models over curves*, to appear in Manuscripta Math.

[Sa] S. Saito, *Motives and filtrations on Chow groups*, Invent. Math. 125 (1996), 149–196.

[Sc] W. Schmid, *Variation of Hodge structure: the singularities of the period mapping*, Invent. Math. 22 (1973), 211–319.

[Sl1] C. Schnell, *Two observations about normal functions*, Clay Math. Proc. 9 (2010), 75–79.

[Sl2] ———, *Local duality and polarized Hodge modules*, arXiv:0904.3480, to appear in Publ. Math. Res. Inst. Sci.

[Sl3] ———, "The boundary behavior of cohomology classes and singularities of normal functions," Ph.D. Thesis, Ohio State University, 2008. Available at OhioLink ETD.

[Sl4] ———, *An alternative construction of the Néron model*, preprint, 2009.

[Sn] C. Schoen, *Varieties dominated by product varieties*, Internat. J. Math. 7 (1996), no. 4, 541–571.

[Sh] T. Shioda, *A note on a theorem of Griffiths on the Abel–Jacobi map*, Invent. Math. 82 (1985), no. 3, 461–465.

[Sp] T. A. Springer, "Linear algebraic groups" (2nd ed.), Birkhäuser, Boston, 1998.

[St] J. Steenbrink, *Limits of Hodge structures*, Inventiones Math. 31 (1976), 229–257.

[SZ] J. Steenbrink and S. Zucker, *Variation of mixed Hodge structure, I*, Invent. Math. 80 (1985), no. 3, 489–542.

[Th] R. Thomas, *Nodes and the Hodge conjecture*, J. Algebraic Geom. 14 (2005), no. 1, 177–185.

[Vo1] C. Voisin, *Hodge loci and absolute Hodge classes*, Compos. Math. 143 (2007), no. 4, 945–958.

[Vo2] ———, "Hodge theory and complex algebraic geometry" (2 vol.), Cambridge Stud. Adv. Math. 76, 77, Cambridge Univ. Press, Cambridge, 2007.

[Yo] A. Young, "Complex analytic Néron models for abelian varieties over higher dimensional parameter spaces", Princeton Univ. Ph.D. thesis.

[Zu1] S. Zucker, *The Hodge conjecture for cubic fourfolds*, Compositio Math. 34 (1977), no. 2, 199–209.

[Zu2] ———, *Generalized intermediate Jacobians and the theorem on normal functions*, Inventiones Math. 33 (1976), no. 2, 185–222.

MATT KERR
DEPARTMENT OF MATHEMATICS
WASHINGTON UNIVERSITY IN ST. LOUIS
CUPPLES I HALL
ONE BROOKINGS DRIVE
ST. LOUIS, MO 63130
UNITED STATES
  matkerr@math.wustl.edu

GREGORY PEARLSTEIN
DEPARTMENT OF MATHEMATICS
MICHIGAN STATE UNIVERSITY
EAST LANSING, MI 48824
UNITED STATES
  gpearl@math.msu.edu

Topology of Stratified Spaces
MSRI Publications
Volume **58**, 2011

Motivic characteristic classes

SHOJI YOKURA

ABSTRACT. Motivic characteristic classes of possibly singular algebraic varieties are homology class versions of motivic characteristics, not classes in the so-called motivic (co)homology. This paper is a survey of them, with emphasis on capturing infinitude finitely and on the motivic nature, in other words, the scissor relation or additivity.

1. Introduction

Characteristic classes are usually cohomological objects defined on real or complex vector bundles. For a smooth manifold, for instance, its characteristic classes are defined through the tangent bundle. For real vector bundles, Stiefel–Whitney classes and Pontraygin classes are fundamental; for complex vector bundles, the Chern class is the fundamental one.

When it comes to a non-manifold space, such as a singular real or complex algebraic or analytic variety, one cannot talk about its cohomological characteristic class, unlike the smooth case, because one cannot define its tangent bundle — although one can define some reasonable substitutes, such as the tangent cone and tangent star cone, which are not vector bundles, but stratified vector bundles.

In the 1960s people started to define characteristic classes on algebraic varieties as homological objects — not through vector bundles, but as higher analogues of geometrically important invariants such as the Euler–Poincaré characteristic, the signature, and so on. I suppose that the theory of characteristic classes of singular spaces starts with Thom's L-class for oriented PL-manifolds

This is an expanded version of the author's talk at the workshop "Topology of Stratified Spaces" held at MSRI, Berkeley, from September 8 to 12, 2008.
Partially supported by Grant-in-Aid for Scientific Research (No. 21540088), the Ministry of Education, Culture, Sports, Science and Technology (MEXT), and JSPS Core-to-Core Program 18005, Japan.

[Thom], whereas Sullivan's Stiefel–Whitney classes and the so-called Deligne–Grothendieck conjecture about the existence of Chern homology classes started the whole story of *capturing characteristic classes of singular spaces as natural transformations*, more precisely as a natural transformation from a certain covariant functor to the homology functor.

The Deligne–Grothendieck conjecture seems to be based on Grothendieck's ideas or Deligne's modification of Grothendieck's conjecture on a *Riemann–Roch type formula* concerning the constructible étale sheaves and Chow rings (see [Grot, Part II, note(87_1), p. 361 ff.]) and was made in its well-known current form by P. Deligne later. R. MacPherson [M1] gave a positive answer to the Deligne–Grothendieck conjecture and, motivated by this solution, P. Baum, W. Fulton and R. MacPherson [BFM1] further established the singular Riemann–Roch Theorem, which is a singular version of Grothendieck–Riemann–Roch, which is a functorial extension of the celebrated Hirzebruch–Riemann–Roch (abbreviated HRR) [Hi]. HRR is the very origin of the Atiyah–Singer Index Theorem.

The main results of [BSY1] (announced in [BSY2]) are the following:

- *"Motivic" characteristic classes of algebraic varieties*, which is a class version of the motivic characteristic. (Note that this "motivic class" is *not* a class in the so-called motivic cohomology in algebraic/arithmetic geometry.)
- Motivic characteristic classes in a sense give rise to *a unification of three well-known important characteristic homology classes*:

  (1) MacPherson's Chern class transformation [M1] (see also [M2; Schw; BrS]);

  (2) Baum, Fulton and MacPherson's Riemann–Roch transformation [BFM1];

  (3) Goresky and MacPherson's L-homology class (see [GM]), or Cappell and Shaneson's L-homology class [CS1] (cf. [CS2]).

This unification result can be understood to be good enough to consider our motivic characteristic classes as a positive solution to the following MacPherson's question or comment, written at the end of his survey paper of 1973 [M2]:

"It remains to be seen whether there is a unified theory of characteristic classes of singular varieties like the classical one outlined above."

The current theory unifies "only three" characteristic classes, but so far it seems to be a reasonble one.

The purpose of this paper is mainly to explain the results from [BSY1] mentioned above (also see [SY]) with emphasis on the "motivic nature" of motivic characteristic classes. In particular, we show that our motivic characteristic class is a very natural class version of the so-called motivic characteristic, just like the way A. Grothendieck extended HRR to Grothendieck –Riemann–Roch. For that, we go back all the way to the natural numbers, which would be thought of as the very origin of a *characteristic* or *characteristic class*.

We naïvely start with the counting of finite sets. Then we want to count infinite sets as if we are still doing the same way of counting finite sets, and want to understand motivic characteristic classes as higher-class versions of this unusual "counting infinite sets", where infinite sets are complex algebraic varieties. (The usual counting of infinite sets, forgetting the structure of a variety at all, lead us into the mathematics of infinity.) The key is Deligne's mixed Hodge structures [De1; De2], or more generally Saito's deep theory of mixed Hodge modules [Sa2], etc.

As to mixed Hodge modules (MHM), in [Sch3] Jörg Schürmann gives a very nice introduction and overview about recent developments on the interaction of theories of characteristic classes and mixed Hodge theory for singular spaces in the complex algebraic context with MHM as a crucial and fundamental key. For example, a study of characteristic classes of the intersection homological Hodge modules has been done in a series of papers by Sylvain Cappell, Anatoly Libgober, Laurentiu Maxim, Jörg Schürmann and Julius Shaneson [CLMS1; CLMS2; CMS1; CMS2; CMSS; MS1; MS2] (in connection with this last one, see also [Y8]).

The very recent book by C. Peters and J. Steenbrink [PS] seems to be a most up-to-date survey on mixed Hodge structures and Saito's mixed Hodge modules. The Tata Lecture Notes by C. Peters [P] (which is a condensed version of [PS]) give a nice introduction to Hodge Theory with more emphasis on the motivic nature.[1]

2. Preliminaries: from natural numbers to genera

We first consider counting the number of elements of finite sets, i.e., natural numbers. Let $\mathcal{FSET}$ be the category of finite sets and maps among them. For an object $X \in \mathcal{FSET}$, let

$$c(X) \in \mathbb{Z}$$

be the number of the elements of X, which is usually denoted by $|X|$ ($\in \mathbb{N}$) and called the cardinal number, or cardinality of X. It satisfies the following four properties on the category $\mathcal{FSET}$ of finite sets:

(1) $X \cong X'$ (bijection or equipotent) $\implies c(X) = c(X')$.
(2) $c(X) = c(X \setminus Y) + c(Y)$ for $Y \subset X$.
(3) $c(X \times Y) = c(X) \cdot c(Y)$.

---

[1] J. Schürmann informed me of the book [PS] and the lecture [P] at the workshop.

(4) $c(pt) = 1$.   (Here pt denotes one point.)

REMARK 2-1. Clearly these four properties characterize the counting $c(X)$. Also note that if $c(X) \in \mathbb{Z}$ satisfies (1)–(3) without (4), then we have $c(pt) = 0$ or $c(pt) = 1$. If $c(pt) = 0$, then it follows from (2) (or (1) and (3)) that $c(X) = 0$ for any finite set X. If $c(pt) = 1$, it follows from (2) that $c(X)$ is the number of elements of a finite set X.

REMARK 2-2. When it comes to infinite sets, cardinality still satisfies properties (1)–(4), but the usual rules of computation no longer work. For natural numbers, $a^2 = a$ implies $a = 0$ or $a = 1$. But the infinite cardinal $\aleph = c(\mathbb{R})$ also has the property that $\aleph^2 = \aleph$; in fact, for any natural number n,

$$c(\mathbb{R}^n) = c(\mathbb{R}), \text{ i.e., } \aleph^n = \aleph.$$

This leads into the *mathematics of infinity*.

One could still imagine counting on the bigger category $\mathcal{SET}$ of sets, where a set can be infinite, and $c(X)$ lies in some integral domain. However, one can see that if for such a counting (1), (2) and (3) are satisfied, it follows automatically that $c(pt) = 0$, contradicting property (4).

In other words: if we consider counting with properties (1)–(3) on the category $\mathcal{SET}$ of all sets, the only possibility is the trivial one: $c(X) = 0$ for any set X!

However, if we consider sets having superstructures on the infrastructure (set) and property (1) is replaced by the invariance of the superstructures, we do obtain more reasonable countings which are finite numbers; thus we can avoid the mysterious "mathematics of infinity" and extend the usual counting $c(X)$ of finite sets very naturally and naïvely. This is exactly what the Euler characteristic, the genus, and many other important and fundamental objects in modern geometry and topology are all about.

Let us consider the following "topological counting" c_{top} on the category $\mathcal{TOP}$ of topological spaces, which assigns to each topological space X a certain integer (or more generally, an element in an integral domain)

$$c_{\text{top}}(X) \in \mathbb{Z}$$

such that it satisfies the following four properties, which are exactly the same as above except for (1):

(1) $X \cong X'$ (homeomorphism = $\mathcal{TOP}$- isomorphism) $\Longrightarrow c_{\text{top}}(X) = c_{\text{top}}(X')$,
(2) $c_{\text{top}}(X) = c_{\text{top}}(X \setminus Y) + c_{\text{top}}(Y)$ for $Y \subset X$ (for the moment no condition),
(3) $c_{\text{top}}(X \times Y) = c_{\text{top}}(X) \cdot c_{\text{top}}(Y)$,
(4) $c_{\text{top}}(pt) = 1$.

REMARK 2-3. As in Remark 2-1, conditions (1) and (3) imply that $c_{top}(pt) = 0$ or 1. If $c(pt) = 0$, it follows from (1) and (3) that $c_{top}(X) = 0$ for any topological space X. Thus the last condition, $c(pt) = 1$, means that c_{top} is a nontrivial counting. Hence, topological counting c_{top} can be regarded as *a nontrivial, multiplicative, additive, topological invariant.*

PROPOSITION 2-4. *If such a c_{top} exists, then*

$$c_{top}(\mathbb{R}^1) = -1, \quad hence \quad c_{top}(\mathbb{R}^n) = (-1)^n.$$

Hence if X is a finite CW-complex with $\sigma_n(X)$ open n-cells, then

$$c_{top}(X) = \sum_n (-1)^n \sigma_n(X) = \chi(X),$$

the Euler–Poincaré characteristic of X.

The equality $c_{top}(\mathbb{R}^1) = -1$ can be seen by considering

$$\mathbb{R}^1 = (-\infty, 0) \sqcup \{0\} \sqcup (0, \infty).$$

Condition (2) implies $c_{top}(\mathbb{R}^1) = c_{top}((-\infty, 0)) + c_{top}(\{0\}) + c_{top}((0, \infty))$, so

$$-c_{top}(\{0\}) = c_{top}((-\infty, 0)) + c_{top}((0, \infty)) - c_{top}(\mathbb{R}^1).$$

Since $\mathbb{R}^1 \cong (-\infty, 0) \cong (0, \infty)$, it follows from (1) and (4) that

$$c_{top}(\mathbb{R}^1) = -c_{top}(\{0\}) = -1.$$

The existence of a counting c_{top} can be shown using ordinary homology/cohomology theory: symbolically,

topological counting c_{top} : ordinary (co)homology theory.

To be more precise, we use Borel–Moore homology theory [BM], the homology theory with closed supports. For a locally compact Hausdorff space X, Borel–Moore homology theory $H_*^{BM}(X; R)$ with a ring coefficient R is isomorphic to the relative homology theory of the pair $(X^c, *)$, with X^c the one-point compactification of X and $*$ the one point added to X:

$$H_*^{BM}(X; R) \cong H_*(X^c, *; R).$$

Hence, if X is compact, Borel–Moore homology theory is the usual homology theory: $H_*^{BM}(X; R) = H_*(X; R)$.

Let $\mathfrak{K}$ be a field, such as $\mathbb{R}$ or $\mathbb{C}$. If the Borel–Moore homology $H_*^{BM}(X; \mathfrak{K})$ is finite-dimensional — say, if X is a finite CW-complex — then the Euler–Poincaré characteristic χ_{BM} using the Borel–Moore homology theory with coefficient field $\mathfrak{K}$, namely

$$\chi_{BM}(X) := \sum_n (-1)^n \dim_{\mathfrak{K}} H_n^{BM}(X; \mathfrak{K}),$$

gives rise to a topological counting χ_{top}, because it satisfies $H_n^{BM}(\mathbb{R}^n, \mathfrak{K}) = \mathfrak{K}$ and $H_k^{BM}(\mathbb{R}^n, \mathfrak{K}) = 0$ for $k \neq n$, and thus

$$\chi_{BM}(\mathbb{R}^n) = (-1)^n.$$

It turns out that for coefficients in a field $\mathfrak{K}$, Borel–Moore homology is *dual* [2] *as a vector space* to the cohomology with compact support, namely

$$H_p^{BM}(X; \mathfrak{K}) = \text{Hom}(H_c^p(X; \mathfrak{K}), \mathfrak{K}).$$

Since $\mathfrak{K}$ is a field, we have

$$H_p^{BM}(X; \mathfrak{K}) \cong H_c^p(X; \mathfrak{K})$$

Hence the Euler-Poincaré characteristic using Borel–Moore homology $\chi_{BM}(X)$ is equal to the Euler-Poincaré characteristic using cohomology with compact support, usually denoted by χ_c:

$$\chi_c(X) = \sum_i (-1)^i \dim_K H_c^i(X; \mathfrak{K}).$$

Since it is quite common to use χ_c, we have

COROLLARY 2-5. *For the category of locally compact Hausdorff spaces,*

$$c_{top} = \chi_c,$$

the Euler–Poincaré characteristic using cohomology with compact support.

REMARK 2-6. This story could be retold as follows: There might be many ways of "topologically counting" on the category $\mathcal{TOP}$ of topological spaces, but they are *all identical to the Euler–Poincaré characteristic with compact support* when restricted to the subcategory of locally compact Hausdorff spaces with finite dimensional Borel–Moore homologies. Symbolically speaking,

$$c_{top} = \chi_c.$$

Next consider the following "algebraic counting" c_{alg} on the category $\mathcal{VAR}$ of *complex* algebraic varieties (of finite type over $\mathbb{C}$), which assigns to each complex algebraic variety X a certain element

$$c_{alg}(X) \in R$$

in a commutative ring R with unity, such that:

(1) $X \cong X'$ ($\mathcal{VAR}$-isomorphism) $\Longrightarrow c_{alg}(X) = c_{alg}(X')$.
(2) $c_{alg}(X) = c_{alg}(X \setminus Y) + c_{alg}(Y)$ for a closed subvariety $Y \subset X$.

---

[2] For an n-dimensional manifold M the Poincaré duality map $\mathcal{PD} : H_c^k(M) \cong H_{n-k}(M)$ is an isomorphism and also $\mathcal{PD} : H^k(M) \cong H_{n-k}^{BM}(M)$ is an isomorphism. Thus they are *Poincaré dual*, but *not dual as vector spaces*.

(3) $c_{\text{alg}}(X \times Y) = c_{\text{alg}}(X) \cdot c_{\text{alg}}(Y)$.

(4) $c_{\text{alg}}(pt) = 1$.

Just like $c(X)$ and $c_{\text{top}}(X)$, the last condition simply means that c_{alg} is a nontrivial counting.

The real numbers $\mathbb{R}$ and in general the Euclidean space $\mathbb{R}^n$ are the most fundamental objects in the category $\mathcal{TOP}$ of topological spaces, and the complex numbers $\mathbb{C}$ and in general complex affine spaces $\mathbb{C}^n$ are the most fundamental objects in the category $\mathcal{VAR}$ of complex algebraic varieties. The decomposition of n-dimensional complex projective space as

$$\mathbb{P}^n = \mathbb{C}^0 \sqcup \mathbb{C}^1 \sqcup \cdots \sqcup \mathbb{C}^{n-1} \sqcup \mathbb{C}^n$$

implies the following:

PROPOSITION 2-7. *If c_{alg} exists, then*

$$c_{\text{alg}}(\mathbb{P}^n) = 1 - y + y^2 - y^3 + \cdots + (-y)^n,$$

where $y := -c_{\text{alg}}(\mathbb{C}^1) \in R$.

REMARK 2-8. Proposition 2-7 already indicates that there could exist infinitely many ways — as many as the elements y — to do algebraic counting c_{alg} on the category $\mathcal{VAR}$ of complex algebraic varieties. This is strikingly different from the topological counting c_{top} and the original counting c, which are uniquely determined. This difference of course lies in the complex structure:

$$\text{set} + \text{topological structure} + \textit{complex structure}.$$

Here there is no question of considering $\mathbb{R}^1$, so the previous argument showing that $c_{\text{top}}(\mathbb{R}^1) = -1$ does not work. In this sense, we should have used the symbol $c_{\text{alg}/\mathbb{C}}$ to emphasize the complex structure, instead of c_{alg}. Since we are dealing with only the category of complex algebraic varieties in this paper, we write just c_{alg}. See Remark 2-11 below for the category of real algebraic varieties.

The existence of a c_{alg} — in fact, of many such ways of algebraically counting — can be shown using *Deligne's theory of mixed Hodge structures* [De1; De2], which comes from the algebraic structure:

$$\text{set} + \text{topological structure} + \textit{complex structure} + \textit{algebraic structure}.$$

Then the Hodge–Deligne polynomial

$$\chi_{u,v}(X) := \sum_{i,p,q \geq 0} (-1)^i (-1)^{p+q} \dim_{\mathbb{C}}(\mathrm{Gr}_F^p \, \mathrm{Gr}_{p+q}^W \, H_c^i(X, \mathbb{C})) u^p v^q$$

satisfies the four properties above with $R = \mathbb{Z}[u, v]$ and $-y := c_{\mathrm{alg}}(\mathbb{C}^1) = uv$, namely any Hodge–Deligne polynomial $\chi_{u,v}$ with $uv = -y$ is a c_{alg}. Here we point out that by Deligne's work only graded terms with $p, q \geq 0$ are nontrivial; otherwise one would have $\chi_{u,v}(X) \in \mathbb{Z}[u, u^{-1}, v, v^{-1}]$.

Similarly one can consider the invariant

$$c_{\mathrm{alg}}(X) := \chi_{y,-1} \in \mathbb{Z}[y],$$

with $c_{\mathrm{alg}}(\mathbb{C}^1) = -y$.

Here we should note that for $(u, v) = (-1, -1)$ we have

$$\chi_{-1,-1}(X) = \chi_c(X) = c_{\mathrm{top}}(X).$$

Further, for a smooth compact variety X, $\chi_{0,-1}(X)$ is the arithmetic genus, while $\chi_{1,-1}(X)$ is the signature. These three cases, $(u, v) = (-1, -1)$, $(0, -1)$ and $(1, -1)$, are very important.

algebraic counting c_{alg}: mixed Hodge theory
$=$ ordinary (co)homology theory $+$ mixed Hodge structures.

REMARK 2-9. (See [DK], for example.) The following description is also fine, but we use the one above in our later discussion on motivic characteristic classes:

$$c_{\mathrm{alg}}(\mathbb{P}^n) = 1 + y + y^2 + y^3 + \cdots + y^n,$$

where $y = c_{\mathrm{alg}}(\mathbb{C}^1) \in \mathbb{Z}[y]$. The Hodge–Deligne polynomial is usually denoted by $E(X; u, v)$ and is defined to be

$$E(X; u, v) := \sum_{i,p,q \geq 0} (-1)^i \dim_{\mathbb{C}}(\mathrm{Gr}_F^p \mathrm{Gr}_{p+q}^W H_c^i(X, \mathbb{C})) u^p v^q.$$

Thus

$$\chi_{u,v}(X) = E(X; -u, -v).$$

The reason why we define $\chi_{u,v}(X)$ to be $E(X; -u, -v)$ rather than $E(X; u, v)$ lies in the definition of Hirzebruch's generalized Todd class and Hirzebruch's χ_y characteristic, which will come below.

The algebraic counting c_{alg} specializes to the topological counting c_{top}. Are there other algebraic countings that specialize to the Hodge–Deligne polynomial $\chi_{u,v}$ (which is sensitive to an algebraic structure)?

CONJECTURE 2-10. *The answer is negative; in other words, there are no extra structures other than Deligne's mixed Hodge structure that contribute more to the algebraic counting c_{alg} of complex algebraic varieties.*

REMARK 2-11. In the category $\mathcal{VAR}(\mathbb{R})$ of *real algebraic varieties*, we can of course consider $c_{\text{alg}/\mathbb{R}}(\mathbb{R}^1)$ of the real line $\mathbb{R}^1$; therefore we might be tempted to the hasty conclusion that in the category of real algebraic varieties the topological counting c_{top}, i.e., χ_c, is sufficient. Unfortunately, the argument for $c_{\text{top}}(\mathbb{R}^1) = -1$ does not work in the category $\mathcal{VAR}(\mathbb{R})$, because $\mathbb{R}^1$ and $(-\infty, 0)$ or $(0, \infty)$ are not isomorphic as real algebraic varieties. Even among compact varieties there do exist real algebraic varieties that are homeomorphic but not isomorphic as real algebraic varieties. For instance (see [MP1, Example 2.7]):

Consider the usual *normal crossing* figure eight curve:

$$\text{F8} = \{(x, y) \mid y^2 = x^2 - x^4\}.$$

The proper transform of F8 under the blowup of the plane at the origin is homeomorphic to a circle, and the preimage of the singular point of F8 is two points.

Next take the *tangential* figure eight curve:

$$t\text{F8} = \{(x, y) \mid ((x+1)^2 + y^2 - 1)((x-1)^2 + y^2 - 1) = 0\},$$

which is the union of two circles tangent at the origin. Here the preimage of the singular point is a single point. Therefore, in contrast to the category of crude topological spaces, in the category of *real algebraic* varieties an "algebraic counting" $c_{\text{alg}/\mathbb{R}}(\mathbb{R}^1)$ is meaningful, i.e., sensitive to the algebraic structure. Indeed, as such a real algebraic counting $c_{\text{alg}/\mathbb{R}}(\mathbb{R}^1)$ there are

the i-th virtual Betti number $\beta_i(X) \in \mathbb{Z}$

and

the virtual Poincaré polynomial $\beta_t(X) = \sum_i \beta_i(X) t^i \in \mathbb{Z}[t]$.

They are both identical to the usual Betti number and Poincaré polynomial on compact nonsingular varieties. For the above two figure eight curves F8 and tF8 we indeed have that

$$\beta_t(\text{F8}) \neq \beta_t(t\text{F8}).$$

For more details, see [MP1] and [To3], and see also Remark 4-12.

Finally, in passing, we also mention the following "cobordism" counting c_{cob} on the category of closed oriented differential manifolds or the category of stably almost complex manifolds:

(1) $X \cong X'$ (cobordant, or bordant) $\implies c_{\text{cob}}(X) = c_{\text{cob}}(X')$.
(2) $c_{\text{cob}}(X \sqcup Y) = c_{\text{cob}}(X) + c_{\text{cob}}(Y)$. (Note: in this case $c_{\text{cob}}(X \setminus Y)$ does not make sense, because $X \setminus Y$ has to be a closed oriented manifold.)
(3) $c_{\text{cob}}(X \times Y) = c_{\text{cob}}(X) \cdot c_{\text{cob}}(Y)$.
(4) $c_{\text{cob}}(pt) = 1$.

As in the cases of the previous countings, (1) and (3) imply $c_{\text{cob}}(pt) = 0$ or $c_{\text{cob}}(pt) = 1$. It follows from (3) that $c_{\text{cob}}(pt) = 0$ implies that $c_{\text{cob}}(X) = 0$ for any closed oriented differential manifolds X. Thus the last condition $c_{\text{cob}}(pt) = 1$ means that our c_{cob} is nontrivial. Such a cobordism counting c_{cob} is nothing but a *genus* such as the signature, the $\hat{A}$-genus, or the elliptic genus. As in Hirzebruch's book, a genus is usually defined as a nontrivial counting satisfying properties (1), (2) and (3). Thus, it is the same as the one given above.

Here is a very simple problem on genera of closed oriented differentiable manifolds or stably almost complex manifolds:

PROBLEM 2-12. *Determine all genera.*

Let $\text{Iso}(G)_n$ be the set of isomorphism classes of smooth closed (and oriented) pure n-dimensional manifolds M for $G = O$ (or $G = SO$), or of pure n-dimensional weakly ("= stably") almost complex manifolds M for $G = U$, i.e., $TM \oplus \mathbb{R}_M^N$ is a complex vector bundle (for suitable N, with $\mathbb{R}_M$ the trivial real line bundle over M). Then

$$\text{Iso}(G) := \bigoplus_n \text{Iso}(G)_n$$

becomes a commutative graded semiring with addition and multiplication given by disjoint union and exterior product, with 0 and 1 given by the classes of the empty set and one point space.

Let $\Omega^G := \text{Iso}(G)/\sim$ be the corresponding *cobordism ring* of closed ($G = O$) and oriented ($G = SO$) or weakly ("= stably") almost complex manifolds ($G = U$) as discussed for example in [Stong]. Here $M \sim 0$ for a closed pure n-dimensional G-manifold M if and only if there is a compact pure $(n+1)$-dimensional G-manifold B with boundary $\partial B \simeq M$. This is indeed a ring with $-[M] = [M]$ for $G = O$ or $-[M] = [-M]$ for $G = SO, U$, where $-M$ has the opposite orientation of M. Moreover, for B as above with $\partial B \simeq M$ one has

$$TB|\partial B \simeq TM \oplus \mathbb{R}_M.$$

This also explains the use of the stable tangent bundle for the definition of a stably or weakly almost complex manifold.

The following structure theorems are fundamental (see [Stong, Theorems on p. 177 and p. 110]):

THEOREM 2-13. (1) (Thom) $\Omega^{SO} \otimes \mathbb{Q} = \mathbb{Q}[\mathbb{P}^2, \mathbb{P}^4, \mathbb{P}^6, \dots, \mathbb{P}^{2n}, \dots]$ *is a polynomial algebra in the classes of the complex even dimensional projective spaces.*

(2) (Milnor) $\Omega_*^U \otimes \mathbb{Q} = \mathbb{Q}[\mathbb{P}^1, \mathbb{P}^2, \mathbb{P}^3, \dots, \mathbb{P}^n, \dots]$ *is a polynomial algebra in the classes of the complex projective spaces.*

So, if we consider a commutative ring R without torsion for a genus $\gamma : \Omega^{SO} \to R$, the genus γ is completely determined by the value $\gamma(\mathbb{P}^{2n})$ of the cobordism class of each even-dimensional complex projective space $\mathbb{P}^{2n}$. Using this value one could consider the related generating "function" or formal power series such as $\sum_n \gamma(\mathbb{P}^{2n})x^n$, or $\sum_n \gamma(\mathbb{P}^{2n})x^{2n}$, and etc. In fact, a more interesting problem is to determine all *rigid* genera such as the signature σ and the A-genus: namely, a genus satisfying the following multiplicativity property stronger than the product property (3):

(3)$_{\text{rigid}}$: $\gamma(M) = \gamma(F)\gamma(B)$ for a fiber bundle $M \to B$ with fiber F and compact connected structure group.

THEOREM 2-14. *Let* $\log_\gamma (x)$ *be the "logarithmic" formal power series in* $R[\![x]\!]$ *given by*

$$\log_\gamma(x) := \sum_n \frac{1}{2n+1}\gamma(\mathbb{P}^{2n})x^{2n+1}.$$

The genus γ *is rigid if and only if it is an elliptic genus; i.e., its logarithm* $\log_\gamma$ *is an elliptic integral; i.e.,*

$$\log_\gamma(x) = \int_0^x \frac{1}{\sqrt{1-2\delta t^2 + \varepsilon t^4}}dt$$

for some $\delta, \varepsilon \in R$.

The "only if" part was proved by S. Ochanine [Oc] and the "if part" was first "physically" proved by E. Witten [Wit] and later "mathematically" proved by C. Taubes [Ta] and also by R. Bott and C. Taubes [BT]. See also B. Totaro's papers [To2; To4].

cobordism counting c_{cob} : Thom's Theorem
rigid genus = elliptic genus : elliptic integral

The oriented cobordism group Ω^{SO} above was extended by M. Atiyah [At] to a generalized cohomology theory, i.e., the oriented cobordism theory $MSO^*(X)$ of a topological space X. The theory $MSO^*(X)$ is defined by the so-called Thom spectra: the infinite sequence of Thom complexes given, for a topological pair (X, Y) with $Y \subset X$, by

$$MSO^k(X, Y) := \lim_{n \to \infty}[\Sigma^{n-k}(X/Y), MSO(n)].$$

Here the homotopy group $[\Sigma^{n-k}(X/Y), MSO(n)]$ is stable.

As a covariant or homology-like version of $MSO^*(X)$, M. Atiyah [At] introduced the bordism theory $MSO_*(X)$ geometrically in quite a simple manner: Let $f_1 : M_1 \to X$, $f_2 : M_2 \to X$ be continuous maps from closed oriented n-dimensional manifolds to a topological space X. f and g are said to be bordant if there exists an oriented manifold W with boundary and a continuous map $g : W \to X$ such that

(1) $g|_{M_1} = f_1$ and $g|_{M_2} = f_2$, and

(2) $\partial W = M_1 \cup -M_2$, where $-M_2$ is M_2 with its reverse orientation.

It turns out that $MSO_*(X)$ is a generalized homology theory and

$$MSO^0(pt) = MSO_0(pt) = \Omega^{SO}.$$

M. Atiyah [At] also showed Poincaré duality for an oriented closed manifold M of dimension n:

$$MSO^k(M) \cong MSO_{n-k}(M).$$

If we replace $SO(n)$ by the other groups $O(n)$, $U(n)$, $\text{Spin}(n)$, we get the corresponding cobordism and bordism theories.

REMARK 2-15 (ELLIPTIC COHOMOLOGY). Given a ring homomorphism $\varphi :$ $MSO^*(pt) \to R$, R is an $MSO^*(pt)$-module and

$$MSO^*(X) \otimes_{MSO^*(pt)} R$$

becomes "almost" a generalized cohomology theory (one not necessarily satisfying the Exactness Axiom). P. S. Landweber [L] gave an algebraic criterion (called the Exact Functor Theorem) for it to become a generalized cohomology theory. Applying this theorem, P. E. Landweber, D. C. Ravenel and R. E. Stong [LRS] showed the following theorem:

THEOREM 2-16. *For the elliptic genus* $\gamma : MSO^*(pt) = MSO_*(pt) = \Omega \to$ $\mathbb{Z}[\frac{1}{2}][\delta, \varepsilon]$, *the following functors are generalized cohomology theories*:

$$MSO^*(X) \otimes_{MSO^*(pt)} \mathbb{Z}[\tfrac{1}{2}][\delta, \varepsilon][\varepsilon^{-1}],$$

$$MSO^*(X) \otimes_{MSO^*(pt)} \mathbb{Z}[\tfrac{1}{2}][\delta, \varepsilon][(\delta^2 - \varepsilon)^{-1}],$$

$$MSO^*(X) \otimes_{MSO^*(pt)} \mathbb{Z}[\tfrac{1}{2}][\delta, \varepsilon][\Delta^{-1}],$$

where $\Delta = \varepsilon(\delta^2 - \varepsilon)^2$.

More generally J. Franke [Fr] showed this:

THEOREM 2-17. *For the elliptic genus* $\gamma : MSO^*(pt) = MSO_*(pt) = \Omega^{SO} \to$ $\mathbb{Z}[\frac{1}{2}][\delta, \varepsilon]$, *the functor*

$$MSO^*(X) \otimes_{MSO^*(pt)} \mathbb{Z}[\tfrac{1}{2}][\delta, \varepsilon][P(\delta, \varepsilon)^{-1}]$$

is a generalized cohomology theory. Here $P(\delta, \varepsilon)$ *is a homogeneous polynomial of positive degree with* $\deg \delta = 4$, $\deg \varepsilon = 8$.

The generalized cohomology theory

$$MSO^*(X) \otimes_{MSO^*(pt)} \mathbb{Z}[\tfrac{1}{2}][\delta, \varepsilon][P(\delta, \varepsilon)^{-1}]$$

is called *elliptic cohomology theory*. It was recently surveyed by J. Lurie [Lu]. It is defined in an algebraic manner, but not in a topologically or geometrically simpler manner than K-theory or the bordism theory $MSO_*(X)$. So, people have been searching for a reasonable geometric or topological construction for elliptic cohomology (cf. [KrSt]).

REMARK 2-18 (MUMBO JUMBO). In the above we see that if you just count points of a variety simply as a set, we get infinity unless it is a finite set or the trivial counting 0, but that if we count it "respecting" the topological and algebraic structures we get a certain reasonable number which is not infinity. Getting carried away, "zeta function-theoretic" formulae such as

$$1 + 1 + 1 + \cdots + 1 + \cdots = -\tfrac{1}{2} = \zeta(0),$$
$$1 + 2 + 3 + \cdots + n + \cdots = -\tfrac{1}{12} = \zeta(-1),$$
$$1^2 + 2^2 + 3^2 + \cdots + n^2 + \cdots = 0 = \zeta(-2),$$
$$1^3 + 2^3 + 3^3 + \cdots + n^3 + \cdots = \tfrac{1}{120} = \zeta(-3)$$

could be considered as based on a counting of infinite sets that respects some kind of "zeta structure" on it, whatever that is. In nature, the equality $1^3 + 2^3 + 3^3 + \cdots + n^3 + \cdots = \tfrac{1}{120}$ is relevant to the *Casimir effect*, named after the Dutch physicist Hendrik B. G. Casimir. (See [Wil, Lecture 7] for the connection.) So, nature perhaps already knows what the zeta structure is. It would be fun, even nonmathematically, to imagine what a zeta structure would be on the natural numbers $\mathbb{N}$, or the integers $\mathbb{Z}$ or the rational numbers $\mathbb{Q}$, or more generally "zeta structured" spaces or varieties. Note that, like the topological counting $c_{\text{top}} = \chi$, zeta-theoretical counting (denoted by c_{zeta} here) was discovered by Euler!

REMARK 2-19. Regarding "counting", one is advised to read Baez [Ba1; Ba2], Baez and Dolan [BD], and Leinster [Lein].

3. Motivic characteristic classes

Any algebraic counting c_{alg} gives rise to the following naïve ring homomorphism to a commutative ring R with unity:

$$c_{\text{alg}} : \text{Iso}(\mathcal{VAR}) \to R \quad \text{defined by } c_{\text{alg}}([X]) := c_{\text{alg}}(X).$$

Here $\text{Iso}(\mathcal{VAR})$ is the free abelian group generated by the isomorphism classes $[X]$ of complex varieties. The additivity relation

$$c_{\text{alg}}([X]) = c_{\text{alg}}([X \setminus Y]) + c_{\text{alg}}([Y]) \text{ for any closed subvariety } Y \subset X$$

—or, in other words,

$$c_{\mathrm{alg}}([X] - [Y] - [X \setminus Y]) = 0 \text{ for any closed subvariety } Y \subset X,$$

induces the following finer ring homomorphism:

$$c_{\mathrm{alg}} : K_0(\mathcal{VAR}) \to R \quad \text{defined by } c_{\mathrm{alg}}([X]) := c_{\mathrm{alg}}(X).$$

Here $K_0(\mathcal{VAR})$ is the Grothendieck ring of complex algebraic varieties, which is $\mathrm{Iso}(\mathcal{VAR})$ modulo the additivity relation

$$[X] = [X \setminus Y] + [Y] \text{ for any closed subvariety } Y \subset X$$

(in other words, $\mathrm{Iso}(\mathcal{VAR})$ modded out by the subgroup generated by elements of the form $[X] - [Y] - [X \setminus Y]$ for any closed subvariety $Y \subset X$).

The equivalence class of $[X]$ in $K_0(\mathcal{VAR})$ should be written as, $[\![X]\!]$, say, but we just use the symbol $[X]$ for simplicity.

More generally, let y be an indeterminate and consider the following homomorphism $c_{\mathrm{alg}} := \chi_y := \chi_{y,-1}$, i.e.,

$$c_{\mathrm{alg}} : K_0(\mathcal{VAR}) \to \mathbb{Z}[y] \quad \text{with } c_{\mathrm{alg}}(\mathbb{C}^1) = -y.$$

This will be called a *motivic characteristic*, to emphasize the fact that its domain is the Grothendieck ring of varieties.

REMARK 3-1. In fact, for the category $\mathcal{VAR}(k)$ of algebraic varieties over any field, the above Grothendieck ring $K_0(\mathcal{VAR}(k))$ can be defined in the same way.

What we want to do is an analogue to the way that Grothendieck extended the celebrated Hirzebruch–Riemann–Roch Theorem (which was the very beginning of the Atiyah–Singer Index Theorem) to the Grothendieck–Riemann–Roch Theorem. In other words, we want to solve the following problem:

PROBLEM 3-2. *Let R be a commutative ring with unity such that $\mathbb{Z} \subset R$, and let y be an indeterminate. Do there exist some covariant functor $\Diamond$ and some natural transformation (here pushforwards are considered for proper maps)*

$$\natural : \Diamond(\ ) \to H_*^{BM}(\ ) \otimes R[y]$$

satisfying conditions (1)–(3) below?

(1) $\Diamond(pt) = K_0(\mathcal{VAR})$.
(2) $\natural(pt) = c_{\mathrm{alg}}$, *i.e.,*

$$\natural(pt) = c_{\mathrm{alg}} : \Diamond(pt) = K_0(\mathcal{VAR}) \to R[y] = H_*^{BM}(pt) \otimes R[y].$$

(3) *For the mapping $\pi_X : X \to pt$ to a point, for a certain distinguished element $\Delta_X \in \Diamond(X)$ we have*

$$\pi_{X*}(\natural(\Delta_X)) = c_{\mathrm{alg}}(X) \in R[y] \quad \text{and} \quad \pi_{X*}(\Delta_X) = [X] \in K_0(\mathcal{VAR}).$$

$$
\begin{array}{ccc}
\Diamond(X) & \xrightarrow{\;\natural(X)\;} & H_*^{BM}(X) \otimes R[y] \\[4pt]
{\scriptstyle \pi_{X*}}\Big\downarrow & & \Big\downarrow {\scriptstyle \pi_{X*}} \\[4pt]
\Diamond(pt) = K_0(\mathcal{VAR}) & \xrightarrow[\natural(pt)=c_{\mathrm{alg}}]{} & R[y].
\end{array}
$$

(If there exist such $\Diamond$ and $\natural$, then $\natural(\Delta_X)$ could be called the *motivic characteristic class* corresponding to the motivic characteristic $c_{\mathrm{alg}}(X)$, just like the Poincaré dual of the total Chern cohomology class $c(X)$ of a complex manifold X corresponds to the Euler–Poincaré characteristic: $\pi_{X*}(c(X) \cap [X]) = \chi(X)$.)

A more concrete one for the Hodge–Deligne polynomial (a prototype of this problem was considered in [Y5]; cf. [Y6]):

PROBLEM 3-3. *Let R be a commutative ring with unity such that $\mathbb{Z} \subset R$, and let u, v be two indeterminates. Do there exist a covariant functor $\Diamond$ and a natural transformation (here pushforwards are considered for proper maps)*

$$\natural : \Diamond(\ ) \to H_*BM(\ ) \otimes R[u, v]$$

satisfying conditions (1)–(3) below?

(1) $\Diamond(pt) = K_0(\mathcal{VAR})$.

(2) $\natural(pt) = \chi_{u,v}$, *i.e.*,

$$\natural(pt) = \chi_{u,v} : \Diamond(pt) = K_0(\mathcal{VAR}) \to R[u, v] = H_*^{BM}(pt) \otimes R[u, v].$$

(3) *For the mapping $\pi_X : X \to pt$ to a point, for a certain distinguished element $\Delta_X \in \Diamond(X)$ we have*

$$\pi_{X*}(\natural(\Delta_X)) = \chi_{u,v}(X) \in R[u, v] \quad \text{and} \quad \pi_{X*}(\Delta_X) = [X] \in K_0(\mathcal{VAR}).$$

One reasonable candidate for the covariant functor $\Diamond$ is the following:

DEFINITION 3-4. (See [Lo2], for example.) *The relative Grothendieck group of X*, denoted by

$$K_0(\mathcal{VAR}/X),$$

is defined to be the free abelian group $\mathrm{Iso}(\mathcal{VAR}/X)$ generated by isomorphism classes $[V \xrightarrow{h} X]$ of morphisms $h : V \to X$ of complex algebraic varieties over X, modulo the additivity relation

$$[V \xrightarrow{h} X] = [V \setminus Z \xrightarrow{h_{|V \setminus Z}} X + [Z \xrightarrow{h_{|Z}} X] \text{ for any closed subvariety } Z \subset V;$$

in other words, $\text{Iso}(\mathcal{VAR}/X)$ modulo the subgroup generated by the elements of the form

$$[V \xrightarrow{h} X] - [Z \xrightarrow{h_{|Z}} X] - [V \setminus Z \xrightarrow{h_{|V \setminus Z}} X]$$

for any closed subvariety $Z \subset V$.

REMARK 3-5. For the category $\mathcal{VAR}(k)$ of algebraic varieties over any field, we can consider the same relative Grothendieck ring $K_0(\mathcal{VAR}(k)/X)$.

NOTE 1. $K_0(\mathcal{VAR}/pt) = K_0(\mathcal{VAR})$.

NOTE 2. $K_0(\mathcal{VAR}/X)^3$ is a covariant functor with the obvious pushforward: for a morphism $f : X \to Y$, the pushforward

$$f_* : K_0(\mathcal{VAR}/X) \to K_0(\mathcal{VAR}/Y)$$

is defined by

$$f_*([V \xrightarrow{h} X]) := [V \xrightarrow{f \circ h} Y].$$

NOTE 3. Although we do not need the ring structure on $K_0(\mathcal{VAR}/X)$ in later discussion, the fiber product gives a ring structure on it:

$$[V_1 \xrightarrow{h_1} X] \cdot [V_2 \xrightarrow{h_2} X] := [V_1 \times_X V_2 \xrightarrow{h_1 \times_X h_2} X].$$

NOTE 4. If $\Diamond(X) = K_0(\mathcal{VAR}/X)$, the distinguished element Δ_X is the isomorphism class of the identity map:

$$\Delta_X = [X \xrightarrow{id_X} X].$$

If we impose one more requirement in Problems 3-2 and 3-3, we can find the answer. The newcomer is the *normalization condition* (or *"smooth condition"*) that for nonsingular X we have

$$\natural(\Delta_X) = c\ell(TX) \cap [X]$$

for a certain normalized multiplicative characteristic class $c\ell$ of complex vector bundles. Note that $c\ell$ is a polynomial in the Chern classes such that it satisfies the normalization condition. Here "normalized" means that $c\ell(E) = 1$ for any trivial bundle E and "multiplicative" means that $c\ell(E \oplus F) = c\ell(E)c\ell(F)$, which is called the *Whitney sum formula*. In connection with the Whitney sum formula, in the analytic or algebraic context, one asks for this multiplicativity for a short exact sequence of vector bundles (which splits only in the topological context):

$$c\ell(E) = c\ell(E')c\ell(E'') \quad \text{for} \quad 1 \to E' \to E \to E'' \to 1.$$

---

[3] According to a recent paper by M. Kontsevich ("Notes on motives in finite characteristic", math.AG/0702206), Vladimir Drinfeld calls an element of $K_0(\mathcal{VAR}/X)$ "poor man's motivic function".

The normalization condition requirement is natural, in the sense that the other well-known/studied characteristic homology classes of possibly singular varieties are formulated as natural transformations satisfying such a normalization condition, as recalled later. Also, as discussed later (see Conjecture 6-1), this seemingly strong requirement of the normalization condition could be eventually dropped.

OBSERVATION 3-6. *Let* $\pi_X : X \to pt$ *be the mapping to a point. It follows from the naturality of* $\natural$ *and the normalization condition that*

$$c_{\mathrm{alg}}([X]) = \natural\big(\pi_{X*}([X \xrightarrow{\mathrm{id}_X} X])\big) = \pi_{X*}\big(\natural([X \xrightarrow{\mathrm{id}_X} X])\big) = \pi_{X*}\big(c\ell(TX) \cap [X]\big).$$

for any nonsingular variety X. Therefore the normalization condition on nonsingular varieties implies that for a nonsingular variety X the algebraic counting $c_{\mathrm{alg}}(X)$ has to be the characteristic number or Chern number [Ful; MiSt]. This is another requirement on c_{alg}, but an inevitable one if we want to capture it functorially (à la Grothendieck–Riemann–Roch) together with the normalization condition above for smooth varieties.

The normalization condition turns out to be essential, and in fact it automatically determines the characteristic class $c\ell$ as follows, if we consider the bigger ring $\mathbb{Q}[y]$ instead of $\mathbb{Z}[y]$:

PROPOSITION 3-7. *If the normalization condition is imposed in Problems 3-2 and 3-3, the multiplicative characteristic class* $c\ell$ *with coefficients in* $\mathbb{Q}[y]$ *has to be the generalized Todd class, or the Hirzebruch class* T_y, *defined as follows: for a complex vector bundle* V,

$$T_y(V) := \prod_{i=1}^{\mathrm{rank}\,V} \left(\frac{\alpha_i(1+y)}{1 - e^{-\alpha_i(1+y)}} - \alpha_i y \right)$$

where the α_i *are the Chern roots of the vector bundle:* $c(V) = \prod_{i=1}^{\mathrm{rank}\,V} (1 + \alpha_i)$.

PROOF. The multiplicativity of $c\ell$ guarantees that if X and Y are smooth compact varieties, then

$$\pi_{X \times Y *}(c\ell(T(X \times Y)) \cap [X \times Y]) = \pi_{X*}(c\ell(TX) \cap [X]) \cdot \pi_{Y*}(c\ell(TY) \cap [Y]).$$

In other words, the Chern number is multiplicative, i.e., it is compatible with the multiplicativity of c_{alg}. Now Hirzebruch's theorem [Hi, Theorem 10.3.1] says that if the multiplicative Chern number defined by a multiplicative characteristic class $c\ell$ with coefficients in $\mathbb{Q}[y]$ satisfies that the corresponding characteristic number of the complex projective space $\mathbb{P}^n$ is equal to $1 - y + y^2 - y^3 + \cdots + (-y)^n$, then the multiplicative characteristic class $c\ell$ has to be the generalized Todd class, i.e., the Hirzebruch class T_y above. $\square$

REMARK 3-8. In other words, in a sense $c_{alg}(\mathbb{C}^1)$ uniquely determines the class version of the motivic characteristic c_{alg}, i.e., the motivic characteristic class. This is very similar to the fact foreseen that $c_{top}(\mathbb{R}^1) = -1$ uniquely determines the "topological counting" c_{top}.

The Hirzebruch class T_y specializes to the following important characteristic classes:

$$y = -1: \quad T_{-1}(V) = c(V) = \prod_{i=1}^{\text{rank } V} (1 + \alpha_i) \quad \text{(total Chern class)}$$

$$y = 0: \quad T_0(X) = td(V) = \prod_{i=1}^{\text{rank } V} \frac{\alpha_i}{1 - e^{-\alpha_i}} \quad \text{(total Todd class)}$$

$$y = 1: \quad T_1(X) = L(V) = \prod_{i=1}^{\text{rank } V} \frac{\alpha_i}{\tanh \alpha_i} \quad \text{(total Thom–Hirzebruch class)}$$

Now we are ready to state our answer to Problem 3-2, which is one of the main theorems of [BSY1]:

THEOREM 3-9 (MOTIVIC CHARACTERISTIC CLASSES). *Let y be an indeterminate.*

(1) *There exists a unique natural transformation*

$$T_{y*} : K_0(\mathcal{VAR}/X) \to H_*^{BM}(X) \otimes \mathbb{Q}[y]$$

satisfying the normalization condition that for a nonsingular variety X

$$T_{y*}([X \xrightarrow{\text{id}_X} X]) = T_y(TX) \cap [X].$$

(2) *For $X = pt$, the transformation $T_{y*} : K_0(\mathcal{VAR}) \to \mathbb{Q}[y]$ equals the Hodge–Deligne polynomial*

$$\chi_{y,-1} : K_0(\mathcal{VAR}) \to \mathbb{Z}[y] \subset \mathbb{Q}[y],$$

namely,

$$T_{y*}([X \to pt]) = \chi_{y,-1}([X]) = \sum_{i, p \geq 0} (-1)^i \dim_{\mathbb{C}}(\text{Gr}_F^p H_c^i(X, \mathbb{C}))(-y)^p.$$

$\chi_{y,-1}(X)$ *is simply denoted by $\chi_y(X)$.*

PROOF. (1) The main part is of course the existence of such a T_{y*}, the proof of which is outlined in a later section. Here we point out only the uniqueness of T_{y*}, which follows from resolution of singularities. More precisely it follows from two results:

(i) Nagata's compactification theorem, or, if we do not wish to use such a fancy result, the projective closure of affine subvarieties. We get the surjective homomorphism

$$A : \text{Iso}^{\text{prop}}(\mathcal{VAR}/X) \twoheadrightarrow K_0(\mathcal{VAR}/X),$$

where $\text{Iso}^{\text{prop}}(\mathcal{VAR}/X)$ is the free abelian group generated by the isomorphism class of *proper* morphisms to X.

(ii) Hironaka's resolution of singularities: it implies, by induction on dimension that any isomorphism class $[Y \xrightarrow{h} X]$ can be expressed as

$$\sum_V a_V [V \xrightarrow{h_V} X],$$

with V nonsingular and $h_V : V \to X$ proper. We get the surjective maps

$$\text{Iso}^{\text{prop}}(\mathcal{SM}/X) \twoheadrightarrow \text{Iso}^{\text{prop}}(\mathcal{VAR}/X);$$

therefore

$$B : \text{Iso}^{\text{prop}}(\mathcal{SM}/X) \twoheadrightarrow K_0(\mathcal{VAR}/X),$$

where $\text{Iso}^{\text{prop}}(\mathcal{SM}/X)$ is the free abelian group generated by the isomorphism class of *proper* morphisms from *smooth varieties* to X.

(iii) The normalization condition ("smooth condition") of page 390.

(iv) The naturality of T_{y*}.

The two surjective homomorphisms A and B also play key roles in the proof of the existence of T_{y*}.

(2) As pointed out in (ii), $K_0(\mathcal{VAR})$ is generated by the isomorphism classes of compact smooth varieties. On a nonsingular compact variety X we have

$$\chi_{y,-1}(X) = \sum_{p,q \geq 0} (-1)^q \dim_{\mathbb{C}} H^q(X; \Omega_X^p) y^p,$$

which is denoted by $\chi_y(X)$ and is called the Hirzebruch χ_y-genus. Next we have the *generalized Hirzebruch–Riemann–Roch Theorem* (gHRR), which says [Hi] that

$$\chi_y(X) = \int_X T_y(TX) \cap [X].$$

Since $\int_X T_y(TX) \cap [X] = \pi_{X*}(T_y(TX) \cap [X]) = T_{y*}([X \to pt])$, we have

$$T_{y*}([X \to pt]) = \chi_{y,-1}([X])$$

on generators of $K_0(\mathcal{VAR})$, and hence on all of $K_0(\mathcal{VAR})$; thus $T_{y*} = \chi_{y,-1}$.

□

REMARK 3-10. Problem 3-3 is slightly more general than Problem 3-2 in the sense that it involves two indeterminates u, v. However, the important keys are the normalization condition for smooth compact varieties and the fact that $\chi_{u,v}(\mathbb{P}^1) = 1 + uv + (uv)^2 + \cdots + (uv)^n$, which automatically implies that $c\ell = T_{-uv}$, as shown in the proof above. In fact, we can say more about u and v: either $u = -1$ or $v = -1$, as shown below (see also [Jo] — the arXiv version). Hence, we can conclude that for Problem 3-3 there is *no* such transformation $\sharp : K_0(\mathcal{VAR}/-) \to H_*^{BM}(-) \otimes R[u, v]$ with both intermediates u and v varying.

To show the claim about u and v, suppose that for X smooth and for a certain multiplicative characteristic class $c\ell$ we have

$$\chi_{u,v}(X) = \pi_{X*}(c\ell(TX) \cap [X]).$$

In particular, consider a smooth elliptic curve E and any d-fold covering

$$\pi : \widetilde{E} \to E$$

with $\widetilde{E}$ a smooth elliptic curve. Note that $T\widetilde{E} = \pi^*TE$ and

$$\chi_{u,v}(E) = \chi_{u,v}(\widetilde{E}) = 1 + u + v + uv = (1+u)(1+v).$$

Hence we have

$$
\begin{aligned}
(1+u)(1+v) = \chi_{u,v}(\widetilde{E}) &= \pi_{\widetilde{E}*}(c\ell(T\widetilde{E}) \cap [\widetilde{E}]) = \pi_{\widetilde{E}*}(c\ell(\pi^*TE) \cap [\widetilde{E}]) \\
&= \pi_{E*}\pi_*(c\ell(\pi^*TE) \cap [\widetilde{E}]) = \pi_{E*}(c\ell(TE) \cap \pi_*[\widetilde{E}]) \\
&= \pi_{E*}(c\ell(TE) \cap d[E]) = d \cdot \pi_{E*}(c\ell(TE) \cap [E]) \\
&= d \cdot \chi_{u,v}(E) = d(1+u)(1+v).
\end{aligned}
$$

Thus we get $(1+u)(1+v) = d(1+u)(1+v)$. Since $d \neq 0$, we must have that $(1+u)(1+v) = 0$, showing that $u = -1$ or $v = -1$.

REMARK 3-11. Note that $\chi_{u,v}(X)$ is symmetric in u and v; thus both special cases $u = -1$ and $v = -1$ give rise to the same $c\ell = T_y$. It suffices to check this for a compact nonsingular variety X. In fact this follows from the Serre duality.

REMARK 3-12. The heart of the mixed Hodge structure is certainly the existence of the weight filtration $W^\bullet$ and the Hodge–Deligne polynomial, i.e., the algebraic counting c_{alg}, involves the mixed Hodge structure, i.e., both the weight filtration $W^\bullet$ and the Hodge filtration $F_\bullet$. However, when one tries to capture c_{alg} *functorially*, only the Hodge filtration $F_\bullet$ gets involved; the weight filtration *does not*, as seen in the Hodge genus χ_y.

DEFINITION 3-13. For a possibly singular variety X, we call

$$T_{y*}(X) := T_{y*}([X \xrightarrow{\text{id}_X} X])$$

the *Hirzebruch class of* X.

COROLLARY 3-14. *The degree of the 0-dimensional component of the Hirze-bruch class of a compact complex algebraic variety X is just the Hodge genus:*

$$\chi_y(X) = \int_X T_{y*}(X).$$

This is another singular analogue of the gHRR theorem ($\chi_y = T_y$), which is a generalization of the famous Hirzebruch–Riemann–Roch Theorem (which was further generalized to the Grothendieck–Riemann–Roch Theorem):

$$\text{Hirzebruch–Riemann–Roch:} \quad p_a(X) = \int_X td(TX) \cap [X],$$

with $p_a(X)$ the arithmetic genus and $td(V)$ the original Todd class. Noticing the above specializations of χ_y and $T_y(V)$, this gHRR is a unification of the following three well-known theorems:

$$y = -1: \quad \chi(X) = \int_X c(X) \cap [X] \quad \text{(Gauss–Bonnet, or Poincaré–Hopf)}$$

$$y = 0: \quad p_a(X) = \int_X td(X) \cap [X] \quad \text{(Hirzebruch–Riemann–Roch)}$$

$$y = 1: \quad \sigma(X) = \int_X L(X) \cap [X] \quad \text{(Hirzebruch's Signature Theorem)}$$

4. Proofs of the existence of the motivic characteristic class T_{y*}

Our motivic characteristic class transformation

$$T_{y*}: K_0(\mathcal{VAR}/X) \to H_*^{BM}(X) \otimes \mathbb{Q}[y]$$

is obtained as the composite

$$T_{y*} = \widetilde{td_{*(y)}^{BFM}} \circ \Lambda_y^{\text{mot}}$$

of the natural transformations

$$\Lambda_y^{\text{mot}}: K_0(\mathcal{VAR}/X) \to G_0(X) \otimes \mathbb{Z}[y]$$

and

$$\widetilde{td_{*(y)}^{BFM}}: G_0(X) \otimes \mathbb{Z}[y] \to H_*^{BM}(X) \otimes \mathbb{Q}[y, (1+y)^{-1}].$$

Here, to describe $\widetilde{td_{*(y)}^{BFM}}$, we need to recall the following Baum–Fulton–MacPherson's Riemann–Roch or Todd class for singular varieties [BFM1]:

THEOREM 4-1. *There exists a unique natural transformation*

$$td_*^{BFM} : G_0(-) \to H_*^{BM}(-) \otimes \mathbb{Q}$$

such that for a smooth X

$$td_*^{BFM}(\mathcal{O}_X) = td(TX) \cap [X].$$

Here $G_0(X)$ is the Grothendieck group of coherent sheaves on X, which is a covariant functor with the pushforward $f_* : G_0(X) \to G_0(Y)$ for a proper morhphism $f : X \to Y$ defined by

$$f_!(\mathcal{F}) = \sum_j (-1)^j [R^j f_* \mathcal{F}].$$

Now set

$$td_*^{BFM}(X) := td_*^{BFM}(\mathcal{O}_X);$$

this is called the Baum–Fulton–MacPherson Todd class of X. Then

$$p_a(X) = \chi(X, \mathcal{O}_X) = \int_X td_*^{BFM}(X) \quad \text{(HRR-type theorem)}.$$

Let

$$td_{*i}^{BFM} : G_0(X) \xrightarrow{td_*^{BFM}} H_*^{BM}(X) \otimes \mathbb{Q} \xrightarrow{\text{projection}} H_{2i}^{BM}(X) \otimes \mathbb{Q}$$

be the i-th (i.e., $2i$-dimensional) component of td_*^{BFM}. Then the above *twisted BFM-Todd class transformation* or *twisted BFM-RR transformation* (cf. [Y4])

$$\widetilde{td_{*(y)}^{BFM}} : G_0(X) \otimes \mathbb{Z}[y] \to H_*^{BM}(X) \otimes \mathbb{Q}[y, (1+y)^{-1}]$$

is defined by

$$\widetilde{td_{*(y)}^{BFM}} := \sum_{i \geq 0} \frac{1}{(1+y)^i} td_{*i}^{BFM}.$$

In this process, $\Lambda_y^{\text{mot}} : K_0(\mathcal{VAR}/X) \to G_0(X) \otimes \mathbb{Z}[y]$ is the key. This object was denoted by mC_* in our paper [BSY1] and called the *motivic Chern class*. In this paper, we use the notation Λ_y^{mot} to emphasize the following property of it:

THEOREM 4-2 ("MOTIVIC" λ_y-CLASS TRANSFORMATION). *There exists a unique natural transformation*

$$\Lambda_y^{\text{mot}} : K_0(\mathcal{VAR}/X) \to G_0(X) \otimes \mathbb{Z}[y]$$

satisfying the normalization condition that for smooth X

$$\Lambda_y^{\text{mot}}([X \xrightarrow{\text{id}} X]) = \sum_{p=0}^{\dim X} [\Omega_X^p] y^p = \lambda_y(T^*X) \otimes [\mathcal{O}_X].$$

Here $\lambda_y(T^*X) = \sum_{p=0}^{\dim X}[\Lambda^p(T^*X)]y^p$ and $\otimes[\mathcal{O}_X] : K^0(X) \cong G_0(X)$ is an isomorphism for smooth X, i.e., taking the sheaf of local sections.

THEOREM 4-3. *The natural transformation*

$$T_{y*} := \widetilde{td^{BFM}_{*(y)}} \circ \Lambda_y^{\mathrm{mot}} : K_0(\mathcal{VAR}/X) \to H_*^{BM}(X) \otimes \mathbb{Q}[y]$$

$$\subset H_*(X) \otimes \mathbb{Q}[y, (1+y)^{-1}]$$

satisfies the normalization condition that for smooth X

$$T_{y*}([X \xrightarrow{\mathrm{id}} X]) = T_y(TX) \cap [X].$$

Hence such a natural transformation is unique.

REMARK 4-4. Why is the image of T_{y*} in $H_*^{BM}(X) \otimes \mathbb{Q}[y]$? Even though the target of

$$\widetilde{td^{BFM}_{*(y)}} : G_0(X) \otimes \mathbb{Z}[y] \to H_*(X) \otimes \mathbb{Q}[y, (1+y)^{-1}]$$

is $H_*^{BM}(X) \otimes \mathbb{Q}[y, (1+y)^{-1}]$, the image of $T_{y*} = \widetilde{td^{BFM}_{*(y)}} \circ \Lambda_y^{\mathrm{mot}}$ is contained in $H_*(X) \otimes \mathbb{Q}[y]$. Indeed, as mentioned, by Hironaka's resolution of singularities, induction on dimension, the normalization condition, and the naturality of T_{y*}, the domain $K_0(\mathcal{VAR}/X)$ is generated by $[V \xrightarrow{h} X]$ with h proper and V smooth. Hence

$$T_{y*}([V \xrightarrow{h} X]) = T_{y*}(h_*[V \xrightarrow{\mathrm{id}_V} V]) = h_*(T_{y*}([V \xrightarrow{\mathrm{id}_V} V]) \in H_*^{BM}(X) \otimes \mathbb{Q}[y].$$

PROOF OF THEOREM 4-3. In [BSY1] we gave a slick way of proving this. Here we give a nonslick, direct one. Let X be smooth.

$$\widetilde{td^{BFM}_{*(y)}} \circ \Lambda_y^{\mathrm{mot}}([X \xrightarrow{\mathrm{id}} X])$$

$$= \widetilde{td^{BFM}_{*(y)}}(\lambda_y(\Omega_X)) = \sum_{i \geq 0} \frac{1}{(1+y)^i} td_{*i}^{BFM}(\lambda_y(\Omega_X))$$

$$= \sum_{i \geq 0} \frac{1}{(1+y)^i}(td_*^{BFM}(\lambda_y(\Omega_X)))_i$$

$$= \sum_{i \geq 0} \frac{1}{(1+y)^i}(td_*^{BFM}(\lambda_y(T^*X) \otimes [\mathcal{O}_X]))_i$$

$$= \sum_{i \geq 0} \frac{1}{(1+y)^i}(ch(\lambda_y(T^*X)) \cap td_*^{BFM}(\mathcal{O}_X))_i$$

$$= \sum_{i \geq 0} \frac{1}{(1+y)^i}(ch(\lambda_y(T^*X)) \cap (td(TX) \cap [X]))_i$$

$$= \sum_{i \geq 0} \frac{1}{(1+y)^i}\left(\prod_{j=1}^{\dim X}(1 + ye^{-\alpha_j}) \prod_{j=1}^{\dim X} \frac{\alpha_j}{1-e^{-\alpha_j}}\right)_{\dim X-i} \cap [X].$$

Furthermore we have

$$\frac{1}{(1+y)^i}\left(\prod_{j=1}^{\dim X}(1+ye^{-\alpha_j})\prod_{j=1}^{\dim X}\frac{\alpha_j}{1-e^{-\alpha_j}}\right)_{\dim X-i}$$

$$=\frac{(1+y)^{\dim X}}{(1+y)^i}\left(\prod_{j=1}^{\dim X}\frac{1+ye^{-\alpha_j}}{1+y}\prod_{j=1}^{\dim X}\frac{\alpha_j}{1-e^{-\alpha_j}}\right)_{\dim X-i}$$

$$=(1+y)^{\dim X-i}\left(\prod_{j=1}^{\dim X}\frac{1+ye^{-\alpha_j}}{1+y}\prod_{j=1}^{\dim X}\frac{\alpha_j}{1-e^{-\alpha_j}}\right)_{\dim X-i}$$

$$=\left(\prod_{j=1}^{\dim X}\frac{1+ye^{-\alpha_j}}{1+y}\prod_{j=1}^{\dim X}\frac{\alpha_j(1+y)}{1-e^{-\alpha_j(1+y)}}\right)_{\dim X-i}$$

$$=\left(\prod_{j=1}^{\dim X}\frac{1+ye^{-\alpha_j}}{1+y}\cdot\frac{\alpha_j(1+y)}{1-e^{-\alpha_j(1+y)}}\right)_{\dim X-i}$$

$$=\left(\prod_{j=1}^{\dim X}\frac{\alpha_j(1+y)}{1-e^{-\alpha_j(1+y)}}-\alpha_j y\right)_{\dim X-i}$$

$$=\left(T_y(TX)\right)_{\dim X-i}.$$

Therefore $\widetilde{td^{BFM}_{*(y)}}\circ\Lambda^{mot}_y([X\xrightarrow{id}X])=T_y(TX)\cap[X]$. □

It remains to show Theorem 4-2. There are at least three proofs, each with its own advantages.

FIRST PROOF (using Saito's theory of mixed Hodge modules [Sa1; Sa2; Sa3; Sa4; Sa5; Sa6]).

Even though Saito's theory is very complicated, this approach turns out to be useful and for example has been used in recent works of Cappell, Libgober, Maxim, Schürmann and Shaneson [CLMS1; CLMS2; CMS1; CMS2; CMSS; MS1; MS2], related to intersection (co)homology. Here we recall only the ingredients which we need to define Λ^{mot}_y:

MHM1 : To X one can associate an abelian category of *mixed Hodge modules* $MHM(X)$, together with a functorial pullback f^* and pushforward $f_!$ on the level of bounded derived categories $D^b(MHM(X))$ for any (not necessarily proper) map. These natural transformations are functors of triangulated categories.

MHM2 : Let $i:Y\to X$ be the inclusion of a closed subspace, with open complement $j:U:=X\setminus Y\to X$. Then one has for $M\in D^bMHM(X)$ a distinguished triangle

$$j_!j^*M\to M\to i_!i^*M\xrightarrow{[1]}.$$

<u>MHM3</u> : For all $p \in \mathbb{Z}$ one has a "filtered De Rham complex" functor of triangulated categories

$$\mathrm{gr}_p^F DR : D^b(MHM(X)) \to D^b_{\mathrm{coh}}(X)$$

commuting with proper pushforward. Here $D^b_{\mathrm{coh}}(X)$ is the bounded derived category of sheaves of $\mathcal{O}_X$-modules with coherent cohomology sheaves. Moreover, $\mathrm{gr}_p^F DR(M) = 0$ for almost all p and $M \in D^b MHM(X)$ fixed.

<u>MHM4</u> : There is a distinguished element $\mathbb{Q}^H_{pt} \in MHM(pt)$ such that

$$\mathrm{gr}^F_{-p} DR(\mathbb{Q}^H_X) \simeq \Omega^p_X[-p] \in D^b_{\mathrm{coh}}(X)$$

for X smooth and pure-dimensional. Here $\mathbb{Q}^H_X := \pi_X^* \mathbb{Q}^H_{pt}$ for $\pi_X : X \to pt$ a constant map, with $\mathbb{Q}^H_{pt}$ viewed as a complex concentrated in degree zero.

The transformations above are functors of triangulated categories; thus they induce functors even on the level of *Grothendieck groups of triangulated categories*, which we denote by the same name. Note that for these *Grothendieck groups* we have isomorphisms

$$K_0(D^b MHM(X)) \simeq K_0(MHM(X)) \quad \text{and} \quad K_0(D^b_{\mathrm{coh}}(X)) \simeq G_0(X)$$

by associating to a complex its alternating sum of cohomology objects.

Now we are ready for the transformations mH and $\mathrm{gr}^F_{-*} DR$. Define

$$mH : K_0(\mathcal{VAR}/X) \to K_0(MHM(X)) \quad \text{by} \quad mH([V \xrightarrow{f} X]) := [f_! \mathbb{Q}^H_V].$$

In a sense $K_0(MHM(X))$ is like the abelian group of "mixed-Hodge-module constructible functions", with the class of $\mathbb{Q}^H_X$ as a "constant function" on X. The well-definedness of mH, i.e., the additivity relation follows from property (MHM2). By (MHM3) we get the following homomorphism commuting with proper pushforward:

$$\mathrm{gr}^F_{-*} DR : K_0(MHM(X)) \to G_0(X) \otimes \mathbb{Z}[y, y^{-1}]$$

defined by

$$\mathrm{gr}^F_{-*} DR([M]) := \sum_p [\mathrm{gr}^F_{-p} DR(M)] \cdot (-y)^p$$

Then we define our Λ_y^{mot} as the composite of these two natural transformations:

$$\Lambda_y^{\mathrm{mot}} := \mathrm{gr}^F_{-*} DR \circ mH : K_0(\mathcal{VAR}/X)$$
$$\xrightarrow{mH} K_0(MHM(X)) \xrightarrow{\mathrm{gr}^F_{-*} DR} G_0(X) \otimes \mathbb{Z}[y].$$

By (MHM4), for X smooth and pure-dimensional we have

$$\mathrm{gr}^F_{-*} DR \circ mH([\mathrm{id}_X]) = \sum_{p=0}^{\dim X} [\Omega^p_X] \cdot y^p \in G_0(X) \otimes \mathbb{Z}[y].$$

Thus we get the unique existence of the "motivic" λ_y-class transformation Λ^{mot}_y. $\square$

SECOND PROOF (using the filtered Du Bois complexes [DB]). Recall the surjective homomorphism

$$A : \mathrm{Iso}^{\mathrm{prop}}(\mathcal{VAR}/X) \twoheadrightarrow K_0(\mathcal{VAR}/X).$$

We can describe its kernel as follows:

THEOREM 4-5. $K_0(\mathcal{VAR}/X)$ is isomorphic to the quotient of $\mathrm{Iso}^{\mathrm{pro}}(\mathcal{VAR}/X)$ modulo the "acyclicity" relation

$$[\varnothing \to X] = 0 \quad and \quad [\tilde{X}' \to X] - [\tilde{Z}' \to X] = [X' \to X] - [Z' \to X], \quad (\mathrm{ac})$$

for any cartesian diagram

$$
\begin{array}{ccc}
\tilde{Z}' & \longrightarrow & \tilde{X}' \\
\downarrow & & \downarrow{\scriptstyle q} \\
Z' & \overset{i}{\longrightarrow} & X' \longrightarrow X,
\end{array}
$$

with q proper, i a closed embedding, and $q : \tilde{X}'\backslash\tilde{Z}' \to X'\backslash Z'$ an isomorphism.

For a proper map $X' \to X$, consider the filtered Du Bois complex

$$(\underline{\Omega}^*_{X'}, F),$$

which has the following properties:

(1) $\underline{\Omega}^*_{X'}$ is a resolution of the constant sheaf $\mathbb{C}$.
(2) $\mathrm{gr}^p_F(\underline{\Omega}^*_{X'}) \in D^b_{\mathrm{coh}}(X')$.
(3) Let $DR(\mathcal{O}_{X'}) = \Omega^*_{X'}$ be the de Rham complex of X' with σ being the stupid filtration. Then there is a filtered morphism

$$\lambda : (\Omega^*_{X'}, \sigma) \to (\underline{\Omega}^*_{X'}, F).$$

If X' is smooth, this is a filtered quasi-isomorphism.

Note that $G_0(X') \cong K_0(D^b_{\mathrm{coh}}(X'))$. Let us define

$$[\mathrm{gr}^p_F(\underline{\Omega}^*_{X'})] := \sum_i (-1)^i H^i(\mathrm{gr}^p_F(\underline{\Omega}^*_{X'})) \in K_0(D^b_{\mathrm{coh}}(X')) = G_0(X').$$

THEOREM 4-6. *The transformation*

$$\Lambda_y^{\mathrm{mot}} : K_0(\mathcal{VAR}/X) \to G_0(X) \otimes \mathbb{Z}[y]$$

defined by

$$\Lambda_y^{\mathrm{mot}}([X' \xrightarrow{h} X]) := \sum_p h_*[\mathrm{gr}_F^p(\underline{\Omega}_{X'}^*)](-y)^p$$

is well-defined and is the unique natural transformation satisfying the normalization condition that for smooth X

$$\Lambda_y^{\mathrm{mot}}([X \xrightarrow{\mathrm{id}_X} X]) = \sum_{p=0}^{\dim X} [\Omega_X^p] y^p = \lambda_y(T^*X) \otimes \mathcal{O}_X.$$

PROOF. The well-definedness follows from the fact that Λ_y^{mot} preserves the acyclicity relation above [DB]. Then uniqueness follows from resolution of singularities and the normalization condition for smooth varieties.   □

REMARK 4-7. When X is smooth, we have

$$[\mathrm{gr}_\sigma^p(\underline{\Omega}_X^*)] = (-1)^p[\Omega_X^p]!$$

That is why we need $(-y)^p$, instead of y^p, in the definition of $\Lambda_y^{\mathrm{mot}}([X' \xrightarrow{h} X])$.

REMARK 4-8. When $y = 0$, we have the natural transformation

$$\Lambda_0^{\mathrm{mot}} : K_0(\mathcal{VAR}/X) \to G_0(X) \quad \text{defined by} \quad \Lambda_0^{\mathrm{mot}}([X' \xrightarrow{h} X]) = h_*[\mathrm{gr}_F^0(\underline{\Omega}_{X'}^*)]$$

satisfying the normalization condition that for a smooth X

$$\Lambda_0^{\mathrm{mot}}([X \xrightarrow{\mathrm{id}_X} X]) = [\mathcal{O}_X].$$ □

THIRD PROOF (using Bittner's theorem on $K_0(\mathcal{VAR}/X)$ [Bi]). Recall the surjective homomorphism

$$B : \mathrm{Iso}^{\mathrm{prop}}(\mathcal{SM}/X) \twoheadrightarrow K_0(\mathcal{VAR}/X).$$

Its kernel is identified by F. Bittner and E. Looijenga as follows [Bi]:

THEOREM 4-9.   *The group $K_0(\mathcal{VAR}/X)$ is isomorphic to the quotient of $\mathrm{Iso}^{\mathrm{prop}}(\mathcal{SM}/X)$ (the free abelian group generated by the isomorphism classes of proper morphisms from smooth varieties to X) by the "blow-up" relation*

$$[\varnothing \to X] = 0 \quad and \quad [\mathrm{Bl}_Y X' \to X] - [E \to X] = [X' \to X] - [Y \to X], \quad \text{(bl)}$$

for any cartesian diagram

$$
\begin{array}{ccc}
E & \xrightarrow{\ i'\ } & \mathrm{Bl}_Y\, X' \\
\downarrow{\scriptstyle q'} & & \downarrow{\scriptstyle q} \\
Y & \xrightarrow{\ i\ } & X' \xrightarrow{\ f\ } X ,
\end{array}
$$

with i a closed embedding of smooth (pure-dimensional) spaces and $f : X' \to X$ proper. Here $\mathrm{Bl}_Y X' \to X'$ is the blow-up of X' along Y with exceptional divisor E. Note that all these spaces over X are also smooth (and pure-dimensional and/or quasiprojective, if this is the case for X' and Y).

The proof of this theorem requires the Weak Factorization Theorem, due to D. Abramovich, K. Karu, K. Matsuki and J. Włodarczyk [AKMW] (see also [Wlo]).   □

COROLLARY 4-10. (1) *Let $B_* : \mathcal{VAR}/k \to \mathcal{AB}$ be a functor from the category* var$/k$ *of (reduced) separated schemes of finite type over* $\mathrm{spec}(k)$ *to the category of abelian groups, which is covariantly functorial for proper morphisms, with $B_*(\varnothing) := \{0\}$. Assume we can associate to any (quasiprojective) smooth space $X \in ob(\mathcal{VAR}/k)$ of pure dimension a distinguished element*

$$
\phi_X \in B_*(X)
$$

such that $h_*(\phi_{X'}) = \phi_X$ for any isomorphism $h : X' \to X$. There exists a unique natural transformation

$$
\Phi : \mathrm{Iso}^{\mathrm{prop}}(\mathcal{SM}/-) \to B_*(-)
$$

satisfying the "normalization" condition that for any smooth X

$$
\Phi([X \xrightarrow{\ \mathrm{id}_X\ } X]) = \phi_X .
$$

(2) *Let $B_* : \mathcal{VAR}/k \to \mathcal{AB}$ and ϕ_X be as above and furthermore we assume that*

$$
q_*(\phi_{\mathrm{Bl}_Y\, X}) - i_* q'_*(\phi_E) = \phi_X - i_*(\phi_Y) \in B_*(X)
$$

for any cartesian blow-up diagram as in the above Bittner's theorem with $f = \mathrm{id}_X$. Then there exists a unique natural transformation

$$
\Phi : K_0(\mathcal{VAR}/-) \to B_*(-)
$$

satisfying the "normalization" condition that for any smooth X

$$
\Phi([X \xrightarrow{\ \mathrm{id}_X\ } X]) = \phi_X .
$$

We will now use Corollary 4-10(2) to conclude our third proof. Consider the coherent sheaf $\Omega_X^p \in G_0(X)$ of a smooth X as the distinguished element ϕ_X of a smooth X. It follows from M. Gros's work [Gr] or the recent work of Guillén and Navarro Aznar [GNA] that it satisfies the blow-up relation

$$q_*(\Omega_{\mathrm{Bl}_Y X}^p) - i_* q_*'(\Omega_E^p) = \Omega_X^p - i_*(\Omega_Y^p) \in G_0(X),$$

which in turn implies a blow-up relation for the λ_y-class:

$$q_*(\lambda_y(\Omega_{\mathrm{Bl}_Y X})) - i_* q_*'(\lambda_y(\Omega_E)) = \lambda_y(\Omega_X) - i_*(\lambda_y(\Omega_Y)) \in G_0(X) \otimes \mathbb{Z}[y].$$

Therefore Corollary 4-10(2) implies this:

THEOREM 4-11. *The transformation*

$$\Lambda_y^{\mathrm{mot}} : K_0(\mathcal{VAR}/X) \to G_0(X) \otimes \mathbb{Z}[y]$$

defined by

$$\Lambda_y^{\mathrm{mot}}([X' \xrightarrow{h} X]) := h_*\left(\sum_{p \geq 0}[\Omega_{X'}^p]y^p\right),$$

where X' is smooth and $h : X' \to X$ is proper, is well-defined and is a unique natural transformation satisying the normalization condition that for smooth X

$$\Lambda_y^{\mathrm{mot}}([X \xrightarrow{\mathrm{id}_X} X]) = \sum_{p=0}^{\dim X}[\Omega_X^p]y^p = \lambda_y(T^*X) \otimes \mathcal{O}_X.$$

REMARK 4-12. The virtual Poincaré polynomial β_t (Remark 2-11) for the category $\mathcal{VAR}(\mathbb{R})$ of real algebraic varieties is the unique homomorphism

$$\beta_t : K_0(\mathcal{VAR}(\mathbb{R})) \to \mathbb{Z}[t] \quad \text{such that } \beta_t(\mathbb{R}^1) = t$$

and $\beta_t(X) = P_t(X)$ is the classical or usual topological Poincaré polynomial for compact nonsingular varieties. The proof of the existence of β_i, thus β_t, also uses Corollary 4-10(2); see [MP1]. Speaking of the Poincaré polynomial $P_t(X)$, we emphasize that this polynoimal cannot be a topological counting at all in the category of topological spaces, simply because the argument in the proof of Proposition 2-4 does not work! The Poincaré polynomial $P_t(X)$ is certainly a *multiplicative* topological invariant, but not an *additive* one.

REMARK 4-13. The virtual Poincaré polynomial $\beta_t : K_0(\mathcal{VAR}(\mathbb{R})) \to \mathbb{Z}[t]$ is the *unique* extension of the Poincaré polynomial $P_t(X)$ to arbitrary varieties. Note that if we consider complex algebraic varieties, the virtual Poincaré polynomial

$$\beta_t : K_0(\mathcal{VAR}) \to \mathbb{Z}[t]$$

is equal to the following motivic characteristic, using only the weight filtration:

$$w\chi(X) = \sum (-1)^i \dim_{\mathbb{C}}\left(\mathrm{Gr}_q^W H_c^i(X, \mathbb{C})\right) t^q,$$

because on any smooth compact complex algebraic variety X they are all the same: $\beta_t(X) = P_t(X) = w\chi(X)$. These last equalities follow from the fact that the Hodge structures on $H^k(X, \mathbb{Q})$ are of pure weight k.

This "weight filtration" motivic characteristic $w\chi(X)$ is equal to the specialization $\chi_{-t,-t}$ of the Hodge–Deligne polynomial for $(u, v) = (-t, -t)$. This observation implies that there is *no class version* of the complex virtual Poincaré polynomial $\beta_t : K_0(\mathcal{VAR}) \to \mathbb{Z}[t]$. In other words, there is no natural transformation

$$\natural : K_0(\mathcal{VAR}/-) \to H_*^{BM}(-) \otimes \mathbb{Z}[t]$$

satisfying the conditions that

- if X is smooth and compact, then $\natural([X \xrightarrow{\mathrm{id}_X} X]) = c\ell(TX) \cap [X]$ for some multiplicative characteristic class of complex vector bundles; and
- $\natural(pt) = \beta_t : K_0(\mathcal{VAR}) \to \mathbb{Z}[t]$.

This is because $\beta_t(X) = \chi_{-t,-t}(X)$ for a smooth compact complex algebraic variety X (hence for all X), and so, as in Remark 3-10, one can conclude that $(-t, -t) = (-1, -1)$. Thus t has to be equal to 1 and cannot be allowed to vary. In other words, the only chance for such a class version is when $t = 1$, which gives the Euler–Poincaré characteristic $\chi : K_0(\mathcal{VAR}) \to \mathbb{Z}$. In that case, we do have the Chern class transformation

$$c_* : K_0(\mathcal{VAR}/-) \to H_*^{BM}(-; \mathbb{Z}).$$

This follows again from Corollary 4-10(2) and the blow-up formula of Chern class [Ful].

REMARK 4-14. The same discussion as in Remark 4-13 can be applied to the context of real algebraic varieties, i.e., the same example for real elliptic curves leads us to the conclusion that $t = 1$ for β_t satisfying the corresponding normalization condition for a normalized multiplicative characteristic class. This class has to be a polynomial in the Stiefel–Whitney classes, and we end up with the Stiefel–Whitney homology class w_*, which also satisfies the corresponding blow-up formula.

REMARK 4-15 (POOR MAN'S MOTIVIC CHARACTERISTIC CLASS). If we use the much simpler covariant functor $\mathrm{Iso}^{\mathrm{prop}}(\mathcal{SM}/X)$ above (the abelian group of "poor man's motivic functions"), we can get the following "poor man's motivic characteristic class" for any characteristic class $c\ell$ of vector bundles: Let $c\ell$ be *any* characteristic class of vector bundles with coefficient ring K. There exists a unique natural transformation

$$cl_* : \mathrm{Iso}^{\mathrm{prop}}(\mathcal{SM}/-) \to H_*^{BM}(-) \otimes K$$

satisfying the normalization condition that for any smooth variety X,

$$cl_*([X \xrightarrow{\mathrm{id}_X} X]) = c\ell(TX) \cap [X].$$

There is a bivariant theoretical version of $\mathrm{Iso}^{\mathrm{prop}}(\mathcal{SM}/X)$ (see [Y7]); a good reference for it is Fulton and MacPherson's AMS memoir [FM].

5. Chern class, Todd class and L-class of singular varieties: towards a unification

Our next task is to describe another main theorem of [BSY1], to the effect that our motivic characteristic class T_{y*} is, in a sense, a unification of MacPherson's Chern class, the Todd class of Baum, Fulton, and MacPherson (discussed in the previous section), and the L-class of singular varieties of Cappell and Shaneson. Let's briefly review these classes:

MacPherson's Chern class [M1]

THEOREM 5-1. *There exists a unique natural transformation*

$$c_*^{\mathrm{Mac}} : F(-) \to H_*^{BM}(-)$$

such that, for smooth X,

$$c_*^{\mathrm{Mac}}(\mathbb{1}_X) = c(TX) \cap [X].$$

Here $F(X)$ is the abelian group of constructible functions, which is a covariant functor with the pushforward $f_* : F(X) \to F(Y)$ for a proper morphism $f : X \to Y$ defined by

$$f_*(\mathbb{1}_W)(y) = \chi_c(f^{-1}(y) \cap W).$$

We call $c_*^{\mathrm{Mac}}(X) := c_*^{\mathrm{Mac}}(\mathbb{1}_X)$ the MacPherson's Chern class of X, or the Chern–Schwartz–MacPherson class. We have

$$\chi(X) = \int_X c_*^{\mathrm{Mac}}(X).$$

The Todd class of Baum, Fulton, and MacPherson [BFM1]

THEOREM 5-2. *There exists a unique natural transformation*

$$td_*^{BFM} : G_0(-) \to H_*^{BM}(-) \otimes \mathbb{Q}$$

such that, for smooth X,

$$td_*^{BFM}(\mathcal{O}_X) = td(TX) \cap [X].$$

Here $G_0(X)$ is the Grothendieck group of coherent sheaves on X, which is a covariant functor with the pushforward $f_* : G_0(X) \to G_0(Y)$ for a proper morphism $f : X \to Y$ defined by

$$f_!(\mathcal{F}) = \sum_j (-1)^j [R^j f_* \mathcal{F}].$$

We call $td_*^{BFM}(X) := td_*^{BFM}(\mathcal{O}_X)$ the Baum–Fulton–MacPherson Todd class of X, and we have

$$p_a(X) = \chi(X, \mathcal{O}_X) = \int_X td_*^{BFM}(X).$$

The L-class of Cappell and Shaneson [CS1; Sh] (cf. [Y4])

THEOREM 5-3. *There exists a unique natural transformation*

$$L_*^{CS} : \Omega(-) \to H_*^{BM}(-) \otimes \mathbb{Q}$$

such that, for smooth X,

$$L_*^{CS}(\mathcal{IC}_X) = L(TX) \cap [X].$$

Here $\Omega(X)$ is the abelian group of Youssin's cobordism classes of self-dual constructible complexes of sheaves on X.

We call $L_*^{GM}(X) := L_*^{CS}(\mathcal{IC}_X)$ the Goresky–MacPherson homology L-class of X. The Goresky–MacPherson theorem [GM] says that

$$\sigma^{GM}(X) = \int_X L_*^{GM}(X).$$

We now explain in what sense our motivic characteristic class transformation

$$T_{y*} : K_0(\mathcal{VAR}/X) \to H_*^{BM}(X) \otimes \mathbb{Q}[y]$$

unifies these three characteristic classes of singular varieties, providing a kind of partial positive answer to MacPherson's question[4] of *whether there is a unified theory of characteristic classes of singular varieties.*

---

[4]Posed in his survey talk [M2] at the Ninth Brazilian Mathematics Colloquium in 1973.

THEOREM 5-4 (UNIFIED FRAMEWORK FOR CHERN, TODD AND HOMOLOGY L-CLASSES OF SINGULAR VARIETIES).

$y = -1$: *There exists a unique natural transformation* $\varepsilon : K_0(\mathcal{VAR}/-) \to F(-)$ *such that, for X nonsingular, $\varepsilon([\mathrm{id} : X \to X]) = \mathbb{1}_X$, and the following diagram commutes*:

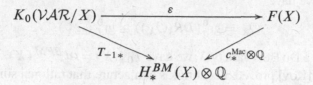

$y = 0$: *There exists a unique natural transformation* $\gamma : K_0(\mathcal{VAR}/-) \to G_0(-)$ *such that, for X nonsingular, $\gamma([\mathrm{id} : X \to X]) = [\mathcal{O}_X]$, and the following diagram commutes*:

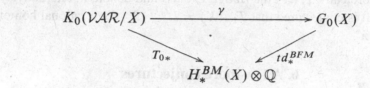

$y = 1$: *There exists a unique natural transformation* $sd : K_0(\mathcal{VAR}/-) \to \Omega(-)$ *such that, for X nonsingular, $sd([\mathrm{id} : X \to X]) = [\mathbb{Q}_X[2\dim X]]$, and the following diagram commutes*:

$$
\begin{array}{ccc}
K_0(\mathcal{VAR}/X) & \xrightarrow{\quad sd \quad} & \Omega(X) \\
& \searrow{\scriptstyle T_{1*}} \quad \swarrow{\scriptstyle L_*^{CS}} & \\
& H_*^{BM}(X) \otimes \mathbb{Q}. &
\end{array}
$$

The first two claims are straightforward; the third, the case $y = 1$, is anything but. In particular, the existence of $sd : K_0(\mathcal{VAR}/-) \to \Omega(-)$ is not obvious at all. The only way we know to prove it is by going through some details of Youssin's work [You] and using Corollary 4-10(2) again. This is done in [BSY1]; see also [BSY2; SY].

REMARK 5-5. $y = -1$: $T_{-1*}(X) = c_*^{\mathrm{Mac}}(X) \otimes \mathbb{Q}$.

$y = 0$: In general, for a singular variety X we have

$$
\Lambda_0^{\mathrm{mot}}([X \xrightarrow{\mathrm{id}_X} X]) \neq [\mathcal{O}_X].
$$

Therefore, in general, $T_{0*}(X) \neq td_*^{BFM}(X)$. So, our $T_{0*}(X)$ shall be called the Hodge–Todd class and denoted by $td_*^H(X)$. However, if X is a Du Bois variety, i.e., every point of X is a Du Bois singularity (note a nonsingular point is also a Du Bois singularity), we DO have

$$\Lambda_0^{mot}([X \xrightarrow{\mathrm{id}_X} X]) = [\mathcal{O}_X].$$

This is because of the definition of Du Bois variety: X is called a Du Bois variety if we have

$$\mathcal{O}_X = \mathrm{gr}_\sigma^0(DR(\mathcal{O}_X)) \cong \mathrm{gr}_F^0(\Omega_X^*).$$

Hence, for a Du Bois variety X we have $T_{0*}(X) = td_*^{BFM}(X)$. For example, S. Kovács [Kov] proved Steenbrink's conjecture that rational singularities are Du Bois, thus for the quotient X of any smooth variety acted on by a finite group we have that $T_{0*}(X) = td_*^{BFM}(X)$.

$y = 1$: In general, $sd([X \xrightarrow{\mathrm{id}_X} X])$ is distinct from $\mathcal{IC}_X$, so $T_{1*}(X) \neq L_*^{GM}(X)$. We therefore call $T_{1*}(X)$ the *Hodge L-class* and denote it, alternatively, by $L_*^H(X)$. It is conjectured that $T_{1*}(X) \neq L_*^{GM}(X)$ for a rational homology manifold X.

6. A few more conjectures

CONJECTURE 6-1. *Any natural transformation*

$$T : K_0(\mathcal{VAR}/X) \to H_*^{BM}(X) \otimes \mathbb{Q}[y]$$

without the normalization condition is a linear combination of components of the form $td_{y*i} : K_0(\mathcal{VAR}/X) \to H_{2i}^{BM}(X) \otimes \mathbb{Q}[y]$:

$$T = \sum_{i \geq 0} r_i(y)\, td_{y*i} \quad (r_i(y) \in \mathbb{Q}[y]).$$

This conjecture means that the normalization condition for smooth varieties imposed to get our motivic characteristic class can be basically *dropped*. This conjecture is motivated by the following theorems:

THEOREM 6-2 [Y1]. *Any natural transformation*

$$T : G_0(-) \to H_*^{BM}(-) \otimes \mathbb{Q}$$

without the normalization condition is a linear combination of components

$$td_{*\,i}^{BFM} : G_0(-) \to H_{2i}^{BM}(-) \otimes \mathbb{Q},$$

that is,

$$T = \sum_{i \geq 0} r_i\, td_{*\,i}^{BFM} \quad (r_i \in \mathbb{Q}).$$

THEOREM 6-3 [KMY]. *Any natural transformation*

$$T : F(-) \to H_*^{BM}(-) \otimes \mathbb{Q}$$

without the normalization condition is a linear combination of components

$$c_{*\,i}^{\mathrm{Mac}} \otimes \mathbb{Q} : G_0(-) \to H_{2i}^{BM}(-) \otimes \mathbb{Q}$$

*of the **rationalized** MacPherson's Chern class $c_*^{\mathrm{Mac}} \otimes \mathbb{Q}$ (i.e., a linear combination of $c_{*\,i}^{\mathrm{Mac}}$ mod torsion):*

$$T = \sum_{i \geq 0} r_i\, c_{*\,i}^{\mathrm{Mac}} \otimes \mathbb{Q} \quad (r_i \in \mathbb{Q}).$$

REMARK 6-4. This theorem certainly implies the uniqueness of such a transformation $c_*^{\mathrm{Mac}} \otimes \mathbb{Q}$ satisfying the normalization. The proof of Theorem 6-3 *does not* appeal to the resolution of singularities at all, therefore modulo torsion the uniqueness of the MacPherson's Chern class transformation c_*^{Mac} is proved without using resolution of singularities. However, in the case of integer coefficients, as shown in [M1], the uniqueness of c_*^{Mac} uses the resolution of singualrities and as far as the author knows, there is no proof available without using this result. Does there exist any mysterious connection between resolution of singularities and finite torsion? (In this connection we quote a comment by J. Schürmann:

> There is indeed a relation between resolution of singularities and torsion information: in [To1] B. Totaro shows by resolution of singularities that the fundamental class $[X]$ of a complex algebraic variety X lies in the image from the complex cobordism $\Omega^U(X) \to H_*(X, \mathbb{Z})$. And this implies some nontrivial topological restrictions: for example, all odd-dimensional elements of the Steenrod algebra vanish on $[X]$ viewed in $H_*(X, \mathbb{Z}_p)$.)

Furthermore, hinted by these two theorems, it would be natural to speculate the following "linearity" on the Cappell–Shaneson L-class also:

CONJECTURE 6-5. *Any natural transformation without the normalization condition*

$$T : \Omega(-) \to H_*^{BM}(-) \otimes \mathbb{Q}$$

is a linear combination of components $L_{*\,i}^{CS} : \Omega(-) \to H_{2i}^{BM}(-) \otimes \mathbb{Q}$:

$$T = \sum_{i \geq 0} r_i\, L_{*\,i}^{CS} \quad (r_i \in \mathbb{Q}).$$

7. Some more remarks

For complex algebraic varieties there is another important homology theory. That is Goresky–MacPherson's *intersection homology theory IH*, introduced in [GM] (see also [KW]). It satisfies all the properties which the ordinary (co)homology theory for nonsingular varieties have, in particular the Poincaré duality holds, in contrast to the fact that in general it fails for the ordinary (co)homology theory of singular varieties. In order that the Poincaré duality theorem holds, one needs to control cycles according to *perversity*, which is sensitive to, or "control", complexity of singularities. M. Saito showed that IH satisfies pure Hodge structure just like the cohomology satisfies the pure Hodge structure for compact smooth manifolds (see also [CaMi1; CaMi2]). In this sense, IH is a convenient gadget for possibly singular varieties, and using the IH, we can also get various invariants which are sensitive to the structure of given possibly singular varieties. For the history of IH, see Kleiman's survey article [Kl], and for L_2-*cohomology* — very closely related to the intersection homology — see [CGM; Go; Lo1; SS; SZ], for example. Thus for the category of compact complex algebraic varieties two competing machines are available:

ordinary (co)homology + mixed Hodge structures

intersection homology + pure Hodge structures

Of course, they are the same for the subcategory of compact smooth varieties.

So, for singular varieties one can introduce the similar invariants using IH; in other words, one can naturally think of the IH-version of the Hirzebruch χ_y genus, because of the pure Hodge structure, denote by χ_y^{IH}: Thus we have invariants χ_y-genus and χ_y^{IH}-genus. As to the class version of these, one should go through the derived category of mixed Hodge modules, because the intersection homology sheaf lives in it. Then obviously the difference between these two genera or between the class versions of these two genera should come from the singularities of the given variety. For this line of investigation, see the articles by Cappell, Libgober, Maxim, and Shaneson [CMS1; CMS2; CLMS1; CLMS2].

The most important result is the *Decomposition Theorem* of Beilinson, Bernstein, Deligne, and Gabber [BBD], which was conjectured by I. M. Gelfand and R. MacPherson. A more geometric proof of this is given in the above mentioned paper [CaMi1] of M. de Cataldo and L. Migliorini.

Speaking of the intersection homology, the general category for IH is the category of pseudomanifolds and the canonical and well-studied invariant for pseudomanifolds is the signature, because of the Poincaré duality of IH. Banagl's monograph [Ba1] is recommended on this topic; see also [Ba2; Ba3; Ba4; BCS; CSW; CW; Wei]. Very roughly, T_{y*} is a kind of deformation or perturbation of Baum–Fulton–MacPherson's Riemann–Roch. It would be interesting to consider a similar kind of deformation of L-class theory defined on the (co)bordism theory of pseudomanifolds. Again we quote J. Schürmann:

A deformation of the L-class theory seems not reasonable. Only the signature = χ_1-genus factorizes over the oriented cobordism ring Ω^{SO}, so that this invariant is of more topological nature related to stratified spaces. For the other desired ("deformation") invariants one needs a complex algebraic or analytic structure. So what is missing up to now is a suitable theory of almost complex stratified spaces.

Finally, since we started the present paper with counting, we end with posing the following question: how about counting pseudomanifolds respecting the structure of pseudomanifolds:

Does "stratified counting" c_{stra} make sense?

For complex algebraic varieties, which are pseudomanifolds, algebraic counting c_{alg} (using mixed Hodge theory = ordinary (co)homology theory + mixed Hodge structure) in fact ignores the stratification. So, in this possible problem, one should consider intersection homology + pure Hodge structure, although intersection homology *is a topological invariant, and hence independent of the stratification.*

J. Schürmann provides one possible answer to the highlighted question above:

One possible answer would be to work in the complex algebraic context with a fixed (Whitney) stratification $X_\bullet$, so that the closure of a stratum S is a union of strata again. Then one can work with the Grothendieck group $K_0(X_\bullet)$ of $X_\bullet$-constructible sets, i.e., those which are a union of such strata. The topological additive counting would be related again to the Euler characteristic and the group $F(X_\bullet)$ of $X_\bullet$-constructible functions. A more sophisticated version is the Grothendieck group $K_0(X_\bullet)$ of $X_\bullet$-constructible sheaves (or sheaf complexes). These are generated by classes $j_!L_S$ for $j : S \to X$, the inclusion of a stratum S, and L_S a local system on S, and also by the intermediate extensions $j_{!*}L_S$, which are perverse sheaves. In relation to signature and duality, one can work with the corresponding cobordism group $\Omega(X_\bullet)$ of Verdier self-dual $X_\bullet$-constructible sheaf complexes. These are generated by $j_{!*}L_S$, with L_S a self-dual local system on S. Finally one can also work with the Grothendieck group $K_0(MHM(X_\bullet))$ of mixed Hodge modules, whose underlying rational complex is $X_\bullet$-constructible. This last group is of course not a topological invariant.

We hope to come back to the problem of a possible "stratified counting" c_{stra}.

Acknowledgements

This paper is based on my talk at the workshop "Topology of Stratified Spaces" held at MSRI, Berkeley, from September 8 to 12, 2008. I thank the organizers (Greg Friedman, Eugénie Hunsicker, Anatoly Libgober, and Laurentiu Maxim) for their invitation to the workshop. I also thank the referee and Jörg Schürmann for their careful reading of the paper and valuable comments and suggestions, and G. Friedmann and L. Maxim for their much-appreciated feedback on an earlier version of the paper. Finally I thank also Silvio Levy, the editor of MSRI Book Series, for his careful reading and editing of the final version.

References

[AKMW] D. Abramovich, K. Karu, K. Matsuki and J. Włodarczyk, *Torification and factorization of birational maps*, J. Amer. Math. Soc. 15:3 (2002), 531–572.

[At] M. Atiyah, *Bordism and cobordism*, Proc. Cambridge Philos. Soc. 57 (1961), 200–208.

[Ba1] J. Baez, *Euler Characteristic versus homotopy cardinality*, lecture at the Fields Institute Program on Homotopy Theory and its Applications (London, ON, 2003), http://math.ucr.edu/home/baez/cardinality/.

[Ba2] J. Baez, *The mysteries of counting: Euler characteristic versus homotopy cardinality*, public lecture at conference "Categories in Algebra, Geometry and Mathematical Physics", (Sydney, 2005), http://math.ucr.edu/home/baez/counting/.

[BD] J. Baez and J. Dolan, *From finite sets to Feynman diagrams*, in *"Mathematics unlimited: 2001 and beyond"*, vol. 1, edited by B. Engquist and W. Schmid, Springer, Berlin, 2001, 29–50.

[Ba1] P. Balmer, *Derived Witt groups of a scheme*, J. Pure Appl. Algebra 141 (1999), 101–129.

[Ba1] M. Banagl, *Topological invariants of stratified spaces*, Springer, Berlin, 2007.

[Ba2] M. Banagl, *Extending intersection homology type invariants to non-Witt spaces*, Memoirs Amer. Math. Soc. no. 760 (2002), 1–83.

[Ba3] M. Banagl, *Computing twisted signatures and L-classes of non-Witt spaces*, Proc. London Math. Soc. (3) 92 (2006), 428–470.

[Ba4] M. Banagl, *The L-class of non-Witt spaces*, Ann. Math. 163:3 (2006), 743–766.

[BCS] M. Banagl, S. E. Cappell and J. L. Shaneson, *Computing twisted signatures and L-classes of stratified spaces*, Math. Ann. 326 (2003), 589–623.

[BFM1] P. Baum, W. Fulton and R. MacPherson, *Riemann–Roch for singular varieties*, Publ. Math. IHES 45 (1975), 101–145.

[BBD] A. A. Beilinson, J. N. Bernstein and P. Deligne, *Faisceaux pervers*, Astérisque 100 (1982),

[Bi] F. Bittner, *The universal Euler characteristic for varieties of characteristic zero*, Comp. Math. 140 (2004), 1011–1032.

[BM] A. Borel and J. C. Moore, Homology theory for locally compact spaces, Michigan Math. J. 7 (1960), 137–159.

[BL1] L. Borisov and A. Libgober, *Elliptic genera for singular varieties*, Duke Math. J. 116 (2003), 319–351.

[BL2] L. Borisov and A. Libgober, *McKay correspondence for elliptic genera*, Ann. Math. 161 (2005), 1521–1569,

[BL3] L. Borisov and A. Libgober, *Higher elliptic genera*, Math. Res. Lett. 15 (2008), 511–520.

[BT] R. Bott and C. Taubes, *On the rigidity theorems of Witten*, J. Amer. Math. Soc. 2 (1989), 137–186.

[BrS] J.-P. Brasselet and M.-H. Schwartz, *Sur les classes de Chern d'une ensemble analytique complexe*, pp. 93–148 in Caractéristique d'Euler–Poincaré, Séminaire E. N. S. 1978–1979, Astérisque 82–83 (1981).

[BSY1] J.-P. Brasselet, J. Schürmann and S. Yokura, *Hirzebruch classes and motivic Chern classes for singular spaces*, J. Topol. Anal. 2:1 (2010), 1–55.

[BSY2] J.-P. Brasselet, J. Schürmann and S. Yokura, *Classes de Hirzebruch et classes de Chern motiviques*, C. R. Acad. Sci. Paris Ser. I 342 (2006), 325–328.

[CLMS1] S. E. Cappell, A. Libgober, L. G. Maxim and J. L. Shaneson, *Hodge genera of algebraic varieties, II*, Math. Ann. 345:4 (2009), 925–972.

[CLMS2] S. E. Cappell, A. Libgober, L. G. Maxim and J. L. Shaneson, *Hodge genera and characteristic classes of complex algebraic varieties*, Electron. Res. Announc. Math. Sci. 15 (2008), 1–7.

[CMSS] S. E. Cappell, L. G. Maxim, J. Schürmann and J. L. Shaneson, *Characteristic classes of complex hypersurfaces*, Adv. Math. 225:5 (2010), 2616–2647.

[CMS1] S. E. Cappell, L. G. Maxim and J. L. Shaneson, *Euler characteristics of algebraic varieties*, Comm. Pure Appl. Math. 61:3, (2008), 409–421.

[CMS2] S. E. Cappell, L. G. Maxim and J. L. Shaneson, *Hodge genera of algebraic varieties, I*, Comm. Pure Appl. Math. 61:3, (2008), 422–449.

[CS1] S. E. Cappell and J. L. Shaneson, *Stratifiable maps and topological invariants*, J. Amer. Math. Soc. 4 (1991), 521–551.

[CS2] S. E. Cappell and J. L. Shaneson, *Genera of algebraic varieties and counting lattice points*, Bull. Amer. Math. Soc. 30 (1994), 62–69.

[CSW] S. E. Cappell, J. L. Shaneson and S. Weinberger, *Classes topologiques caractéristiques pour les actions de groupes sur les espaces singuliers*, C. R. Acad. Sci. Paris Sér. I. Math. 313 (1991), 293–295.

[CW] S. E. Cappell and S. Weinberger, *Classification de certains espaces stratifiés*, C. R. Acad. Sci. Paris Sér. I. Math. 313 (1991), 399–401.

[CGM] J. Cheeger, M. Goresky and R. MacPherson, L_2 *cohomology and intersection homology for singular algebraic varieties*, in *Seminar on differential geometry*, Annals Math. Studies 102 (1982), 303–340

[DK] V. I. Danilov and A. G. Khovanskii, *Newton polyhedra and an algorithm for computing Hodge–Deligne numbers*, Izv. Akad. Nauk SSSR Ser. Mat. 50:5 (1986), 925–945; In Russian; translated in Bull. Amer. Math. Soc. 30 (1994), 62–69.

[CaMi1] M. de Cataldo and L. Migliorini, *The Hodge theory of algebraic maps*, Ann. Scient. Éc. Norm. Sup. (4) 38 (2005), 693–750.

[CaMi2] M. de Cataldo and L. Migliorini, *The Decomposition Theorem and the topology of algebraic maps*, to appear in Bull. Amer. Math. Soc.

[De1] P. Deligne, *Théorie de Hodge II*, Publ. Math. IHES 40 (1971), 5–58.

[De2] P. Deligne, *Théorie de Hodge III*, Publ. Math. IHES 44 (1974), 5–78.

[DB] Ph. Du Bois, *Complexe de de Rham filtré d'une variété singulière*, Bull. Soc. Math. France 109 (1981), 41–81.

[Fr] J. Franke, *On the construction of elliptic cohomology*, Math. Nachr. 158 (1992), 43–65.

[Fri] G. Friedman, *Intersection homology of stratified fibrations and neighborhoods*, Adv. in Math. 215 (2007), 24–65.

[Ful] W. Fulton, *Intersection theory*, Springer (1984).

[FL] W. Fulton and S. Lang, *Riemann–Roch algebra*, Springer (1985).

[FM] W. Fulton and R. MacPherson, *Categorical framework for the study of singular spaces*, Mem. Amer. Math. Soc. 243 (1981).

[FuM] J. Fu and C. McCrory, *Stiefel–Whitney classes and the conormal cycle of a singular variety*, Trans. Amer. Math. Soc. 349 (1997), 809–835.

[Go] M. Goresky, L_2-*cohomology is intersection cohomology*, preprint, available at http://www.math.ias.edu/~goresky/pdf/zucker.pdf.

[GM] M. Goresky and R. MacPherson, *Intersection homology theory*, Topology 149 (1980), 155–162.

[Gr] M. Gros, *Classes de Chern et classes de cycles en cohomologie de Hodge–Witt logarithmique*, Bull. Soc. Math. France Mem. 21 (1985).

[Grot] A. Grothendieck, *Récoltes et semailles: réflexions et témoignages sur un passé de mathématicien*, preprint, 1985.

[GNA] F. Guillén and V. Navarro Aznar, *Un critère d'extension des foncteurs définis sur les schémas lisses*, Publ. Math. IHES 95 (2002), 1–91.

[Hi] F. Hirzebruch, *Topological methods in algebraic geometry*, Springer, New York (1966).

[HBJ] F. Hirzebruch, T. Berger and R. Jung, *Manifolds and modular forms*, Aspects Math. E.20, Vieweg (1992).

[Jo] D. Joyce, *Constructible functions on Artin stacks*, math. AG/0403305v4. A revised version appeared in J. London Math. Soc. 74 (2006), 583–606.

[KS] M. Kashiwara and P. Schapira, *Sheaves on manifolds*, Springer, Berlin (1990).

[Ken] G. Kennedy, *MacPherson's Chern classes of singular varieties*, Com. Algebra. 9 (1990), 2821–2839.

[KMY] G. Kennedy, C. McCrory and S. Yokura, *Natural transformations from constructible functions to homology*, C. R. Acad. Sci. Paris I 319 (1994), 969–973,

[KW] F. Kirwan and J. Woolf, *An introduction to intersection homology theory* (2nd ed.), Chapman Hall/CRC, 2006.

[Kl] S. Kleiman, *The development of intersection homology theory*, in "A century of mathematics in America", Part II, Hist. Math. 2, Amer. Math. Soc., (1989), 543–585.

[Kov] S. J. Kovács, *Rational, log canonical, Du Bois singularities: on the conjectures of Kollár and Steenbrink*, Comp. Math. 118 (1999), 123–133.

[KrSt] M. Kreck and S. Stolz, HP^2-*bundles and elliptic homology*, Acta Math., 171 (1993), 231–261.

[L] P. S. Landweber, *Homological properties of comodules over $MU_* MU$ and $BP_* BP$*, Amer. J. Math., 98 (1976), 591–610.

[LRS] P. S. Landweber, D. C. Ravenel and R. E. Stong, *Periodic cohomology theories defined by elliptic curves*, Contemp. Math., 181 (1995), 317–337.

[Lein] T. Leinster, *The Euler characteristic of a category*, Documenta Mathematica 13 (2008), 21–49.

[LM] M. Levine and F. Morel, *Algebraic cobordism*, Springer (2006).

[Lo1] E. Looijenga, L_2-*cohomology of locally symmetric varieties*, Compositio Math. 67:1 (1988), 3–20.

[Lo2] E. Looijenga, *Motivic measures*, Séminaire Bourbaki 874, Astérisque 276 (2002), 267–297.

[Lu] J. Lurie, *A survey of elliptic cohomology*, available at http://www-math.mit.edu/~lurie/papers/survey.pdf, 2007.

[M1] R. MacPherson, *Chern classes for singular algebraic varieties*, Ann. Math. 100 (1974), 423–432.

[M2] R. MacPherson, *Characteristic classes for singular varieties*, Proceedings of the 9th Brazilian Mathematical Colloquium (Poços de Caldas, 1973) Vol. II, Instituto de Matemática Pura e Aplicada, Rio de Janeiro, (1977), 321–327.

[MS1] L. Maxim and J. Schürmann, *Hodge-theoretic Atiyah–Meyer formulae and the stratified multiplicative property*, in "Singularities I: Algebraic and Analytic Aspects", Contemporary Math. 474 (2008), 145–167.

[MS2] L. Maxim and J. Schürmann, *Twisted genera of symmetric products*, preprint, arXiv:0906.1264v1.

[MP1] C. McCrory and A. Parusiński, *Virtual Betti numbers of real algebraic varieties*, C. R. Acad. Sci. Paris, Ser. I 336 (2003), 763–768.

[MP2] C. McCrory and A. Parusiński, *The weight filtration for real algebraic varieties*, in this volume.

[Mi-Ra] *Elliptic cohomology: geometry, applications, and higher chromatic analogues*, edited by H. R. Miller and D. Ravenel, London Math. Soc. Lecture Note Series 342 (2007).

[MiSt] J. W. Milnor and J. D. Stasheff, *Characteristic classes*, Ann. Math. Studies 76, Princeton Univ. Press (1974).

[Oc] S. Ochanine, *Sur les genres multiplicatifs définis par des intégrales elliptiques*, Topology 26 (1987), 143–151.

[P] C. A. M. Peters, *Tata Lectures on motivic aspects of Hodge theory*, Lecture Notes, Tata Institute, Mumbai, December 2007.

[PS] C. A. M. Peters and J. H. M. Steenbrink, *Mixed Hodge structures*, Springer, Berlin, 2008.

[Sa1] M. Saito, *Modules de Hodge polarisables*, Publ. RIMS 24 (1988), 849–995.

[Sa2] M. Saito, *Mixed Hodge Modules*, Publ. RIMS 26 (1990), 221–333.

[Sa3] M. Saito, *Induced D-modules and differential complexes*, Bull. Soc. Math. France 117 (1989), 361–387.

[Sa4] M. Saito, *D-modules on analytic spaces*, Publ. RIMS 27 (1991), 291–332.

[Sa5] M. Saito, *Mixed Hodge complexes on algebraic varieties*, Math. Ann. 316 (2000), 283–331.

[Sa6] M. Saito, *Arithmetic mixed sheaves*, Inv. Math. 144 (2001), 533–569.

[SS] L. Saper and M. Stern, L_2-*cohomology of arithmetic varieties*, Ann. Math. (2) 132:1 (1990), 1–69.

[SZ] L. Saper and S. Zucker, *An introduction to L_2-cohomology*, in *Several Complex Variables and Complex Geometry* (edited by E. Bedford et al.), Proc. Symp. Pure Math., Vol. 52, Part 2, Amer. Math. Soc. (1991), 519–534.

[Si] P. H. Siegel, *Witt spaces: A geometric cycle theory for KO-homology at odd primes*, Amer. J. Math. 105 (1983), 1067–1105.

[Sch1] J. Schürmann, *Topology of singular spaces and constructible sheaves*, Monografie Matematyczne 63 (New Series), Birkhäuser, Basel (2003).

[Sch2] J. Schürmann, *A generalized Verdier-type Riemann-Roch theorem for Chern-Schwartz-MacPherson classes*, preprint, arXiv:math.AG/0202175

[Sch3] J. Schürmann, *Characteristic classes of mixed Hodge modules*, in this volume.

[SY] J. Schürmann and S. Yokura, *A survey of characteristic classes of singular spaces*, in "Singularity Theory: Dedicated to Jean-Paul Brasselet on his 60th birthday", edited by Denis Chéniot et al., World Scientific (2007), 865–952.

[Schw] M.-H. Schwartz, *Classes caractéristiques définies par une stratification d'une variété analytique complexe*, C. R. Acad. Sci. Paris 260 (1965), 3262–3264 and 3535–3537.

[Sh] J. L. Shaneson, *Characteristic classes, lattice points, and Euler–MacLaurin formulae*, Proc. Intern. Congress Math., Zürich 1994, vol. 1 (1995), 612–624

[Sr] V. Srinivas, *The Hodge characteristic*, Lectures in *Jet Schemes Seminar, MSRI, December 2002*, Manuscript preprint (2002).

[Stong] R. E. Stong, *Notes on Cobordism Theory*, Princeton Univ. Press, 1968.

[Su] D. Sullivan, *Combinatorial invariants of analytic spaces*, Lect. Notes Math. 192 (1970), 165–168.

[Ta] C. Taubes, S^1 *actions and elliptic genera*, Comm. Math. Phys., 122 (1989), 455–526.

[Thom] R. Thom, *Les classes caractéristiques de Pontrjagin des variétés triangulées*, Symp. Intern. de Topologia Algebraica, Unesco (1958).

[To1] B. Totaro, *Torsion algebraic cycles and complex cobordism*, J. Amer. Math. Soc. 10 (1997), 467–493.

[To2] B. Totaro, *Chern numbers for singular varieties and elliptic homology*, Ann. Math. 151 (2000), 757–791.

[To3] B. Totaro, *Topology of singular algebraic varieties*, Proc. Int. Cong. Math. Beijing, Vol. 1 (2002), 533–541.

[To4] B. Totaro, *The elliptic genus of a singular variety*, pp. 360–364 in [Mi-Ra] (2007).

[Wei] S. Weinberger, *The topological classification of stratified spaces*, University of Chicago Press, Chicago, 1994.

[Wil] Floyd L. Williams, *Lectures on zeta functions, L-functions and modular forms with some physical applications*, in *A window into zeta and modular physics* (edited by Klaus Kirsten and Floyd L. Williams), MSRI Publications vol. 57, Cambridge Univ. Press, New York (2010), 7–100.

[Wit] E. Witten, *The index of the Dirac operator in loop spaces*, pp. 161–181 in [La].

[Wlo] J. Włodarczyk, *Toroidal varieties and the weak factorization theorem*, Inv. Math. 154 (2003), 223–331.

[Y1] S. Yokura, *On the universality of Baum-Fulton-MacPherson's Riemann-Roch for singular varieties*, Proc. Japan. Acad. 68 (1992), 119–122.

[Y2] S. Yokura, *A generalized Grothendieck-Riemann-Roch theorem for Hirzebruch's χ_y-characteristic and T_y-characteristic*, Publ. Res. Inst. Math. Sci. 30 (1994), 603–610.

[Y2] S. Yokura, *On Cappell-Shaneson's homology L-class of singular algebraic varieties*, Trans. Amer. Math. Soc. 347 (1995), 1005–1012.

[Y4] S. Yokura, *A singular Riemann-Roch theorem for Hirzebruch characteristics.*, Banach Center Publ. 44 (1998), 257–268.

[Y5] S. Yokura, *Constructible functions and Hodge polynomials*, unpublished note, 2003.

[Y6] S. Yokura, *Characteristic classes of proalgebraic varieties and motivic measures*, arXiv:math/0606352v2.

[Y7] S. Yokura, *Oriented bivariant theory, I*, International J. Math. 20:10 (2009), 1305–1334.

[Y8] S. Yokura, *Motivic Milnor classes*, J. Singularities, 1 (2010), 39–59.

[You] B. Youssin, *Witt groups of derived categories*, K-Theory 11 (1997), 373–395.

SHOJI YOKURA
DEPARTMENT OF MATHEMATICS AND COMPUTER SCIENCE
FACULTY OF SCIENCE
KAGOSHIMA UNIVERSITY
21-35 KORIMOTO 1-CHOME
KAGOSHIMA 890-0065
JAPAN
  yokura@sci.kagoshima-u.ac.jp

Characteristic classes of mixed Hodge modules

JÖRG SCHÜRMANN

ABSTRACT. This paper is an extended version of an expository talk given at
the workshop "Topology of Stratified Spaces" at MSRI in September 2008.
It gives an introduction and overview about recent developments on the interaction of the theories of characteristic classes and mixed Hodge theory for
singular spaces in the complex algebraic context.

It uses M. Saito's deep theory of mixed Hodge modules as a black box,
thinking about them as "constructible or perverse sheaves of Hodge structures", having the same functorial calculus of Grothendieck functors. For
the "constant Hodge sheaf", one gets the "motivic characteristic classes" of
Brasselet, Schürmann, and Yokura, whereas the classes of the "intersection
homology Hodge sheaf" were studied by Cappell, Maxim, and Shaneson. The
classes associated to "good" variation of mixed Hodge structures where studied
in connection with understanding the monodromy action by these three authors
together with Libgober, and also by the author.

There are two versions of these characteristic classes. The K-theoretical
classes capture information about the graded pieces of the filtered de Rham
complex of the filtered D-module underlying a mixed Hodge module. Application of a suitable Todd class transformation then gives classes in homology.
These classes are functorial for proper pushdown and exterior products, together with some other properties one would expect for a satisfactory theory
of characteristic classes for singular spaces.

For "good" variation of mixed Hodge structures they have an explicit classical description in terms of "logarithmic de Rham complexes". On a point
space they correspond to a specialization of the Hodge polynomial of a mixed
Hodge structure, which one gets by forgetting the weight filtration.

We also indicate some relations with other subjects of the conference, like
index theorems, signature, L-classes, elliptic genera and motivic characteristic
classes for singular spaces.

CONTENTS

1. Introduction

This paper gives an introduction and overview about recent developments on the interaction of the theories of characteristic classes and mixed Hodge theory for singular spaces in the complex algebraic context. The reader is not assumed to have a background on any of these subjects, and the paper can also be used as a bridge for communication between researchers on one of these subjects.

General references for the theory of characteristic classes of singular spaces are the survey [48] and the paper [55] in these proceedings. As references for mixed Hodge theory one can use [38; 52], as well as the nice paper [37] for explaining the motivic viewpoint to mixed Hodge theory. Finally as an introduction to M. Saito's deep theory of mixed Hodge modules one can use [38, Chapter 14], [41] as well as the introduction [45].

The theory of mixed Hodge modules is used here more or less as a black box; we think about them as constructible or perverse sheaves of Hodge structures, having the same functorial calculus of Grothendieck functors. The underlying theory of constructible and perverse sheaves can be found in [7; 30; 47].

For the "constant Hodge sheaf" $\mathbb{Q}_Z^H$ one gets the "motivic characteristic classes" of Brasselet, Schürmann, and Yokura [9], as explained in these proceedings [55]. The classes of the "intersection homology Hodge sheaf" IC_Z^H were studied by Cappell, Maxim, and Shaneson in [10; 11]. Also, the classes associated to "good" variation of mixed Hodge structures where studied via Atiyah–Meyer type formulae by Cappell, Libgober, Maxim, and Shaneson in [12; 13]. For a summary compare also with [35].

There are two versions of these characteristic classes, the *motivic Chern class transformation* MHC_y and the *motivic Hirzebruch class transformation* MHT_{y*}. The K-theoretical classes MHC_y capture information about the graded pieces of the filtered de Rham complex of the filtered D-module underlying a mixed Hodge module. Application of a suitable twisting $td_{(1+y)}$ of the Todd class transformation td_* of Baum, Fulton, and MacPherson [5; 22] then gives the classes $MHT_{y*} = td_{(1+y)} \circ MHC_y$ in homology. It is the *motivic Hirzebruch class transformation* MHT_{y*}, which unifies three concepts:

($y = -1$) the (rationalized) *Chern class transformation* c_* of MacPherson [34];
($y = 0$) the *Todd class transformation* td_* of Baum–Fulton–MacPherson [5];
($y = 1$) the *L-class transformation* L_* of Cappell and Shaneson [14].

(Compare with [9; 48] and also with [55] in these proceedings.) But in this paper we focus on the K-theoretical classes MHC_y, because these imply then also the corresponding results for MHT_{y*} just by application of the (twisted) Todd class transformation. So the *motivic Chern class transformation* MHC_y studied here is really the basic one!

Here we explain the functorial calculus of these classes, first stating in a very precise form the key results used from Saito's theory of mixed Hodge modules, and then explaining how to get from this the basic results about the motivic Chern class transformation MHC_y. These results are illustrated by many interesting examples. For the convenience of the reader, the most general results are only stated near the end of the paper. In fact, while most of the paper is a detailed survey of the K-theoretical version of the theory as developed in [9; 12; 13; 35], it is this last section that contains new results on the important functorial properties of these characteristic classes. The first two sections do not use mixed Hodge modules and are formulated in the now classical language of (variation of) mixed Hodge structures. Here is the plan of the paper:

SECTION 2 introduces pure and mixed Hodge structures and the corresponding Hodge genera, such as the E-polynomial and the χ_y-genus. These are suitable generating functions of Hodge numbers with χ_y using only the Hodge filtration F, whereas the E-polynomial also uses the weight filtration. We also carefully explain why only the χ_y-genus can be further generalized to characteristic classes, i.e., why one has to forget the weight filtration for applications to characteristic classes.

SECTION 3 motivates and explains the notion of a variation (or family) of pure and mixed Hodge structures over a smooth (or maybe singular) base. Basic examples come from the cohomology of the fibers of a family of complex algebraic varieties. We also introduce the notion of a "good" variation of mixed Hodge structures on a complex algebraic manifold M, to shorten the notion for a graded polarizable variation of mixed Hodge structures on M that is *admissible* in the sense of Steenbrink and Zucker [50] and Kashiwara [28], with *quasi-unipotent monodromy* at infinity, i.e., with respect to a compactification $\overline{M}$ of M by a compact complex algebraic manifold $\overline{M}$, with complement $D := \overline{M} \setminus M$ a normal crossing divisor with smooth irreducible components. Later on these will give the basic example of so-called "smooth" mixed Hodge modules. And for these good variations we introduce a simple *cohomological* characteristic class transformation MHC^y, which behaves nicely with respect to smooth pullback, duality and (exterior) products. As a first approximation to more general mixed Hodge modules and their characteristic classes, we also study in detail functorial properties of the canonical Deligne extension across a normal crossing divisor D at infinity (as above), leading to *cohomological* characteristic classes $MHC^y(j_*(\cdot))$ defined in terms of "logarithmic de Rham complexes". These classes of good variations have been studied in detail in [12; 13; 35], and most results described here are new functorial reformulations of the results from these sources.

SECTION 4 starts with an introduction to Saito's functorial theory of algebraic mixed Hodge modules, explaining its power in many examples, including how to get a pure Hodge structure on the global intersection cohomology $IH^*(Z)$ of a compact complex algebraic variety Z. From this we deduce the basic calculus of Grothendieck groups $K_0(MHM(\cdot))$ of mixed Hodge modules needed for our motivic Chern class transformation MHC_y. We also explain the relation to the motivic viewpoint coming from relative Grothendieck groups of complex algebraic varieties.

SECTION 5.1 is devoted to the definition of our motivic characteristic *homology class* transformations MHC_y and MHT_{y*} for mixed Hodge modules. By Saito's theory they commute with push down for proper morphisms, and on a compact space one gets back the corresponding χ_y-genus by pushing down to a point, i.e., by taking the degree of these characteristic homology classes.

SECTIONS 5.2 AND 5.3 finally explain other important functoriality properties:

(1) multiplicativity for exterior products;

(2) the behavior under smooth pullback given by a Verdier Riemann–Roch formula;

(3) a "going up and down" formula for proper smooth morphisms;

(4) multiplicativity between MHC^y and MHC_y for a suitable (co)homological pairing in the context of a morphism with smooth target (as special cases one gets interesting Atiyah and Atiyah–Meyer type formulae, as studied in [12; 13; 35]);

(5) the relation between MHC_y and duality, i.e., the Grothendieck duality transformation for coherent sheaves and Verdier duality for mixed Hodge modules;

(6) the identification of MHT_{-1*} with the (rationalized) Chern class transformation $c_* \otimes \mathbb{Q}$ of MacPherson for the underlying constructible sheaf complex or function.

Note that such a functorial calculus is expected for any good theory of functorial characteristic classes of singular spaces (compare [9; 48]):

- for MacPherson's Chern class transformation c_* compare with [9; 31; 34; 48];
- for the Baum–Fulton–MacPherson Todd class transformation td_* compare with [5; 6; 9; 22; 24; 48];
- for Cappell and Shaneson's L-class transformation L_* compare with [2; 3; 4; 9; 14; 48; 49; 54].

The counterpart of mixed Hodge modules in these theories are constructible functions and sheaves (for c_*), coherent sheaves (for td_*) and selfdual perverse or constructible sheaf complexes (for L_*). The cohomological counterpart of the smooth mixed Hodge modules (i.e., good variation of mixed Hodge structures) are locally constant functions and sheaves (for c^*), locally free coherent sheaves or vector bundles (for the Chern character ch^*) and selfdual local systems (for a twisted Chern character of the KO-classes of Meyer [36]).

In this paper we concentrate mainly on pointing out the relation and analogy to the L-class story related to important signature invariants, because these are the subject of many other talks from the conference given in more topological terms. Finally also some relations to other themes of the conference, like index theorems, L^2-cohomology, elliptic genera and motivic characteristic classes for singular spaces, will be indicated.

2. Hodge structures and genera

2A. Pure Hodge structures. Let M be a compact *Kähler manifold* (e.g., a complex projective manifold) of complex dimension m. By classical Hodge theory one gets the decomposition (for $0 \le n \le 2m$)

$$H^n(M, \mathbb{C}) = \bigoplus_{p+q=n} H^{p,q}(M) \tag{2-1}$$

of the complex cohomology of M into the spaces $H^{p,q}(M)$ of harmonic forms of type (p, q). This decomposition doesn't depend on the choice of a Kähler form (or metric) on M, and for a complex algebraic manifold M it is of algebraic nature. Here it is more natural to work with the *Hodge filtration*

$$F^i(M) := \bigoplus_{p \ge i} H^{p,q}(M) \tag{2-2}$$

so that $H^{p,q}(M) = F^p(M) \cap \overline{F^q(M)}$, with $\overline{F^q(M)}$ the complex conjugate of $F^q(M)$ with respect to the real structure $H^n(M, \mathbb{C}) = H^n(M, \mathbb{R}) \otimes \mathbb{C}$. If

$$\Omega_M^{\bullet} = [\mathcal{O}_M \xrightarrow{d} \cdots \xrightarrow{d} \Omega_M^m]$$

denotes the usual holomorphic de Rham complex (with $\mathcal{O}_M$ in degree zero), then one gets

$$H^*(M, \mathbb{C}) = H^*(M, \Omega_M^\bullet)$$

by the holomorphic Poincaré lemma, and the Hodge filtration is induced from the "stupid" decreasing filtration

$$F^p \Omega_M^\bullet = [0 \longrightarrow \cdots 0 \longrightarrow \Omega_M^p \xrightarrow{d} \cdots \xrightarrow{d} \Omega_M^m]. \quad (2\text{-}3)$$

More precisely, the corresponding *Hodge to de Rham spectral sequence* degenerates at E_1, with

$$E_1^{p,q} = H^q(M, \Omega_M^p) \simeq H^{p,q}(M). \quad (2\text{-}4)$$

The same results are true for a compact complex manifold M that is only *bimeromorphic to a Kähler manifold* (compare [38, Corollary 2.30], for example). This is especially true for a compact complex algebraic manifold M. Moreover in this case one can calculate by Serre's GAGA theorem $H^*(M, \Omega_M^\bullet)$ also with the algebraic (filtered) de Rham complex in the Zariski topology.

Abstracting these properties, one can say the $H^n(M, \mathbb{Q})$ gets an induced *pure Hodge structure of weight n* in the following sense:

DEFINITION 2.1. Let V be a finite-dimensional rational vector space. A (rational) Hodge structure of weight n on V is a decomposition

$$V_\mathbb{C} := V \otimes_\mathbb{Q} \mathbb{C} = \bigoplus_{p+q=n} V^{p,q}, \quad \text{with } V^{q,p} = \overline{V^{p,q}} \quad \text{(Hodge decomposition)}.$$

In terms of the (decreasing) *Hodge filtration* $F^i V_\mathbb{C} := \bigoplus_{p \geq i} V^{p,q}$, this is equivalent to the condition

$$F^p V \cap \overline{F^q V} = \{0\} \quad \text{whenever } p + q = n + 1 \quad (n\text{-opposed filtration}).$$

Then $V^{p,q} = F^p \cap \overline{F^q}$, with $h^{p,q}(V) := \dim(V^{p,q})$ the corresponding *Hodge number*.

If V, V' are rational vector spaces with Hodge structures of weight n and m, then $V \otimes V'$ gets an induced Hodge structure of weight $n + m$, with Hodge filtration

$$F^k(V \otimes V')_\mathbb{C} := \bigoplus_{i+j=k} F^i V_\mathbb{C} \otimes F^j V'_\mathbb{C}. \quad (2\text{-}5)$$

Similarly the dual vector space $V^\vee$ gets an induced Hodge structure of weight $-n$, with

$$F^k(V_\mathbb{C}^\vee) := (F^{-k} V_\mathbb{C})^\vee. \quad (2\text{-}6)$$

A basic example is the *Tate Hodge structure* of weight $-2n \in \mathbb{Z}$ given by the one-dimensional rational vector space

$$\mathbb{Q}(n) := (2\pi i)^n \cdot \mathbb{Q} \subset \mathbb{C}, \quad \text{with } \mathbb{Q}(n)_{\mathbb{C}} = (\mathbb{Q}(n)_{\mathbb{C}})^{-n,-n}.$$

Then integration defines an isomorphism

$$H^2(P^1(\mathbb{C}), \mathbb{Q}) \simeq \mathbb{Q}(-1),$$

with $\mathbb{Q}(-n) = \mathbb{Q}(-1)^{\otimes n}$, $\mathbb{Q}(1) = \mathbb{Q}(-1)^{\vee}$ and $\mathbb{Q}(n) = \mathbb{Q}(1)^{\otimes n}$ for $n > 0$.

DEFINITION 2.2. A *polarization* of a rational Hodge structure V of weight n is a rational $(-1)^n$-symmetric bilinear form S on V such that

$$S(F^p, F^{n-p+1}) = 0 \quad \text{for all } p$$

and

$$i^{p-q} S(u, \bar{u}) > 0 \quad \text{for all nonzero } u \in V^{p,q}.$$

So for n *even* one gets in particular

$$(-1)^{p-n/2} S(u, \bar{u}) > 0 \quad \text{for all } q \text{ and all nonzero } u \in V^{p,q}. \tag{2-7}$$

V is called *polarizable* if such a polarization exists.

For example, the cohomology $H^n(M, \mathbb{Q})$ of a projective manifold is polarizable by the choice of a suitable Kähler form! Also note that a polarization of a rational Hodge structure V of weight n induces an isomorphism of Hodge structures (of weight n):

$$V \simeq V^{\vee}(-n) := V^{\vee} \otimes_{\mathbb{Q}} \mathbb{Q}(-n).$$

So if we choose the isomorphism of rational vector spaces

$$\mathbb{Q}(-n) = (2\pi i)^{-n} \cdot \mathbb{Q} \simeq \mathbb{Q},$$

then a polarization induces a $(-1)^n$-symmetric duality isomorphism $V \simeq V^{\vee}$.

2B. Mixed Hodge structures. The cohomology (with compact support) of a singular or noncompact complex algebraic variety, denoted by $H^n_{(c)}(X, \mathbb{Q})$, can't have a pure Hodge structure in general, but by Deligne's work [20; 21] it carries a canonical functorial (graded polarizable) *mixed Hodge structure* in the following sense:

DEFINITION 2.3. A finite-dimensional rational vector space V has a mixed Hodge structure if there is a (finite) increasing *weight filtration* $W = W_{\bullet}$ on V (by rational subvector spaces), and a (finite) decreasing Hodge filtration $F = F^{\bullet}$ on $V_{\mathbb{C}}$, such that F induces a Hodge structure of weight n on $Gr_n^W V :=$ $W_n V / W_{n-1} V$ for all n. Such a mixed Hodge structure is called (*graded*) *polarizable* if each graded piece $Gr_n^W V$ is polarizable.

A morphism of mixed Hodge structures is just a homomorphism of rational vector spaces compatible with both filtrations. Such a morphism is then *strictly* compatible with both filtrations, so that the category $mHs^{(p)}$ of (graded polarizable) mixed Hodge structures is an abelian category, with Gr_*^W, Gr_F^* and $Gr_F^* Gr_*^W$ preserving short exact sequences. The category $mHs^{(p)}$ is also endowed with a tensor product $\otimes$ and a duality $(\cdot)^\vee$, where the corresponding Hodge and weight filtrations are defined as in (2-5) and (2-6). So for a complex algebraic variety X one can consider its cohomology class

$$[H_{(c)}^*(X)] := \sum_i (-1)^i \cdot [H_{(c)}^i(X, \mathbb{Q})] \in K_0(mHs^{(p)})$$

in the Grothendieck group $K_0(mHs^{(p)})$ of (graded polarizable) mixed Hodge structures. The functoriality of Deligne's mixed Hodge structure means, in particular, that for a closed complex algebraic subvariety $Y \subset X$, with open complement $U = X \backslash Y$, the corresponding long exact cohomology sequence

$$\cdots H_c^i(U, \mathbb{Q}) \to H_c^i(X, \mathbb{Q}) \to H_c^i(Y, \mathbb{Q}) \to \cdots \qquad (2\text{-}8)$$

is an exact sequence of mixed Hodge structures. Similarly, for complex algebraic varieties X, Z, the Künneth isomorphism

$$H_c^*(X, \mathbb{Q}) \otimes H_c^*(Z, \mathbb{Q}) \simeq H_c^*(X \times Z, \mathbb{Q}) \qquad (2\text{-}9)$$

is an isomorphism of mixed Hodge structures. Let us denote by $K_0(\text{var}/pt)$ the Grothendieck group of complex algebraic varieties, i.e., the free abelian group of isomorphism classes $[X]$ of such varieties divided out by the *additivity relation*

$$[X] = [Y] + [X \backslash Y]$$

for $Y \subset X$ a closed complex subvariety. This is then a commutative ring with addition resp. multiplication induced by the disjoint union resp. the product of varieties. So by (2-8) and (2-9) we get an induced ring homomorphism

$$\chi_{\text{Hdg}} : K_0(\text{var}/pt) \to K_0(mHs^{(p)}); \ [X] \mapsto [H_c^*(X)]. \qquad (2\text{-}10)$$

2C. Hodge genera. The *E-polynomial*

$$E(V) := \sum_{p,q} h^{p,q}(V) \cdot u^p v^q \in \mathbb{Z}[u^{\pm 1}, v^{\pm 1}] \qquad (2\text{-}11)$$

of a rational mixed Hodge structure V with *Hodge numbers*

$$h^{p,q}(V) := \dim_{\mathbb{C}} Gr_F^p Gr_{p+q}^W(V_{\mathbb{C}}),$$

induces a *ring* homomorphism

$$E : K_0(mHs^{(p)}) \to \mathbb{Z}[u^{\pm 1}, v^{\pm 1}], \quad \text{with } E(\mathbb{Q}(-1)) = uv.$$

Note that $E(V)(u, v)$ is *symmetric* in u and v, since $h(V) = \sum_n h(W_n V)$ and $V^{q,p} = \overline{V^{p,q}}$ for a pure Hodge structure. With respect to *duality* one has in addition the relation

$$E(V^\vee)(u, v) = E(V)(u^{-1}, v^{-1}). \tag{2-12}$$

Later on we will be mainly interested in the specialized ring homomorphism

$$\chi_y := E(-y, 1) : K_0(mHs^{(p)}) \to \mathbb{Z}[y^{\pm 1}], \quad \text{with } \chi_y(\mathbb{Q}(-1)) = -y,$$

defined by

$$\chi_y(V) := \sum_{p} \dim_{\mathbb{C}}(Gr_F^p(V_{\mathbb{C}})) \cdot (-y)^p. \tag{2-13}$$

So here one uses only the Hodge and forgets the weight filtration of a mixed Hodge structure. With respect to *duality* one has then the relation

$$\chi_y(V^\vee) = \chi_{1/y}(V). \tag{2-14}$$

Note that $\chi_{-1}(V) = \dim(V)$ and for a pure polarized Hodge structure V of weight n one has by $\chi_1(V) = (-1)^n \chi_1(V^\vee) = (-1)^n \chi_1(V)$ and (2-7):

$$\chi_1(V) = \begin{cases} 0 & \text{for } n \text{ odd,} \\ \operatorname{sgn} V & \text{for } n \text{ even,} \end{cases}$$

where $\operatorname{sgn} V$ is the *signature* of the induced symmetric bilinear form $(-1)^{n/2} S$ on V. A similar but deeper result is the famous *Hodge index theorem* (compare [52, Theorem 6.3.3], for example):

$$\chi_1([H^*(M)]) = \operatorname{sgn}(H^m(M, \mathbb{Q}))$$

for M a compact Kähler manifold of even complex dimension $m = 2n$. Here the right side denotes the signature of the symmetric intersection pairing

$$H^m(M, \mathbb{Q}) \times H^m(M, \mathbb{Q}) \xrightarrow{\cup} H^{2m}(M, \mathbb{Q}) \simeq \mathbb{Q}.$$

The advantage of χ_y compared to E (and the use of $-y$ in the definition) comes from the following question:

Let $E(X) := E([H^*(X)])$ for X a complex algebraic variety. For M a compact complex algebraic manifold one gets by (2-4):

$$E(M) = \sum_{p,q \geq 0} (-1)^{p+q} \cdot \dim_{\mathbb{C}} H^q(M, \Omega_M^p) \cdot u^p v^q.$$

Is there a (normalized multiplicative) characteristic class

$$cl^* : \operatorname{Iso}(\mathbb{C} - VB(M)) \to H^*(M)[u^{\pm 1}, v^{\pm 1}]$$

of complex vector bundles such that the E-polynomial is a characteristic number in the sense that

$$E(M) = \sharp(M) := \deg(cl^*(TM) \cap [M]) \in H^*(pt)[u^{\pm 1}, v^{\pm 1}] \quad (2\text{-}15)$$

for any compact complex algebraic manifold M with fundamental class $[M]$?

So the cohomology class $cl^*(V) \in H^*(M)[u^{\pm 1}, v^{\pm 1}]$ should only depend on the isomorphism class of the complex vector bundle V over M and commute with pullback. Multiplicativity says

$$cl^*(V) = cl^*(V') \cup cl^*(V'') \in H^*(M)[u^{\pm 1}, v^{\pm 1}]$$

for any short exact sequence $0 \to V' \to V \to V'' \to 0$ of complex vector bundles on M. Finally cl^* is normalized if $cl^*(\text{trivial}) = 1 \in H^*(M)$ for any trivial vector bundle. Then the answer to the question is *negative*, because there are unramified coverings $p : M' \to M$ of elliptic curves M, M' of (any) degree $d > 0$. Then $p^*TM \simeq TM'$ and $p_*([M']) = d \cdot [M]$, so the projection formula would give for the topological characteristic numbers the relation

$$\sharp(M') = d \cdot \sharp(M).$$

But one has

$$E(M) = (1-u)(1-v) = E(M') \neq 0,$$

so the equality $E(M) = \sharp(M)$ is not possible! Here we don't need to ask cl^* to be multiplicative or normalized. But if we use the invariant $\chi_y(X) := \chi_y([H^*(X)])$, then $\chi_y(M) = 0$ for an elliptic curve, and $\chi_y(M)$ is a characteristic number in the sense above by the famous *generalized Hirzebruch Riemann–Roch theorem* [27]:

THEOREM 2.4 (GHRR). *There is a unique normalized multiplicative characteristic class*

$$T_y^* : \mathrm{Iso}(\mathbb{C} - VB(M)) \to H^*(M, \mathbb{Q})[y]$$

such that

$$\chi_y(M) = \deg(T_y^*(TM) \cap [M]) = \langle T_y^*(TM), [M] \rangle \in \mathbb{Z}[y] \subset \mathbb{Q}[y]$$

for any compact complex algebraic manifold M. Here $\langle \cdot, \cdot \rangle$ is the Kronecker pairing between cohomology and homology.

The *Hirzebruch class* T_y^* and χ_y-genus unify the following (total) characteristic classes and numbers:

y	T_y^*	[name]	χ_y	[name]
-1	c^*	Chern class	χ	Euler characteristic
0	td^*	Todd class	χ_a	arithmetic genus
1	L^*	L-class	sgn	signature

In fact, gHRR is just a cohomological version of the following K-theoretical calculation. Let M be a compact complex algebraic manifold, so that

$$\chi_y(M) = \sum_{p,q \geq 0} (-1)^{p+q} \cdot \dim_{\mathbb{C}} H^q(M, \Omega_M^p) \cdot (-y)^p$$

$$= \sum_{p \geq 0} \chi(H^*(M, \Omega_M^p)) \cdot y^p. \qquad (2\text{-}16)$$

Let us denote by $K_{an}^0(Y)$ (or $G_0^{an}(Y)$) the Grothendieck group of the exact (or abelian) category of holomorphic vector bundles (or coherent $\mathcal{O}_Y$-module sheaves) on the complex variety Y, i.e., the free abelian group of isomorphism classes V of such vector bundles (or sheaves), divided out by the relation

$$[V] = [V'] + [V''] \quad \text{for any short exact sequence } 0 \to V' \to V \to V'' \to 0.$$

Then $G_0^{an}(Y)$ (or $K_{an}^0(Y)$) is of (co)homological nature, with

$$f_* : G_0^{an}(X) \to G_0^{an}(Y), \quad [\mathcal{F}] \mapsto \sum_{i \geq 0} (-1)^i [R^i f_* \mathcal{F}]$$

the functorial pushdown for a proper holomorphic map $f : X \to Y$. In particular, for X compact, the constant map $k : X \to pt$ is proper, with

$$\chi(H^*(X, \mathcal{F})) = k_*([\mathcal{F}]) \in G_0^{an}(pt) \simeq K_{an}^0(pt) \simeq \mathbb{Z}.$$

Moreover, the tensor product $\otimes_{\mathcal{O}_Y}$ induces a natural pairing

$$\cap = \otimes : K_{an}^0(Y) \times G_0^{an}(Y) \to G_0^{an}(Y),$$

where we identify a holomorphic vector bundle V with its locally free coherent sheaf of sections $\mathcal{V}$. So for X compact we can define a *Kronecker pairing*

$$K_{an}^0(X) \times G_0^{an}(X) \to G_0^{an}(pt) \simeq \mathbb{Z}; \quad \langle [\mathcal{V}], [\mathcal{F}] \rangle := k_*([\mathcal{V} \otimes_{\mathcal{O}_X} \mathcal{F}]).$$

The *total λ-class* of the dual vector bundle

$$\lambda_y(V^\vee) := \sum_{i \geq 0} \Lambda^i(V^\vee) \cdot y^i$$

defines a multiplicative characteristic class

$$\lambda_y((\cdot)^\vee) : K_{an}^0(Y) \to K_{an}^0(Y)[y].$$

And for a compact complex algebraic manifold M one gets the equality

$$\chi_y(M) = \sum_{i \geq 0} k_*[\Omega_M^i] \cdot y^i$$

$$= \langle \lambda_y(T^*M), [\mathcal{O}_M] \rangle \in G_0^{an}(pt)[y] \simeq \mathbb{Z}[y]. \qquad (2\text{-}17)$$

3. Characteristic classes of variations of mixed Hodge structures

This section explains the definition of *cohomological* characteristic classes associated to good variations of mixed Hodge structures on complex algebraic and analytic manifolds. These were previously considered in [12; 13; 35] in connection with Atiyah–Meyer type formulae of Hodge-theoretic nature. Here we also consider important functorial properties of these classes.

3A. Variation of Hodge structures. Let $f : X \to Y$ be a *proper smooth* morphism of complex algebraic varieties or a *projective smooth* morphism of complex analytic varieties. Then the higher direct image sheaf $L = L^n :=$ $R^n f_* \mathbb{Q}_X$ is a *locally constant sheaf* on Y with finite-dimensional stalks

$$L_y = (R^n f_* \mathbb{Q}_X)_y = H^n(\{f = y\}, \mathbb{Q})$$

for $y \in Y$. Let $\mathcal{L} := L \otimes_{\mathbb{Q}_Y} \mathcal{O}_Y \simeq R^n f_*(\Omega^\bullet_{X/Y})$ denote the corresponding holomorphic vector bundle (or locally free sheaf), with $\Omega^\bullet_{X/Y}$ the *relative holomorphic de Rham complex*. Then the stupid filtration of $\Omega^\bullet_{X/Y}$ determines a decreasing filtration F of $\mathcal{L}$ by holomorphic subbundles $F^p\mathcal{L}$, with

$$Gr_F^p((R^{p+q} f_* \mathbb{Q}_X) \otimes_{\mathbb{Q}_Y} \mathcal{O}_Y) \simeq R^q f_*(\Omega^p_{X/Y}), \tag{3-1}$$

inducing for all $y \in Y$ the Hodge filtration F on the cohomology

$$H^n(\{f = y\}, \mathbb{Q}) \otimes \mathbb{C} \simeq \mathcal{L}|_y$$

of the compact and smooth algebraic fiber $\{f = y\}$ (compare [38, Chapter 10]). If Y (and therefore also X) is smooth, then $\mathcal{L}$ gets an induced *integrable Gauss–Manin connection*

$$\nabla : \mathcal{L} \to \mathcal{L} \otimes_{\mathcal{O}_Y} \Omega^1_Y, \quad \text{with } L \simeq \ker \nabla \text{ and } \nabla \circ \nabla = 0,$$

satisfying the *Griffiths transversality condition*

$$\nabla(F^p\mathcal{L}) \subset F^{p-1}\mathcal{L} \otimes_{\mathcal{O}_Y} \Omega^1_Y \quad \text{for all } p. \tag{3-2}$$

This motivates the following notion:

DEFINITION 3.1. A *holomorphic family* (L, F) of Hodge structures of weight n on the reduced complex space Y is a local system L with rational coefficients and finite-dimensional stalks on Y, and a decreasing filtration F of $\mathcal{L} = L \otimes_{\mathbb{Q}_Y} \mathcal{O}_Y$ by holomorphic subbbundles $F^p\mathcal{L}$ such that F determines by $L_y \otimes_{\mathbb{Q}} \mathbb{C} \simeq \mathcal{L}|_y$ a pure Hodge structure of weight n on each stalk L_y ($y \in Y$).

If Y is a smooth complex manifold, then such a holomorphic family (L, F) is called a *variation* of Hodge structures of weight n if, in addition, Griffiths transversality (3-2) holds for the induced connection $\nabla : \mathcal{L} \to \mathcal{L} \otimes_{\mathcal{O}_Y} \Omega^1_Y$.

Finally a *polarization* of (L, F) is a pairing of local systems $S : L \otimes_{\mathbb{Q}_Y} L \to \mathbb{Q}_Y$ that induces a polarization of Hodge structures on each stalk L_y $(y \in Y)$.

For example in the geometric case above, one can get such a polarization on $L = R^n f_* \mathbb{Q}_X$ for $f : X \to Y$ a *projective smooth* morphism of complex algebraic (or analytic) varieties. The existence of a polarization is needed for example for the following important result of Schmid [46, Theorem 7.22]:

THEOREM 3.2 (RIGIDITY). *Let Y be a connected complex manifold Zarisky open in a compact complex analytic manifold $\overline{Y}$, with (L, F) a polarizable variation of pure Hodge structures on Y. Then $H^0(Y, L)$ gets an induced Hodge structure such that the evaluation map $H^0(Y, L) \to L_y$ is an isomorphism of Hodge structures for all $y \in Y$. In particular the variation (L, F) is constant if the underlying local system L is constant.*

3B. Variation of mixed Hodge structures. If one considers a morphism $f : X \to Y$ of complex algebraic varieties with Y smooth, which is a topological fibration with possible singular or noncompact fiber, then the locally constant direct image sheaves $L = L^n := R^n f_* \mathbb{Q}_X$ $(n \geq 0)$ are *variations of mixed Hodge structures* in the sense of the following definitions.

DEFINITION 3.3. Let Y be a reduced complex analytic space. A *holomorphic family of mixed Hodge structures* on Y consists of

(1) a local system L of rational vector spaces on Y with finite-dimensional stalks,
(2) a finite decreasing *Hodge filtration* F of $\mathcal{L} = L \otimes_{\mathbb{Q}_Y} \mathcal{O}_Y$ by holomorphic subbundles $F^p \mathcal{L}$,
(3) a finite increasing *weight filtration* W of L by local subsystems $W_n L$,

such that the induced filtrations on $\mathcal{L}_y \simeq L_y \otimes_{\mathbb{Q}} \mathbb{C}$ and L_y define a mixed Hodge structure on all stalks L_y $(y \in Y)$.

If Y is a smooth complex manifold, such a holomorphic family (L, F, W) is called a *variation of mixed Hodge structures* if, in addition, Griffiths transversality (3-2) holds for the induced connection $\nabla : \mathcal{L} \to \mathcal{L} \otimes_{\mathcal{O}_Y} \Omega_Y^1$.

Finally, (L, F, W) is called *graded polarizable* if the induced family (or variation) of pure Hodge structures $Gr_n^W L$ (with the induced Hodge filtration F) is polarizable for all n.

With the obvious notion of morphisms, the two categories $FmHs^{(p)}(Y)$ and $VmHs^{(p)}(Y)$ of (graded polarizable) families and variations of mixed Hodge structures on Y become abelian categories with a tensor product $\otimes$ and duality $(\cdot)^\vee$. Again, any such morphism is strictly compatible with the Hodge and weight filtrations. Moreover, one has for a holomorphic map $f : X \to Y$ (of complex manifolds) a functorial pullback

$$f^* : FmHs^{(p)}(Y) \to FmHs^{(p)}(X) \quad (\text{or } f^* : VmHs^{(p)}(Y) \to VmHs^{(p)}(X)),$$

commuting with tensor product $\otimes$ and duality $(\cdot)^\vee$. On a point space pt one just gets back the category

$$FmHs^{(p)}(pt) = VmHs^{(p)}(pt) = mHs^{(p)}$$

of (graded polarizable) mixed Hodge structures. Using the pullback under the constant map $k : Y \to pt$, we get the constant family (or variation) of Tate Hodge structures $\mathbb{Q}_Y(n) := k^*\mathbb{Q}(n)$ on Y.

3C. Cohomological characteristic classes.

The Grothendieck group $K^0_{an}(Y)$ of holomorphic vector bundles on the complex variety Y is a commutative ring with multiplication induced by $\otimes$ and has a duality involution induced by $(\cdot)^\vee$. For a holomorphic map $f : X \to Y$ one has a functorial pullback f^* of rings with involutions. The situation is similar for $K^0_{an}(Y)[y^{\pm 1}]$, if we extend the duality involution by

$$([V] \cdot y^k)^\vee := [V^\vee] \cdot (1/y)^k.$$

For a family (or variation) of mixed Hodge structures (L, F, W) on Y let us introduce the characteristic class

$$MHC^y((L, F, W)) := \sum_p [Gr_F^p(\mathcal{L})] \cdot (-y)^p \in K^0_{an}(Y)[y^{\pm 1}]. \qquad (3\text{-}3)$$

Because morphisms of families (or variations) of mixed Hodge structures are strictly compatible with the Hodge filtrations, we get induced group homomorphisms of Grothendieck groups:

$$MHC^y : K_0(FmHs^{(p)}(Y)) \to K^0_{an}(Y)[y^{\pm 1}],$$
$$MHC^y : K_0(VmHs^{(p)}(Y)) \to K^0_{an}(Y)[y^{\pm 1}].$$

Note that $MHC^{-1}((L, F, W)) = [\mathcal{L}] \in K^0_{an}(Y)$ is just the class of the associated holomorphic vector bundle. And for $Y = pt$ a point, we get back the χ_y-genus:

$$\chi_y = MHC^y : K_0(mHs^{(p)}) = K_0(FmHs^{(p)}(pt)) \to K^0_{an}(pt)[y^{\pm 1}] = \mathbb{Z}[y^{\pm 1}].$$

THEOREM 3.4. *The transformations*

$$MHC^y : K_0(FmHs^{(p)}(Y)) \to K^0_{an}(Y)[y^{\pm 1}],$$
$$MHC^y : K_0(VmHs^{(p)}(Y)) \to K^0_{an}(Y)[y^{\pm 1}],$$

are contravariant functorial, and are transformations of commutative rings with unit, i.e., they commute with products and respect units: $MHC^y([\mathbb{Q}_Y(0)]) = [\mathcal{O}_Y]$. *Similarly they respect duality involutions:*

$$MHC^y([(L, F, W)^\vee]) = \sum_p [(Gr_F^{-p}(\mathcal{L}))^\vee] \cdot (-y)^p = \left(MHC^y([(L, F, W)])\right)^\vee.$$

EXAMPLE 3.5. Let $f : X \to Y$ be a *proper smooth* morphism of complex algebraic varieties or a *projective smooth* morphism of complex analytic varieties, so that the higher direct image sheaf $L^n := R^n f_* \mathbb{Q}_X$ $(n \geq 0)$ with the induced Hodge filtration as in (3-1) defines a holomorphic family of pure Hodge structures on Y. If m is the complex dimension of the fibers, then $L_n = 0$ for $n > 2m$, so one can define

$$[Rf_* \mathbb{Q}_X] := \sum_{n=0}^{2m} (-1)^n \cdot [(R^n f_* \mathbb{Q}_X, F)] \in K_0(FmHs(Y)).$$

Then one gets, by (3-1),

$$MHC^y([Rf_* \mathbb{Q}_X]) = \sum_{p,q \geq 0} (-1)^{p+q} \cdot [R^q f_* \Omega_{X/Y}^p] \cdot (-y)^p$$

$$= \sum_{p \geq 0} f_*[\Omega_{X/Y}^p] \cdot y^p$$

$$=: f_* \left(\lambda_y(T_{X/Y}^*)\right) \in K_{an}^0(Y)[y]. \qquad (3\text{-}4)$$

Assume moreover that

(a) Y is a connected complex manifold Zarisky open in a compact complex analytic manifold $\overline{Y}$, and

(b) all direct images sheaves $L^n := R^n f_* \mathbb{Q}_X$ $(n \geq 0)$ are *constant*.

Then one gets by the *rigidity theorem* 3.2 (for $z \in Y$):

$$f_* \left(\lambda_y(T_{X/Y}^*)\right) = \chi_y(\{f = z\}) \cdot [\mathcal{O}_Y] \in K_{an}^0(Y)[y].$$

COROLLARY 3.6 (MULTIPLICATIVITY). *Let $f : X \to Y$ be a smooth morphism of compact complex algebraic manifolds, with Y connected. Let $T_{X/Y}^*$ be the relative holomorphic cotangent bundle of the fibers, fitting into the short exact sequence*

$$0 \to f^* T^* Y \to T^* X \to T_{X/Y}^* \to 0.$$

Assume all direct images sheaves $L^n := R^n f_* \mathbb{Q}_X$ $(n \geq 0)$ are constant, i.e., $\pi_1(Y)$ acts trivially on the cohomology $H^*(\{f = z\})$ of the fiber. Then one *gets the multiplicativity of the χ_y-genus (with $k : Y \to pt$ the constant map):*

$$\begin{aligned}
\chi_y(X) &= (k \circ f)_* [\lambda_y(T^*X)] \\
&= k_* f_* \left([\lambda_y(T^*_{X/Y})] \otimes f^*[\lambda_y(T^*Y)] \right) \\
&= k_* \left(\chi_y(\{f = z\}) \cdot [\lambda_y(T^*Y)] \right) \\
&= \chi_y(\{f = z\}) \cdot \chi_y(Y).
\end{aligned} \tag{3-5}$$

REMARK 3.7. The multiplicativity relation (3-5) specializes for $y = 1$ to the classical multiplicativity formula

$$\operatorname{sgn}(X) = \operatorname{sgn}(\{f = z\}) \cdot \operatorname{sgn}(Y)$$

of Chern, Hirzebruch, and Serre [16] for the signature of an oriented fibration of smooth coherently oriented compact manifolds, if $\pi_1(Y)$ acts trivially on the cohomology $H^*(\{f = z\})$ of the fiber. So it is a Hodge theoretic counterpart of this. Moreover, the corresponding Euler characteristic formula for $y = -1$

$$\chi(X) = \chi(\{f = z\}) \cdot \chi(Y)$$

is even true *without* $\pi_1(Y)$ acting trivially on the cohomology $H^*(\{f = z\})$ of the fiber!

The Chern–Hirzebruch–Serre signature formula was motivational for many subsequent works which studied monodromy contributions to invariants (genera and characteristic classes). See, for exmaple, [1; 4; 10; 11; 12; 13; 14; 35; 36].

Instead of working with holomorphic vector bundles, we can of course also use only the underlying topological complex vector bundles, which gives the forgetful transformation

$$For : K^0_{an}(Y) \to K^0_{top}(Y).$$

Here the target can also be viewed as the even part of $\mathbb{Z}_2$-graded topological complex K-cohomology. Of course, the forgetful transformation is contravariant functorial and commutes with product $\otimes$ and with duality $(\cdot)^\vee$. This duality induces a $\mathbb{Z}_2$-grading on $K^0_{top}(Y)[\frac{1}{2}]$ by splitting into the (anti-)invariant part, and similarly for $K^0_{an}(Y)[\frac{1}{2}]$. Then the (anti-)invariant part of $K^0_{top}(Y)[\frac{1}{2}]$ can be identified with the even part of $\mathbb{Z}_4$-graded topological real K-theory $KO^0_{top}(Y)[\frac{1}{2}]$ (and $KO^2_{top}(Y)[\frac{1}{2}]$).

Assume now that (L, F) is a holomorphic family of pure Hodge structures of weight n on the complex variety Y, with a polarization $S : L \otimes_{\mathbb{Q}_Y} L \to \mathbb{Q}_Y$. This induces an isomorphism of families of pure Hodge structures of weight n:

$$L \simeq L^\vee(-n) := L^\vee \otimes \mathbb{Q}_Y(-n).$$

So if we choose the isomorphism of rational local systems

$$\mathbb{Q}_Y(-n) = (2\pi i)^{-n} \cdot \mathbb{Q}_Y \simeq \mathbb{Q}_Y,$$

the polarization induces a $(-1)^n$-symmetric duality isomorphism $L \simeq L^\vee$ of the underlying local systems. And for such an (anti)symmetric selfdual local system L Meyer [36] has introduced a KO-characteristic class

$$[L]_{KO} \in KO_{\text{top}}^0(Y)[\tfrac{1}{2}] \oplus KO_{\text{top}}^2(Y)[\tfrac{1}{2}]) = K_{\text{top}}^0(Y)[\tfrac{1}{2}],$$

so that for Y a compact oriented manifold of even real dimension $2m$ the following *twisted signature formula* is true:

$$\text{sgn}(H^m(Y, L)) = \langle ch^*(\Psi^2([L]_{KO})), L^*(TM) \cap [M] \rangle. \quad (3\text{-}6)$$

Here $H^m(Y, L)$ gets an induced (anti)symmetric duality, with $\text{sgn}(H^m(Y, L))$ defined as 0 in case of an antisymmetric pairing. Moreover ch^* is the Chern character, Ψ^2 the second Adams operation and L^* is the Hirzebruch–Thom L-class.

We now explain that $[L]_{KO}$ agrees up to some universal signs with

$$For(MHC^1((L, F)).$$

The underlying topological complex vector bundle of $\mathcal{L}$ has a natural real structure, so that, as a topological complex vector bundle, one gets an orthogonal decomposition

$$\mathcal{L} = \bigoplus_{p+q=n} \mathcal{H}^{p,q}, \quad \text{with } \mathcal{H}^{p,q} = F^p\mathcal{L} \cap \overline{F^q\mathcal{L}} = \overline{\mathcal{H}^{q,p}},$$

with

$$For(MHC^1((L, F)) = \sum_{\substack{p \text{ even} \\ q}} [\mathcal{H}^{p,q}] - \sum_{\substack{p \text{ odd} \\ q}} [\mathcal{H}^{p,q}]. \quad (3\text{-}7)$$

If n is even, both sums on the right are invariant under conjugation. And, by (2-7), $(-1)^{-n/2} \cdot S$ is positive definite on the corresponding real vector bundle $(\bigoplus_{p \text{ even},q} \mathcal{H}^{p,q})_\mathbb{R}$, and negative definite on $(\bigoplus_{p \text{ odd},q} \mathcal{H}^{p,q})_\mathbb{R}$. So if we choose the pairing $(-1)^{n/2} \cdot S$ for the isomorphism $L \simeq L^\vee$, then this agrees with the splitting introduced by Meyer [36] in the definition of his KO-characteristic class $[L]_{KO}$ associated to this *symmetric* duality isomorphism of L:

$$For(MHC^1((L, F)) = [L]_{KO} \in KO_{\text{top}}^0(Y)[\tfrac{1}{2}].$$

Similarly, if n is odd, both sums of the right hand side in (3-7) are exchanged under conjugation. If we choose the pairing $(-1)^{(n+1)/2} \cdot S$ for the isomorphism $L \simeq L^\vee$, then this agrees by Definition 2.2 with the splitting introduced by

Meyer [36] in the definition of his KO-characteristic class $[L]_{KO}$ associated to this *antisymmetric* duality isomorphism of L:

$$For(MHC^1((L, F)) = [L]_{KO} \in KO^2_{top}(Y)[\tfrac{1}{2}].$$

COROLLARY 3.8. *Let (L, F) be a holomorphic family of pure Hodge structures of weight n on the complex variety Y, with a polarization S chosen. The class $[L]_{KO}$ introduced in [36] for the duality isomorphism coming from the pairing $(-1)^{n(n+1)/2} \cdot S$ is equal to*

$$For(MHC^1((L, F)) = [L]_{KO} \in KO^0_{top}(Y)[\tfrac{1}{2}] \oplus KO^2_{top}(Y)[\tfrac{1}{2}] = K^0_{top}(Y)[\tfrac{1}{2}].$$

It is therefore independent of the choice of the polarization S. Moreover, this identification is functorial under pullback and compatible with products (as defined in [36, p. 26] for (anti)symmetric selfdual local systems).

There are Hodge theoretic counterparts of the twisted signature formula (3-6). Here we formulate a corresponding K-theoretical result. Let (L, F, W) be a variation of mixed Hodge structures on the m-dimensional complex manifold M. Then

$$H^n(M, L) \simeq H^n(M, DR(\mathcal{L}))$$

gets an induced (decreasing) F filtration coming from the filtration of the holomorphic de Rham complex of the vector bundle $\mathcal{L}$ with its integrable connection ∇:

$$DR(\mathcal{L}) = [\mathcal{L} \xrightarrow{\nabla} \cdots \xrightarrow{\nabla} \mathcal{L} \otimes_{\mathcal{O}_M} \Omega^m_M]$$

(with $\mathcal{L}$ in degree zero), defined by

$$F^p DR(\mathcal{L}) = [F^p \mathcal{L} \xrightarrow{\nabla} \cdots \xrightarrow{\nabla} F^{p-m}\mathcal{L} \otimes_{\mathcal{O}_M} \Omega^m_M]. \qquad (3\text{-}8)$$

Note that here we are using Griffiths transversality (3-2)!

The following result is due to Deligne and Zucker [56, Theorem 2.9, Lemma 2.11] in the case of a compact Kähler manifold, whereas the case of a compact complex algebraic manifold follows from Saito's general results as explained in the next section.

THEOREM 3.9. *Assume M is a compact Kähler manifold or a compact complex algebraic manifold, with (L, F, W) a graded polarizable variation of mixed (or pure) Hodge structures on M. Then $H^n(M, L) \simeq H^n(M, DR(\mathcal{L}))$ gets an induced mixed (or pure) Hodge structure with F the Hodge filtration. Moreover, the corresponding Hodge to de Rham spectral sequence degenerates at E_1 so that*

$$Gr^p_F(H^n(M, L)) \simeq H^n(M, Gr^p_F DR(\mathcal{L})) \quad \text{for all } n, p.$$

Therefore one gets as a corollary (compare [12; 13; 35]):

$$
\begin{aligned}
\chi_y(H^*(M, L)) &= \sum_{n,p} (-1)^n \cdot \dim_{\mathbb{C}} \left(H^n(M, Gr_F^p DR(\mathcal{L})) \right) \cdot (-y)^p \\
&= \sum_p \chi \left(H^*(M, Gr_F^p DR(\mathcal{L})) \right) \cdot (-y)^p \\
&= \sum_{p,i} (-1)^i \cdot \chi \left(H^*(M, Gr_F^{p-i}(\mathcal{L}) \otimes_{\mathcal{O}_M} \Omega_M^i) \right) \cdot (-y)^p \\
&= k_* \left(MHC^y(L) \otimes \lambda_y(T^*M) \right) \\
&=: \langle MHC^y(L), \lambda_y(T^*M) \cap [\mathcal{O}_M] \rangle \in \mathbb{Z}[y^{\pm 1}].
\end{aligned}
\tag{3-9}
$$

3D. Good variation of mixed Hodge structures.

DEFINITION 3.10 (GOOD VARIATION). Let M be a complex algebraic manifold. A graded polarizable variation of mixed Hodge structures (L, F, W) on M is called good if it is *admissible* in the sense of Steenbrink and Zucker [50] and Kashiwara [28], with *quasi-unipotent monodromy* at infinity, i.e., with respect to a compactification $\overline{M}$ of M by a compact complex algebraic manifold $\overline{M}$, with complement $D := \overline{M} \setminus M$ a normal crossing divisor with smooth irreducible components.

EXAMPLE 3.11 (PURE AND GEOMETRIC VARIATIONS). Two important examples for such a good variation of mixed Hodge structures are the following:

(i) A polarizable variation of *pure* Hodge structures is always admissible by a deep theorem of Schmid [46, Theorem 6.16]. So it is good precisely when it has quasi-unipotent monodromy at infinity.

(ii) Consider a morphism $f : X \to Y$ of complex algebraic varieties with Y smooth, which is a topological fibration with possible singular or noncompact fiber. The locally constant direct image sheaves $R^n f_* \mathbb{Q}_X$ and $R^n f_! \mathbb{Q}_X$ ($n \geq 0$) are good variations of mixed Hodge structures (compare Remark 4.4).

This class of good variations on M is again an abelian category $VmHs^g(M)$ stable under tensor product $\otimes$, duality $(\cdot)^\vee$ and pullback f^* for f an algebraic morphism of complex algebraic manifolds. Moreover, in this case all vector bundles $F^p L$ of the Hodge filtration carry the structure of a unique underlying complex algebraic vector bundle (in the Zariski topology), so that the characteristic class transformation MHC^y can be seen as a natural contravariant transformation of rings with involution

$$
MHC^y : K_0(VmHs^g(M)) \to K_{\mathrm{alg}}^0(M)[y^{\pm 1}].
$$

438 JÖRG SCHÜRMANN

In fact, consider a (partial) compactification $\overline{M}$ of M as above, with $D :=$ $\overline{M} \setminus M$ a normal crossing divisor with smooth irreducible components and $j :$ $M \to \overline{M}$ the open inclusion. Then the holomorphic vector bundle $\mathcal{L}$ with integrable connection ∇ corresponding to L has a unique *canonical Deligne extension* $(\overline{\mathcal{L}}, \overline{\nabla})$ to a holomorphic vector bundle $\overline{\mathcal{L}}$ on $\overline{M}$, with *meromorphic* integrable connection

$$\overline{\nabla} : \overline{\mathcal{L}} \to \overline{\mathcal{L}} \otimes_{\mathcal{O}_{\overline{M}}} \Omega^1_{\overline{M}}(\log(D)) \tag{3-10}$$

having *logarithmic poles* along D. Here the *residues* of $\overline{\nabla}$ along D have real eigenvalues, since L has *quasi-unipotent monodromy* along D. And the canonical extension is characterized by the property that all these eigenvalues are in the half-open interval $[0, 1)$. Moreover, also the Hodge filtration F of $\mathcal{L}$ extends uniquely to a filtration F of $\overline{\mathcal{L}}$ by holomorphic subvector bundles

$$F^p \overline{\mathcal{L}} := j_*(F^p \mathcal{L}) \cap \overline{\mathcal{L}} \subset j_* \mathcal{L},$$

since L is *admissible* along D. Finally, Griffiths transversality extends to

$$\overline{\nabla}(F^p \overline{\mathcal{L}}) \subset F^{p-1} \overline{\mathcal{L}} \otimes_{\mathcal{O}_{\overline{M}}} \Omega^1_{\overline{M}}(\log(D)) \quad \text{for all } p. \tag{3-11}$$

For more details see [19, Proposition 5.4] and [38, § 11.1, 14.4].

If we choose $\overline{M}$ as a compact algebraic manifold, then we can apply Serre's GAGA theorem to conclude that $\overline{\mathcal{L}}$ and all $F^p \overline{\mathcal{L}}$ are *algebraic* vector bundles, with $\overline{\nabla}$ an *algebraic* meromorphic connection.

REMARK 3.12. The canonical Deligne extension $\overline{\mathcal{L}}$ (as above) with its Hodge filtration F has the following compabilities (compare [19, Part II]):

SMOOTH PULLBACK: Let $f : \overline{M}' \to \overline{M}$ be a smooth morphism so that $D' :=$ $f^{-1}(D)$ is also a normal crossing divisor with smooth irreducible components on $\overline{M}'$ with complement M'. Then one has

$$f^*(\overline{\mathcal{L}}) \simeq \overline{f^* \mathcal{L}} \quad \text{and} \quad f^*(F^p \overline{\mathcal{L}}) \simeq F^p \overline{f^* \mathcal{L}} \text{ for all } p. \tag{3-12}$$

EXTERIOR PRODUCT: Let L and L' be two good variations on M and M'. Then their canonical Deligne extensions satisfy

$$\overline{\mathcal{L} \boxtimes_{\mathcal{O}_{M \times M'}} \mathcal{L}'} \simeq \overline{\mathcal{L}} \boxtimes_{\mathcal{O}_{\overline{M} \times \overline{M}'}} \overline{\mathcal{L}'},$$

since the residues of the corresponding meromorphic connections are compatible. Then one has for all p

$$F^p(\overline{\mathcal{L} \boxtimes_{\mathcal{O}_{M \times M'}} \mathcal{L}'}) \simeq \bigoplus_{i+k=p} (F^i \overline{\mathcal{L}}) \boxtimes_{\mathcal{O}_{\overline{M} \times \overline{M}'}} (F^k \overline{\mathcal{L}'}). \tag{3-13}$$

TENSOR PRODUCT: In general the canonical Deligne extensions of two good variations L and L' on M are *not* compatible with tensor products, because of the choice of different residues for the corresponding meromorphic connections. This problem doesn't appear if one of these variations, lets say L', is already defined on $\overline{M}$. Let L and L' be a good variation on M and $\overline{M}$, respectively. Then their canonical Deligne extensions satisfy

$$\overline{\mathcal{L} \otimes_{\mathcal{O}_M} (\mathcal{L}'|M)} \simeq \overline{\mathcal{L}} \otimes_{\mathcal{O}_{\overline{M}}} \mathcal{L}',$$

and one has for all p:

$$F^p(\overline{\mathcal{L} \otimes_{\mathcal{O}_M} (\mathcal{L}'|M)}) \simeq \bigoplus_{i+k=p} (F^i \overline{\mathcal{L}}) \otimes_{\mathcal{O}_{\overline{M}}} (F^k \mathcal{L}'). \qquad (3\text{-}14)$$

Let $\overline{M}$ be a partial compactification of M as before, i.e., we don't assume that $\overline{M}$ is compact, with $m := \dim_{\mathbb{C}}(M)$. Then the *logarithmic de Rham complex*

$$DR_{\log}(\overline{\mathcal{L}}) := [\overline{\mathcal{L}} \xrightarrow{\overline{\nabla}} \cdots \xrightarrow{\overline{\nabla}} \overline{\mathcal{L}} \otimes_{\mathcal{O}_{\overline{M}}} \Omega^m_{\overline{M}}(\log(D))]$$

(with $\overline{\mathcal{L}}$ in degree zero) is by [19] quasi-isomorphic to $Rj_* L$, so that

$$H^*(M, L) \simeq H^*(\overline{M}, DR_{\log}(\overline{\mathcal{L}})).$$

So these cohomology groups get an induced (decreasing) F-filtration coming from the filtration

$$F^p DR_{\log}(\overline{\mathcal{L}}) = [F^p \overline{\mathcal{L}} \xrightarrow{\overline{\nabla}} \cdots \xrightarrow{\overline{\nabla}} F^{p-m} \overline{\mathcal{L}} \otimes_{\mathcal{O}_{\overline{M}}} \Omega^m_{\overline{M}}(\log(D))].$$
$$(3\text{-}15)$$

For $\overline{M}$ a compact algebraic manifold, this is again the Hodge filtration of an induced mixed Hodge structure on $H^*(M, L)$ (compare with Corollary 4.7).

THEOREM 3.13. *Assume $\overline{M}$ is a smooth algebraic compactification of the algebraic manifold M with the complement D a normal crossing divisor with smooth irreducible components. Let (L, F, W) be a good variation of mixed Hodge structures on M. Then $H^n(M, L) \simeq H^*(\overline{M}, DR_{\log}(\overline{\mathcal{L}}))$ gets an induced mixed Hodge structure with F the Hodge filtration. Moreover, the corresponding Hodge to de Rham spectral sequence degenerates at E_1 so that*

$$Gr_F^p(H^n(M, L)) \simeq H^n(M, Gr_F^p DR_{\log}(\overline{\mathcal{L}})) \quad \text{for all } n, p.$$

Therefore one gets as a corollary (compare [12; 13; 35]):

$$\chi_y(H^*(M, L)) = \sum_{n,p} (-1)^n \cdot \dim_{\mathbb{C}} \left(H^n \left(M, Gr_F^p DR_{\log}(\bar{\mathcal{L}}) \right) \right) \cdot (-y)^p$$

$$= \sum_p \chi \left(H^* \left(M, Gr_F^p DR_{\log}(\bar{\mathcal{L}}) \right) \right) \cdot (-y)^p$$

$$= \sum_{p,i} (-1)^i \chi \left(H^* \left(M, Gr_F^{p-i}(\bar{\mathcal{L}}) \otimes_{\mathcal{O}_{\overline{M}}} \Omega^i_{\overline{M}}(\log(D)) \right) \right) (-y)^p$$

$$=: \left\langle MHC^y(Rj_*L), \lambda_y \left(\Omega^1_{\overline{M}}(\log(D)) \right) \cap [\mathcal{O}_{\overline{M}}] \right\rangle \in \mathbb{Z}[y^{\pm 1}].$$
$$(3\text{-}16)$$

Here we use the notation

$$MHC^y(Rj_*L) := \sum_p [Gr_F^p(\bar{\mathcal{L}})] \cdot (-y)^p \in K^0_{\mathrm{alg}}(\overline{M})[y^{\pm 1}]. \qquad (3\text{-}17)$$

Remark 3.12 then implies:

COROLLARY 3.14. *Let $\overline{M}$ be a smooth algebraic partial compactifiction of the algebraic manifold M with the complement D a normal crossing divisor with smooth irreducible components. Then $MHC^y(Rj_*(\cdot))$ induces a transformation*

$$MHC^y(j_*(\cdot)) : K_0(VmHs^g(M)) \to K^0_{\mathrm{alg}}(\overline{M})[y^{\pm 1}].$$

(1) *This is contravariant functorial for a smooth morphism $f : \overline{M}' \to \overline{M}$ of such partial compactifications, i.e.,*

$$f^* \left(MHC^y(j_*(\cdot)) \right) \simeq MHC^y \left(j'_*(f^*(\cdot)) \right).$$

(2) *It commutes with exterior products for two good variations L, L':*

$$MHC^y \left((j \times j')_*[(L \boxtimes_{\mathbb{Q}_{M \times M'}} L')] \right) = MHC^y(j_*[L]) \boxtimes MHC^y(j'_*[(L')]).$$

(3) *Let L be a good variation on M, and L' one on $\overline{M}$. Then $MHC^y(j_*[\cdot])$ is multiplicative in the sense that*

$$MHC^y \left(j_*[(L \otimes_{\mathbb{Q}_M} (L'|M)]) \right) = MHC^y(j_*[L]) \otimes MHC^y([L']).$$

4. Calculus of mixed Hodge modules

4A. Mixed Hodge modules. Before discussing extensions of the characteristic cohomology classes MHC^y to the singular setting, we need to briefly recall some aspects of Saito's theory [39; 40; 41; 43; 44] of algebraic mixed Hodge modules, which play the role of singular extensions of good variations of mixed Hodge structures.

To each complex algebraic variety Z, Saito associated a category $MHM(Z)$ of *algebraic mixed Hodge modules* on Z (cf. [39; 40]). If Z is smooth, an object of this category consists of an algebraic (regular) holonomic D-module $(\mathcal{M}, F)$ with a good filtration F together with a perverse sheaf K of rational vector spaces, both endowed a finite increasing filtration W such that

$$\alpha : DR(\mathcal{M})^{\mathrm{an}} \simeq K \otimes_{\mathbb{Q}_Z} \mathbb{C}_Z \quad \text{is compatible with } W$$

under the Riemann–Hilbert correspondence coming from the (shifted) analytic de Rham complex (with α a chosen isomorphism). Here we use left D-modules, and the sheaf $\mathcal{D}_Z$ of algebraic differential operators on Z has the increasing filtration F with $F_i \mathcal{D}_Z$ given by the differential operators of order $\leq i$ ($i \in \mathbb{Z}$). Then a *good* filtration F of the algebraic holonomic D-module $\mathcal{M}$ is given by a bounded from below, increasing and exhaustive filtration $F_p \mathcal{M}$ by *coherent* algebraic $\mathcal{O}_Z$-modules such that

$$F_i \mathcal{D}_Z (F_p \mathcal{M}) \subset F_{p+i} \mathcal{M} \quad \text{for all } i, p,$$
$$\text{and this is an equality for } i \text{ big enough.} \tag{4-1}$$

In general, for a singular variety Z one works with suitable local embeddings into manifolds and corresponding filtered D-modules supported on Z. In addition, these objects are required to satisfy a long list of complicated properties (not needed here). The *forgetful* functor rat is defined as

$$\mathrm{rat} : MHM(Z) \to \mathrm{Perv}(\mathbb{Q}_Z), \quad (\mathcal{M}(F), K, W) \mapsto K.$$

THEOREM 4.1 (M. SAITO). *$MHM(Z)$ is an abelian category with*

$$\mathrm{rat} : MHM(Z) \to \mathrm{Perv}(\mathbb{Q}_Z)$$

exact and faithful. It extends to a functor

$$\mathrm{rat} : D^b MHM(Z) \to D^b_c(\mathbb{Q}_Z)$$

to the derived category of complexes of $\mathbb{Q}$-sheaves with algebraically constructible cohomology. There are functors

$$f_*, \ f_!, \ f^*, \ f^!, \ \otimes, \ \boxtimes, \ \mathcal{D} \quad \text{on } D^b MHM(Z),$$

which are "lifts" via rat *of the similar (derived) functors defined on $D^b_c(\mathbb{Q}_Z)$, with (f^*, f_*) and $(f_!, f^!)$ also pairs of adjoint functors. One has a natural map $f_! \to f_*$, which is an isomorphism for f proper. Here $\mathcal{D}$ is a duality involution $\mathcal{D}^2 \simeq \mathrm{id}$ "lifting" the Verdier duality functor, with*

$$\mathcal{D} \circ f^* \simeq f^! \circ \mathcal{D} \quad \text{and} \quad \mathcal{D} \circ f_* \simeq f_! \circ \mathcal{D}.$$

Compare with [40, Theorem 0.1 and §4] for more details (as well as with [43] for a more general formal abstraction). The usual truncation $\tau_{\leq}$ on $D^b MHM(Z)$ corresponds to the *perverse truncation* $^p\tau_{\leq}$ on $D^b_c(Z)$ so that

$$\text{rat} \circ H = {}^p\mathcal{H} \circ \text{rat},$$

where H stands for the cohomological functor in $D^b MHM(Z)$ and $^p\mathcal{H}$ denotes the perverse cohomology (always with respect to the self-dual middle perversity).

EXAMPLE 4.2. Let M be a complex algebraic manifold of pure complex dimension m, with (L, F, W) a good variation of mixed Hodge structures on M. Then $\mathcal{L}$ with its integrable connection ∇ is a holonomic (left) D-module with $\alpha : DR(\mathcal{L})^{\text{an}} \simeq L[m]$, where this time we use the shifted de Rham complex

$$DR(\mathcal{L}) := [\mathcal{L} \xrightarrow{\nabla} \cdots \xrightarrow{\nabla} \mathcal{L} \otimes_{\mathcal{O}_M} \Omega_M^m]$$

with $\mathcal{L}$ in degree $-m$, so that $DR(\mathcal{L})^{\text{an}} \simeq L[m]$ is a perverse sheaf on M. The filtration F induces by Griffiths transversality (3-2) a good filtration $F_p(\mathcal{L}) := F^{-p}\mathcal{L}$ as a filtered D-module. As explained before, this comes from an underlying algebraic filtered D-module. Finally α is compatible with the induced filtration W defined by

$$W^i(L[m]) := W^{i-m}L[m] \quad \text{and} \quad W^i(\mathcal{L}) := (W^{i-m}L) \otimes_{\mathbb{Q}_M} \mathcal{O}_M.$$

And this defines a mixed Hodge module $\mathcal{M}$ on M, with $\text{rat}(\mathcal{M})[-m]$ a local system on M.

A mixed Hodge module $\mathcal{M}$ on the pure m-dimensional complex algebraic manifold M is called *smooth* if $\text{rat}(\mathcal{M})[-m]$ is a local system on M. Then this example corresponds to [40, Theorem 0.2], whereas the next theorem corresponds to [40, Theorem 3.27 and remark on p. 313]:

THEOREM 4.3 (M. SAITO). *Let M be a pure m-dimensional complex algebraic manifold. Associating to a good variation of mixed Hodge structures $\mathbb{V} = (L, F, W)$ on M the mixed Hodge module $\mathcal{M} := \mathbb{V}_H$ as in Example 4.2 defines an equivalence of categories*

$$MHM(M)_{sm} \simeq VmHs^g(M)$$

between the categories of smooth mixed Hodge modules $MHM(M)_{sm}$ and good variation of mixed Hodge structures on M. This commutes with exterior product $\boxtimes$ and with the pullbacks

$$f^* : VmHs^g(M) \to VmHs^g(M') \quad \text{and} \quad f^*[m'-m] : MHM(M) \to MHM(M')$$

for an algebraic morphism of smooth algebraic manifolds M, M' of dimension m, m'. For $M = pt$ a point, one gets in particular an equivalence

$$MHM(pt) \simeq mHs^p.$$

REMARK 4.4. These two theorems explain why a geometric variation of mixed Hodge structures as in Example 3.11(2) is good.

By the last identification of the theorem, there exists a unique Tate object

$$\mathbb{Q}^H(n) \in MHM(pt)$$

such that $\text{rat}(\mathbb{Q}^H(n)) = \mathbb{Q}(n)$ and $\mathbb{Q}^H(n)$ is of type $(-n, -n)$:

$$MHM(pt) \ni \mathbb{Q}^H(n) \simeq \mathbb{Q}(n) \in mHs^p.$$

For a complex variety Z with constant map $k : Z \to pt$, define

$$\mathbb{Q}_Z^H(n) := k^*\mathbb{Q}^H(n) \in D^b MHM(Z), \quad \text{with } \text{rat}(\mathbb{Q}_Z^H(n)) = \mathbb{Q}_Z(n).$$

So tensoring with $\mathbb{Q}_Z^H(n)$ defines the Tate twist $\cdot(n)$ of mixed Hodge modules. To simplify the notation, let $\mathbb{Q}_Z^H := \mathbb{Q}_Z^H(0)$. If Z is *smooth* of complex dimension n then $\mathbb{Q}_Z[n]$ is perverse on Z, and $\mathbb{Q}_Z^H[n] \in MHM(Z)$ is a single mixed Hodge module, explicitly described by

$$\mathbb{Q}_Z^H[n] = ((\mathcal{O}_Z, F), \mathbb{Q}_Z[n], W), \quad \text{with } gr_i^F = 0 = gr_{i+n}^W \text{ for all } i \neq 0.$$

It follows from the definition that every $\mathcal{M} \in MHM(Z)$ has a finite increasing *weight filtration* W so that the functor $M \to Gr_k^W M$ is exact. We say that $\mathcal{M} \in D^b MHM(Z)$ has *weights* $\leq n$ (resp. $\geq n$) if $Gr_j^W H^i M = 0$ for all $j > n+i$ (resp. $j < n+i$). $\mathcal{M}$ is called *pure of weight n* if it has weights both $\leq n$ and $\geq n$. For the following results compare with [40, Proposition 2.26 and (4.5.2)]:

PROPOSITION 4.5. *If f is a map of algebraic varieties, then $f_!$ and f^* preserve weight $\leq n$, and f_* and $f^!$ preserve weight $\geq n$. If f is smooth of pure complex fiber dimension m, then $f^! \simeq f^*[2m](m)$ so that $f^*, f^!$ preserve pure objects for f smooth. Moreover, if $\mathcal{M} \in D^b MHM(X)$ is pure and $f : X \to Y$ is proper, then $f_*\mathcal{M} \in D^b MHM(Y)$ is pure of the same weight as $\mathcal{M}$.*

Similarly the duality functor $\mathcal{D}$ exchanges "weight $\leq n$" and "weight $\geq -n$", in particular it preserves pure objects. Finally let $j : U \to Z$ be the inclusion of a Zariski open subset. Then the intermediate extension *functor*

$$j_{!*} : MHM(U) \to MHM(Z) : \mathcal{M} \mapsto \text{Im}(H^0(j_!\mathcal{M}) \to H^0(j_*(\mathcal{M}))) \quad (4\text{-}2)$$

preserves weight $\leq n$ and $\geq n$, and so preserves pure objects (of weight n).

We say that $\mathcal{M} \in D^b MHM(Z)$ is supported on $S \subset Z$ if and only if $\mathrm{rat}(\mathcal{M})$ is supported on S. There are the abelian subcategories $MH(Z, k)^p \subset MHM(Z)$ of pure Hodge modules of weight k, which in the algebraic context are assumed to be polarizable (and extendable at infinity).

For each $k \in \mathbb{Z}$, the abelian category $MH(Z, k)^p$ is semisimple, in the sense that every pure Hodge module on Z can be uniquely written as a finite direct sum of pure Hodge modules with strict support in irreducible closed subvarieties of Z. Let $MH_S(Z, k)^p$ denote the subcategory of *pure Hodge modules of weight k with strict support in S*. Then every $\mathcal{M} \in MH_S(Z, k)^p$ is generically a good variation of Hodge structures $\mathbb{V}_U$ of weight $k - d$ (where $d = \dim S$) on a Zariski dense smooth open subset $U \subset S$; i.e., $\mathbb{V}_U$ is polarizable with quasi-unipotent monodromy at infinity. This follows from Theorem 4.3 and the fact that a perverse sheaf is generically a shifted local system on a smooth dense Zariski open subset $U \subset S$. Conversely, every such good variation of Hodge structures $\mathbb{V}$ on such an U corresponds by Theorem 4.3 to a pure Hodge module $\mathbb{V}_H$ on U, which can be extended in an unique way to a pure Hodge module $j_{!*}\mathbb{V}_H$ on S with strict support (here $j : U \to S$ is the inclusion). Under this correspondence, for $M \in MH_S(Z, k)^p$ we have that

$$\mathrm{rat}(\mathcal{M}) = IC_S(\mathbb{V})$$

is the *twisted intersection cohomology complex* for $\mathbb{V}$ the corresponding variation of Hodge structures. Similarly

$$\mathcal{D}(j_{!*}\mathbb{V}_H) \simeq j_{!*}(\mathbb{V}_H^\vee)(d). \tag{4-3}$$

Moreover, a *polarization* of $\mathcal{M} \in MH_S(Z, k)^p$ corresponds to an isomorphism of Hodge modules (compare [38, Definition 14.35, Remark 14.36])

$$S : \mathcal{M} \simeq \mathcal{D}(\mathcal{M})(-k), \tag{4-4}$$

whose restriction to U gives a polarization of $\mathbb{V}$. In particular it induces a self-duality isomorphism

$$S : \mathrm{rat}(\mathcal{M}) \simeq \mathcal{D}(\mathrm{rat}(\mathcal{M}))(-k) \simeq \mathcal{D}(\mathrm{rat}(\mathcal{M}))$$

of the underlying twisted intersection cohomology complex, if an isomorphism $\mathbb{Q}_U(-k) \simeq \mathbb{Q}_U$ is chosen.

So if U is smooth of pure complex dimension n, then $\mathbb{Q}_U^H[n]$ is a pure Hodge module of weight n. If moreover $j : U \hookrightarrow Z$ is a Zariski-open dense subset in Z, then the *intermediate extension* $j_{!*}$ for mixed Hodge modules (cf. also with [7]) preserves the weights. This shows that if Z is a complex algebraic variety of pure dimension n and $j : U \hookrightarrow Z$ is the inclusion of a smooth Zariski-open dense subset then the intersection cohomology module $IC_Z^H := j_{!*}(\mathbb{Q}_U^H[n])$ is pure of weight n, with underlying perverse sheaf $\mathrm{rat}(IC_Z^H) = IC_Z$.

Note that the stability of a pure object $\mathcal{M} \in MHM(X)$ under a proper morphism $f : X \to Y$ implies the famous *decomposition theorem* of [7] in the context of pure Hodge modules [40, (4.5.4) on p. 324]:

$$f_* \mathcal{M} \simeq \bigoplus_i H^i f_* \mathcal{M}[-i], \quad \text{with } H^i f_* \mathcal{M} \text{ semisimple for all } i. \tag{4-5}$$

Assume Y is pure-dimensional, with $f : X \to Y$ a *resolution of singularities*, i.e., X is smooth with f a proper morphism, which generically is an isomorphism on some Zariski dense open subset U. Then $\mathbb{Q}_X^H$ is pure, since X is smooth, and IC_Y^H has to be the direct summand of $H^0 f_* \mathbb{Q}_X^H$ which corresponds to $\mathbb{Q}_U^H$.

COROLLARY 4.6. *Assume Y is pure-dimensional, with $f : X \to Y$ a resolution of singularities. Then IC_Y^H is a direct summand of $f_* \mathbb{Q}_X^H \in D^b MHM(Y)$.*

Finally we get the following results about the existence of a mixed Hodge structure on the cohomology (with compact support) $H_{(c)}^i(Z, \mathcal{M})$ for $\mathcal{M} \in D^b MHM(Z)$.

COROLLARY 4.7. *Let Z be a complex algebraic variety with constant map $k : Z \to pt$. Then the cohomology (with compact support) $H_{(c)}^i(Z, \mathcal{M})$ of $\mathcal{M} \in D^b MHM(Z)$ gets an induced graded polarizable mixed Hodge structure:*

$$H_{(c)}^i(Z, \mathcal{M}) = H^i(k_{*(!)} \mathcal{M}) \in MHM(pt) \simeq mHs^p.$$

In particular:

(1) *The rational cohomology (with compact support) $H_{(c)}^i(Z, \mathbb{Q})$ of Z gets an induced graded polarizable mixed Hodge structure by*

$$H^i(Z, \mathbb{Q}) = \text{rat}(H^i(k_* k^* \mathbb{Q}^H)) \quad \text{and} \quad H_c^i(Z, \mathbb{Q}) = \text{rat}(H^i(k_! k^* \mathbb{Q}^H)).$$

(2) *Let $\mathbb{V}_U$ be a good variation of mixed Hodge structures on a smooth pure n-dimensional complex variety U, which is Zariski open and dense in a variety Z, with $j : U \to Z$ the open inclusion. Then the global twisted intersection cohomology (with compact support)*

$$IH_{(c)}^i(Z, \mathbb{V}) := H_{(c)}^i(Z, IC_Z(\mathbb{V})[-n])$$

gets a mixed Hodge structure by

$$IH_{(c)}^i(Z, \mathbb{V}) = H^i(k_{*(!)} IC_Z(\mathbb{V})[-n]) = H^i(k_{*(!)} j_{!*}(\mathbb{V})[-n]).$$

If Z is compact, with $\mathbb{V}$ a polarizable variation of pure Hodge structures of weight w, then also $IH^i(Z, \mathbb{V})$ has a (polarizable) pure Hodge structure of weight $w + i$.

(3) *Let $\mathbb{V}$ be a good variation of mixed Hodge structures on a smooth (pure-dimensional) complex manifold M, which is Zariski open and dense in complex algebraic manifold $\overline{M}$, with complement D a normal crossing divisor with smooth irreducible components. Then $H^i(M, \mathbb{V})$ gets a mixed Hodge structure by*

$$H^i(M, \mathbb{V}) \simeq H^i(\overline{M}, j_*\mathbb{V}) \simeq H^i(k_*j_*\mathbb{V}),$$

with $j : U \to Z$ the open inclusion.

REMARK 4.8. Here are important properties of these mixed Hodge structures:

(1) By a deep theorem of Saito [44, Theorem 0.2, Corollary 4.3], the mixed Hodge structure on $H^i_{(c)}(Z, \mathbb{Q})$ defined as above coincides with the classical mixed Hodge structure constructed by Deligne ([20; 21]).

(2) Assume we are in the context of Corollary 4.7(3) with $Z = \overline{M}$ projective and $\mathbb{V}$ a good variation of pure Hodge structures on $U = M$. Then the pure Hodge structure of (2) on the global intersection cohomology $IH^i(Z, \mathbb{V})$ agrees with that of [15; 29] defined in terms of L^2-cohomology with respect to a Kähler metric with Poincaré singularities along D (compare [40, Remark 3.15]). The case of a 1-dimensional complex algebraic curve $Z = \overline{M}$ due to Zucker [56, Theorem 7.12] is used in the work of Saito [39, (5.3.8.2)] in the proof of the stability of pure Hodge modules under projective morphisms [39, Theorem 5.3.1] (compare also with the detailed discussion of this 1-dimensional case in [45]).

(3) Assume we are in the context of Corollary 4.7(3) with $\overline{M}$ compact. Then the mixed Hodge structure on $H^i(M, \mathbb{V})$ is the one of Theorem 3.13, whose Hodge filtration F comes from the filtered logarithmic de Rham complex (compare [40, § 3.10, Proposition 3.11]).

4B. Grothendieck groups of algebraic mixed Hodge modules. In this section, we describe the functorial calculus of Grothendieck groups of algebraic mixed Hodge modules. Let Z be a complex algebraic variety. By associating to (the class of) a complex the alternating sum of (the classes of) its cohomology objects, we obtain the following identification (compare, for example, [30, p. 77] and [47, Lemma 3.3.1])

$$K_0(D^b MHM(Z)) = K_0(MHM(Z)). \qquad (4\text{-}6)$$

In particular, if Z is a point, then

$$K_0(D^b MHM(pt)) = K_0(mHs^p), \qquad (4\text{-}7)$$

and the latter is a commutative ring with respect to the tensor product, with unit $[\mathbb{Q}^H]$. Then we have, for any complex $\mathcal{M}^\bullet \in D^b MHM(Z)$, the identification

$$[\mathcal{M}^\bullet] = \sum_{i \in \mathbb{Z}} (-1)^i [H^i(\mathcal{M}^\bullet)] \in K_0(D^b MHM(Z)) \cong K_0(MHM(Z)). \quad (4\text{-}8)$$

In particular, if for any $\mathcal{M} \in MHM(Z)$ and $k \in \mathbb{Z}$ we regard $\mathcal{M}[-k]$ as a complex concentrated in degree k, then

$$[\mathcal{M}[-k]] = (-1)^k [\mathcal{M}] \in K_0(MHM(Z)). \quad (4\text{-}9)$$

All the functors f_*, $f_!$, f^*, $f^!$, $\otimes$, $\boxtimes$, $\mathcal{D}$ induce corresponding functors on $K_0(MHM(\cdot))$. Moreover, $K_0(MHM(Z))$ becomes a $K_0(MHM(pt))$-module, with the multiplication induced by the exact exterior product with a point space:

$$\boxtimes : MHM(Z) \times MHM(pt) \to MHM(Z \times \{pt\}) \simeq MHM(Z).$$

Also note that

$$\mathcal{M} \otimes \mathbb{Q}_Z^H \simeq \mathcal{M} \boxtimes \mathbb{Q}_{pt}^H \simeq \mathcal{M}$$

for all $\mathcal{M} \in MHM(Z)$. Therefore, $K_0(MHM(Z))$ is a unitary $K_0(MHM(pt))$-module. The functors f_*, $f_!$, f^*, $f^!$ commute with exterior products (and f^* also commutes with the tensor product $\otimes$), so that the induced maps at the level of Grothendieck groups $K_0(MHM(\cdot))$ are $K_0(MHM(pt))$-linear. Similarly $\mathcal{D}$ defines an involution on $K_0(MHM(\cdot))$. Moreover, by the functor

$$\text{rat} : K_0(MHM(Z)) \to K_0(D_c^b(\mathbb{Q}_Z)) \simeq K_0(\text{Perv}(\mathbb{Q}_Z)),$$

all these transformations lift the corresponding transformations from the (topological) level of Grothendieck groups of constructible (or perverse) sheaves.

REMARK 4.9. The Grothendieck group $K_0(MHM(Z))$ has two different types of generators:

(1) It is generated by the classes of pure Hodge modules $[IC_S(\mathbb{V})]$ with strict support in an irreducible complex algebraic subset $S \subset Z$, with $\mathbb{V}$ a good variation of (pure) Hodge structures on a dense Zariski open smooth subset U of S. These generators behave well under duality.

(2) It is generated by the classes $f_*[j_*\mathbb{V}]$, with $f : \overline{M} \to Z$ a proper morphism from the smooth complex algebraic manifold $\overline{M}$, $j : M \to \overline{M}$ the inclusion of a Zariski open and dense subset M, with complement D a normal crossing divisor with smooth irreducible components, and $\mathbb{V}$ a good variation of mixed (or if one wants also pure) Hodge structures on M. These generators will be used in the next section about characteristic classes of mixed Hodge modules.

Here (1) follows from the fact that a mixed Hodge module has a finite weight filtration, whose graded pieces are pure Hodge modules, i.e., are finite direct sums of pure Hodge modules $IC_S(\mathbb{V})$ with strict support S as above. The claim in (2) follows by induction from resolution of singularities and from the existence of a "standard" distinguished triangle associated to a closed inclusion.

Let $i : Y \to Z$ be a closed inclusion of complex algebraic varieties with open complement $j : U = Z \backslash Y \to Z$. Then one has by Saito's work [40, (4.4.1)] the following functorial distinguished triangle in $D^b MHM(Z)$:

$$j_! j^* \xrightarrow{\ ad_j\ } \mathrm{id} \xrightarrow{\ ad_i\ } i_* i^* \xrightarrow{\ [1]\ } . \tag{4-10}$$

Here the maps ad are the adjunction maps, with $i_* = i_!$ since i is proper. If $f : Z \to X$ is a complex algebraic morphism, then we can apply $f_!$ to get another distinguished triangle

$$f_! j_! j^* \mathbb{Q}_Z^H \xrightarrow{\ ad_j\ } f_! \mathbb{Q}_Z^H \xrightarrow{\ ad_i\ } f_! i_! i^* \mathbb{Q}_Z^H \xrightarrow{\ [1]\ } . \tag{4-11}$$

On the level of Grothendieck groups, we get the important *additivity relation*

$$f_![\mathbb{Q}_Z^H] = (f \circ j)_![\mathbb{Q}_U^H] + (f \circ i)_![\mathbb{Q}_Y^H]$$
$$\in K_0(D^b MHM(X)) = K_0(MHM(X)). \tag{4-12}$$

COROLLARY 4.10. *One has a natural group homomorphism*

$$\chi_{\mathrm{Hdg}} : K_0(\mathrm{var}\,/X) \to K_0(MHM(X)); [f : Z \to X] \mapsto [f_! \mathbb{Q}_Z^H],$$

which commutes with pushdown $f_!$, exterior product $\boxtimes$ and pullback g^*. For $X = pt$ this corresponds to the ring homomorphism (2-10) under the identification $MHM(pt) \simeq mHs^p$.

Here $K_0(\mathrm{var}\,/X)$ is the motivic *relative Grothendieck group* of complex algebraic varieties over X, i.e., the free abelian group generated by isomorphism classes $[f] = [f : Z \to X]$ of morphisms f to X, divided out be the *additivity relation*

$$[f] = [f \circ i] + [f \circ j]$$

for a closed inclusion $i : Y \to Z$ with open complement $j : U = Z \backslash Y \to Z$. The pushdown $f_!$, exterior product $\boxtimes$ and pullback g^* for these relative Grothendieck groups are defined by composition, exterior product and pullback of arrows. The fact that χ_{Hdg} commutes with exterior product $\boxtimes$ (or pullback g^*) follows then from the corresponding Künneth (or base change) theorem for the functor

$$f_! : D^b MHM(Z) \to D^b MHM(X)$$

(contained in Saito's work [43] and [40, (4.4.3)]).

Let $\mathbb{L} := [\mathbb{A}_{\mathbb{C}}^1] \in K_0(\mathrm{var}/pt)$ be the class of the affine line, so that

$$\chi_{\mathrm{Hdg}}(\mathbb{L}) = [H^2(P^1(\mathbb{C}), \mathbb{Q})] = [\mathbb{Q}(-1)] \in K_0(MHM(pt)) = K_0(mHs^p)$$

is the Lefschetz class $[\mathbb{Q}(-1)]$. This class is invertible in $K_0(MHM(pt)) = K_0(mHs^p)$ so that the transformation χ_{Hdg} of Corollary 4.10 factorizes over the localization

$$M_0(\mathrm{var}/X) := K_0(\mathrm{var}/X)[\mathbb{L}^{-1}].$$

Altogether we get the following diagram of natural transformations commuting with $f_!$, $\boxtimes$ and g^*:

$$
\begin{array}{ccccc}
F(X) & \xleftarrow{\;\;\mathrm{can}\;\;} & M_0(\mathrm{var}/X) & \longleftarrow & K_0(\mathrm{var}/X) \\
\big\uparrow{\scriptstyle \chi_{\mathrm{stalk}}} & & \big\downarrow{\scriptstyle \chi_{\mathrm{Hdg}}} & & \\
K_0(D_c^b(X)) & \xleftarrow[\;\;\mathrm{rat}\;\;]{} & K_0(MHM(X)). & &
\end{array}
\qquad (4\text{-}13)
$$

Here $F(X)$ is the group of algebraically constructible functions on X, which is generated by the collection $\{1_Z\}$, for $Z \subset X$ a closed complex algebraic subset, with χ_{stalk} given by the Euler characteristic of the stalk complexes (compare [47, §2.3]). The pushdown $f_!$ for algebraically constructible functions is defined for a morphism $f : Y \to X$ by

$$f_!(1_Z)(x) := \chi\left(H_c^*(Z \cap \{f = x\}, \mathbb{Q})\right) \quad \text{for } x \in X,$$

so that the horizontal arrow marked "can" is given by

$$\mathrm{can} : [f : Y \to X] \mapsto f_!(1_Y), \quad \text{with } \mathrm{can}(\mathbb{L}) = 1_{pt}.$$

The advantage of $M_0(\mathrm{var}/X)$ compared to $K_0(\mathrm{var}/X)$ is that it has an induced *duality* involution $\mathcal{D} : M_0(\mathrm{var}/X) \to M_0(\mathrm{var}/X)$ characterized uniquely by the equality

$$\mathcal{D}([f : M \to X]) = \mathbb{L}^{-m} \cdot [f : M \to X]$$

for $f : M \to X$ a proper morphism with M smooth and pure m-dimensional (compare [8]). This "motivic duality" $\mathcal{D}$ commutes with pushdown $f_!$ for proper f, so that χ_{Hdg} also commutes with duality by

$$
\begin{aligned}
\chi_{\mathrm{Hdg}}(\mathcal{D}[\mathrm{id}_M]) &= \chi_{\mathrm{Hdg}}\left(\mathbb{L}^{-m} \cdot [\mathrm{id}_M]\right) = [\mathbb{Q}_M^H(m)] \\
&= [\mathbb{Q}_M^H[2m](m)] = [\mathcal{D}(\mathbb{Q}_M^H)] = \mathcal{D}\left(\chi_{\mathrm{Hdg}}([\mathrm{id}_M])\right)
\end{aligned}
\qquad (4\text{-}14)
$$

for M smooth and pure m-dimensional. In fact by resolution of singularities and "additivity", $K_0(\mathrm{var}/X)$ is generated by such classes $f_![\mathrm{id}_M] = [f : M \to X]$.

Then all the transformations in the diagram (4-13) *commute with duality*, were $K_0(D_c^b(X))$ gets this involution from Verdier duality, and $\mathcal{D} = \mathrm{id}$ for algebraically constructible functions by $\mathrm{can}([\mathbb{Q}(-1)]) = 1_{pt}$ (compare also with

[47, §6.0.6]). Similarly they commute with f_* and $g^!$ defined by the relations (compare [8]):

$$\mathcal{D} \circ g^* = g^! \circ \mathcal{D} \quad \text{and} \quad \mathcal{D} \circ f_* = f_! \circ \mathcal{D}.$$

For example for an open inclusion $j : M \to \overline{M}$, one gets

$$\chi_{\text{Hdg}} (j_*[\text{id}_M]) = j_*[\mathbb{Q}_M^H]. \tag{4-15}$$

5. Characteristic classes of mixed Hodge modules

5A. Homological characteristic classes. In this section we explain the theory of K-theoretical characteristic homology classes of mixed Hodge modules based on the following result of Saito (compare with [39, §2.3] and [44, §1] for the first part, and with [40, §3.10, Proposition 3.11]) for part (2)):

THEOREM 5.1 (M. SAITO). *Let Z be a complex algebraic variety. Then there is a functor of triangulated categories*

$$Gr_p^F DR : D^b MHM(Z) \to D_{\text{coh}}^b(Z) \tag{5-1}$$

commuting with proper push-down, with $Gr_p^F DR(\mathcal{M}) = 0$ for almost all p and $\mathcal{M}$ fixed, where $D_{\text{coh}}^b(Z)$ is the bounded derived category of sheaves of algebraic $\mathcal{O}_Z$-modules with coherent cohomology sheaves. If M is a (pure m-dimensional) complex algebraic manifold, then one has in addition:

(1) *Let $\mathcal{M} \in MHM(M)$ be a single mixed Hodge module. Then $Gr_p^F DR(\mathcal{M})$ is the corresponding complex associated to the de Rham complex of the underlying algebraic left D-module $\mathcal{M}$ with its integrable connection ∇:*

$$DR(\mathcal{M}) = [\mathcal{M} \xrightarrow{\nabla} \cdots \xrightarrow{\nabla} \mathcal{M} \otimes_{\mathcal{O}_M} \Omega_M^m]$$

with $\mathcal{M}$ in degree $-m$, filtered by

$$F_p DR(\mathcal{M}) = [F_p\mathcal{M} \xrightarrow{\nabla} \cdots \xrightarrow{\nabla} F_{p+m}\mathcal{M} \otimes_{\mathcal{O}_M} \Omega_M^m].$$

(2) *Let $\overline{M}$ be a smooth partial compactification of the complex algebraic manifold M with complement D a normal crossing divisor with smooth irreducible components, with $j : M \to \overline{M}$ the open inclusion. Let $\mathbb{V} = (L, F, W)$ be a good variation of mixed Hodge structures on M. Then the filtered de Rham complex*

$$(DR(j_*\mathbb{V}), F) \quad \text{of} \quad j_*\mathbb{V} \in MHM(\overline{M})[-m] \subset D^b MHM(\overline{M})$$

is filtered quasi-isomorphic to the logarithmic de Rham complex $DR_{\log}(\mathcal{L})$ with the increasing filtration $F_{-p} := F^p$ ($p \in \mathbb{Z}$) associated to the decreasing

F-filtration (3-15). *In particular* $Gr_{-p}^F DR(j_* \mathbb{V})$ ($p \in \mathbb{Z}$) *is quasi-isomorphic to*

$$Gr_F^p DR_{\log}(\overline{\mathcal{L}}) = [Gr_F^p \overline{\mathcal{L}} \xrightarrow{Gr \, \overline{\nabla}} \cdots \xrightarrow{Gr \, \overline{\nabla}} Gr_F^{p-m} \overline{\mathcal{L}} \otimes_{\mathcal{O}_{\overline{M}}} \Omega_{\overline{M}}^m (\log(D))].$$

Here the filtration $F_p DR(\mathcal{M})$ of the de Rham complex is well defined, since the action of the integrable connection ∇ is given in local coordinates $(z_1, \ldots, z_m)$ by

$$\nabla(\cdot) = \sum_{i=1}^m \frac{\partial}{\partial z_i}(\cdot) \otimes dz_i, \quad \text{with } \frac{\partial}{\partial z_i} \in F_1 \mathcal{D}_M,$$

so that $\nabla(F_p \mathcal{M}) \subset F_{p+1} \mathcal{M}$ for all p by (4-1). For later use, let us point that the maps $Gr \, \nabla$ and $Gr \, \overline{\nabla}$ in the complexes

$$Gr_p^F DR(\mathcal{M}) \quad \text{and} \quad Gr_F^p DR_{\log}(\overline{\mathcal{L}})$$

are $\mathcal{O}$-linear!

EXAMPLE 5.2. Let M be a pure m-dimensional complex algebraic manifold. Then

$$Gr_{-p}^F DR(\mathbb{Q}_M^H) \simeq \Omega_M^p[-p] \in D_{\mathrm{coh}}^b(M)$$

if $0 \leq p \leq m$, and $Gr_{-p}^F DR(\mathbb{Q}_M^H) \simeq 0$ otherwise. Assume in addition that $f : M \to Y$ is a resolution of singularities of the pure-dimensional complex algebraic variety Y. Then IC_Y^H is a direct summand of $f_* \mathbb{Q}_M^H \in D^b MHM(Y)$ so that by functoriality $gr_{-p}^F DR(IC_Y^H)$ is a direct summand of $Rf_* \Omega_M^p[-p] \in D_{\mathrm{coh}}^b(Y)$. In particular

$$Gr_{-p}^F DR(IC_Y^H) \simeq 0 \quad \text{for } p < 0 \text{ or } p > m.$$

The transformations $Gr_p^F DR$ ($p \in \mathbb{Z}$) induce functors on the level of Grothendieck groups. Therefore, if $G_0(Z) \simeq K_0(D_{\mathrm{coh}}^b(Z))$ denotes the Grothendieck group of coherent *algebraic* $\mathcal{O}_Z$-sheaves on Z, we get group homomorphisms

$$Gr_p^F DR : K_0(MHM(Z)) = K_0(D^b MHM(Z)) \to K_0(D_{\mathrm{coh}}^b(Z)) \simeq G_0(Z).$$

DEFINITION 5.3. The *motivic Hodge Chern class transformation*

$$MHC_y : K_0(MHM(Z)) \to G_0(Z) \otimes \mathbb{Z}[y^{\pm 1}]$$

is defined by

$$[\mathcal{M}] \mapsto \sum_{i,p} (-1)^i [\mathcal{H}^i(Gr_{-p}^F DR(\mathcal{M}))] \cdot (-y)^p. \tag{5-2}$$

So this characteristic class captures information from the graded pieces of the filtered de Rham complex of the filtered D-module underlying a mixed Hodge module $\mathcal{M} \in MHM(Z)$, instead of the graded pieces of the filtered D-module itself (as more often studied). Let $p' = \min\{p \mid F_p\mathcal{M} \neq 0\}$. Using Theorem 5.1(1) for a local embedding $Z \hookrightarrow M$ of Z into a complex algebraic manifold M of dimension m, one gets

$$Gr_p^F DR(\mathcal{M}) = 0 \quad \text{for } p < p' - m,$$

and

$$Gr_{p'-m}^F DR(\mathcal{M}) \simeq (F_{p'}\mathcal{M}) \otimes_{\mathcal{O}_M} \omega_M$$

is a coherent $\mathcal{O}_Z$-sheaf independent of the local embedding. Here we are using left D-modules (related to variation of Hodge structures), whereas for this question the corresponding filtered right D-module (as used in [42])

$$\mathcal{M}^r := \mathcal{M} \otimes_{\mathcal{O}_M} \omega_M \quad \text{with} \quad F_p\mathcal{M}^r := (F_{p+m}\mathcal{M}) \otimes_{\mathcal{O}_M} \omega_M$$

would work better. Then the coefficient of the "top-dimensional" power of y in

$$MHC_y([\mathcal{M}]) = [F_{p'}\mathcal{M} \otimes_{\mathcal{O}_M} \omega_M] \otimes (-y)^{m-p'} + \sum_{i < m-p'} (\cdots) \cdot y^i \in G_0(Z)[y^{\pm 1}] \tag{5-3}$$

is given by the class $[F_{p'}\mathcal{M} \otimes_{\mathcal{O}_M} \omega_M] \in G_0(Z)$ of this coherent $\mathcal{O}_Z$-sheaf (up to a sign). Using resolution of singularities, one gets for example for an m-dimensional complex algebraic variety Z that

$$MHC_y([\mathbb{Q}_Z^H]) = [\pi_*\omega_M] \cdot y^m + \sum_{i < m} (\cdots) \cdot y^i \in G_0(Z)[y^{\pm 1}],$$

with $\pi : M \to Z$ any resolution of singularities of Z (compare [44, Corollary 0.3]). More generally, for an irreducible complex variety Z and $\mathcal{M} = IC_Z^H(\mathcal{L})$ a pure Hodge module with strict support Z, the corresponding coherent $\mathcal{O}_Z$-sheaf

$$S_Z(\mathcal{L}) := F_{p'} IC_Z^H(\mathcal{L}) \otimes_{\mathcal{O}_M} \omega_M$$

only depends on Z and the good variation of Hodge structures $\mathcal{L}$ on a Zariski open smooth subset of Z, and it behaves much like a dualizing sheaf. Its formal properties are studied in Saito's proof given in [42] of a conjecture of Kollár. So the "top-dimensional" power of y in $MHC_y([IC_Z^H(\mathcal{L})])$ exactly picks out (up to a sign) the class $[S_Z(\mathcal{L})] \in G_0(Z)$ of this interesting coherent sheaf $S_Z(\mathcal{L})$ on Z.

Let $td_{(1+y)}$ be the *twisted Todd transformation*

$$td_{(1+y)} : G_0(Z) \otimes \mathbb{Z}[y^{\pm 1}] \to H_*(Z) \otimes \mathbb{Q}[y^{\pm 1}, (1+y)^{-1}] ;$$

$$[\mathcal{F}] \mapsto \sum_{k \geq 0} td_k([\mathcal{F}]) \cdot (1+y)^{-k}, \tag{5-4}$$

where $H_*(\cdot)$ stands either for the Chow homology groups $CH_*(\cdot)$ or for the Borel–Moore homology groups $H_{2*}^{BM}(\cdot)$ (in even degrees), and td_k is the degree k component in $H_k(Z)$ of the *Todd class transformation* $td_* : G_0(Z) \to H_*(Z) \otimes \mathbb{Q}$ of Baum, Fulton, and MacPherson [5], which is linearly extended over $\mathbb{Z}[y^{\pm 1}]$. Compare also with [22, Chapter 18] and [24, Part II].

DEFINITION 5.4. The (un)normalized *motivic Hirzebruch class transformations* MHT_{y*} (and $\widetilde{MHT}_{y*}$) are defined by the composition

$$MHT_{y*} := td_{(1+y)} \circ MHC_y : K_0(MHM(Z)) \to H_*(Z) \otimes \mathbb{Q}[y^{\pm 1}, (1+y)^{-1}] \tag{5-5}$$

and

$$\widetilde{MHT}_{y*} := td_* \circ MHC_y : K_0(MHM(Z)) \to H_*(Z) \otimes \mathbb{Q}[y^{\pm 1}]. \tag{5-6}$$

REMARK 5.5. By precomposing with the transformation χ_{Hdg} from Corollary 4.10 one gets similar transformations

$$mC_y := MHC_y \circ \chi_{\text{Hdg}}, \quad T_{y*} := MHT_{y*} \circ \chi_{\text{Hdg}}, \quad \tilde{T}_{y*} := \widetilde{MHT}_{y*} \circ \chi_{\text{Hdg}}$$

defined on the relative Grothendieck group of complex algebraic varieties as studied in [9]. Then it is the (normalized) motivic Hirzebruch class transformation T_{y*}, which, as mentioned in the Introduction, "unifies" in a functorial way

($y = -1$) the (rationalized) Chern class transformation c_* of MacPherson [34];
($y = 0$) the Todd class transformation td_* of Baum–Fulton–MacPherson [5];
($y = 1$) the L-class transformation L_* of Cappell and Shaneson [14].

(Compare with [9; 48] and also with [55] in these proceedings.)

In this paper we work most the time only with the more important K-theoretical transformation MHC_y. The corresponding results for MHT_{y*} follow from this by the known properties of the Todd class transformation td_* (compare [5; 22; 24]).

EXAMPLE 5.6. Let $\mathbb{V} = (V, F, W) \in MHM(pt) = mHs^p$ be a (graded polarizable) mixed Hodge structure. Then:

$$MHC_y([\mathbb{V}]) = \sum_p \dim_{\mathbb{C}}(Gr_F^p V_{\mathbb{C}}) \cdot (-y)^p = \chi_y([\mathbb{V}]) \in \mathbb{Z}[y^{\pm 1}]$$
$$= G_0(pt) \otimes \mathbb{Z}[y^{\pm 1}]. \tag{5-7}$$

So over a point the transformation MHC_y coincides with the χ_y-genus ring homomorphism $\chi_y : K_0(mHs^p) \to \mathbb{Z}[y^{\pm 1}]$ (and similarly for $\widetilde{MHT}_{y*}$ and MHT_{y*}).

The *motivic Chern class* $C_y(Z)$ and the *motivic Hirzebruch class* $T_{y*}(Z)$ of a complex algebraic variety Z are defined by

$$C_y(Z) := MHC_y([\mathbb{Q}_Z^H]) \quad \text{and} \quad T_{y*}(Z) := MHT_{y*}([\mathbb{Q}_Z^H]). \tag{5-8}$$

Similarly, if U is a pure n-dimensional complex algebraic manifold and L is a local system on U underlying a good variation of mixed Hodge structures $\mathcal{L}$, we define the *twisted motivic Chern and Hirzebruch characteristic classes* by (compare [12; 13; 35])

$$C_y(U; \mathcal{L}) := MHC_y([\mathcal{L}^H]) \quad \text{and} \quad T_{y*}(U; \mathcal{L}) := MHT_{y*}([\mathcal{L}^H]), \tag{5-9}$$

where $\mathcal{L}^H[n]$ is the smooth mixed Hodge module on U with underlying perverse sheaf $L[n]$. Assume, in addition, that U is dense and Zariski open in the complex algebraic variety Z. Let $IC_Z^H, IC_Z^H(L) \in MHM(Z)$ be the (twisted) intersection homology (mixed) Hodge module on Z, whose underlying perverse sheaf is IC_Z or $IC_Z(L)$, as the case may be. Then we define *intersection characteristic classes* as follows (compare [9; 11; 13; 35]):

$$\begin{aligned} IC_y(Z) &:= MHC_y([IC_Z^H[-n]]), \\ IT_{y*}(Z) &:= MHT_{y*}([IC_Z^H[-n]]), \end{aligned} \tag{5-10}$$

and, similarly,

$$\begin{aligned} IC_y(Z; \mathcal{L}) &:= MHC_y([IC_Z^H(\mathcal{L})[-n]]), \\ IT_{y*}(Z; \mathcal{L}) &:= MHT_{y*}([IC_Z^H(\mathcal{L})[-n]]). \end{aligned} \tag{5-11}$$

By definition and Theorem 5.1, the transformations MHC_y and MHT_{y*} *commute with proper push-forward*. The following *normalization* property holds (compare [9]): If M is smooth, then

$$C_y(Z) = \lambda_y(T^*M) \cap [\mathcal{O}_M] \quad \text{and} \quad T_{y*}(Z) = T_y^*(TM) \cap [M], \tag{5-12}$$

where $T_y^*(TM)$ is the cohomology Hirzebruch class of M as in Theorem 2.4.

EXAMPLE 5.7. Let Z be a compact (possibly singular) complex algebraic variety, with $k : Z \to pt$ the proper constant map to a point. Then for $\mathcal{M} \in D^b MHM(Z)$ the pushdown

$$k_*(MHC_y(\mathcal{M})) = MHC_y(k_*\mathcal{M}) = \chi_y([H^*(Z, \mathcal{M})])$$

is the Hodge genus

$$\chi_y([H^*(Z, \mathcal{M})]) = \sum_{i,p} (-1)^i \dim_{\mathbb{C}}(Gr_F^p H^i(Z, \mathcal{M})) \cdot (-y)^p. \tag{5-13}$$

In particular:

(1) If Z is smooth, then

$$k_* C_y(Z) = \chi_y(Z) := \chi_y\left([H^*(Z, \mathbb{Q})]\right)$$
$$k_* C_y(Z; \mathcal{L}) = \chi_y(Z; \mathcal{L}) := \chi_y\left([H^*(Z, \mathcal{L})]\right).$$

(2) If Z is pure-dimensional, then

$$k_* IC_y(Z) = I\chi_y(Z) := \chi_y\left([IH^*(Z, \mathbb{Q})]\right)$$
$$k_* IC_y(Z; \mathcal{L}) = I\chi_y(Z; \mathcal{L}) := \chi_y\left([IH^*(Z, \mathcal{L})]\right).$$

Note that, for Z compact,

$$I\chi_{-1}(Z) = \chi([IH^*(Z; \mathbb{Q})]$$

is the *intersection (co)homology Euler characteristic* of Z, whereas, for Z projective,

$$I\chi_1(Z) = \text{sgn}\left(IH^*(Z, \mathbb{Q})\right)$$

is the *intersection (co)homology signature* of Z, introduced by Goresky and MacPherson [25]. In fact this follows as in the smooth context from Saito's relative version of the Hodge index theorem for intersection cohomology [39, Theorem 5.3.2]. Finally $\chi_0(Z)$ and $I\chi_0(Z)$ are two possible extensions to singular varieties of the *arithmetic genus*. Here it makes sense to take $y = 0$, since one has, by Example 5.2,

$$k_* IC_y(Z) = I\chi_y(Z) \in \mathbb{Z}[y].$$

It is conjectured that, for a pure n-dimensional compact variety Z,

$$IT_{1*}(Z) \stackrel{?}{=} L_*(Z) \in H_{2*}(Z, \mathbb{Q})$$

is the Goresky–MacPherson homology L-class [25] of the Witt space Z; see [9, Remark 5.4]. Similarly one should expect for a pure-dimensional compact variety Z that

$$\alpha(IC_1(Z)) \stackrel{?}{=} \Delta(Z) \in KO_0^{\text{top}}(Z)[\tfrac{1}{2}] \oplus KO_2^{\text{top}}(Z)[\tfrac{1}{2}] \simeq K_0^{\text{top}}(Z)[\tfrac{1}{2}], \quad (5\text{-}14)$$

where $\alpha : G_0(Z) \to K_0^{\text{top}}(Z)$ is the K-theoretical Riemann–Roch transformation of Baum, Fulton, and MacPherson [6], and $\Delta(Z)$ is the *Sullivan class* of the Witt space Z (compare with [3] in these proceedings). These conjectured equalities are true for a smooth Z, or more generally for a pure n-dimensional compact complex algebraic variety Z with a *small resolution* of singularities $f : M \to Z$, in which case one has $f_*(\mathbb{Q}_M^H) = IC_Z^H[-n]$, so that

$$IT_{1*}(Z) = f_* T_{1*}(M) = f_* L_*(M) = L_*(Z)$$

and

$$\alpha\left(IC_1(Z)\right) = f_*\left(\alpha(C_1(M))\right) = f_* \Delta(M) = \Delta(Z).$$

Here the functoriality $f_* L_*(M) = L_*(Z)$ and $f_* \Delta(M) = \Delta(Z)$ for a small resolution follows, for instance, from [54], which allows one to think of the characteristic classes L_* and Δ as covariant functors for suitable Witt groups of selfdual constructible sheaf complexes.

In particular, the classes $f_* C_1(M)$ and $f_* T_{1*}(M)$ do not depend on the choice of a small resolution. In fact the same functoriality argument applies to

$$IC_y(Z) = f_* C_y(M) \in G_0(Z) \otimes \mathbb{Z}[y],$$
$$IT_{y*}(Z) = f_* T_{y*}(M) \in H_{2*}(Z) \otimes \mathbb{Q}[y, (1+y)^{-1}];$$

compare [11; 35]. Note that in general a complex variety Z doesn't have a small resolution, and even if it exists, it is in general not unique. This type of independence question were discussed by Totaro [51], pointing out the relation to the famous *elliptic genus and classes* (compare also with [32; 53] in these proceedings). Note that we get such a result for the K-theoretical class

$$IC_y(Z) = f_* C_y(M) \in G_0(Z) \otimes \mathbb{Z}[y] \, !$$

5B. Calculus of characteristic classes. So far we only discussed the functoriality of MHC_y with respect to proper push down, and the corresponding relation to Hodge genera for compact Z coming from the push down for the proper constant map $k : Z \to pt$. Now we explain some other important functoriality properties. Their proof is based on the following (see [35, (4.6)], for instance):

EXAMPLE 5.8. Let $\overline{M}$ be a smooth partial compactification of the complex algebraic manifold M with complement D a normal crossing divisor with smooth irreducible components, with $j : M \to \overline{M}$ the open inclusion. Let $\mathbb{V} = (L, F, W)$ be a good variation of mixed Hodge structures on M. Then the filtered de Rham complex

$$(DR(j_* \mathbb{V}), F) \quad \text{of} \quad j_* \mathbb{V} \in MHM(\overline{M})[-m] \subset D^b MHM(\overline{M})$$

is by Theorem 5.1(2) filtered quasi-isomorphic to the logarithmic de Rham complex $DR_{\log}(\mathcal{L})$ with the increasing filtration $F_{-p} := F^p$ ($p \in \mathbb{Z}$) associated to the decreasing F-filtration (3-15). Then

$$MHC_y(j_* \mathbb{V}) = \sum_{i,p} (-1)^i [\mathcal{H}^i (Gr_F^p DR_{\log}(\mathcal{L}))] \cdot (-y)^p$$

$$= \sum_p [Gr_F^p DR_{\log}(\mathcal{L})] \cdot (-y)^p$$

$$\overset{(*)}{=} \sum_{i,p} (-1)^i [Gr_F^{p-i}(\mathcal{L}) \otimes_{\mathcal{O}_{\overline{M}}} \Omega_{\overline{M}}^i(\log(D))] \cdot (-y)^p$$

$$= MHC^y(Rj_* L) \cap \left(\lambda_y \left(\Omega_{\overline{M}}^1(\log(D)) \right) \cap [\mathcal{O}_{\overline{M}}] \right). \quad (5\text{-}15)$$

In particular for $j = \mathrm{id} : M \to M$ we get the following *Atiyah–Meyer type formula* (compare [12; 13; 35]):

$$MHC_y(\mathbb{V}) = MHC^y(\mathcal{L}) \cap \left(\lambda_y(T^*M) \cap [\mathcal{O}_M]\right). \tag{5-16}$$

REMARK 5.9. The formula (5-15) is a class version of the formula (3-16) of Theorem 3.13, which one gets back from (5-15) by pushing down to a point for the proper constant map $k : \overline{M} \to pt$ on the compactification $\overline{M}$ of M.

Also note that in the equality $(*)$ in (5-15) we use the fact that the complex $Gr_F^p DR_{\log}(\mathcal{L})$ has coherent (locally free) objects, with $\mathcal{O}_{\overline{M}}$-linear maps between them.

The formula (5-15) describes a *splitting* of the characteristic class $MHC_y(j_*\mathbb{V})$ into two terms:

(coh) a cohomological term $MHC^y(Rj_*\mathcal{L})$, capturing the information of the good variation of mixed Hodge structures $\mathcal{L}$, and

(hom) the homological term $\lambda_y\left(\Omega_{\overline{M}}^1(\log(D))\right) \cap [\mathcal{O}_{\overline{M}}] = MHC_y(j_*\mathbb{Q}_M^H)$, capturing the information of the underlying space or embedding $j : M \to \overline{M}$.

By Corollary 3.14, the term $MHC^y(Rj_*L)$ has good functorial behavior with respect to exterior and suitable tensor products, as well as for smooth pullbacks. For the exterior products one gets similarly (compare [19, Proposition 3.2]):

$$\Omega_{\overline{M \times M'}}^1(\log(D \times M' \cup M \times D')) \simeq \left(\Omega_{\overline{M}}^1(\log(D))\right) \boxtimes \left(\Omega_{\overline{M'}}^1(\log(D'))\right)$$

so that

$$\lambda_y\left(\Omega_{\overline{M \times M'}}^1(\log(D \times M' \cup M \times D'))\right) \cap [\mathcal{O}_{\overline{M} \times \overline{M'}}]$$
$$= \left(\lambda_y\left(\Omega_{\overline{M}}^1(\log(D))\right) \cap [\mathcal{O}_{\overline{M}}]\right) \boxtimes \left(\lambda_y\left(\Omega_{\overline{M'}}^1(\log(D'))\right) \cap [\mathcal{O}_{\overline{M'}}]\right)$$

for the product of two partial compactifications as in example 5.8. But the Grothendieck group $K_0(MHM(Z))$ of mixed Hodge modules on the complex variety Z is generated by classes of the form $f_*(j_*[\mathbb{V}])$, with $f : \overline{M} \to Z$ proper and $M, \overline{M}, \mathbb{V}$ as before. Finally one also has the multiplicativity

$$(f \times f')_* = f_* \boxtimes f'_*$$

for the push down for proper maps $f : \overline{M} \to Z$ and $f' : \overline{M'} \to Z'$ on the level of Grothendieck groups $K_0(MHM(\cdot))$ as well as for $G_0(\cdot) \otimes \mathbb{Z}[y^{\pm 1}]$. Then one gets the following result from Corollary 3.14 and Example 5.8 (as in [9, Proof of Corollary 2.1(3)]):

COROLLARY 5.10 (MULTIPLICATIVITY FOR EXTERIOR PRODUCTS). *The motivic Chern class transformation MHC_y commutes with exterior products*:

$$MHC_y([M \boxtimes M']) = MHC_y([M] \boxtimes [M'])$$
$$= MHC_y([M]) \boxtimes MHC_y([M']) \quad (5\text{-}17)$$

for $M \in D^b MHM(Z)$ and $M' \in D^b MHM(Z')$.

Next we explain the behavior of MHC_y for smooth pullbacks. Consider a cartesian diagram of morphisms of complex algebraic varieties

$$\begin{array}{ccc} \overline{M}' & \xrightarrow{g'} & \overline{M} \\ f' \downarrow & & \downarrow f \\ Z' & \xrightarrow{g} & Z, \end{array}$$

with g smooth, f proper and $M, \overline{M}, \mathbb{V}$ as before. Then g' too is smooth and f' is proper, and one has the *base change isomorphism*

$$g^* f_* = f'_* g'^*$$

on the level of Grothendieck groups $K_0(MHM(\cdot))$ as well as for $G_0(\cdot) \otimes \mathbb{Z}[y^{\pm 1}]$. Finally for the induced partial compactification $\overline{M}'$ of $M' := g'^{-1}(M)$, with complement D' the induced normal crossing divisor with smooth irreducible components, one has a short exact sequence of vector bundles on $\overline{M}'$:

$$0 \to g'^*\big(\Omega^1_{\overline{M}}(\log(D))\big) \to \Omega^1_{\overline{M}'}(\log(D')) \to T^*_{g'} \to 0,$$

with $T^*_{g'}$ the relative cotangent bundle along the fibers of the smooth morphism g'. And by base change one has $T^*_{g'} = f'^*(T^*_g)$. So for the corresponding lambda classes we get

$$\lambda_y\big(\Omega^1_{\overline{M}'}(\log(D'))\big) = \big(g'^*\lambda_y\big(\Omega^1_{\overline{M}}(\log(D))\big)\big) \otimes \lambda_y(T^*_{g'})$$
$$= \big(g'^*\lambda_y\big(\Omega^1_{\overline{M}}(\log(D))\big)\big) \otimes f'^*\lambda_y(T^*_g). \quad (5\text{-}18)$$

Finally (compare also with [9, Proof of Corollary 2.1(4)]), by using the *projection formula*

$$\lambda_y(T^*_g) \otimes f'_*(\cdot) = f'_*\big(f'^*\lambda_y(T^*_g) \otimes (\cdot)\big):$$
$$G_0(\overline{M}') \otimes \mathbb{Z}[y^{\pm 1}] \to G_0(Z') \otimes \mathbb{Z}[y^{\pm 1}]$$

one gets from Corollary 3.14 and Example 5.8 the following consequence:

COROLLARY 5.11 (VRR FOR SMOOTH PULLBACKS). *For a smooth morphism*
$g : Z' \to Z$ *of complex algebraic varieties one has for the motivic Chern class*
transformation the following Verdier Riemann–Roch formula:

$$\lambda_y(T_g^*) \cap g^* MHC_y([\mathcal{M}]) = MHC_y(g^*[\mathcal{M}]) = MHC_y([g^*\mathcal{M}]) \qquad (5\text{-}19)$$

for $\mathcal{M} \in D^b MHM(Z)$. *In particular*

$$g^* MHC_y([\mathcal{M}]) = MHC_y(g^*[\mathcal{M}]) = MHC_y([g^*\mathcal{M}]) \qquad (5\text{-}20)$$

for g an étale morphism (i.e., a smooth morphism with zero dimensional fibers),
or in more topological terms, for g an unramified covering. The most important
special case is that of an open embedding.

If moreover g is also proper, then one gets from Corollary 5.11 and the projection
formula the following result:

COROLLARY 5.12 (GOING UP AND DOWN). *Let* $g : Z' \to Z$ *be a smooth and*
proper morphism of complex algebraic varieties. Then one has for the motivic
Chern class transformation the following going up und down formula:

$$
\begin{aligned}
MHC_y(g_* g^*[\mathcal{M}]) &= g_* MHC_y(g^*[\mathcal{M}]) \\
&= g_* \left(\lambda_y(T_g^*) \cap g^* MHC_y([\mathcal{M}]) \right) \\
&= \left(g_* \lambda_y(T_g^*) \right) \cap MHC_y([\mathcal{M}]) \qquad (5\text{-}21)
\end{aligned}
$$

for $\mathcal{M} \in D^b MHM(Z)$, *with*

$$g_* \left(\lambda_y(T_g^*) \right) := \sum_{p,q \geq 0} (-1)^q \cdot [R^q g_*(\Omega_{Z'/Z}^p)] \cdot y^p \in K_{alg}^0(Z)[y]$$

the algebraic cohomology class being given (as in Example 3.5) by

$$MHC^y([Rg_* \mathbb{Q}_{Z'}]) = \sum_{p,q \geq 0} (-1)^q \cdot [R^q g_*(\Omega_{Z'/Z}^p)] \cdot y^p.$$

Note that all higher direct image sheaves $R^q g_*(\Omega_{Z'/Z}^p)$ *are locally free in this*
case, since g is a smooth and proper morphism of complex algebraic varieties
(compare with [18]). *In particular*

$$g_* C_y(Z') = \left(g_* \lambda_y(T_g^*) \right) \cap C_y(Z),$$

and

$$g_* IC_y(Z') = \left(g_* \lambda_y(T_g^*) \right) \cap IC_y(Z)$$

for Z and Z' pure-dimensional. If, in addition, Z and Z' are compact, with
$k : Z \to pt$ *the constant proper map, then*

$$\chi_y(g^*[\mathcal{M}]) = k_* g_* MHC_y(g^*[\mathcal{M}]) = \langle g_* \lambda_y(T_g^*), MHC_y([\mathcal{M}]) \rangle. \qquad (5\text{-}22)$$

In particular,

$$\chi_y(Z') = \langle g_* \lambda_y(T_g^*), C_y(Z) \rangle \quad and \quad I\chi_y(Z') = \langle g_* \lambda_y(T_g^*), IC_y(Z) \rangle.$$

The result of this corollary can also be seen form a different viewpoint, by making the "going up and down" calculation already on the level of Grothendieck groups of mixed Hodge modules, where this time one only needs the assumption that $f : Z' \to Z$ is proper (to get the projection formula):

$$f_* f^*[\mathcal{M}] = [f_* f^* \mathcal{M}] = [f_*(\mathbb{Q}_{Z'}^H \otimes f^* \mathcal{M})] = [f_* \mathbb{Q}_{Z'}^H] \otimes [\mathcal{M}] \in K_0(MHM(Z))$$

for $\mathcal{M} \in D^b MHM(Z)$. The problem for a singular Z is then that we do not have a precise relation between

$$[f_* \mathbb{Q}_{Z'}^H] \in K_0(MHM(Z)) \quad and \quad [Rf_* \mathbb{Q}_{Z'}] \in K_0(FmHs^p(Z)).$$

REMARK 5.13. What is missing up to now is the right notion of a good variation (or family) of mixed Hodge structures on a *singular* complex algebraic variety Z! This class should contain at least

(1) the higher direct image local systems $R^i f_* \mathbb{Q}_{Z'}$ ($i \in \mathbb{Z}$) for a smooth and proper morphism $f : Z' \to Z$ of complex algebraic varieties, and
(2) the pullback $g^* \mathcal{L}$ of a good variation of mixed Hodge structures $\mathcal{L}$ on a smooth complex algebraic manifold M under an algebraic morphism $g : Z \to M$.

At the moment we have to assume that Z is smooth (and pure-dimensional), so as to use Theorem 4.3.

Nevertheless, in case (2) above we can already prove the following interesting result (compare with [35, §4.1] for a similar result for MHT_{y*} in the case when f is a closed embedding):

COROLLARY 5.14 (MULTIPLICATIVITY). *Let* $f : Z \to N$ *be a morphism of complex algebraic varieties, with* N *smooth and pure n-dimensional. Then one has a natural pairing*

$$f^*(\cdot) \cap (\cdot) : K_0(VmHs^g(N)) \times K_0(MHM(Z)) \to K_0(MHM(Z)),$$
$$([\mathcal{L}], [\mathcal{M}]) \mapsto [f^*(\mathcal{L}^H) \otimes \mathcal{M}].$$

Here $\mathcal{L}^H[m]$ *is the smooth mixed Hodge module on* N *with underlying perverse sheaf* $L[m]$. *One also has a similar pairing on (co)homological level:*

$$f^*(\cdot) \cap (\cdot) : K_{alg}^0(N) \otimes \mathbb{Z}[y^{\pm 1}] \times G_0(Z) \otimes \mathbb{Z}[y^{\pm 1}] \to G_0(Z) \otimes \mathbb{Z}[y^{\pm 1}],$$
$$([\mathcal{V}] \cdot y^i, [\mathcal{F}] \cdot y^j) \mapsto [f^*(\mathcal{V}) \otimes \mathcal{F}] \cdot y^{i+j}.$$

And the motivic Chern class transformations MHC^y and MHC$_y$ commute with these natural pairings:

$$MHC_y([f^*(\mathcal{L}^H) \otimes \mathcal{M}]) = MHC^y([f^*\mathcal{L}]) \cap MHC_y([\mathcal{M}])$$
$$= f^*(MHC^y([\mathcal{L}])) \cap MHC_y([\mathcal{M}]) \qquad (5\text{-}23)$$

for $\mathcal{L} \in VmHs^g(N)$ and $\mathcal{M} \in D^bMHM(Z)$.

For the proof we can once more assume $\mathcal{M} = g_*j_*\mathbb{V}$ for $g : \overline{M} \to Z$ proper, with $\overline{M}$ a pure-dimensional smooth complex algebraic manifold, $j : M \to \overline{M}$ a Zariski open inclusion with complement D a normal crossing divisor with smooth irreducible components, and finally $\mathbb{V}$ a good variation of mixed Hodge structures on M. Using the projection formula, it is then enough to prove

$$MHC_y([g^*f^*(\mathcal{L}^H) \otimes j_*\mathbb{V}]) = MHC^y([g^*f^*\mathcal{L}]) \cap MHC_y([j_*\mathbb{V}]).$$

But $g^*f^*\mathcal{L}$ is a good variation of mixed Hodge structures on $\overline{M}$. Therefore, by Example 5.8 and Corollary 3.14(3), both sides are equal to

$$(MHC^y(g^*f^*\mathcal{L}) \otimes MHC^y(j_*\mathbb{V})) \cap (\lambda_y(\Omega^1_{\overline{M}}(\log(D))) \cap [\mathcal{O}_{\overline{M}}]).$$

As an application of the very special case where $f = \mathrm{id} : Z \to N$ is the identity of a complex algebraic manifold Z, with

$$MHC_y([\mathbb{Q}_Z^H]) = \lambda_y(T^*Z) \cap [\mathcal{O}_Z],$$

one gets the Atiyah–Meyer type formula (5-16) as well as the following result (cf. [12; 13; 35]):

EXAMPLE 5.15 (ATIYAH TYPE FORMULA). Let $g : Z' \to Z$ be a proper morphism of complex algebraic varieties, with Z smooth and connected. Assume that for a given $\mathcal{M} \in D^bMHM(Z')$ all direct image sheaves

$$R^i g_* \mathrm{rat}(\mathcal{M}) \quad (i \in \mathbb{Z}) \quad \text{are locally constant} :$$

for instance, g may be a locally trivial fibration and $\mathcal{M} = \mathbb{Q}_{Z'}^H$, or $\mathcal{M} = IC_{Z'}^H$ (for Z' pure-dimensional), so that they all underlie a good variation of mixed Hodge structures. Then one can define

$$[Rg_* \mathrm{rat}(\mathcal{M})] := \sum_{i \in \mathbb{Z}} (-1)^i \cdot [R^i g_* \mathrm{rat}(\mathcal{M})] \in K_0(VmHs^g(Z)),$$

with

$$g_*MHC_y([\mathcal{M}]) = MHC_y(g_*[\mathcal{M}])$$
$$= MHC^y([Rg_* \mathrm{rat}(\mathcal{M})]) \otimes (\lambda_y(T^*Z) \cap [\mathcal{O}_Z]). \qquad (5\text{-}24)$$

Here is a final application:

EXAMPLE 5.16 (FORMULA OF ATIYAH–MEYER TYPE FOR INTERSECTION COHOMOLOGY). Let $f : Z \to N$ be a morphism of complex algebraic varieties, with N smooth and pure n-dimensional (e.g., a closed embedding). Assume also Z is pure m-dimensional. Then one has for a good variation of mixed Hodge structures $\mathcal{L}$ on N the equality

$$IC_Z^H(f^*\mathcal{L})[-m] \simeq f^*\mathcal{L}^H \otimes IC_Z^H[-m] \in MHM(Z)[-m] \subset D^b MHM(Z),$$

so that

$$IC_y(Z; f^*\mathcal{L}) = MHC^y(f^*\mathcal{L}) \cap IC_y(Z) = f^*\left(MHC^y(\mathcal{L})\right) \cap IC_y(Z). \quad (5\text{-}25)$$

If in addition Z is also compact, then one gets by pushing down to a point:

$$I\chi_y(Z; f^*\mathcal{L}) = \langle MHC^y(f^*\mathcal{L}), IC_y(Z) \rangle. \quad (5\text{-}26)$$

REMARK 5.17. This example should be seen as a Hodge-theoretical version of the corresponding result of Banagl, Cappell, and Shaneson [4] for the L-classes $L_*(IC_Z(L))$ of a selfdual *Poincaré local system* L on all of Z. The special case of Example 5.16 for f a closed inclusion was already explained in [35, §4.1].

Finally note that all the results of this section can easily be applied to the (un)normalized *motivic Hirzebruch class transformation* MHT_{y*} (and $\widehat{MHT}_{y*}$), because the *Todd class transformation* $td_* : G_0(\cdot) \to H_*(\cdot) \otimes \mathbb{Q}$ of Baum, Fulton, and MacPherson [5] has the following properties (compare also with [22, Chapter 18] and [24, Part II]):

FUNCTORIALITY: The Todd class transformation td_* commutes with push-down f_* for a proper morphism $f : Z \to X$:

$$td_*(f_*([\mathcal{F}])) = f_*(td_*([\mathcal{F}])) \quad \text{for } [\mathcal{F}] \in G_0(Z).$$

MULTIPLICATIVITY FOR EXTERIOR PRODUCTS: The Todd class transformation td_* commutes with exterior products:

$$td_*([\mathcal{F} \boxtimes \mathcal{F}']) = td_*([\mathcal{F}]) \boxtimes td_*([\mathcal{F}']) \quad \text{for } [\mathcal{F}] \in G_0(Z) \text{ and } [\mathcal{F}'] \in G_0(Z').$$

VRR FOR SMOOTH PULLBACKS: For a smooth morphism $g : Z' \to Z$ of complex algebraic varieties one has for the Todd class transformation td_* the following Verdier Riemann–Roch formula:

$$td^*(T_g) \cap g^* td_*([\mathcal{F}]) = td_*(g^*[\mathcal{F}]) = td_*([g^*\mathcal{F}]) \quad \text{for } [\mathcal{F}] \in G_0(Z).$$

MULTIPLICATIVITY: Let $ch^* : K^0_{\text{alg}}(\cdot) \to H^*(\cdot) \otimes \mathbb{Q}$ be the cohomological *Chern character* to the cohomology $H^*(\cdot)$ given by the operational Chow ring $CH^*(\cdot)$ or the usual cohomology $H^{2*}(\cdot, \mathbb{Z})$ in even degrees. Then one has the multiplicativity relation

$$td_*([\mathcal{V} \otimes \mathcal{F}]) = ch^*([\mathcal{V}]) \cap td_*([\mathcal{F}])$$

for $[\mathcal{V}] \in K_{\mathrm{alg}}^0(Z)$ and $[\mathcal{F}] \in G_0(Z)$, with Z a (possible singular) complex algebraic variety.

5C. Characteristic classes and duality. In this final section we explain the characteristic class version of the duality formula (2-14) for the χ_y-genus. We also show that the specialization of MHT_{y*} for $y = -1$ exists and is equal to the rationalized MacPherson Chern class c_* of the underlying constructible sheaf complex. The starting point is the following result [39, §2.4.4]:

THEOREM 5.18 (M. SAITO). *Let M be a pure m-dimensional complex algebraic manifold. Then one has for $\mathcal{M} \in D^b MHM(M)$ the duality result (for $j \in \mathbb{Z}$)*

$$Gr_j^F(DR(\mathcal{D}\mathcal{M})) \simeq \mathcal{D}(Gr_{-j}^F DR(\mathcal{M})) \in D_{\mathrm{coh}}^b(M). \tag{5-27}$$

Here $\mathcal{D}$ on the left side is the duality of mixed Hodge modules, wheres $\mathcal{D}$ on the right is the Grothendieck duality

$$\mathcal{D} = \mathrm{Rhom}(\cdot, \omega_M[m]) : D_{\mathrm{coh}}^b(M) \to D_{\mathrm{coh}}^b(M),$$

with $\omega_M = \Omega_M^m$ the canonical sheaf of M.

A priori this is a duality for the corresponding analytic (cohomology) sheaves. Since $\mathcal{M}$ and $DR(\mathcal{M})$ can be extended to smooth complex algebraic compactification $\overline{M}$, one can apply Serre's GAGA theorem to get the same result also for the underlying algebraic (cohomology) sheaves.

COROLLARY 5.19 (CHARACTERISTIC CLASSES AND DUALITY). *Let Z be a complex algebraic variety with* dualizing complex $\omega_Z^{\bullet} \in D_{\mathrm{coh}}^b(Z)$, *so that the* Grothendieck duality transformation $\mathcal{D} = \mathrm{Rhom}(\cdot, \omega_Z^{\bullet})$ *induces a duality involution*

$$\mathcal{D} : G_0(Z) \to G_0(Z).$$

Extend this to $G_0(Z) \otimes \mathbb{Z}[y^{\pm 1}]$ by $y \mapsto 1/y$. Then the motivic Hodge Chern class transformation MHC_y commutes with duality $\mathcal{D}$:

$$MHC_y(\mathcal{D}(\cdot)) = \mathcal{D}(MHC_y(\cdot)) : K_0(MHM(Z)) \to G_0(Z) \otimes \mathbb{Z}[y^{\pm 1}]. \tag{5-28}$$

Note that for $Z = pt$ a point this reduces to the duality formula (2-14) for the χ_y-genus. For dualizing complexes and (relative) Grothendieck duality we refer to [26; 17; 33] as well as [24, Part I, §7]). Note that for M smooth of pure dimension m, one has

$$\omega_M[m] \simeq \omega_M^{\bullet} \in D_{\mathrm{coh}}^b(M).$$

Moreover, for a proper morphism $f : X \to Z$ of complex algebraic varieties one has the relative Grothendieck duality isomorphism

$$Rf_* \left(\mathrm{Rhom}(\mathcal{F}, \omega_X^\bullet) \right) \simeq \mathrm{Rhom}(Rf_* \mathcal{F}, \omega_Z^\bullet) \quad \text{for } \mathcal{F} \in D_{\mathrm{coh}}^b(X),$$

so that the duality involution

$$\mathcal{D} : G_0(Z) \otimes \mathbb{Z}[y^{\pm 1}] \to G_0(Z) \otimes \mathbb{Z}[y^{\pm 1}]$$

commutes with proper push down. Since $K_0(MHM(Z))$ is generated by classes $f_*[\mathcal{M}]$, with $f : M \to Z$ proper morphism from a pure dimensional complex algebraic manifold M (and $\mathcal{M} \in MHM(M)$), it is enough to prove (5-28) in the case $Z = M$ a pure dimensional complex algebraic manifold, in which case it directly follows from Saito's result (5-27).

For a systematic study of the behavior of the Grothendieck duality transformation $\mathcal{D} : G_0(Z) \to G_0(Z)$ with respect to exterior products and smooth pullback, we refer to [23] and [24, Part I, §7], where a corresponding "bivariant" result is stated. Here we only point out that the dualities $(\cdot)^\vee$ and $\mathcal{D}$ commute with the *pairings* of Corollary 5.14:

$$f^* \left((\cdot)^\vee \right) \cap (\mathcal{D}(\cdot)) = \mathcal{D} \left(f^*(\cdot) \cap (\cdot) \right) :$$
$$K_{\mathrm{alg}}^0(N) \otimes \mathbb{Z}[y^{\pm 1}] \times G_0(Z) \otimes \mathbb{Z}[y^{\pm 1}] \to G_0(Z) \otimes \mathbb{Z}[y^{\pm 1}], \tag{5-29}$$

and similarly

$$f^* \left((\cdot)^\vee \right) \cap (\mathcal{D}(\cdot)) = \mathcal{D} \left(f^*(\cdot) \cap (\cdot) \right) :$$
$$K_0(VmHs^g(N)) \times K_0(MHM(Z)) \to K_0(MHM(Z)). \tag{5-30}$$

Here the last equality needs only be checked for classes $[IC_S(\mathcal{L})]$, with $S \subset Z$ irreducible of dimension d and $\mathcal{L}$ a good variation of pure Hodge structures on a Zariski dense open smooth subset U of S, and $\mathbb{V}$ a good variation of pure Hodge structures on N. But then the claim follows from

$$f^*(\mathbb{V}) \otimes IC_S(\mathcal{L}) \simeq IC_S(f^*(\mathbb{V})|U \otimes \mathcal{L})$$

and (4-3) in the form

$$\mathcal{D}(IC_S(f^*(\mathbb{V})|U \otimes \mathcal{L})) \simeq IC_S((f^*(\mathbb{V})|U \otimes \mathcal{L})^\vee)(d)$$
$$\simeq IC_S(f^*(\mathbb{V}^\vee)|U \otimes \mathcal{L}^\vee)(d).$$

REMARK 5.20. The Todd class transformation $td_* : G_0(\cdot) \to H_*(\cdot) \otimes \mathbb{Q}$, too, commutes with duality (compare with [22, Example 18.3.19] and [24, Part I, Corollary 7.2.3]) if the duality involution $\mathcal{D} : H_*(\cdot) \otimes \mathbb{Q} \to H_*(\cdot) \otimes \mathbb{Q}$ in homology is defined as $\mathcal{D} := (-1)^i \cdot \mathrm{id}$ on $H_i(\cdot) \otimes \mathbb{Q}$. So also the unnormalized Hirzebruch class transformation $\widetilde{MHT}_{y*}$ commutes with duality, if this duality in homology is extended to $H_*(\cdot) \otimes \mathbb{Q}[y^{\pm 1}]$ by $y \mapsto 1/y$.

As a final result of this paper, we have:

PROPOSITION 5.21. *Let Z be a complex algebraic variety, and consider $[\mathcal{M}] \in K_0(MHM(Z))$. Then*

$$MHT_{y*}([\mathcal{M}]) \in H_*(Z) \otimes \mathbb{Q}[y^{\pm 1}] \subset H_*(Z) \otimes \mathbb{Q}[y^{\pm 1}, (1+y)^{-1}],$$

so that the specialization $MHT_{-1*}([\mathcal{M}]) \in H_*(Z) \otimes \mathbb{Q}$ for $y = -1$ is defined. Then

$$MHT_{-1*}([\mathcal{M}]) = c_*([\mathrm{rat}(\mathcal{M})]) =: c_*(\chi_{\mathrm{stalk}}([\mathrm{rat}(\mathcal{M})])) \in H_*(Z) \otimes \mathbb{Q} \quad (5\text{-}31)$$

is the rationalized MacPherson Chern class of the underlying constructible sheaf complex $\mathrm{rat}(\mathcal{M})$ (or the constructible function $\chi_{\mathrm{stalk}}([\mathrm{rat}(\mathcal{M})])$). In particular

$$MHT_{-1*}(\mathcal{D}[\mathcal{M}]) = MHT_{-1*}([\mathcal{D}\mathcal{M}]) = MHT_{-1*}([\mathcal{M}]). \quad (5\text{-}32)$$

Here χ_{stalk} is the transformation form the diagram (4-13). Similarly, all the transformations from this diagram (4-13), like χ_{stalk} and rat, commute with duality $\mathcal{D}$. This implies already the last claim, since $\mathcal{D} = \mathrm{id}$ for algebraically constructible functions (compare [47, §6.0.6]). So we only need to prove the first part of the proposition. Since MHT_{-1*} and c_* both commute with proper push down, we can assume $[\mathcal{M}] = [j_*\mathbb{V}]$, with $Z = \overline{M}$ a smooth pure-dimensional complex algebraic manifold, $j : M \to \overline{M}$ a Zariski open inclusion with complement D a normal crossing divisor with smooth irreducible components, and $\mathbb{V}$ a good variation of mixed Hodge structures on M. So

$$\widetilde{MHT}_{y*}([j_*\mathbb{V}]) = ch^* \left(MHC^y(Rj_*\mathcal{L}) \right) \cap \widetilde{MHT}_{y*}([j_*\mathbb{Q}_M^H]) \in H_*(\overline{M}) \otimes \mathbb{Q}[y^{\pm 1}]$$

by (5-15) and the *multiplicativity* of the Todd class transformation td_*. Introduce the *twisted Chern character*

$$ch^{(1+y)} : K_{\mathrm{alg}}^0(\cdot) \otimes \mathbb{Q}[y^{\pm 1}] \to H^*(\cdot) \otimes \mathbb{Q}[y^{\pm 1}],$$

$$[\mathcal{V}] \cdot y^j \mapsto \sum_{i \geq 0} ch^i([\mathcal{V}]) \cdot (1+y)^i \cdot y^j, \quad (5\text{-}33)$$

with $ch^i([\mathcal{V}]) \in H^i(\cdot) \otimes \mathbb{Q}$ the i-th component of ch^*. Then one easily gets

$$MHT_{y*}([j_*\mathbb{V}]) = ch^{(1+y)} \left(MHC^y(Rj_*\mathcal{L}) \right) \cap MHT_{y*}([j_*\mathbb{Q}_M^H])$$

$$\in H_*(\overline{M}) \otimes \mathbb{Q}[y^{\pm 1}, (1+y)^{-1}].$$

But $[j_*\mathbb{Q}_M^H] = \chi_{\mathrm{Hdg}}(j_*[\mathrm{id}_M])$ is by (4-15) in the image of

$$\chi_{\mathrm{Hdg}} : M_0(\mathrm{var}/\overline{M}) = K_0(\mathrm{var}/\overline{M})[\mathbb{L}^{-1}] \to K_0(MHM(\overline{M})).$$

So for $MHT_{y*}([j_*\mathbb{Q}_M^H])$ we can apply the following special case of Proposition 5.21:

LEMMA 5.22. *The transformation*

$$T_{y*} = MHT_{y*} \circ \chi_{\mathrm{Hdg}} : M_0(\mathrm{var}/Z) \to H_*(Z) \otimes \mathbb{Q}[y^{\pm 1}, (1+y)^{-1}]$$

takes values in $H_*(Z) \otimes \mathbb{Q}[y^{\pm 1}] \subset H_*(Z) \otimes \mathbb{Q}[y^{\pm 1}, (1+y)^{-1}]$, *with*

$$T_{-1*} = T_{-1*} \circ \mathcal{D} = c_* \circ \mathrm{can} : M_0(\mathrm{var}/Z) \to H_*(Z) \otimes \mathbb{Q}.$$

Assuming this lemma, we can derive from the following commutative diagram that the specialization $MHT_{-1*}([j_*\mathbb{V}])$ for $y = -1$ exists:

$$
\begin{array}{ccc}
H^*(\cdot) \otimes \mathbb{Q}[y^{\pm 1}] \times H_*(\cdot) \otimes \mathbb{Q}[y^{\pm 1}, (1+y)^{-1}] & \xrightarrow{\cap} & H_*(\cdot) \otimes \mathbb{Q}[y^{\pm 1}, (1+y)^{-1}] \\
\scriptstyle{incl.} \uparrow & & \uparrow \scriptstyle{incl.} \\
H^*(\cdot) \otimes \mathbb{Q}[y^{\pm 1}] \times H_*(\cdot) \otimes \mathbb{Q}[y^{\pm 1}] & \xrightarrow{\cap} & H_*(\cdot) \otimes \mathbb{Q}[y^{\pm 1}] \\
\scriptstyle{y=-1} \downarrow & & \downarrow \scriptstyle{y=-1} \\
H^*(\cdot) \otimes \mathbb{Q} \times H_*(\cdot) \otimes \mathbb{Q} & \xrightarrow{\cap} & H_*(\cdot) \otimes \mathbb{Q}.
\end{array}
$$

Moreover $ch^{(1+y)}(MHC^y(Rj_*\mathcal{L}))$ specializes for $y = -1$ just to

$$\mathrm{rk}(L) = ch^0([\bar{\mathcal{L}}]) \in H^0(\overline{M}) \otimes \mathbb{Q},$$

with $\mathrm{rk}(L)$ the rank of the local system L on M. So we get

$$MHT_{-1*}([j_*\mathbb{V}]) = \mathrm{rk}(L) \cdot c_*(j_*1_M) = c_*(\mathrm{rk}(L) \cdot j_*1_M) \in H_*(\overline{M}) \otimes \mathbb{Q},$$

with $\mathrm{rk}(L) \cdot j_*1_M = \chi_{\mathrm{stalk}}(\mathrm{rat}([j_*\mathbb{V}]))$.

It remains to prove Lemma 5.22. But all transformations — T_{y*}, $\mathcal{D}$, c_* and can — commute with pushdown for proper maps. Moreover, by resolution of singularities and additivity, $M_0(\mathrm{var}/Z)$ is generated by classes $[f : N \to Z] \cdot \mathbb{L}^k$ ($k \in \mathbb{Z}$), with N smooth pure n-dimensional and f proper. So it is enough to prove that $T_{y*}([\mathrm{id}_N] \cdot \mathbb{L}^k) \in H_*(N) \otimes \mathbb{Q}[y^{\pm 1}]$, with

$$T_{y*}([\mathrm{id}_N] \cdot \mathbb{L}^k) = T_{y*}(\mathcal{D}([\mathrm{id}_N] \cdot \mathbb{L}^k)) = c_*(\mathrm{can}([\mathrm{id}_N] \cdot \mathbb{L}^k)).$$

But by the *normalization condition* for our characteristic class transformations one has (compare [9]):

$$T_{y*}([\mathrm{id}_N]) = T_y^*(TN) \cap [N] \in H_*(N) \otimes \mathbb{Q}[y],$$

with $T_{-1*}([\mathrm{id}_N]) = c^*(TN) \cap [N] = c_*(1_N)$. Similarly

$$T_{y*}([\mathbb{L}]) = \chi_y([\mathbb{Q}(-1)]) = -y \quad \text{and} \quad \mathrm{can}([\mathbb{L}]) = 1_{pt},$$

so the multiplicativity of T_{y*} for exterior products (with a point space) yields

$$T_{y*}([\mathrm{id}_N] \cdot \mathbb{L}^k) \in H_*(N) \otimes \mathbb{Q}[y^{\pm 1}].$$

Moreover

$$T_{-1*}([\mathrm{id}_N] \cdot \mathbb{L}^k) = c_*(1_N) = c_*(\mathrm{can}([\mathrm{id}_N] \cdot \mathbb{L}^k)).$$

Finally $\mathcal{D}([\mathrm{id}_N] \cdot \mathbb{L}^k) = [\mathrm{id}_N] \cdot \mathbb{L}^{k-n}$ by the definition of $\mathcal{D}$, so that

$$T_{-1*}([\mathrm{id}_N] \cdot \mathbb{L}^k) = T_{-1*}\big(\mathcal{D}([\mathrm{id}_N] \cdot \mathbb{L}^k)\big).$$

Acknowledgements

This paper is an extended version of an expository talk given at the workshop "Topology of Stratified Spaces" at MSRI in September 2008. I thank the organizers (G. Friedman, E. Hunsicker, A. Libgober and L. Maxim) for the invitation to this workshop. I also would like to thank S. Cappell, L. Maxim and S. Yokura for some discussions on the subject of this paper.

References

[1] M. F. Atiyah, *The signature of fiber bundles*, pp. 73–84 in *Global analysis: papers in honor of K. Kodaira*, Univ. Tokyo Press, Tokyo, 1969.

[2] M. Banagl, *Topological invariants of stratified spaces*, Springer, Berlin, 2007.

[3] M. Banagl, *The signature of singular spaces and its refinements to generalized homology theories*, pp. 227–263 in this volume.

[4] M. Banagl, S. E. Cappell, J. L. Shaneson, *Computing twisted signatures and L-classes of stratified spaces*, Math. Ann. 326 (2003), 589–623.

[5] P. Baum, W. Fulton, R. MacPherson, *Riemann–Roch for singular varieties*, Publ. IHES 45 (1975), 101–145.

[6] P. Baum, W. Fulton, R. MacPherson, *Riemann–Roch and topological K-theory for singular varieties*, Acta Math. 143 (1979), 155–192.

[7] A. A. Beilinson, J. Bernstein, P. Deligne, *Faisceaux pervers*, Astérisque 100, Soc. Math. de France, Paris, 1982.

[8] F. Bittner, *The universal Euler characteristic for varieties of characteristic zero*, Comp. Math. 140 (2004), 1011–1032.

[9] J. P. Brasselet, J. Schürmann, S. Yokura, *Hirzebruch classes and motivic Chern classes of singular spaces*, J. Topol. Anal. 2 (2010), 1–55.

[10] S. E. Cappell, L. G. Maxim, J. L. Shaneson, *Euler characteristics of algebraic varieties*, Comm. Pure Appl. Math. 61 (2008), 409–421.

[11] S. E. Cappell, L. G. Maxim, J. L. Shaneson, *Hodge genera of algebraic varieties, I*, Comm. Pure Appl. Math. 61 (2008), 422–449.

[12] S. E. Cappell, A. Libgober, L. G. Maxim, J. L. Shaneson, *Hodge genera of algebraic varieties, II*, Math. Annalen 345 (2009), 925–972.

[13] S. E. Cappell, A. Libgober, L. G. Maxim, J. L. Shaneson, *Hodge genera and characteristic classes of complex algebraic varieties*, Electron. Res. Announc. Math. Sci. 15 (2008), 1–7.

[14] S. E. Cappell, J. L. Shaneson, *Stratifiable maps and topological invariants*, J. Amer. Math. Soc. 4 (1991), 521–551

[15] E. Cattani, A. Kaplan, W. Schmid, L^2 *and intersection cohomologies for a polarizable variation of Hodge structure*, Inv. Math. 87 (1987), 217–252.

[16] S. S. Chern, F. Hirzebruch, J.-P. Serre, *On the index of a fibered manifold*, Proc. Amer. Math. Soc. 8 (1957), 587–596.

[17] B. Conrad, *Grothendieck Duality and Base Change*, Lecture Notes in Mathematics, Vol. 1750. Springer, 2000.

[18] P. Deligne, *Théorème de Lefschetz et critères de dégénérescence de suites spectrales*, Publ. Math. IHES 35 (1968), 107–126.

[19] P. Deligne, *Equation différentielles a point singular régulier*, Springer, Berlin, 1969.

[20] P. Deligne, *Théorie de Hodge II*, Publ. Math. IHES 40 (1971), 5–58.

[21] P. Deligne, *Théorie de Hodge III*, Publ. Math. IHES 44 (1974), 5–78.

[22] W. Fulton, *Intersection theory*, Springer, 1981.

[23] W. Fulton, L. Lang, *Riemann–Roch algebra*, Springer, New York, (1985).

[24] W. Fulton, R. MacPherson, *Categorical framework for the study of singular spaces*, Memoirs Amer. Math. Soc. 243 (1981).

[25] M. Goresky, R. MacPherson, *Intersection homology II*, Invent. Math. 71 (1983), 77–129.

[26] R. Hartshorne, *Residues and duality*, Lecture Notes in Mathematics, Vol. 20. Springer, New York, 1966.

[27] F. Hirzebruch, *Topological methods in algebraic geometry*, Springer, Berlin, 1966.

[28] M. Kashiwara, *A study of a variation of mixed Hodge structures*, Publ. RIMS 22 (1986), 991–1024.

[29] M. Kashiwara, T. Kawai, *The Poincaré lemma for variations of polarized Hodge structures*, Publ. RIMS 23 (1987), 345–407.

[30] M. Kashiwara, P. Schapira, *Sheaves on manifolds*, Springer, Berlin, 1990.

[31] G. Kennedy, *MacPherson's Chern classes of singular varieties*, Comm. Alg. 18 (1990), 2821–2839.

[32] A. Libgober, *Elliptic genera, real algebraic varieties and quasi-Jacobi forms*, pp. 99–125 in these proceedings.

[33] J. Lipmann, M. Hashimoto, *Foundations of Grothendieck Duality for Diagrams of Schemes*, Lecture Notes in Mathematics, Vol. 1960. Springer, 2009.

[34] R. MacPherson, *Chern classes for singular algebraic varieties*, Ann. of Math. (2) 100 (1974), 423–432.

[35] L. Maxim, J. Schürmann, *Hodge-theoretic Atiyah–Meyer formulae and the stratified multiplicative property*, pp. 145–167 in "Singularities I: algebraic and analytic aspects", Contemp. Math. 474 (2008).

[36] W. Meyer, *Die Signatur von lokalen Koeffizientensystemen und Faserbündeln*, Bonner Mathematische Schriften 53, (Universität Bonn), 1972.

[37] C. Peters, *Tata lectures on motivic aspects of Hodge theory*, Lecture Notes of the Tata Institute of Fundamental Research, 2010.

[38] C. Peters, J. Steenbrink, *Mixed Hodge structures*, Ergebnisse der Math. (3) Vol. 52, Springer, Berlin, 2008.

[39] M. Saito, *Modules de Hodge polarisables*, Publ. RIMS 24 (1988), 849–995.

[40] M. Saito, *Mixed Hodge modules*, Publ. RIMS 26 (1990), 221–333.

[41] M. Saito, *Introduction to mixed Hodge modules*, pp. 145–162 in Actes du Colloque de Théorie de Hodge (Luminy, 1987), Astérisque Vol. 179–180 (1989).

[42] M. Saito, *On Kollar's Conjecture*, Proceedings of Symposia in Pure Mathematics 52, Part 2 (1991), 509–517.

[43] M. Saito, *On the formalism of mixed sheaves*, preprint, arXiv:math/0611597.

[44] M. Saito, *Mixed Hodge complexes on algebraic varieties*, Math. Ann. 316 (2000), 283–331.

[45] C. Sabbah, *Hodge theory, singularities and D-modules*, preprint (2007), homepage of the author.

[46] W. Schmid, *Variation of Hodge structures: the singularities of the period mapping*, Inv. Math. 22 (1973), 211–319.

[47] J. Schürmann, *Topology of singular spaces and constructible sheaves*, Monografie Matematyczne, Vol.63, Birkhäuser, Basel, 2003.

[48] J. Schürmann, S. Yokura, *A survey of characteristic classes of singular spaces*, pp. 865–952 in "Singularity theory" (Marseille, 2005), edited by D. Chéniot et al., World Scientific, Singapore, 2007.

[49] P. H. Siegel, *Witt spaces: A geometric cycle theory for KO-homology at odd primes*, Amer. J. Math. 105 (1983), 1067–1105.

[50] J. Steenbrink, S. Zucker, *Variations of mixed Hodge structures*, Inv. Math. 80 (1983).

[51] B. Totaro, *Chern numbers for singular varieties and elliptic homology*, Ann. of Math. (2) 151 (2000), 757–791.

[52] C. Voisin, *Hodge theory and complex algebraic geometry I*, Cambridge Studies in Advanced Mathematics 76, Cambridge University Press, 2002.

[53] R. Waelder, *Rigidity of differential operators and Chern numbers of singular spaces*, pp. 35–54 in this volume.

[54] J. Woolf, *Witt groups of sheaves on topological spaces*, Comment. Math. Helv. 83 (2008), 289–326.

[55] S. Yokura, *Motivic characteristic classes*, pp. 375–418 in these proceedings.

[56] S. Zucker, *Hodge theory with degenerating coefficients: L_2-cohomology in the Poincaré metric*, Ann. Math. 109 (1979), 415–476.

JÖRG SCHÜRMANN
MATHEMATISCHE INSTITUT
UNIVERSITÄT MÜNSTER
EINSTEINSTR. 62
48149 MÜNSTER
GERMANY
  jschuerm@math.uni-muenster.de

Topology of Stratified Spaces
MSRI Publications
Volume 58, 2011

Workshop on the Topology of Stratified Spaces
Open Problems

The following open problems were suggested by the participants both during and following the Workshop.

1. L^2 Hodge and signature theorems; Signature theory on singular spaces

(a) (suggested by Eugénie Hunsicker)

Consider a pseudomanifold X as in Cheeger, [7]. Cheeger proves in this paper that if the smooth part of X, X_{reg}, is endowed with an iterated cone metric, and if X is a Witt space, then the L^2 cohomology of X_{reg} is isomorphic to the middle perversity intersection cohomology of X (which is also unique due to the Witt condition). This implies in turn that the space of L^2 harmonic forms for the maximal extension is isomorphic to the middle perversity intersection cohomology, and from this we get that the operator $d + \delta$ is essentially self-adjoint on X_{reg}, which in turn means that this operator has a unique closed extension to $L^2(X_{\text{reg}})$. Thus in the setting of Witt spaces and conical metrics, there is a clear and simple relationship between harmonic forms, L^2-cohomology and intersection cohomology.

If the Witt condition is dropped, then there is not generally a unique middle perversity intersection cohomology, and $d + \delta$ generally has different possible extensions, and in particular, can have different possible self-adjoint extensions. In the case of a pseudomanifold with only one singular stratum endowed again with a conical metric, the non-Witt case was studied in [14]. In this paper, it is shown that the operator $d + \delta$ on X_{reg} endowed with a cone metric has self-adjoint extensions whose kernels are isomorphic to the upper and to the lower middle perversity intersection cohomologies on X, and the kernel of the minimal extension of $d + \delta$ (for an appropriately chosen cone metric) is isomorphic to the image of lower middle in upper middle perversity intersection cohomology.

Versions of L^2 cohomology also can relate to more general perversities. For example, Nagase showed in [18] that for each standard perversity $\overline{p}$ greater than or equal to the upper middle perversity on a pseudomanifold X, there exists an incomplete metric on the regular set of X for which the L^2 cohomology associated to the maximal extension of d is isomorphic to $IH^{\overline{p}}(X)$. This in particular implies that there exists a self-dual L^2 extension of $d + \delta$ for this metric whose kernel is also isomorphic to $IH^{\overline{p}}(X)$.

It seems likely that these phenomena are part of a larger relationship among L^2 extensions of the geometric operator $d + \delta$ on the regular set of a pseudomanifold for various metrics, weighted L^2 cohomology for these metrics and intersection cohomologies on X with various perversities. It would be interesting to explore this further. In particular, consider the metrics constructed in [18]. What other closed L^2 extensions of $d + \delta$ exist for these metrics, and which perversity intersection cohomologies will their kernels be isomorphic to? Further, are there generalizations of intersection cohomology that are isomorphic to the kernel of some such extensions? Finally, can we understand the interesting extensions of $d + \delta$ using analytic approaches to singularities, such as the Melrose b-calculus or Schulze or Boutet de Monvel calculi?

(b) (suggested by Paolo Piazza)

Consider a non-Witt pseudomanifold X that has a Lagrangian structure à la Banagl. See for example Chapter 9 in the book [4]. For such spaces one can define a signature and an L-class.

(i) Is there an analytic description of the signature? More precisely: endow X with an iterated conic metric. Is there an extension of the signature operator which is Fredholm and such that its index is equal to the above signature?

For Witt spaces this is a well known result due to Cheeger and recently re-established by Albin, Leichtnam, Mazzeo and Piazza in the preprint [1]. In the latter preprint the extension of the full signature package from closed manifolds to Witt spaces, leading to the definition and the homotopy invariance of higher signatures on Witt spaces, is discussed. Notice that the higher signatures on Witt spaces involve the L-class of Goresky–MacPherson.

(ii) What part of the signature package on closed manifolds and Witt spaces can be extended to these non-Witt spaces?

(c) (suggested by Shmuel Weinberger)

What kind of elliptic operator theory "tracks" (i.e. contains a signature operator for G-manifolds) the cosheaf homology term in stratified surgery for M/G? Things are easy when G acts locally freely so the quotient is an orbifold but this looks interesting in general.

(d) (suggested by Shmuel Weinberger; clarifications by Les Saper and Paolo Piazza)

The entrance of sheaves with nontrivial local cohomology whose global vanishing is important for global self-duality in Saper's talk suggests that compactifications have global "index invariants" in $L(\mathbb{R}P)$ or $K(\mathbb{C}P)$ but do not localize, i.e. pullback to $K(B\Gamma)$ (when Γ has torsion). For Γ with torsion, a "pullback" probably would be accidental nonsense. This is like what happens (for a different reason) in Fowler's talk on uniform lattices with torsion.

(e) (suggested by David Trotman)

If P and P' are homeomorphic PL Witt spaces (i.e. you think of them as different triangulations of the same object), it is a consequence of Goresky–MacPherson II and Siegel (or other combinations) that these are cobordant in the Witt sense. How elementary is this fact? Is there a *direct* proof?

2. Topology of algebraic varieties

(a) (suggested by Anatoly Libgober)

Does there exist a cobordism theory of pairs (X, D) such that for D log-terminal, $\mathcal{E}ll(X, D)$ is invariant under such cobordisms? See [6] for a discussion of elliptic genus of pairs and results related to this question.

(b) (suggested by Clint McCrory)

(i) Define intersection homology for real algebraic varieties. This question appears on Goresky and MacPherson's 1994 problem list [9]. Interesting work has been done by van Hamel [21]; see also [3; 20].

(ii) Simplify Akbulut and King's conjectural topological characterization of real algebraic varieties [2], and compute the bordism ring of real algebraic varieties. Invariants are "Akbulut–King numbers" [15].

(iii) What is the topology of the weight filtration of a real algebraic variety [17]? How can the filtration vary within a homeomorphism type? Is the filtration trivial for $\mathbb{Z}_2$ homology manifolds? Is it a bi-Lipshitz invariant (Trotman)?

(iv) Prove that the Stiefel–Whitney homology classes of a real algebraic variety are topologically invariant (*cf.* [8]).

(v) Which real toric varieties X are maximal, that is, when is

$$\dim H_*(X(\mathbb{R}); \mathbb{Z}_2) = \dim H_*(X(\mathbb{C}); \mathbb{Z}_2)?$$

See Hower's counterexample [12].

(vi) If a complex algebraic variety is defined over $\mathbb{R}$, what is the relation between the Deligne weight filtration of the cohomology (or homology) of the complex points and the weight filtration defined by Totaro [20] and McCrory and Parusiński [17] for the real points? The weight filtration of the homology of a complex variety can be defined with arbitrary coefficients. What is the relation between the weight filtration of the homology of the complex points with $\mathbb{Z}_2$ coefficients and the weight filtration for the real points?

(vii) Are there motivic characteristic classes for real varieties analogous to those defined by Brasselet, Schürmann, and Yokura [22] for complex varieties? The virtual Betti numbers β_q of real algebraic varieties [16] satisfy the "scissor relations"

$$\beta_q(X) = \beta_q(Y) + \beta_q(X \setminus Y)$$

for Y a closed subvariety of X. Can the virtual Betti numbers be extended to characteristic classes of real varieties?

(c) (suggested by David Trotman)

(i) Is it true that every topologically conical complex stratification of a complex analytic variety is Whitney (A)-regular? (This is not true for real algebraic varieties.)

(ii) Does every Whitney C^k stratified set admit a C^k triangulation such that the open simplices are strata of a Whitney stratification? The same question replacing "Whitney" by "Bekka."

(iii) It is known that families of (germs of) complex hypersurfaces with an isolated singularity have constant Milnor number if an only if they have constant topological type (except for "only if" for surfaces where it is an open question). Could it be true that having constant topological type is equivalent to the family being Bekka C-regular over the parameter space?

(iv) Can Goresky–MacPherson's Morse theory be made to work for tame Bekka stratifications instead of tame Whitney stratifications?

(v) It is known (Noirel -1996) that every abstract stratified space (Thom–Mather space) can be embedded in some $\mathbb{R}^n$ as a semi-algebraic Whitney stratified set (even Verdier regular) with semi-algebriac control data without refining the original stratification. Is there such a Mostowski stratified embedding, or at least locally bi-Lipschitz trivial semi-algebraic stratification without refinement? (By theorems of Parusiński (1992) or Valette (2005), there are refinements with these properties.)

(vi) Suppose 2 germs of complex analytic functions on $\mathbb{C}^n$ with isolated singularities at 0 are topologically equivalent, i.e. there exists a homeomorphism $h : (\mathbb{C}^n, 0) \to (\mathbb{C}^n, 0)$ over $f : (\mathbb{C}^n, 0) \to \mathbb{C}$ and $g : (\mathbb{C}^n, 0) \to \mathbb{C}$ such that $f = gh$. Can one find another such homeomorphism h_1 such that h_1 preserves distance to the origin, i.e. $\|h_1(z)\| = \|z\|$ for z near $0 \in \mathbb{C}^n$? (A positive answer would solve Zariski's 1970 problem about the topological invariance of the multiplicity.)

3. Mixed Hodge theory and singularities

(suggested by Matt Kerr and Gregory Pearlstein)

The period domain classifying Hodge structures of type $(h^{n,0}, h^{n-1,1}, \ldots, h^{0,n})$, (say all > 0), $n > 1$ odd, is a *non*-locally-symmetric homogeneous space. Understand the L^2-cohomology groups

$$H^q_{(2)}\big(\Gamma \backslash D, \otimes_k ((\Lambda^{h^{k,n-k}})\mathcal{H}^{k,n-k})^{\otimes a_k}\big)$$

and the role played by these in algebraizing images of period maps. (Note that Γ is an arithmetic subgroup and $K_{\Gamma \backslash D}$ is a line bundle of the form shown.) Possible reference (somewhat outdated): [11].

4. Characteristic class theories for singular varieties

(a) (suggested by Jörg Schürmann)

We work in the algebraic context over $\mathbb{C}$. Find a pure-dimensional variety X such that the class of the intersection cohomology complex IC_X in the Grothendieck group of complex algebraically constructible sheaves

$$[IC_X] \in K_0(D^b_c(X))$$

is *not* in the subgroup generated by $[Rf_* \mathbb{Q}_Z]$ with Z smooth pure-dimensional and $f : Z \to X$ proper.

Note that it is important to take only the classes of the total direct images. If one asks the same question for the subgroup generated by *direct summands* of $[Rf_* \mathbb{Q}_Z]$ with Z smooth pure-dimensional and $f : Z \to X$ proper, then $[IC_X]$ belongs to this subgroup by the *decomposition theorem* (compare [19, Corollary 4.6], for example).

A positive answer to the question above, stated in the "topological context" of algebraically constructible sheaves, would also give an example such that the class

$$[IC_X^H] \in K_0(MHM(X))$$

of the corresponding (pure) intersection Hodge module IC_X^H in the Grothendieck group of algebraic mixed Hodge modules is *not* in the image of the natural group homomorphism

$$\chi_{\text{Hdg}} : K_0(\text{var}/X) \to K_0(MHM(X))$$

from the *motivic relative Grothendieck group* of complex algebraic varieties over X (compare [19, Section 4.2]).

This fact would further justify the study of characteristic classes of mixed Hodge modules in the works of Cappell, Libgober, Maxim, Schürmann, and Shaneson; see [19] and the references therein.

(b) (suggested by Shmuel Weinberger)

Are the elliptic genera, etc. part of an integral theory the way L-classes come back from $KO(M)$ in index theory? Schürmann and Yokura know something about this but with too few variables.

References

[1] Pierre Albin, Eric Leichtnam, Rafe Mazzeo, Paolo Piazza, *The signature package on Witt spaces, I. Index classes*, http://arxiv.org/abs/0906.1568

[2] S. Akbulut, H. King, *Topology of real algebraic sets*, MSRI Publ. **25**, Springer Verlag, New York, 1992.

[3] M. Banagl, *The signature of singular spaces and its refinements to generalized homology theories*, in this volume.

[4] M. Banagl, *Topological invariants of stratified spaces*, Springer Monographs in Mathematics, Springer, New York, 2006

[5] Jean-Paul Brasselet, Joerg Schürmann, Shoji Yokura, *Hirzebruch classes and motivic Chern classes for singular spaces*, J. Topol. Anal. 2:1 (2010), 1–55.

[6] Lev Borisov, Anatoly Libgober, *Elliptic genera of singular varieties*, Duke Math. J. 116 (2003), 319–351.

[7] Cheeger, Jeff, *On the Hodge theory of Riemannian pseudomanifolds*. Geometry of the Laplace operator (Proc. Sympos. Pure Math., Univ. Hawaii, Honolulu, Hawaii, 1979), pp. 91–146, Proc. Sympos. Pure Math., XXXVI, Amer. Math. Soc., 1980.

[8] J. Fu, C. McCrory, *Stiefel–Whitney classes and the conormal cycle of a real analytic variety*, Trans. Amer. Math. Soc. **349** (1997), 809–835.

[9] M. Goresky, R. MacPherson, *Problems and bibliography on intersection homology*, in *Intersection cohomology*, A. Borel *et al.*, Birkhäuser 1994, 221–229.

[10] P. Griffiths, M. Green, M. Kerr, *Some enumerative global properties of variations of Hodge structure*, to appear in Moscow Math. J.

[11] P. Griffiths, W. Schmid, *Locally homogeneous complex manifolds*, Acta Math. **123** (1969), 253-302.

[12] V. Hower, *A counterexample to the maximality of toric varieties*, Proc. Amer. Math. Soc., **136** (2008), 4139–4142.

[13] E. Hunsicker, *Hodge and signature theorems for a family of manifolds with fibre bundle boundary*, Geom. Topol. 11 (2007), 1581–1622.

[14] E. Hunsicker, R. Mazzeo, *Harmonic forms on manifolds with edges*, Int. Math. Res. Not. (2005) .

[15] C. McCrory, A. Parusiński, *The topology of real algebraic sets of dimension 4: necessary conditions*, Topology **39** (2000), 495–523.

[16] C. McCrory, A. Parusiński, *Virtual Betti numbers of real algebraic varieties*, Comptes Rendus Acad. Sci. Paris, Ser. I, **336** (2003), 763–768.

[17] C. McCrory, A. Parusiński, *The weight filtration for real algebraic varieties*, in this volume.

[18] M. Nagase, L^2-*cohomology and intersection homology of stratified spaces*, Duke Math. J. 50 (1983).

[19] J. Schürmann, *Characteristic classes of mixed Hodge modules*, in this volume.

[20] B. Totaro, *Topology of singular algebraic varieties*, Proc. Int. Cong. Math. Beijing (2002), 533-541.

[21] J. van Hamel, *Towards an intersection homology theory for real algebraic varieties*, Int. Math. Research Notices **25** (2003), 1395–1411.

[22] S. Yokura, *Motivic characteristic classes*, in this volume.

Printed in the United States
By Bookmasters